Safety Data, Analysis, and Evaluation 2018

Transportation Research Record: Journal of the Transportation Research Board Volume 2672, Issue 38

ISBN: 978-1-5108-8246-1

Printed from e-media with permission by:

Curran Associates, Inc.
57 Morehouse Lane
Red Hook, NY 12571

Some format issues inherent in the e-media version may also appear in this print version.

Published by Sage Publications USA

ISBN (Print) 978-1-5108-8246-1
ISBN 2018 Printed Set (All Issues) 978-1-5108-7735-1

Printed by Curran Associates, Inc. (2019)

For permission requests, please contact sagepub.com/journals-permissions

Additional copies of this publication are available from:

Curran Associates, Inc.
57 Morehouse Lane
Red Hook, NY 12571 USA
Phone: 845-758-0400
Fax: 845-758-2633
Email: curran@proceedings.com
Web: www.proceedings.com

Volume 2672 Number 38 December 2018

TRR
JOURNAL OF THE TRANSPORTATION RESEARCH BOARD

Contents

Safety Data, Analysis, and Evaluation 2018

Articles

TRB

TRANSPORTATION RESEARCH BOARD

CORE SPONSORS

Transportation Departments of the 50 States and the District of Columbia

Federal Government

U.S. Department of Transportation
- Federal Aviation Administration (FAA)
- Federal Highway Administration
- Federal Motor Carrier Safety Administration (FMCSA)
- Federal Railroad Administration (FRA)
- Federal Transit Administration
- National Highway Traffic Safety Administration (NHTSA)
- Office of the Assistant Secretary of Transportation for Research and Technology (OST-R)

U.S. Air Force Civil Engineer Center (AFCEC)
U.S. Coast Guard (USCG)
U.S. Department of Energy
U.S. Department of the Interior (DOI)
U.S. Environmental Protection Agency (EPA)

Nongovernmental Organizations

American Public Transportation Association (APTA)
Association of American Railroads (AAR)
California Air Resources Board

The National Academies of

SCIENCES • ENGINEERING • MEDICINE

The **National Academy of Sciences** was established in 1863 by an Act of Congress, signed by President Lincoln, as a private, nongovernmental institution to advise the nation on issues related to science and technology. Members are elected by their peers for outstanding contributions to research. Dr. Marcia McNutt is president.

The **National Academy of Engineering** was established in 1964 under the charter of the National Academy of Sciences to bring the practices of engineering to advising the nation. Members are elected by their peers for extraordinary contributions to engineering. Dr. C. D. Mote, Jr., is president.

The **National Academy of Medicine** (formerly the Institute of Medicine) was established in 1970 under the charter of the National Academy of Sciences to advise the nation on medical and health issues. Members are elected by their peers for distinguished contributions to medicine and health. Dr. Victor J. Dzau is president.

The three Academies work together as the National Academies of Sciences, Engineering, and Medicine to provide independent, objective analysis and advice to the nation and conduct other activities to solve complex problems and inform public policy decisions. The Academies also encourage education and research, recognize outstanding contributions to knowledge, and increase public understanding in matters of science, engineering, and medicine.

Learn more about the National Academies of Sciences, Engineering, and Medicine at **www.national-academies.org.**

The **Transportation Research Board** is one of seven major programs of the National Academies of Sciences, Engineering, and Medicine. The mission of the Transportation Research Board is to increase the benefits that transportation contributes to society by providing leadership in transportation innovation and progress through research and information exchange, conducted within a setting that is objective, interdisciplinary, and multimodal. The Board's varied activities annually engage about 7,000 engineers, scientists, and other transportation researchers and practitioners from the public and private sectors and academia, all of whom contribute their expertise in the public interest. The program is supported by state transportation departments, federal agencies including the component administrations of the U.S. Department of Transportation, and other organizations and individuals interested in the development of transportation.

Learn more about the Transportation Research Board at **www.TRB.org**.

Article

TRR
JOURNAL OF THE TRANSPORTATION RESEARCH BOARD

Transportation Research Record
2018, Vol. 2672(38) 1–11

DOI: 10.1177/0361198118790861
journals.sagepub.com/home/trr
SAGE

Assessing Surrogate Safety Measures using a Safety Pilot Model Deployment Dataset

Zhaoxiang He[1], Xiao Qin[1], Pan Liu[2], and Md Abu Sayed[1]

Abstract

Emerging data sources such as Safety Pilot Model Deployment (SPMD) provide a great opportunity to gain a better understanding of collision mechanisms and to develop novel safety metrics. The SPMD program was a comprehensive data collection effort under real-world conditions in Ann Arbor, Michigan, covering over 73 lane-miles and including approximately 3,000 pieces of onboard vehicle equipment and 30 pieces of roadside equipment. In-vehicle data (e.g., speed, location) collected by the SPMD program can potentially be an important supplement to traditional crash data-oriented safety analysis. The goal of this study was to assess roadway link-level surrogate safety measures using the vehicle trajectory data from SPMD. The study's objectives included: 1) developing a framework to process the SPMD dataset using big-data analytics; 2) converting raw vehicle motion data from SPMD to surrogate safety measures; and 3) analyzing the statistical relationship between crash records and the calculated safety index. The statistical models showed that modified time to collision (MTTC) outperforms time to collision (TTC) and deceleration rate to avoid collision (DRAC) with respect to its goodness of fit. The findings are promising in that augmenting safety analysis with surrogate measures and vehicle performance (e.g., speed and brake duration from connected vehicles) improves the overall model performance. Such information is vital for safety analysis, especially in the absence of detailed roadway and traffic data.

Traffic deaths in 2015 set the record for the largest annual increase in the United States since 1966 (*1*). The preliminary information from the National Highway Traffic Safety Administration (NHTSA) shows that traffic fatalities increased by 10.4% from 2015 to 2016 (*1*). On average, more than 6 million crashes are reported annually, resulting in more than 30 thousand fatalities and 2 million injuries every year on the U.S. highways and streets (*1*). Because of the tremendous loss caused by traffic accidents, there is a keen interest in developing better safety metrics and seeking solutions to reduce crashes.

Crash history has long been considered by researchers and safety professionals as a reliable performance measure for road safety, and possibly the most straightforward measure. However, the shortcoming of using crash history is that sites without crashes cannot be properly evaluated. Moreover, collecting crash data for a valid statistical evaluation usually takes a long time, and crash data often are biased because of issues like underreporting and reporting thresholds. The predictive methods in the Highway Safety Manual (HSM) provide scientific approaches to calculating the expected annual average crashes, given a series of contributing variables, but the requirement for detailed roadway and traffic data makes the application impractical in many local agencies.

The traffic conflict technique (TCT), which supports surrogate safety measures for road safety, provides an alternative to evaluate road safety. Traffic conflict is defined as "an observable situation in which two or more road users approach each other in time and space for such an extent that there is a risk of collision if their movements remain unchanged" (*2*). This type of measurement is usually derived from the vehicle kinematic characteristics before possible conflicts using the collision theory. TCT not only offers an objective view of collision likelihood based on a vehicle's motion, but also considers

[1]Department of Civil and Environmental Engineering, University of Wisconsin-Milwaukee, Milwaukee, WI
[2]Jiangsu Key Laboratory of Urban ITS, Southeast University, Jiangsu Province Collaborative Innovation Center of Modern Urban Traffic Technologies, Nanjing, China

Corresponding Author:
Address correspondence to Xiao Qin: qinx@uwm.edu

driver behaviors such as speed, acceleration/deceleration, and following distance. Traditionally, traffic microsimulation models and field observations, either from observers or by fixed videos, can help to measure conflicts. Nowadays, emerging data from initiatives such as Safety Pilot Model Deployment (SPMD) present new opportunities to test, validate, and evaluate surrogate safety measures with high-resolution vehicle trajectory data.

The main use of the surrogate measures is to evaluate road safety in the absence of crash data or when crash data is limited. Surrogate measures can be used to measure safety performance of a highway facility or to estimate effectiveness of safety treatment. These measures can provide rapid evaluation for innovative intersection designs or new traffic control strategies that usually require a longer period of time to accumulate an adequate number of sites and crash history. They can measure the safety improvements for rare crash types such as pedestrian or bicycle crashes. Moreover, surrogate safety measures can be used to evaluate "what if" scenarios in microscopic traffic simulation models.

SPMD is a research initiative that demonstrates connected vehicle safety technologies in real-world implementation. Data are collected from vehicles equipped with vehicle-to-vehicle (V2V) and vehicle-to-infrastructure (V2I) communication devices in Ann Arbor, Michigan (*3*). Approximately 3,000 vehicles are instrumented with data acquisition systems (DAS) and V2V communication devices (*3*). In-vehicle data such as speed, location, and direction are collected and communicated through basic safety messages (BSMs) by dedicated short range communications (DSRC) 10 times per second (*3*). The goal of this study is to retrieve proper information from vehicle trajectory data in SPMD, construct several surrogate safety measures, and assess their performance in a safety evaluation.

Literature Review

Surrogate safety measures originate from quantifying the kinematics of a conflict in traffic microsimulation models (*2*). Broadly, surrogate safety measures can be categorized as either temporal or nontemporal (*2*). Time to collision (TTC), "the time to collide if two vehicles continue at their present speed and along the same path," is a commonly used temporal surrogate measure that has been widely used to evaluate road safety in different traffic conditions (*4–7*). TTC is usually measured for each time stamp, and a threshold is used to determine whether a collision will happen if the current speed and direction are maintained (*5*). Dijkstra used TTC to measure traffic conflicts with various levels of risk, identifying the relationship between conflicts and observed crashes (*4*). Laureshyn used TTC to analyze conflicts between turning movements and through movements near the intersection (*7*). TTC threshold values have been proposed for different conditions (*8*, *9*), and a threshold of 4 s has been used to differentiate between safe and uncomfortable situations (*8*). Hydén and Linderholm use 1.5 s as the TTC threshold to detect a severe conflict (*9*). Note that the severity of a conflict or the value of a surrogate measure is to estimate the crash risk rather than the crash severity.

Time Exposed TTC (TET) and Time Integrated TTC (TIT) are two modified TTC indicators that can be used when TTC is below the threshold value (*10*). TET is the summation of all times that a driver approaches the front vehicle with a TTC below the threshold, and TIT is the integration of the TTC profile when TTC is below the threshold (*10*). TTC assumes a constant speed and ignores possible conflicts caused by the change of speed; therefore, modified time to collision (MTTC) is suggested because it considers the vehicle's acceleration or deceleration (*11*). Charly and Mathew used MTTC to identify mid-block conflicts under mixed traffic conditions, and evaluated the temporal and spatial correlation between conflicts and observed mid-block crashes (*12*). Other temporal indicators include post-encroachment time (PET), where PET is the time between the first vehicle leaving a common spatial area and the second arriving at the area (*13*).

Distance-related and deceleration-related are two types of nontemporal safety indicators. The proportion of stopping distance (PSD), "the ratio of the remaining distance to the point of collision to the minimum acceptable stopping distance," is commonly used with the assumption of the maximum available deceleration rate (MADR) (*14*). An unsafe situation is detected if PSD is less than 1, when a collision cannot be avoided even under MADR. The conventional deceleration-based index is deceleration rate to avoid collision (DRAC), which is the deceleration required to avoid a crash (*15*). The Crash Potential Index (CPI) is defined as the probability that a given vehicle's DRAC exceeds its MADR during a given time interval (*16*). Evaluation should be conducted for candidate surrogate measures, as there are no current benchmarks (*17*).

Surrogate safety measures can be calculated from traffic microsimulation models which assume that drivers do not practice "unsafe" behaviors; however, driver error contributes tremendously to observed conflicts (e.g., crash or near crash) (*17*). Field observations or video data can also provide non-simulation–based safety surrogate measures. The latest automated video-based techniques make video data processing more efficient (*18*, *19*). Saunier and Sayed proposed an automated traffic conflict detection method to assess road safety based on video data with a feature-based vehicle tracking algorithm for intersection (*18*). Zangenehpour developed a new video-based methodology to extract different road users in traffic videos automatically, and then evaluated the relationship between calculated PET and historical crash data (*19*). Vehicle trajectory information can be derived from video data, which help calculate surrogate safety measures for different situations (*20–22*). Astarita used observed vehicle tracking data from a fixed video to calculate TTC and DRAC in an attempt to measure the behavior of drivers approaching an intersection (*20*). Meng and Weng explored the risk of rear-end crashes in work zones by using the trajectory data to calculate DRAC (*21*). Oh and Kim also used the vehicle trajectory to calculate TTC for rear-end crashes (*22*).

Emerging data sources now offer new opportunities and insights for traffic safety evaluation. SHRP2 Naturalistic Driving Study (NDS) provides a detailed examination of the role of driver performance and behavior in safety. The study focuses on 1) analyzing the statistical relationship between surrogate safety measures of collisions (e.g., conflicts, critical incidents, near-collisions) and actual collisions, and 2) using surrogate measures to formulate exposure-based risk measures (*23–26*). By using the in-vehicle NDS data supplemented by driver characteristics (e.g., gender, age, driver experience, speed selection), researchers captured surrogate measures to understand driver behavior during traffic conflicts (*25*). Montgomery et al. found a statistically significant difference in TTC at braking for different gender and age groups by using the 100-car NDS dataset (*27*). Although these studies show promising results, more work is required to refine methodologies and validate surrogate measures (*24*, *25*, *27*).

Another emerging data source is SPMD, which is collected from vehicles equipped with vehicle-to-vehicle (V2V) and vehicle-to-infrastructure (V2I) communication devices under the real-world conditions in Ann Arbor, Michigan (*3*). The deployment included approximately 3,000 pieces of onboard vehicle equipment and 30 pieces of roadside equipment in an area of over 73 lane-miles. SPMD data includes driving data (DAS data), BSM data, roadside equipment data, weather data, and network traffic volume data (*3*). Driving data provide vehicles' kinematic and geographic information, and BSM data contain vehicles' position and motion with the status of components such as brakes, turning lights, wipers, and so on (*3*). Compared with video image based data in the NDS, the data such as DAS and BSM from SPMD are relatively easy to process. Moreover, SPMD data can provide real-time safety evaluation in a connected vehicle environment.

SPMD data have been applied in studies on transportation planning, traffic volume/congestion estimation, and extreme events identification (*28–32*). Deering processed spatial aggregation of trips into origin and destination zones for transportation planning by organizing SPMD data (basic safety message and driving data) into a trip-level dataset (*28*). Vasudevan et al. predicted congestion states from BSMs by using big-data graph analytics (*31*), and Zheng et al. used SPMD data to estimate traffic volumes for signalized intersections (*32*). The authors in Zheng's study developed an approach to estimate traffic volume using GPS trajectory data from connected vehicle (CV) devices under low market penetration rates (*32*). Liu et al. used data analytics to extract critical information (e.g., extreme event) embedded in BSMs, providing drivers instantaneous feedback about dangers in surrounding roadway environments (*30*). Extreme events were identified if instantaneous acceleration (the sum of motion vectors of longitudinal and lateral accelerations) exceeded the 95th percentile thresholds which change with speed (*30*). The identified extreme events were then connected to the vehicle maneuvering status (e.g., brake, turn signal) and driving context (e.g., number of objects, distance to the closest objects) to explore their relationship (*30*). Another study predicted the risky behavior of drivers (speed) at the point of curvature (PC) for different monitoring periods using the BSMs (*29*).

The challenge of using SPMD data is its size and complexity. The raw data tables are very large; even the 2-month BSM files are more than 400 GB. Deering found that during data processing, the implementation framework affected computation (*28*). A distributed computing framework like Hadoop had significantly reduced computation time when processing the CV data (*28*). Other challenges include understanding the data format and dealing with null values and outliers (*28*).

Data Collection

Because of the complexity of crash events and associated surrogate measures, only rear-end crashes in the mid-block were considered for this study. Road network data and crash data for Washtenaw County were collected from the Southeast Michigan Council of Governments Open Data Portal (http://maps-semcog.opendata.arcgis.com/). There were 52,386 crashes in the County from 2011 to 2015, including 17,103 rear-end crashes. According to Michigan's definition of intersection-related crashes, the 75-ft radius was used to remove rear-end crashes at the intersection (*33*), and the rest were kept as mid-block crashes. Figure 1 shows the road network and the crash points in Washtenaw County.

Complete speed limit information is not available for the road network, and an estimation was therefore made based on the road's functional class. A primary road has a value of 55 mph, a secondary road has a value of

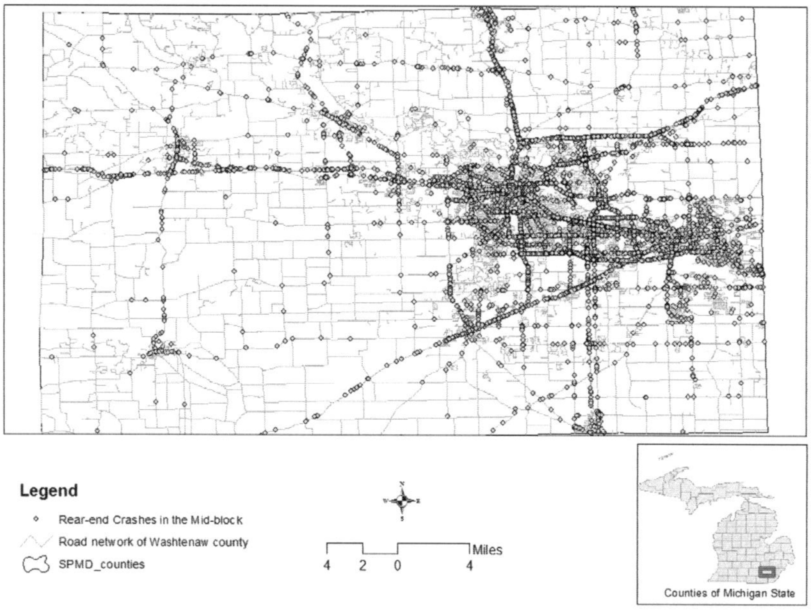

Figure 1. Road network and crash points.

35 mph, and a local road or city road has a value of 25 mph. Road functional class information can be retrieved from "TIGER/Line Shapefiles" (https://www.census.gov/geo/maps-data/data/tiger-line.html). Annual average daily traffic (AADT) can be obtained from the same Governments Open Data Portal (http://maps-semcog.opendata.arcgis.com/), but only state highways are included.

Around 2,800 vehicles participated in the SPMD program, including 2,450 equipped with a vehicle awareness device (VAD), 300 with aftermarket safety devices (ASD), 19 with retrofit safety devices (RSD), and 67 with integrated safety devices (ISD). Among the vehicles with VAD, 2,350 are cars, 60 are trucks and 85 are transit buses. Vehicles with VAD can only transmit BSM. ASD can transmit and receive BSM but is only installed on cars. Similar to ASD but designed for freight and transit, RSD is installed on 16 trucks and three transit buses. Compared with ASD or RSD, ISD can not only send and receive BSM but also connect to the vehicle data processing system. Generally, drivers who reported the most driving within the SPMD area were selected to the SPMD program. For the 64 cars with ISD, participants were selected based on age and gender to ensure a similar number of drivers in each group.

Two months of the SPMD dataset (October 2012 and April 2013) are free to download in the transportation data sharing platform, or Research Data Exchange (RDE) (https://www.its-rde.net/index.php/rdedataenvironment/10018#). This study used the DataWsu file (12 GB) and the DataFrontTargets file (4.34 GB) in the driving dataset DAS1 collected from around 100 equipped vehicles. DataWsu, from the wireless safety unit (WSU), includes mainly GPS-based data elements (e.g., speed, longitudinal acceleration) and the state of some components such as brake status and headlamp status. DataFrontTargets, which was populated mainly with the aid of Mobileye's vision-based Advanced Driver Assistance Systems (http://www.mobileye.com/), collects front target information such as distance to the front target and relative speed for the front target. In this study, common data fields "Device," "Trip," and "Time" were used to link the two datasets. Table 1 contains the primary data elements in the DataWsu file and the DataFrontTargets file, along with a brief description of each.

Methodology

TTC, MTTC, and DRAC were chosen as the three safety surrogate measures because of their popularity. Methods for calculating link-based surrogate safety measures were developed using information extracted from the real-world SPMD data set. Vehicle kinematics from the SPMD data set were used to calculate vehicle-level safety surrogate measures at each timestamp, which were aggregated into trip-level safety surrogate measures for each link. The link-level indexes combine trip-level indexes for

Table 1. Major Data Elements in the DataWsu File and DataFrontTargets File

DataWsu			
Field name	Type	Units	Description
Device	Integer	none	A unique numeric ID assigned to each DAS. This ID also doubles as a vehicle's ID
Trip	Integer	none	Count of ignition cycles—each ignition cycle commences when the ignition is in the on position and ends when it is in the off position
Time	Integer	centiseconds	Time in centiseconds since DAS started, which (generally) starts when the ignition is in the on position
GpsValidWsu	Integer	none	Communicates whether a GPS data point is valid or not
GpsTimeWsu	Integer	ms	Epoch GPS time received from the remote vehicle that has been targeted by the host vehicle's WSU
LatitudeWsu	Float	degrees	Latitude from WSU receiver
LongitudeWsu	Float	degrees	Longitude from WSU receiver
AltitudeWsu	Real	m	Altitude from WSU receiver
GpsHeadingWsu	Real	degrees	Heading from WSU GPS receiver
GpsSpeedWsu	Real	m/s	Speed from WSU GPS receiver
SpeedWsu	Real	km/h	Speed from vehicle CAN Bus via WSU
TurnSngRWsu	Integer	none	Right turn signal from vehicle CAN Bus via WSU
TurnSngLWsu	Integer	none	Left turn signal from vehicle CAN Bus via WSU
BrakeAbsTcsWsu	Integer	none	Brake, ABS, and traction control from vehicle CAN Bus via WSU
AxWsu	Real	m/s^2	Longitudinal acceleration from vehicle CAN Bus via WSU
PrndlWsu	Integer	none	Current transmission state (Park, Reverse, Neutral, Drive, Low) from vehicle CAN Bus via WSU
HeadlampWsu	Integer	none	Headlamp state from vehicle CAN Bus via WSU
WiperWsu	Integer	none	Wiper state from vehicle CAN Bus via WSU
ThrottleWsu	Real	none	Throttle position from vehicle CAN Bus via WSU
SteerWsu	Real	degrees	Steering angle/position from vehicle CAN Bus via WSU

DataFrontTargets			
Field name	Type	Units	Description
TargetId	Integer	none	Numeric ID assigned by the Mobileye sensor to distinguish between the different objects being tracked; the closest obstacle is given a TargetId value of 1
ObstacleId	Integer	none	ID of new obstacle, as assigned by the Mobileye sensor, and its value will be the last used free ID
Range	Integer	m	Longitudinal position of an object, typically the closest object, relative to a reference point on the host vehicle, according to the Mobileye sensor
RangeRate	Real	m/s	Longitudinal velocity of an object, typically the closest object, relative to the host vehicle, according to the Mobileye sensor
Transversal	Real	m	The lateral position of the obstacle, as determined by the Mobileye sensor
TargetType	Integer	none	Classification of an identified obstacle/target as a car, truck, pedestrian, and so on
Status	Integer	none	Classification of the motion (kinematic state) of an identified obstacle/target as stopped, moving, and so on
CIPV	Integer	none	Field communicating whether an obstacle is the closest in a vehicle's path

each trip on the link. The statistical relationships between surrogate safety measures and rear-end crashes were developed. Additional safety information such as vehicle maneuvering decisions (e.g., excess speed and brake time duration) and traffic volume were also included in the regression models.

Vehicle-Level Safety Surrogate Measures

Safety surrogate measures are usually measured at each timestamp between vehicles that are interacting with each other. In Figure 2, it is assumed that if the distance to the front vehicle is larger than 250 m, or the lateral difference is greater than 3 m, there will be no conflict at this time-spatial point (*14*). Let V_F be the front vehicle speed, and V_S be the subject vehicle speed; the relative speed equals ΔV and is $(V_S - V_F)$. Let a_F be the front vehicle acceleration and a_S be the subject vehicle acceleration; the relative acceleration Δa equals $(a_S - a_F)$. D represents the distance between the two vehicles. The SPMD dataset provides all of this information except for a_F, which can be calculated from the front vehicle speed information between two consecutive time stamps. Equations 1–3 detail surrogate safety measures TTC, MTTC, and DRAC, and the assumptions.

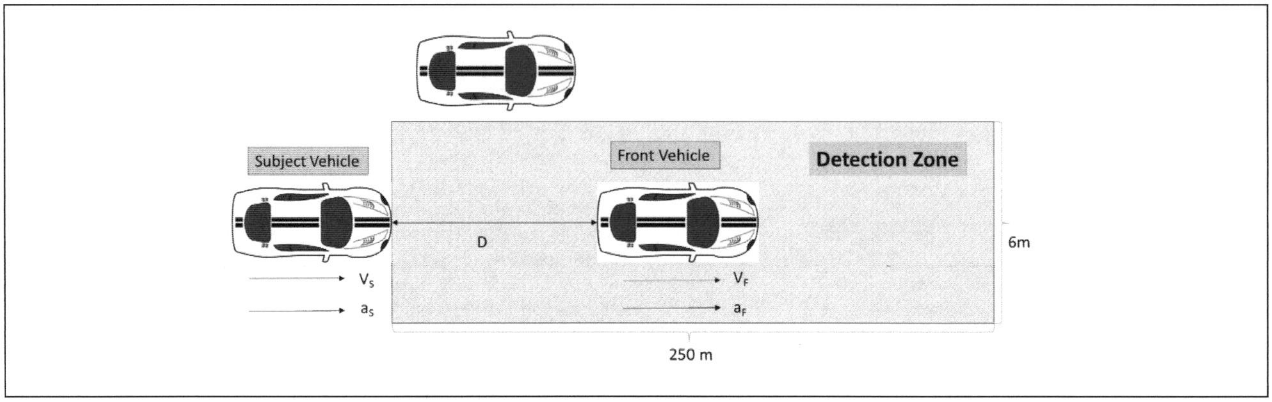

Figure 2. Illustration of vehicle motion in SPMD dataset.

- TTC is defined as "the time that remains until a collision between two vehicles would have occurred if the collision course and speed difference are maintained" (*2*):

$$\text{TTC} = \frac{D}{\Delta V} \tag{1}$$

When $\Delta V \leq 0$, there is no risk of collision at that moment. Let L be the link length. The upper limit of TTC is set to be $\frac{L}{V_s}$, ensuring that if there is a conflict, it is within the roadway segment link.

- MTTC considers the trajectory parameters of the two consecutive vehicles, including their relative distance, speed, and acceleration (*11*). The threshold of MTTC is also assumed to be $\frac{L}{V_s}$:

$$\text{MTTC} = \begin{cases} \max(t_1, t_2), & \text{if } \Delta a > 0 \\ \min(t_1, t_2), & \text{if } \Delta a < 0 \text{ and } \Delta V > 0 \\ t_3, & \text{if } \Delta a = 0 \text{ and } \Delta V > 0 \\ \text{na}, & \text{if } \Delta a \leq 0 \text{ and } \Delta V \leq 0 \end{cases} \tag{2}$$

where

$$t_1 = \frac{-\Delta V + \sqrt{\Delta V^2 + 2\Delta a D}}{\Delta a}$$

$$t_2 = \frac{-\Delta V - \sqrt{\Delta V^2 + 2\Delta a D}}{\Delta a}$$

$$t_3 = \frac{D}{\Delta V}$$

- DRAC is "the minimum deceleration rate required by the following vehicle to avoid a crash with the leading vehicle if the speed of leading vehicle is unchanged during the process" (*15*):

$$\text{DRAC} = \frac{V_S{}^2 - V_F{}^2}{2(D - V_S{*}\text{PRT})} \tag{3}$$

where PRT is the perception–reaction time. The default PRT value used is 0.92 s, which is adopted from Triggs and Harris's study for rear-end collision (*34*). When $V_S - V_F \leq 0$, there is no risk of collision at that moment. When $D \leq V_S$*PRT, the index can be treated as positive infinity, meaning the collision is certain to happen.

Trip-Level Safety Surrogate Measures

A trip is defined as the time a vehicle traverses a roadway segment link. After the vehicle-level safety surrogate measures are available, they will be aggregated into safety surrogate measures for each vehicle trip. During a nonstop long trip, a vehicle may traverse the same link multiple times. Thus vehicle-level safety surrogate measures with the same "Trip" field value can be aggregated into multiple trip-level safety indexes. For the same vehicle, if there are multiple front targets at the same timestamp, the closest one ("CIPV = 1") is kept.

It is assumed that the safety surrogate measure for each time interval equals the value for the end timestamp of that interval. For trip i in a link, the four safety indexes (SI) are formulated in Equations 4–7 to compute the trip-level safety surrogate measures on a link. For the sake of generality, TTC, MTTC, DRAC, and their aggregations are called SI.

- Time Duration (SI_{1i}) is the total time when the surrogate measures exist. Note that the time duration should be less than or equal to the link travel time $\frac{L}{V_s}$:

$$\text{SI}_{1i} = \sum_P (t_{i,j} - t_{i,j-1}) \tag{4}$$

- Average Index (SI_{2i}) is the weighted average of surrogate measures over time:

$$SI_{2i} = \frac{\sum_P (t_{i,j} - t_{i,j-1})*\text{Index}_{i,j}}{\sum_P (t_{i,j} - t_{i,j-1})} \quad (5)$$

- Median Index (SI_{3i}) is the median value of surrogate measures over time:

$$SI_{3i} = \underset{P}{\text{median}}\{\text{Index}_{i,j}\} \quad (6)$$

- Extreme Index (SI_{4i}) is either the minimum TTC or MTTC, or the maximum DRAC, representing the most dangerous situation:

$$SI_{4i} = \min_P\{\text{Index}_{i,j}\}\text{or}\max_P\{\text{Index}_{i,j}\} \quad (7)$$

where P = {time intervals when safety surrogate measures are available}, and j is the index of time stamp.

Link-Level Safety Surrogate Measures

One link could include multiple trips: different vehicles can travel on the same link or one vehicle can travel the same link multiple times. Trip-level safety surrogate measures need to be aggregated to a link-level safety surrogate measure. The proposed link-level safety surrogate measures for each link are presented in Equations 8–11.

- Time Duration (SI_1) is the average length of time of all trips in a link:

$$SI_1 = \frac{\sum_1^N SI_{1i}}{N} \quad (8)$$

- Average Index (SI_2) is the average index of all trips in a link:

$$SI_2 = \frac{\sum_1^N (SI_{2i}*SI_{1i})}{\sum_1^N SI_{1i}} \quad (9)$$

- Median Index (SI_3) is the median index of all trips in a link:

$$SI_3 = \underset{1 \le i \le N}{\text{median}}\{SI_{3i}\} \quad (10)$$

- Extreme Index (SI_4) is the minimum or maximum index of all trips in a link:

$$SI_4 = \min_{1 \le i \le N}\{SI_{4i}\}\text{or}\max_{1 \le i \le N}\{SI_{4i}\} \quad (11)$$

Data Processing

The DataWsu data and DataFrontTargets data were used in this study to calculate SI. Figure 3 shows the framework of data processing and safety index calculation in this study, along with the size and volume of the database after each process.

Initially, each dataset was examined using Python programming language to check the data type and data organization. The datasets were then imported into Hadoop to conduct a query using Apache Hive, a query language that is built on top of Apache Hadoop. Unfortunately, because of the complicated relationship among columns, the aggregation of two big datasets in Hadoop was extremely slow. Moreover, current Hadoop-GIS does not support advanced spatial analysis such as spatial join, so the two datasets were exported into small files to be joined in the PostgreSQL database. The combined data were then imported into ArcGIS to integrate link and intersection information. The combined data were imported back into the PostgreSQL database, and 75-ft buffer zones were created to remove points around the intersections. Finally, a combined dataset was generated for the target type of "car" or "truck."

Vehicle-level SI were calculated within the "range" based on the combined dataset with driving information and front vehicle information (i.e., distance to the front vehicle is less than 250 m and the lateral distance is less than 3 m). Then, safety surrogate measures were calculated for each link. Rear-end crashes in the mid-block and the links with safety surrogate measures, are shown in Figure 4.

Results

The statistical relationship between the link safety surrogate measures and mid-block rear-end crashes was developed using the negative binomial (NB) model, of which the mean is estimated as a log-linear function of the explanatory variables. Only links with observed surrogate measures were selected. Of the 2,772 selected links, the average link length is 352 m, the median length is 213 m, the minimum length is 35 m, and the maximum length is 3,635 m. The surrogate safety measures and other independent variables include vehicle maneuvering actions such as average speed and brake duration and segment link length. The link length is treated as an independent variable rather than as an exposure variable to a crash because the length is used to set the upper limit for a safety index. Variance inflation factors (VIFs) were calculated for each independent variable to examine the possibility of multicollinearity. In this study, all VIFs are less than 10 except for the average index and median index. Small VIF values mean no high correlations exist among the independent variables after excluding average index and median index. Traffic volume such as AADT is typically included as the traffic exposure to crashes. However, during the variable selection process, AADT

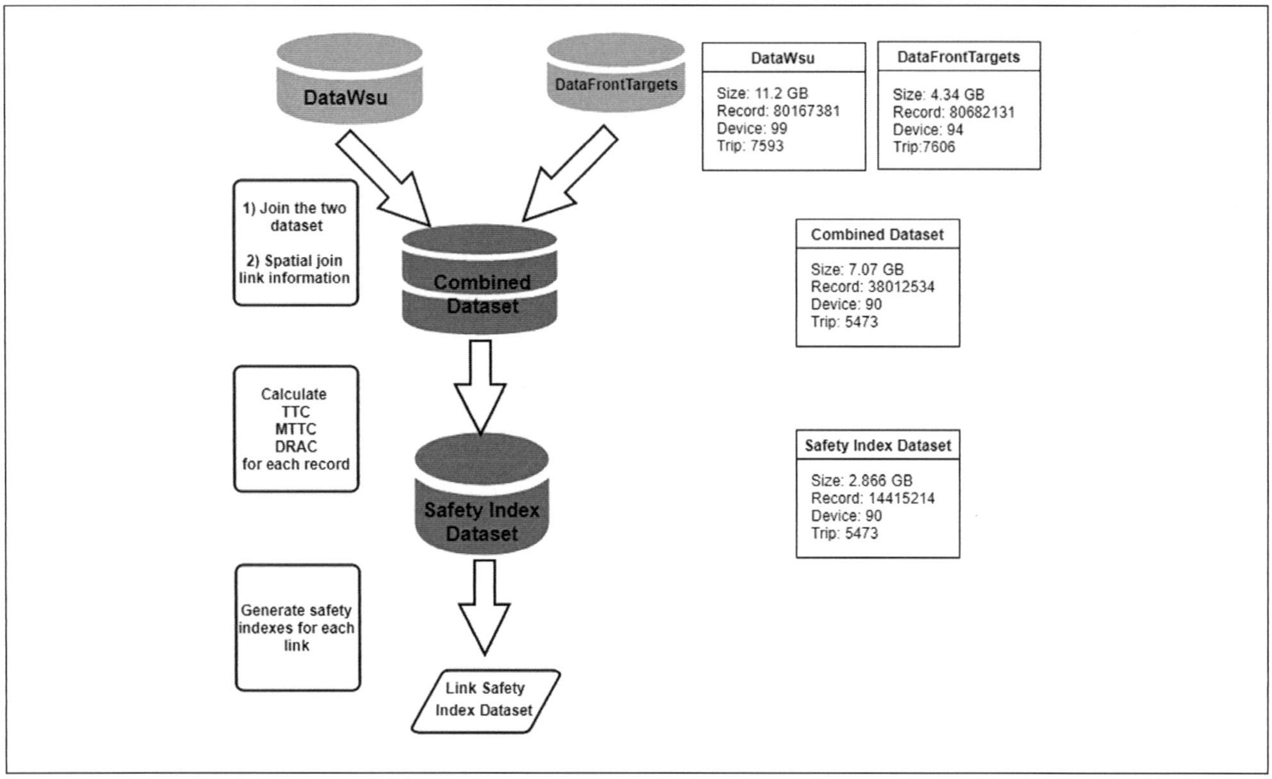

Figure 3. Data processing framework.

was not statistically significant for the selected links and, therefore, AADT was excluded from the final NB models. The results of the models are shown in Table 2.

For TTC, the variables of time duration, extreme index, average speed, brake duration, and link length are statistically significant at the 1% level or lower. Time duration and extreme index are two SI that affect crash frequency at a statistically significant level. The positive sign of time duration means that the longer the dangerous situation lasts, the more frequently a crash will occur. The negative sign of extreme index suggests a larger minimum TTC is associated with a smaller number of crashes. The positive impact of average speed on crashes indicates that a faster speed results in more crashes. The longer brake time within the link means a worse link safety condition (higher crash frequency). The positive effect of link length on crashes simply suggests that more crashes may happen if the link is longer.

When compared with TTC, the differences in parameter estimates for MTTC include the lower time duration value and larger extreme index value. It is found that the mean value of time duration for MTTC is larger than for TTC (1.22 s versus 0.56 s) and the mean value of extreme index for MTTC is lower than for TTC (1.86 s versus 3.43 s), suggesting an MTTC-based conflict is more easily detected compared with a TTC-based conflict. As an acceleration-based surrogate measure, all SI are statistically significant for DRAC. The positive sign of extreme index for DRAC means the increase of the maximum DRAC leads to more crashes. Average speed and brake duration are always statistically significant with the similar parameter estimates, irrespective of the SI. Excess speed, reflecting the driver's aggressiveness in choosing speed, was calculated and included in the model. The excess speed equals zero when the average speed is less than the speed limit and equals the difference between the two if the average speed is greater than the speed limit. Model results show insignificant SI when replacing average speed with excess speed.

The dispersion parameter was estimated in the NB model to measure the data overdispersion. The respective values of 0.399, 0.419, and 0.372 in the models for TTC, MTTC, and DRAC justify the choice of the NB model. Goodness-of-fit measures such as Akaike Information Criterion (AIC) and pseudo R-squared were used to compare the model performance. AIC uses the maximum log-likelihood function with a penalizing term related to the number of variables. A lower AIC value indicates a better fit. McFadden pseudo R-squared, analogous to the R-squared value for linear regression models, equals

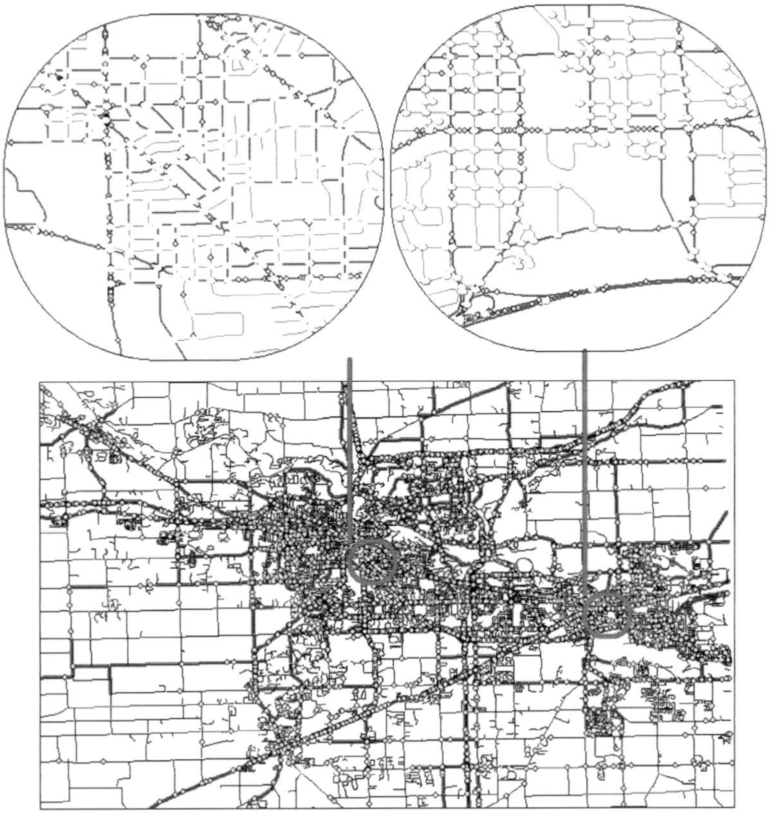

Figure 4. Maps with observed safety index and crash points.

Table 2. NB Models for Mid-Block Rear-End Crashes

	TTC		MTTC		DRAC	
	Estimate	*p*-value	Estimate	*p*-value	Estimate	*p*-value
(Intercept)	–0.990***	$<2e^{-16}$	–0.624***	$<2e^{-16}$	–1.237***	$<2e^{-16}$
Time duration	0.222**	0.002	0.147***	$<2e^{-16}$	0.063***	$<2e^{-16}$
Extreme index	–0.100***	$<2e^{-16}$	–0.518***	$<2e^{-16}$	4×10^{-6}**	0.007
Average speed	0.082***	$<2e^{-16}$	0.083***	$<2e^{-16}$	0.077***	$<2e^{-16}$
Brake duration	1.136***	$<2e^{-16}$	1.040***	$<2e^{-16}$	1.390***	$<2e^{-16}$
Link length	0.001***	$<2e^{-16}$	0.001***	$<2e^{-16}$	0.001***	$<2e^{-16}$
Number of observations	2,772		2,772		2,772	
Dispersion parameter	0.399		0.419		0.372	
2 log-likelihood	–9615.862		–9542.735		–9701.151	
AIC	9629.900		9556.700		9715.200	
McFadden pseudo *R*-squared	0.066		0.075		0.054	

Note: AIC = Akaike Information Criterion.
***p = .001. **p = .01. *p = .05.

1 minus the ratio of the log-likelihood of the full model to the log-likelihood of the intercept-only model. The pseudo R-squared takes a value between 0 and 1, and a higher value indicates better model performance. The relatively low R-squared values mean that if crash data are considered as ground truth, it cannot be concluded with confidence which surrogate measure is the best safety metric. Among the three indexes, MTTC is considered as the most statistically significant surrogate safety measure for crashes because it has the highest pseudo R-squared value and the lowest AIC value. In summary, the proposed safety information can be useful for evaluating segment link safety when roadway geometric and traffic information is limited.

Conclusion

An SPMD dataset was used to evaluate rear-end crashes in the mid-block. Three main objectives were accomplished: 1) develop a framework to process the SPMD big data; 2) construct surrogate safety measures from SPMD data; and 3) analyze the statistical relationship between crash records and a calculated safety index.

Unlike other studies that adopted surrogate safety measures to identify traffic conflicts, this study attempted to evaluate segment safety using surrogate safety measures. Aggregated surrogate measures were developed for a trip-level and a link-level safety index, as surrogate measures are usually taken at the vehicle level by measuring the time and space between a pair of vehicles. Surrogate measures are treated as SI, which quantify the severity of a potential traffic conflict. For example, a higher value of TTC means a higher safety index. In the real world, however, a TTC of 3 s may result in a crash whereas a TTC of 1 s may not; this is dependent on the driver. Thus, arbitrary threshold values were not used in the study to preserve the integrity of the information. The logic of this study is to keep the calculated surrogate safety measures for all the time points and then provide indicators (e.g., time duration, average index, median, or minimum/maximum index) to summarize these SI on the same trip for each link.

The NB models for mid-block rear-end crashes show the expected impact of explanatory variables on crashes. Among the three models, MTTC has a better goodness of fit when compared with TTC and DRAC. The findings show that augmenting safety analysis with surrogate measures and vehicle performance (i.e., speed and brake duration from CVs) improves the overall model performance. Such information can be vital when detailed roadway and traffic data are absent.

Some abnormal numbers were produced and removed, and the measurement errors in the dataset were unknown. The study is also less comprehensive because there is no record of the dataset in some columns (e.g. right turn or left turn signal). The complexity of the dataset means that some of the assumptions or data processing approaches used in this study may not be optimal in all situations; thus, future studies should search for other effective approaches. Future studies can be expanded into the comparison of other safety surrogate measures such as PET, Delta-V, and extended Delta-V. New and emerging safety surrogate measures can be developed with the rich information provided through CV safety technologies. In addition, future studies can also take advantage of the high-resolution vehicle kinematics and Signal Phasing and Timing (SPaT) data in the SPMD program to study risky driver behaviors such as red light running.

Author Contributions

The authors confirm contribution to the paper as follows: study conception and design: Qin, He; data collection: He, Sayed; analysis and interpretation of results: He, Qin, Sayed; draft manuscript preparation: He, Qin, Liu, Sayed. All authors reviewed the results and approved the final version of the manuscript.

References

1. NHTSA. *Crash Stats: Early Estimate of Motor Vehicle Traffic Fatalities for the First Half, January to June of 2016*. https://crashstats.nhtsa.dot.gov/Api/Public/ViewPublication/812332.
2. Gettman, D., and L. Head. Surrogate Safety Measures from Traffic Simulation Models. *Transportation Research Record: Journal of the Transportation Research Board*, 2003. 1840: 104–115.
3. Henclewood, D., M. Abramovich, and B. Yelchuru. *Safety Pilot Model Deployment—One Day Sample Data Environment Data Handbook, V. 1.2*. USDOT Research and Technology Innovation Administrations, Washington, D.C., 2014.
4. Dijkstra, A., P. Marchesini, F. Bijleveld, V. Kars, H. Drolenga, and M. van Maarseveen. Do Calculated Conflicts in Microsimulation Model Predict Number of Crashes? *Transportation Research Record: Journal of the Transportation Research Board*, 2010. 2147: 105–112.
5. Hayward, J. C. Near Miss Determination through Use of a Scale of Danger. *Highway Research Record*, Vol. 384, 1972, pp. 24–34.
6. Kuang, Y., and X. Qu. A Review of Crash Surrogate Events. In M. Beer, S.-K. Au, & J. W. Hall (eds.), *Vulnerability, Uncertainty, and Risk: Quantification, Mitigation, and Management*. American Society of Civil Engineers, Reston, VA, 2014. pp. 2254–2264.
7. Laureshyn, A., Å. Svensson, and C. Hydén. Evaluation of Traffic Safety, Based on Micro-Level Behavioural Data: Theoretical Framework and First Implementation. *Accident Analysis & Prevention*, Vol. 42, No. 6, 2010, pp. 1637–1646.

8. Horst, R. Time-to-Collision as a Cue for Decision-Making in Braking. *Vision in Vehicles*, Vol. 3, 1991, pp. 10–26.
9. Hydén, C., and L. Linderholm. The Swedish Traffic-Conflicts Technique. In *International Calibration Study of Traffic Conflict Techniques*, Springer, Berlin Heidelberg, 1984, pp. 133–139.
10. Minderhoud, M. M., and P. H. Bovy. Extended Time-to-Collision Measures for Road Traffic Safety Assessment. *Accident Analysis & Prevention*, Vol. 33, No. 1, 2001, pp. 89–97.
11. Ozbay, K., H. Yang, B. Bartin, and S. Mudigonda. Derivation and Validation of New Simulation-Based Surrogate Safety Measure. *Transportation Research Record: Journal of the Transportation Research Board*, 2008. 2083: 105–113.
12. Charly, A., and T. V. Mathew. Estimation of Modified Time To Collision as Surrogate For Mid-Block Crashes under Mixed Traffic Conditions. Presented at 96th Annual Meeting of the Transportation Research Board, Washington, D.C., 2017.
13. Gettman, D., L. Pu, T. Sayed, and S. G. Shelby. *Surrogate Safety Assessment Model and Validation: Final Report*. Federal Highway Administration, Washington, D.C., 2008.
14. Allen, B. L., B. T. Shin, and P. J. Cooper. Analysis of Traffic Conflicts and Collisions. *Transportation Research Record: Journal of the Transportation Research Board*, 1978. 667: 67–74.
15. Almqvist, S., C. Hydén, and R. Risser. Use of Speed Limiters in Cars for Increased Safety and a Better Environment. *Transportation Research Record: Journal of the Transportation Research Board*, 1991. 1318: 34–39.
16. Cunto, F. J. C., and F. F. Saccomanno. Microlevel Traffic Simulation Method for Assessing Crash Potential at Intersections. Presented at 86th Annual Meeting of the Transportation Research Board, Washington, D.C., 2007.
17. Young, W., A. Sobhani, M. G. Lenné, and M. Sarvi. Simulation of Safety: A Review of the State of the Art in Road Safety Simulation Modelling. *Accident Analysis & Prevention*, Vol. 66, 2014, pp. 89–103.
18. Saunier, N., and T. Sayed. Automated Analysis of Road Safety with Video Data. *Transportation Research Record: Journal of the Transportation Research Board*, 2007. 2019: 57–64.
19. Zangenehpour, S. *A Video-Based Methodology for Extracting Microscopic Data and Evaluating Safety Countermeasures at Intersections Using Surrogate Safety Indicators*. Doctoral dissertation. McGill University, Montreal, Québec, Canada, 2016.
20. Astarita, V., G. Guido, A. Vitale, and V. Giofré. A New Microsimulation Model for the Evaluation of Traffic Safety Performances. *European Transport\Trasporti Europei*, No. 51, 2012, pp. 1–16.
21. Meng, Q., and J. Weng. Evaluation of Rear-End Crash Risk at Work Zone Using Work Zone Traffic Data. *Accident Analysis & Prevention*, Vol. 43, No. 4, 2011, pp. 1291–1300.
22. Oh, C., and T. Kim. Estimation of Rear-End Crash Potential Using Vehicle Trajectory Data. *Accident Analysis & Prevention*, Vol. 42, No. 6, 2010, pp. 1888–1893.
23. Campbell, K. L. The SHRP 2 Naturalistic Driving Study: Addressing Driver Performance and Behavior in Traffic Safety. *TR News*, No. 282, 2012, pp. 30–35.
24. Tarko, A. P. Use of Crash Surrogates and Exceedance Statistics to Estimate Road Safety. *Accident Analysis & Prevention*, Vol. 45, 2012, pp. 230–240.
25. Wu, K. F., and P. P. Jovanis. Crashes and Crash-Surrogate Events: Exploratory Modeling with Naturalistic Driving Data. *Accident Analysis & Prevention*, Vol. 45, 2012, pp. 507–516.
26. Markkula, G., J. Engström, J. Lodin, J. Bärgman, and T. Victor. A Farewell to Brake Reaction Times? Kinematics-Dependent Brake Response in Naturalistic Rear-End Emergencies. *Accident Analysis & Prevention*, Vol. 95, 2016, pp. 209–226.
27. Montgomery, J., K. D. Kusano, and H. C. Gabler. Age and Gender Differences in Time to Collision at Braking from the 100-Car Naturalistic Driving Study. *Traffic Injury Prevention*, Vol. 15, No. Suppl. 1, 2014, pp. S15–S20.
28. Deering, A. M. *A Framework for Processing Connected Vehicle Data in Transportation Planning Applications*. Doctoral dissertation. The University of Texas at Austin, Austin, TX, 2017.
29. Ghanipoor Machiani, S., A. Jahangiri, V. Balali, and C. Belt. Predicting Driver Risky Behavior for Curve Speed Warning Systems Using Real Field Connected Vehicle Data. Presented at 96th Annual Meeting of the Transportation Research Board, Washington, D.C., 2017.
30. Liu, J., and A. J. Khattak. Delivering Improved Alerts, Warnings, and Control Assistance Using Basic Safety Messages Transmitted between Connected Vehicles. *Transportation Research Part C: Emerging Technologies*, Vol. 68, 2016, pp. 83–100.
31. Vasudevan, M., D. Negron, M. Feltz, J. Mallette, and K. Wunderlich. Predicting Congestion States from Basic Safety Messages by Using Big-Data Graph Analytics. *Transportation Research Record: Journal of the Transportation Research Board*, 2015. 2500: 59–66.
32. Zheng, J., and H. X. Liu. Estimating Traffic Volumes for Signalized Intersections Using Connected Vehicle Data. *Transportation Research Part C: Emerging Technologies*, Vol. 79, 2017, pp. 347–362.
33. Abdel-Aty, M., and X. Wang. Identifying Intersection-Related Traffic Crashes for Accurate Safety Representation. *Institute of Transportation Engineers. ITE Journal*, Vol. 79, No. 12, 2009, p. 38.
34. Triggs, T. J., and W. G. Harris. *Reaction Time of Drivers to Road Stimuli*. Monash University Accident Research Centre, Victoria, Australia, 1982.

The Safety Section (ANB00) peer-reviewed this paper (18-04931).

Article

TRR
JOURNAL OF THE TRANSPORTATION RESEARCH BOARD

Transportation Research Record
2018, Vol. 2672(38) 12–21

DOI: 10.1177/0361198118791379
journals.sagepub.com/home/trr
SAGE

Predicting Imminent Crash Risk with Simulated Traffic from Distant Sensors

Zhi Chen[1], Xiao Qin[1], Renxin Zhong[2], Pan Liu[3], and Yang Cheng[4]

Abstract

The aim of this research was to investigate the performance of simulated traffic data for real-time crash prediction when loop detector stations are distant from the actual crash location. Nearly all contemporary real-time crash prediction models use traffic data from physical detector stations; however, the distance between a crash location and its nearest detector station can vary considerably from site to site, creating inconsistency in detector data retrieval and subsequent crash prediction. Moreover, large distances between crash locations and detector stations imply that traffic data from these stations may not truly reflect crash-prone conditions. Crash and noncrash events were identified for a freeway section on I-94 EB in Wisconsin. The cell transmission model (CTM), a macroscopic simulation model, was applied in this study to instrument segments with virtual detector stations when physical stations were not available near the crash location. Traffic data produced from the virtual stations were used to develop crash prediction models. A comparison revealed that the predictive accuracy of models developed with virtual station data was comparable to those developed with physical station data. The finding demonstrates that simulated traffic data are a viable option for real-time crash prediction given distant detector stations. The proposed approach can be used in the real-time crash detection system or in a connected vehicle environment with different settings.

A driver must constantly respond to changes in traffic and other roadway conditions by changing speed, switching lanes, or even changing directions. The inevitable and frequent changes in driving conditions can result in driver errors. The early detection of crash-prone traffic conditions can alert the driver to make necessary evasive maneuvers, and it can also lead to appropriate traffic control strategies to mitigate imminent crash risk. The factors contributing to a crash are directly related to or result from the prevailing traffic conditions before the event. The wide deployment of Advanced Traveler Information Systems (ATIS) has made the collection, storage, and processing of real-time traffic data readily available. Researchers can now gather real-time information related to crash occurrence. Among different types of traffic sensors, inductive loop detectors have been a popular data source for real-time crash prediction. Numerous studies have investigated the relationship between crash risk and prevailing traffic conditions using real-time traffic information collected from loop detectors; therefore, it would be beneficial to develop a real-time crash prediction model that can detect crash-prone traffic patterns.

The lead time before a crash has been thoroughly studied in cases in which freeway crashes were predicted using loop detector data at different time slices (*1*). However, studies regarding how the space between detector stations and crash locations affects crash prediction accuracy are rare. It is expected that the traffic conditions near a crash location would better reflect the hazardous conditions leading to the crash. In nearly all previous studies, crash prediction models were developed using the traffic data from immediate upstream and downstream loop detector stations; however, crashes may take place anywhere between two detector stations, and the distance between a crash location and the nearest detector station can vary considerably from site to site. The variation creates inconsistency in detector data retrieval and subsequent crash prediction, and also casts

[1]Department of Civil and Environmental Engineering, University of Wisconsin-Milwaukee, Milwaukee, WI
[2]School of Engineering, Sun Yat-Sen University, Guangzhou, China
[3]Jiangsu Key Laboratory of Urban ITS, Southeast University, Jiangsu Province Collaborative Innovation Center of Modern Urban Traffic Technologies, Nanjing, China
[4]Department of Civil & Environmental Engineering, University of Wisconsin-Madison, Madison, WI

Corresponding Author:
Address correspondence to Xiao Qin: qinx@uwm.edu

doubt on how well the traffic data from these stations reflects crash-prone conditions.

The goal of this study was to examine the performance of simulated traffic data in detecting crash-prone traffic patterns given large distances between loop detector stations. The cell transmission model (CTM) was applied to simulate traffic conditions close to a crash location to accommodate the challenges of spatial disparities in loop detectors and discrepancies in crash prediction.

Literature Review

Ideally, traffic conditions at the location and time of a crash should be used in capturing and recording what exactly happened before the crash. The reality, however, is that traffic data often are observed from physical detectors which are not likely to be installed anywhere near the crash. Alternatively, virtual detectors may be exploited at any location through the development of traffic simulation models. CTM is a representative macroscopic traffic flow simulation model proposed by Daganzo (*2*). In a CTM, a highway is partitioned into a series of cells, and the cell length can be user-defined. The traffic density and other characteristics in each cell can be calculated following the traffic flow–density relationship defined by the fundamental diagram (FD). The methodology is detailed in the next section. CTM is very compatible with traffic flow data collected from loop detectors, which is a clear advantage for real-time crash prediction. CTM and its extensions, including switching-mode model (SMM) and stochastic CTM, have shown promising results in accurately predicting traffic flow characteristics using loop detector data as inputs (*3–5*). Studies show that simulated traffic densities and flow rates may achieve an error rate as low as 7.9% (*5*) and 4% (*4*).

Although CTM is a powerful model for simulating traffic flow characteristics, the existence of an underlying relationship between traffic flow and crash risk warrants the validity of a real-time crash prediction model. Many efforts have been made to define and quantify such a relationship. Some studies used data mining techniques including the Kohonen clustering algorithm, neural networks, and the Bayesian network (*6–8*). Although data mining methods are capable of accounting for correlation within speed, flow, and occupancy (*6*), they cannot identify explicit relationships between crash risk and traffic flow variables. Other studies have used statistical regression models because they provide a clear connection between crash probability and traffic flow variables. Among all methods, the case-control design has been the most popular in real-time crash prediction studies (*1, 6–13*) because the design controls for exogenous factors such as locations and roadway geometries, and provides more accurate estimates by using both crash and noncrash traffic information (*1*). However, the case-control method assumes that each stratum of a crash event and its matched noncrash events have different constant terms, meaning that only the odds ratio of crash probability is predicted. In other words, the crash probability cannot be directly predicted given the explanatory variables. An alternative method is the binary logistic regression. The binary logistic model is used to estimate the probability of a binary response, such as a crash event based on one or more predictors, meaning the crash probability can be directly predicted given the explanatory variables (*14–16*).

The relationship between crash probability and traffic flow variables (e.g., mean, standard deviation, and coefficient of variation of traffic flow, speed, and occupancy) is under intense scrutiny (*1, 8, 9, 17–19*). Roshandel et al. conducted a comprehensive review of the relationship between real-time traffic conditions and freeway crashes using the meta-analysis of past literature (*20*). The authors identified some statistically significant crash-contributing factors such as average speed and speed variation which have been consistently reported in past studies. The authors also pointed out a limitation of using loop detector data, which is that researchers must use data from loop detectors that could be far from crash locations (*20*). The spacings between loop detector stations vary significantly within and across studies, rendering findings that are unreliable within the same study and inconsistent between studies.

Methodology

This section covers the methodology for simulating spatial and temporal traffic during the period before a crash, and also details the regression method for predicting crash occurrence. The traffic variables are simulated with CTM, and the binary logistic regression model is used to estimate the probability of a crash based on simulated traffic conditions.

Cell Transmission Model (CTM)

CTM is a macroscopic traffic simulation model that predicts macroscopic traffic behavior such as flow and density at a finite number of cells at different time steps on a given highway corridor (*2*). It is a powerful simulation tool that can capture many important traffic phenomena such as queue formation and dissipation as well as shockwave propagation (*2*). Compared with microscopic simulation models, CTM is computationally efficient and easier to configure and calibrate. CTM also operates sufficiently with aggregated traffic flow data from detector stations. The core component of CTM is the FD that

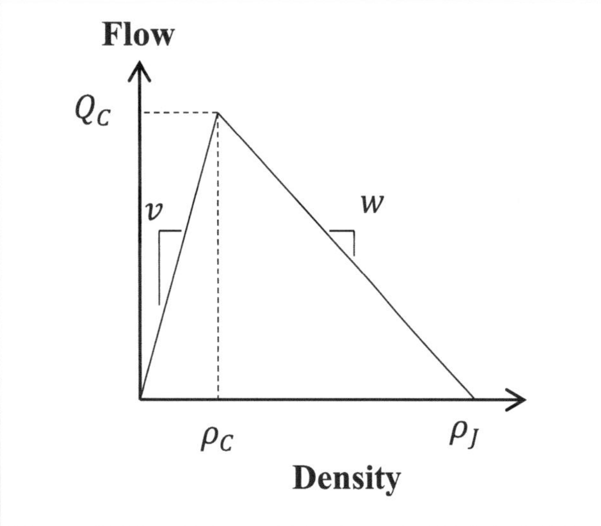

Figure 1. Triangular fundamental diagram.

defines the flow–density relationship and governs how the traffic state in each cell evolves over time. Figure 1 shows a triangular FD for developing CTM.

In CTM, a highway segment is divided into a series of cells. The density of each cell evolves following the conservation law of vehicles. Assume that Cell i is characterized by the triangular FD in Figure 1, where Q_C is the capacity flow, ρ_C is the critical density, ρ_J is the jam density, v is the free-flow speed, and w is the shockwave speed. The density for Cell i without on- or off-ramps is determined by

$$\rho_i(k+1) = \rho_i(k) + \frac{T}{l_i}(q_i(k+1) - q_i(k)) \quad (1)$$

where

k is the time step index,

$\rho_i(k)$ is the density of Cell i during the kth time step,

T is the length of the time step,

l_i is the length of Cell i, and

$q_i(k)$ is the flow rate into Cell i during the kth time step. The flow rate is determined by the sending and receiving functions. For Cell i, the sending function $S_i(k)$ represents the maximum flow that may be supplied during the kth time step, and the receiving function $R_i(k)$ represents the maximum flow that may be received. The two functions are determined in Equations 2 and 3, respectively, as

$$S_i(k) = \min(v_i\rho_i(k), Q_{C,i}) \quad (2)$$

$$R_i(k) = \min\left(Q_{C,i}, w_i\left(\rho_{J,i} - \rho_i(k)\right)\right) \quad (3)$$

The flow rate, $q_i(k)$, is determined by

$$q_i(k) = \min(S_{i-1}(k), R_i(k)) \quad (4)$$

Binary Logistic Regression Model

In a binary logistic regression model, the probability of a crash event can be formulated as

$$p(\boldsymbol{X_i}) = \frac{1}{1 + e^{-g(X_i)}} \quad (5)$$

where $p(\boldsymbol{X_i})$ represents the crash probability given $\boldsymbol{X_i} = (x_{i,1}, x_{i,2}, \ldots, x_{i,k})$, a set of k explanatory variables for sample i, and $g(X_i)$ is a linear combination of the following variable set

$$g(X_i) = \beta_0 + \beta_1{*}x_{i,1} + \beta_2{*}x_{i,2} + \ldots + \beta_k{*}x_{i,k} \quad (6)$$

where $(\boldsymbol{\beta}_0, \beta_1, \beta_2, \ldots, \beta_k)$ are the corresponding coefficients for $(x_{i,1}, x_{i,2}, \ldots, x_{i,k})$.

The parameters $\boldsymbol{\beta} = (\beta_0, \beta_1, \beta_2, \ldots, \beta_k)$ can be estimated by maximizing the following log-likelihood function:

$$lnL(\boldsymbol{\beta}, \boldsymbol{X_i}) = \sum_{i=1}^{n}\left[(\beta_0 + \beta_1{*}x_{i,1} + \ldots + \beta_k{*}x_{i,k}) - ln\left(1 + e^{-(\beta_0 + \beta_1{*}x_{i,1} + \ldots + \beta_k{*}x_{i,k})}\right)\right] \quad (7)$$

Data Description

The dataset used in this study consists of 3 years (2012–2014) of crash data, roadway characteristics, and traffic data for a 4.15-mi I-94 EB corridor in Wisconsin. The corridor has three lanes with one on-ramp and one off-ramp. There are seven mainline loop detector stations and one off-ramp loop detector station on the EB and WB corridor, and there is no detector station on the on-ramp. The layout of the corridor and detector stations is shown in Figure 2.

The 3rd and 6th detector stations were assumed to be absent and were not used in this study in order to test the concept of using simulated traffic data for crash prediction in the setting of large detector spacing. The remaining stations were named as S1, S2, S3, S4, and S5 as shown in Figure 2. The CTM model would provide inaccurate simulated traffic data given missing on-ramp flow, so one segment between S1 and S3 and one segment between S4 and S5 were included in place of the missing on-ramp detector station between stations S3 and S4.

Spacings between S1 and S2, S2 and S3, and S4 and S5 are 1.00 mi, 1.06 mi, and 1.38 mi, respectively. Segments of interest include one 2.06 mi-long segment between S1 and S3, and a 1.38 mi-long segment between S4 and S5. The two segments are divided into 13 cells for CTM simulation. Cells have a uniform length of 0.25 mi with a few exceptions that range from 0.26 mi to 0.29 mi.

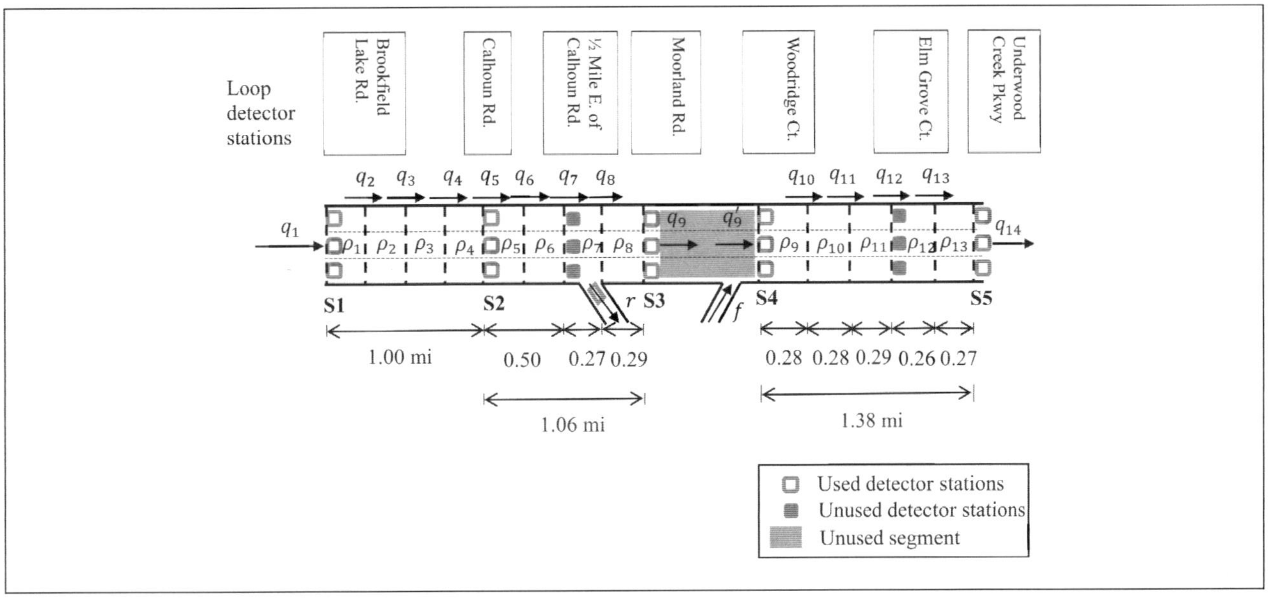

Figure 2. Layout of physical loop detector stations and CTM cells.

In Figure 2, the default cell length is 0.25 mi if no other length is specified. The average cell length is 0.264 mi with a standard deviation of 0.016 mi.

Crash data were retrieved from the web-based query and retrieval facility for Wisconsin Department of Transportation (DOT) crash data as well as from reports archived in the WisTransPortal data management system. Crashes occurring at the study site from 2012 to 2014 were included in the study. Any other crash that happened within 2 h after a crash occurrence and within 2 mi of the crash location was considered a secondary crash and was subsequently removed (*21*). Crash time is required to build up the stratum, so crashes with missing times were excluded. Weather information such as snow or rain was collected from the Wisconsin DOT's Road Weather Information System (RWIS), and traffic information in a 1-min time interval was extracted from the V-SPOC (Volume, Speed, and Occupancy) application suite (*22*). All inaccurate data were eliminated from the raw 1-min data based on the following criteria: 1) occupancy < 0 or > 100; 2) speed < 0 or > 100; 3) volume < 0 or > 50 in 1 min; 4) volume > 0 with speed $= 0$ or speed > 0 with volume $= 0$ (*23*).

For each crash in the dataset, 1-min traffic data from the 0–5 min or 5–10 min intervals before the crash occurred were collected from immediately upstream and downstream physical loop detector stations. For example, if a crash occurred at 10:00 a.m., the traffic data were extracted from 9:55 to 10:00 a.m. if the 0–5 min interval was of interest, or from 9:50 to 9:55 a.m. if the 5–10 min interval was of interest. Each crash has a geolocation, which determines the cell in which the crash occurred. Traffic data that are not affected by or associated with crash occurrence, also called "noncrash events," are also required to develop the crash prediction model. Ten noncrash events were selected for each crash by randomly selecting a time among 1,578,240 1-min intervals in 2012–2014 (60 min × 24 h × 1096 days in 2012–2014) and a cell among 13 cells. Noncrash events were selected in a way that ensured none of the times was within 2 h of any crash. The 5-min traffic data consisting of data from five 1-min intervals was retrieved from physical detector stations for noncrash events in the same way it was for crash events.

CTM Setup and Calibration

An FD is required to operate the CTM simulation. Differences in roadway characteristics such as distances to on-/off-ramps could lead to cells having different traffic patterns and therefore different FDs. The calibration algorithm proposed by Zhong et al. (*24*) was applied for calibrating FDs. The algorithm found the optimal FD parameters to minimize the discrepancy between CTM simulated traffic data and observed traffic data. That study can be referenced for more details.

Traffic data must be collected from detector stations in order to calibrate FDs. Traffic data from five detector stations were collected between 4:00 a.m. and 12:00 p.m. on May 6, 2013. Free flow, which is the onset and offset of congestion, was observed at all five stations during this time,, which is ideal for FD calibration. It is not computer-efficient to solve for optimal FDs for each cell when the cell number is relatively large and the station

Table 1. FD Parameters

Cell*	v	ρ_C	ρ_J	Q_C	w
1, 2	58.74	108.04	398.65	6347	21.84
3, 4	58.22	112.96	545.34	6577	15.21
5, 6	66.38	100.16	409	6649	21.53
7	60.52	107.76	569.44	6522	14.13
8	56.5	99.26	544.03	5608	12.61
9	63.42	92.6	370.42	5873	21.14
10, 11, 12	73.32	79.77	644.5	5848	10.36
13	67.64	86.6	359.95	5858	21.43

Note: * denotes the cells sharing the same FD.

number is small. Therefore, it was determined that several adjacent cells might share one FD. Table 1 presents the FD parameters for different cells. The first column shows the cells with the same FD. For example, Cell 1 and 2 have the same FD. Note that Cell 7 has its own FD as it has an off-ramp and could have a very different FD.

The validity of proposed FD parameters was examined by comparing simulated traffic variables with observed traffic variables at the two physical stations that were not used for calibrating CTM. Traffic data from physical detector stations in Cells 7 and 12 were collected from 4:00 a.m. to 12:00 p.m. on May 6, 2013. During the same time period, traffic data were simulated from virtual stations by CTM in the same cells. The mean absolute percentage error (MAPE) is used to measure the difference:

$$\text{MAPE} = \frac{1}{N}\sum\nolimits_{i=1}^{N}\frac{|\bar{y}(i) - y(i)|}{y(i)} \tag{8}$$

where N is the count of 1-min traffic records, $\bar{y}(i)$ is the ith simulated traffic volume or density, and $y(i)$ is the ith observed traffic flow rate or traffic density. In Cell 7, MAPEs are 5.76% and 7.03% for the flow rate and density at the first physical station. In Cell 12, MAPEs are 5.95% and 11.46% for the flow and density at the second physical station. The small MAPE values show that CTM with proposed FD parameters can provide reliable measurements of traffic flow behavior at these virtual stations.

CTM Simulation

In the simulation, the Courant–Friedrichs–Lewy (CFL) condition (*25*) needs to be fulfilled to guarantee a feasible CTM. . The CFL condition occurs when a vehicle cannot travel across more than one cell during one simulation step, that is, $v_i{*}T \leq l_i$ where v_i is the free-flow speed, T is the simulation time step, and l_i is the cell length. The CFL constraint prohibits the use of a time step as large as 1 min based on the constructed cell length. Therefore, 5 s was used as the time step (T = 5 s) to guarantee the feasibility of the CTM.

The cell and segment within which the crash occurred, as well as its upstream and downstream detector stations, were identified to simulate crash location traffic data. S1 and S3 are considered as the upstream and downstream stations for the first segment, and S4 and S5 are the two stations for the second segment. The flow data from both stations during the 0–5 min or 5–10 min period before a crash were used as the in-flow and out-flow of the segment consisting of the cells in between. A 0th-order interpolation was applied to generate the in-flow/out-flow data in 5 s. A CTM was then run to simulate how traffic data in those cells evolve at each time step within the 5-min time interval.

Virtual detector stations were set up at the beginning of each cell for the CTM simulation. The spacing between virtual stations was equal to the cell length, and was therefore consistent in distance. The virtual stations were expected to capture traffic conditions at locations closer to the crash site where traffic conditions should be more related to the crash than those collected from more distant physical detector stations. Similar to physical detector stations, virtual stations were set up to measure flow, speed, and density. Because the time step was set to be 5 s for the CTM simulation, virtual stations would detect 5-s traffic data. To be comparable to the 1-min traffic data from physical stations, 5-s simulated traffic data were aggregated to 1-min traffic data.

A virtual station k is located at the beginning of Cell k, and measures flow, q_k, density, ρ_k and speed, s_k. The model developed with traffic data from the two virtual upstream and two virtual downstream stations is called a two-station setting. The model developed with traffic data from one virtual upstream and one virtual downstream station is called a one-station setting.

Figure 3 shows that for crashes occurring in Cell k, the two-station setting would include virtual station $k-1$ and k as two virtual upstream stations, and virtual

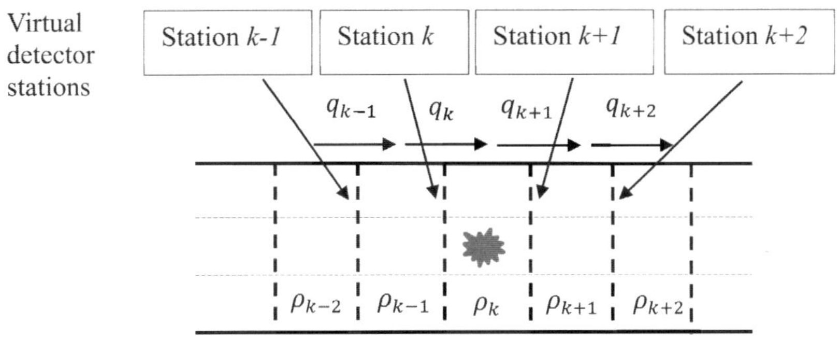

Figure 3. Layout of virtual loop detector stations.

station $k + 1$ and $k + 2$ as two virtual downstream stations; the one-station setting would only include virtual station k as one virtual upstream station, and virtual station $k + 1$ as one virtual downstream station. For each crash/noncrash that occurred in the remaining cells, the virtual upstream and downstream detector stations were identified and traffic data were collected from these virtual detector stations for developing crash prediction models.

Analysis and Discussion

The traffic variables such as mean, standard deviation, and coefficient of variation of flow, density, and speed were calculated for physical or virtual detector stations for each event. Traffic data from one immediate upstream and one immediate downstream station were used to generate variables for physical stations.

Several studies have considered the impact of various traffic states on safety. Abdel-Aty et al. assessed crash risk at two different traffic states categorized by speed (*15*). The authors discovered different crash-contributing traffic parameters in the high- and low-speed regimes. Li et al. (*10*) found that different sets of statistically significant traffic variables affect crash probability in distinct traffic states when expanding from two to four traffic states based on the speed from upstream and downstream detector stations (*26*). Traffic state was added to this study to improve model performance per the previous study's findings. The average density was used to measure the level of traffic congestion (*26*); traffic is congested only if the average density is greater than the critical density. Traffic state was determined based on the combination of the upstream and downstream traffic conditions:

- Free Flow (FF): when both upstream state and downstream state are free flow;
- Bottleneck front (BN): when upstream is congested and downstream is free flow;
- Back of queue (BQ): when upstream is free flow and downstream is congested; and
- Congested traffic (CT): when both upstream and downstream are congested.

Table 2 displays the candidate variables including traffic flow variables, weather condition and ramp presence. Three models were developed using different data sources to generate traffic variables: physical stations, virtual stations in a one-station setting, and virtual stations in a two-station setting. The three models are referred to as Model P, V1, and V2. Two time intervals were tested for all three models: 0–5 min before a crash and 5–10 min before a crash. A crash/noncrash dataset suitable for all models was identified so that all six models could be compared; Events with missing physical detector data were removed, and the final dataset consists of 531 events, including 66 crashes and 465 noncrashes.

Crash prediction models were developed using the binary logistic regression model to identify the relationship between crash risk and explanatory variables. The stepwise variable selection method was applied to identify the significant variables providing the best goodness-of-fit. Estimation results were obtained by fitting the model with those selected variables. The prediction accuracy of models was checked by conducting the leave-one-out cross-validation (LOOCV) with selected significant variables from each model. The LOOCV method uses one observation as the validation dataset and all the remaining observations as the training dataset. A model was fitted, given the training dataset, and was then used to predict the crash probability of that single observation in the validation dataset. This procedure was then repeated for all observations in the dataset. Based on the LOOCV results, ROC (receiver operating characteristic) curves for all six models are plotted in Figure 4 and the AUC (area under curve) values are presented in Table 3.

Table 2. Candidate Variables for Model Development

Variable	Description
UFA	Average flow at the upstream station(s) (vph)
UFS	SD of flow at the upstream station(s) (vph)
UFCVS	Coefficient of variation of flow at the upstream station(s) (vph)
UDA	Average density at the upstream station(s) (vpm)
UDS	SD of density at the upstream station(s) (vpm)
UDCVS	Coefficient of variation of density at the upstream station(s) (vpm)
USA	Average speed at the upstream station(s) (mph)
USS	SD of speed at the upstream station(s) (mph)
USCVS	Coefficient of variation of speed at the upstream station(s) (mph)
DFA	Average flow at the downstream station(s) (vph)
DFS	SD of flow at the downstream station(s) (vph)
DFCVS	Coefficient of variation of flow at the downstream station(s) (vph)
DDA	Average density at the downstream station(s) (vpm)
DDS	SD of density at the downstream station(s) (vpm)
DDCVS	Coefficient of variation of density at the downstream station(s) (vpm)
DSA	Average speed at the downstream station(s) (mph)
DSS	SD of speed at the downstream station(s) (mph)
DSCVS	Coefficient of variation of speed at the downstream station(s) (mph)
Diff_FA	Average absolute difference in flow between upstream and downstream stations
Diff_DA	Average absolute difference in density between upstream and downstream stations
Diff_SA	Average absolute difference in speed between upstream and downstream stations
Diff_FS	SD of absolute difference in flow between upstream and downstream stations
Diff_DS	SD of absolute difference in density between upstream and downstream stations
Diff_SS	SD of absolute difference in speed between upstream and downstream stations
State	Traffic state including four types: FF, BN, BQ, CT[a]
Ramp	1 = if there is a ramp between upstream and downstream stations; 0 = otherwise
Weather	Weather condition including three types: Normal, Rain, Snow

Note: vph = vehicles per hour; SD = standard deviation; vpm = vehicles per mile.
[a]BQ state was not observed in the dataset.

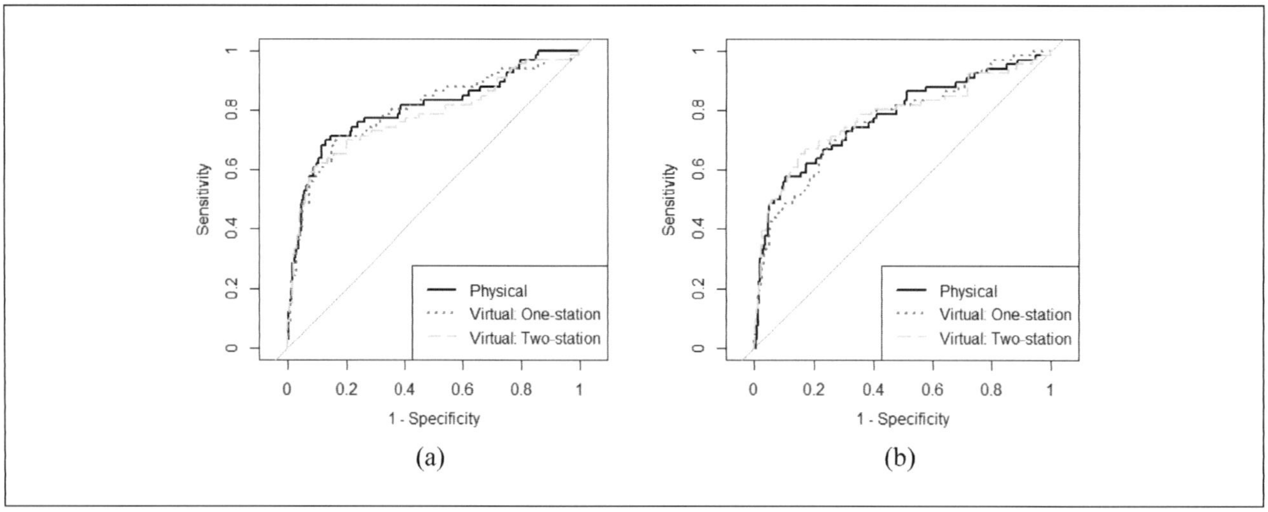

Figure 4. ROC curves for three models with different data sources from two time intervals: (*a*) 0–5 min before the crash occurrence and (*b*) 5–10 min before the crash occurrence.

ROC curves in Figure 4 do not show evident differences between three models at two time intervals as differences in AUC values are marginal. A larger AUC indicates better accuracy. Model P provides the best accuracy for the 0–5 min interval, but Model V1 provides almost the same accuracy rate. For the 5–10 min interval, Model V2 outperformed Model P and V1. Generally, virtual station data provides comparable accuracy to the physical station data. The satisfactory performance of virtual station data indicates that the CTM with well-

Table 3. AUC for Three Models at Two Time Intervals

Time interval	Physical (P)	Virtual: one-station (V1)	Virtual: two-station (V2)
0–5 min	0.8068	0.8017	0.7798
5–10 min	0.7751	0.7643	0.7778

Table 4. Model Comparison for 5–10 min Interval

Physical (P)		Virtual: two-station (V2)	
Variable	Estimate	Variable	Estimate
Constant	−3.887	Constant	−2.765
DFA	0.000525	DFA	0.00235
Diff_FA	−0.000771	UFA	−0.00231
Diff_DA	0.0266	UDS	0.0716
DSCVS	4.869	Ramp	−1.129
		State: FF[a]	–
		State: BQ	2.946
		State: CT	1.666

[a]FF state is the base level.

calibrated FDs can sufficiently capture the traffic conditions within a long segment. The comparison between AUCs also revealed that the accuracy of the 0–5 min interval is always better than that of the 5–10 min interval. It is expected that the crash occurrence should be more related to the traffic conditions as they become closer in time to the crash occurrence.

At the current stage, traffic agencies may need buffer time to detect crash risk, disseminate warnings, and implement traffic control strategies. Therefore, it is worthwhile to investigate the models for the 5–10 min interval. Because Model V2 provides the best performance in this interval, it was compared with Model P. Table 4 shows that the two models have very different sets of statistically significant variables. It should be noted that the two models were developed from different data sources. Both models include DFA as a significant variable, and both show consistent signs on the estimate coefficients. The results from Model P suggest that crash risk increases with an increase in the average flow and speed variation at the downstream station. However, crash risk decreases with an increase in the absolute flow difference between two stations. Model V2 suggests that crash risk increases with an increase in the average flow at the downstream station and the density variation at the upstream station. However, crash risk decreases with an increase in the average flow at the upstream station. Model V2 also identifies the impact of traffic states: the BQ and CT states are more crash-prone than the FF state. The small size of cells means that CTM may capture local state transitions which cannot be captured by physical stations located far away. The negative sign of the ramp presence is counterintuitive, as it suggests that the presence of off-ramp would reduce crash risk. It is possible that the actual impact of the ramp is confounded by the traffic state.

Conclusion

Many real-time crash prediction studies have used traffic data collected from detector stations. The traffic data from these stations may not represent the actual traffic conditions contributing to crashes if the stations are located far from the actual crash location. Moreover, the spacing between stations could vary considerably from site to site, which compromises crash prediction accuracy when multiple sites are considered. This critical issue may render unreliable findings based on the traffic data collected directly from detector stations.

This paper attempted to exploit simulated traffic data and predict crash occurrence in a real-time fashion given distant stations. A macroscopic simulation model CTM was applied to instrument the freeway corridor with virtual detector stations that can be placed anywhere. It is expected that this simulated traffic data from virtual detector stations would yield more accurate prediction results than the traffic data from distant physical detector stations. Two time intervals were tested—0–5 min and 5–10 min before a crash occurrence—and both a one-station and a two-station setting were applied to determine which setting is more suitable in two different time interval choices. Models developed with physical station data were compared with those developed with virtual station data, leading to the conclusion that models developed with virtual station data provided a prediction accuracy rate similar to those developed with physical station data. The results suggest that simulated traffic data can be used when the spacing between detectors is large.

The proposed CTM approach can be applied in a real-time crash detection system. The application can vary according to the constraints and accuracy between one-station and two-station settings. The one-station setting can be used when a buffer time is needed for agencies to detect crash risk, take corresponding actions, and inform drivers. The two-station setting can be used when a buffer time is not necessary, such as in a connected and autonomous driving environment in which vehicles have the ability to make immediate maneuvers.

Future research should be directed toward furthering the development of the CTM approach. More importantly, the methodological advancement should be measured by improved flexibility to model traffic dynamics that correlate to crash risk, higher crash prediction accuracy, and reliability. One potential improvement could be calibrating separate FDs under some safety-related circumstances such as inclement weather, low light conditions, or work zones. Another direction for future research is establishing the relationship between macroscopic traffic characteristics correlated with crash occurrence and microscopic traffic characteristics that correspond to safety surrogate measures (e.g. traffic conflicts). The empirical relationship may be explored and measured via simulation models, but theoretical development will require in-depth knowledge of traffic flow theory and its application in highway safety.

Author Contributions

The authors confirm contribution to the paper as follows: study conception and design: Xiao Qin, Zhi Chen; data collection: Zhi Chen, Yang Cheng; analysis and interpretation of results: Zhi Chen, Xiao Qin, Renxin Zhong; draft manuscript preparation: Zhi Chen, Xiao Qin, Renxin Zhong, Pan Liu. All authors reviewed the results and approved the final version of the manuscript.

References

1. Abdel-Aty, M., N. Uddin, A. Pande, F. Abdalla, and L. Hsia. Predicting Freeway Crashes from Loop Detector Data by Matched Case-Control Logistic Regression. *Transportation Research Record: Journal of the Transportation Research Board*, 2004. 1897: 88–95.
2. Daganzo, C. F. The Cell Transmission Model: Network Traffic. *Transportation Research Part B: Methodological*, Vol. 29, No. 2, 1994, pp. 79–93.
3. Muñoz, L., X. Sun, R. Horowitz, and L. Alvarez. Traffic Density Estimation with the Cell Transmission Model. *Proc., 2003 American Control Conference*, Vol. 5, IEEE, Denver, Colo., 2003, pp. 3750–3755.
4. Muñoz, L., X. Sun, R. Horowitz, and L. Alvarez. Piecewise-Linearized Cell Transmission Model and Parameter Calibration Methodology. *Transportation Research Record: Journal of the Transportation Research Board*, 2006. 1965: pp. 183–191.
5. Sumalee, A., R. X. Zhong, T. L. Pan, and W. Y. Szeto. Stochastic Cell Transmission Model (SCTM): A Stochastic Dynamic Traffic Model for Traffic State Surveillance and Assignment. *Transportation Research Part B: Methodological*, Vol. 45, No. 3, 2011, pp. 507–533.
6. Hossain, M., and Y. Muromachi. A Bayesian Network Based Framework for Real-Time Crash Prediction on the Basic Freeway Segments of Urban Expressways. *Accident Analysis & Prevention*, Vol. 45, 2012, pp. 373–381.
7. Sun, J., and J. Sun. A Dynamic Bayesian Network Model for Real-Time Crash Prediction Using Traffic Speed Conditions Data. *Transportation Research Part C: Emerging Technologies*, Vol. 54, 2015, pp. 176–186.
8. Pande, A., and M. Abdel-Aty. Comprehensive Analysis of The Relationship Between Real-Time Traffic Surveillance Data and Rear-End Crashes on Freeways. *Transportation Research Record: Journal of the Transportation Research Board*, 2006. 1953: 31–40.
9. Abdel-Aty, M. A., and R. Pemmanaboina. Calibrating a Real-Time Traffic Crash-Prediction Model Using Archived Weather and ITS Traffic Data. *IEEE Transactions on Intelligent Transportation Systems*, Vol. 7, No. 2, 2006, pp. 167–174.
10. Li, Z. B., W. Wang, R. Y. Chen, P. Liu, and C. C. Xu. Evaluation of the Impacts of Speed Variation on Freeway Traffic Collisions in Various Traffic States. *Traffic Injury Prevention*, Vol. 14, No. 8, 2013, pp. 861–866.
11. Kwak, H.-C., and S. Kho. Predicting Crash Risk and Identifying Crash Precursors on Korean Expressways Using Loop Detector Data. *Accident Analysis & Prevention*, Vol. 88, 2016, pp. 9–19.
12. Xu, C., P. Liu, and W. Wang. Evaluation of the Predictability of Real-Time Crash Risk Models. *Accident Analysis & Prevention*, Vol. 94, 2016, pp. 207–215.
13. Chen, Z., X. Qin, and M. R. R. Shaon. Modeling Lane-Change Related Crashes with Lane-Specific Real-Time Traffic and Weather Data. *Journal of Intelligent Transportation Systems*, 2017, pp. 1–10.
14. Xu, C., W. Wang, and P. Liu. A Genetic Programming Model for Real-Time Crash Prediction on Freeways. *IEEE Transactions on Intelligent Transportation Systems*, Vol. 14, No. 2, 2013, pp. 574–586.
15. Abdel-Aty, M., N. Uddin, and A. Pande. Split Models for Predicting Multivehicle Crashes During High-Speed and Low-Speed Operating Conditions on Freeways. *Transportation Research Record: Journal of the Transportation Research Board*, 2005. 1908: 51–58.
16. Zheng, Z., S. Ahn, and C. M. Monsere. Impact of Traffic Oscillations on Freeway Crash Occurrences. *Accident Analysis & Prevention*, Vol. 42, No. 2, 2010, pp. 626–636.
17. Hourdos, J., V. Garg, P. Michalopoulos, and G. Davis. Real-Time Detection of Crash-Prone Conditions at Freeway High-Crash Locations. *Transportation Research Record: Journal of the Transportation Research Board*, 2006. 1968: 83–91.
18. Lee, C., M. Abdel-Aty, and L. Hsia. Potential Real-Time Indicators of Sideswipe Crashes on Freeways. *Transportation Research Record: Journal of the Transportation Research Board*, 2006. 1953: 41–49.
19. Lee, C., B. Hellinga, and F. Saccomanno. Proactive Freeway Crash Prevention Using Real-Time Traffic Control. *Canadian Journal of Civil Engineering*, Vol. 30, No. 6, 2003, pp. 1034–1041.
20. Roshandel, S., Z. Zheng, and S. Washington. Impact of Real-Time Traffic Characteristics on Freeway Crash Occurrence: Systematic Review and Meta-Analysis. *Accident Analysis & Prevention*, Vol. 79, 2015, pp. 198–211.
21. Moore, J. E., G. Giuliano, and S. Cho. Secondary Accident Rates on Los Angeles Freeways. *Journal of Transportation Engineering*, Vol. 130, No. 3, 2004, pp. 280–285.
22. Parker, S. T., and Y. Tao. WisTransPortal: A Wisconsin Traffic Operations Data Hub. *Proc., Ninth International Conference on Applications of Advanced Technology in Transportation (AATT)*, ASCE, New York, 2006, pp. 611–616.

23. Al-Deek, H. M., C. Venkata, and S. Ravi Chandra. New Algorithms for Filtering and Imputation of Real-Time and Archived Dual-Loop Detector Data in I-4 Data Warehouse. *Transportation Research Record: Journal of the Transportation Research Board*, 2004. 1867: 116–126.
24. Zhong, R., C. Chen, A. H. Chow, T. Pan, F. Yuan, and Z. He. Automatic Calibration of Fundamental Diagram for First-Order Macroscopic Freeway Traffic Models. *Journal of Advanced Transportation*, Vol. 50, No. 3, 2016, pp. 363–385.
25. Courant, R., K. Friedrichs, and H. Lewy. On the Partial Difference Equations of Mathematical Physics. *IBM journal of Research and Development*, Vol. 11, No. 2, 1967, pp. 215–234.
26. Yeo, H., K. Jang, A. Skabardonis, and S. Kang. Impact of Traffic States on Freeway Crash Involvement Rates. *Accident Analysis and Prevention*, Vol. 50, 2013, pp. 713–723.

The Safety Section (ANB00) peer-reviewed this paper (18-05725).

Article

TRR
JOURNAL OF THE TRANSPORTATION RESEARCH BOARD

Transportation Research Record
2018, Vol. 2672(38) 22–34

DOI: 10.1177/0361198118790617
journals.sagepub.com/home/trr
SAGE

Exploration of Contributing Factors Related to Driver Errors on Highway Segments

Mohammad Razaur Rahman Shaon[1], Xiao Qin[1], Zhi Chen[1], and Jian Zhang[2]

Abstract

A significant portion of crashes occurred on highway segments, with more than 90% of crashes associated with driving errors. To avoid a crash, a driver needs to detect a hazard, decide the safest driving maneuvers, and execute them properly. Driver errors at any of these sequential phases may lead to a crash; therefore, it is necessary to identify the contributing factors and assess their influence on driver behavior. To assist this investigation, a multinomial probit model was employed to study driver errors reported in crashes in rural and urban areas. The modeling results identified many highway geometric features, traffic conditions, roadway events, and driver characteristics as statistically correlated to different types of driver error. Following the extensive list, the impacts of error-contributing factors were discussed within each error category. This exercise helps to gain a better understanding of similar or varying effects of explanatory variables across different error categories. The broad and insightful information will help researchers and safety professionals to better understand when, where, and how the driver error may lead to a crash and to develop cost-effective preventive countermeasures.

Highway safety analysis is mostly focused on analyzing crash occurrence or severity, where highway and traffic engineering-related data such as roadway geometric characteristics and traffic conditions are used as explanatory variables. It is well known that human factors probably contribute to over 90% of crashes. According to the National Motor Vehicle Crash Causation Survey (NMVCCS), almost 94% of crashes are caused by driver errors (*1*). Without specifically considering driver factors, crash modeling results may be biased due to the absence of human factors. Thus, understanding what contributes to driver error and how to incorporate driver behavior into crash prediction have become increasingly important topics among safety researchers.

Crashes are complex events, as reflected by the 110 data elements recommended in the Model Minimum Uniform Crash Criteria (*2*). Most information can be obtained directly by reviewing detailed crash reports, including police officers' judgment on driver factors contributing to the occurrence of a crash. Crash information can be augmented by socioeconomic, demographic, land use, and traffic pattern information to substantiate the knowledge of how a driver interacts with—and how his/her behavior is influenced by—roadway design, traffic conditions, and other contextual factors.

For a driver, there is a four-phase process of seeing and reacting to a hazard; that is, perception, intellection, emotion, and volition, or "PIEV." An error can happen during any of these four phases. This study is particularly focused on understanding when, where, and how drivers make mistakes that contribute to a crash, drawing cues from a comprehensive list of variables ranging from highway geometry, traffic, roadway, weather, and lighting conditions, events such as construction zone and debris on roadway, as well as driver information such as age, gender, vehicle types, and so forth. Specifically, the categorization method in the NMVCCS study was followed; this groups driver errors into four categories: recognition error, decision error, performance error, and non-performance error. As each error type is specific and unique, the relating explanatory variables identified through statistical models may be more informative to safety professionals. The new insight into the circumstances leading to driver errors and a better

[1]Department of Civil and Environmental Engineering, University of Wisconsin-Milwaukee, Milwaukee, WI
[2]Research Center for Internet of Mobility, School of Transportation, Southeast University, Nanjing, Jiangsu, PR China

Corresponding Author:
Address correspondence to Xiao Qin: qinx@uwm.edu

understanding of possible causes will support the development of tangible, practical, and, more importantly, targeted and cost-effective enforcement strategies, driver education and training programs, engineering solutions, and vehicle safety technologies.

Literature Review

Crash occurrence may be attributed to errors by drivers or the interaction between driver behavior and roadway design features (*3*). As indicated by police records, driver errors can range from a traffic infraction in which the driver is not paying attention, to an intentional traffic violation such as failure to yield or significantly exceeding the speed limit. However, according to the Human Factors Guidelines for Road Systems: "*Road users cannot be expected to solve either highway design or traffic engineering problems without making mistakes and/or compromising operational efficiency and safety*" (*4*).

Understanding the interaction between driver errors and roadway geometric and contextual features is crucial. It has been well established from crash count models that highway design, traffic conditions, and contextual factors such as weather events are related to crash occurrence and have an effect on driver behavior. The American Automobile Association (AAA) Foundation for Traffic Safety estimated that 56% of the fatal crashes that occurred between 2003 and 2007 involved potentially aggressive driving behavior, in which speeding was the most common aggressive action, making up about 31% of total fatal crashes. Hauer noted that the speed at which people choose to travel is affected by roadway design and vehicle characteristics (*5*). Tate and Turner investigated the relationship between observed travel speed, road geometry, and crashes in New Zealand (*6*) and concluded that drivers' speed choices were more strongly related to curve radius than curve design speed, and that the approach speed environment has a significant impact on speed choice. Liu and Chen documented that "driving too fast for conditions" was more likely to occur on roads with higher speed limits (50+ mph) than other crashes (*7*). The authors also noted that a significant proportion of speeding-related crashes occurred on adverse road surface conditions such as "Snowy/Slushy/Icy/Slippery" and "Wet" road pavement compared with other crashes.

Distracted driving is another major driver error that contributes to crashes. Novice drivers appeared to be prone to distraction while driving (*8*). Naturalistic driving studies showed that talking on a cell phone raises the risk of collision by more than 30%, and drivers who text are at 23 times higher crash risk compared with non-distracted drivers (*9*). Results from the National Occupant Protection Use Survey, conducted annually by the National Center for Statistics and Analysis of the National Highway Traffic Safety Administration (NHTSA), showed that females from all age groups are more prone to use electronic devices while driving (*10*). Electronic device use percentage was found to be similar in age brackets from 16 to 69 years. This suggests that gender might be a more important variable than age cohort in distracted driving error.

Work zone is a roadway event that has been reported to increase crash rate, according to previous literature (*11–13*). Drivers in a construction/work zone encounter a complex array of warning signs, barrels, pylons, construction equipment, and machines, which can create hazards for drivers. The new traffic patterns and challenging roadway configurations in work zones, such as stopping or slow traffic, trucks merging from the ramp, uneven pavement, narrowed lanes, and absence of shoulders, require drivers to operate their vehicles with extra caution and impose considerable stress on their driving tasks.

Previously, impaired driving has been identified as a contributing factor to driver error (*14*). Use of alcohol can significantly affect a driver's decision-making process. Blomberg et al. conducted a case-control study to explore the relationship of blood alcohol concentration (BAC) with relative crash risk (*14*). Results showed that elevated relative risk beginning at 0.05–0.06% BAC, with an accelerating increase in risk at BACs greater than 0.10%. In 2015 in the U.S., 41% of drivers killed in roadway speeding-related crashes had a BAC of 0.08 g/dl or higher in their blood (*15*). Besides alcohol impairment, drug-impaired driving has a significant effect on driving behavior (*16–18*). The Governors Highway Safety Association (GHSA) sponsored a study that found that fatalities caused by drugged driving surpass those caused by alcohol-impaired driving in the United States (*19*). In 2015, 43% of motorists who died in a road accident had drugs in their systems, whereas 37% of motorists who died tested positive for alcohol.

Several studies have investigated driver error for segment and intersection-related crashes (*7*, *20–23*), most of which discussed factors contributing to driver error for specific types of crashes, such as speed. An overall discussion of the factors contributing to driver error is rare. Wang and Qin investigated the factors contributing to driver errors at uncontrolled, sign-controlled, and signal-controlled intersections (*23*). The authors categorized driver error based on traffic violations recorded in the crash report, roadway characteristics (presence of curve, visibility, speed limit), driver characteristics (age, gender, driving under the influence), environmental characteristics (weather condition, roadway condition, lighting condition), and vehicle type (passenger car, light truck, heavy truck) to predict different types of driver error. Sign-controlled intersections are found to have the highest percentage of driver error, followed by signalized, and then uncontrolled intersections. Drivers are more prone to serious errors if their vision is obscured. Adverse environmental characteristics, such as snow or ice on the pavement, negatively affect the severity of driver error. Driver age, gender, and alcohol or drug use greatly influence the severity of error outcome. The findings confirmed that driver errors are the outcome not only of a driver's psychological behavior but also of the interaction with external factors during the driver's decision making.

Based on previous research, this study is an attempt to investigate the relationship between driver errors and observable factors on highway segments in both rural and urban areas. More information on these contributing factors would help researchers and safety professionals to develop cost-effective countermeasures.

Methodology

The multinomial probit model (MNP) is a discrete outcome model that considers a response variable with three or more levels without accounting for order between levels. Two popular choices to model multiclass categorical variables are multinomial logit (MNL) and MNP models. The MNL model is built on the independence of irrelevant alternatives assumption, meaning adding or deleting an alternative will not change the ratio between the probability of any pair of existing alternatives. In simple words, MNL cannot account for correlation between any pair of existing alternatives. This may not be always true if the dependent variable categories are correlated. The MNP model relaxes the independence assumption *(24)*. The driver error categories defined in this study are not independent as they are categorized based on sequential events. From a practical perspective, it is obvious that performance error depends on the decision of activity the driver tends to execute, and decision error depends on the recognition of hazardous situation perceived by the driver. Considering dependency between driver error categories, the MNP model was an appropriate choice for model development in this study. The utility function of the MNP model that determines the preference or possible value of attaining the outcome i (i = 1, 2, . . ., I) for observation n can be written as *(25)*:

$$U_{in} = \beta_i X_{in} + \varepsilon_{in} \tag{1}$$

$$[\varepsilon_{1n}, \varepsilon_{2n}, \varepsilon_{3n}, \ldots\ldots, \varepsilon_{in}] \sim MVN(0, \Sigma)$$

where,

X_{in} = vector of independent variables for nth observation with ith outcome

β_i = vector of corresponding unknown coefficients

ε_{in} = disturbance term that accounts for unobserved effects and random errors

The disturbance term ε_{in} for ith driver error type has a mean of zero and is correlated among different error types. Thus, the disturbance vector is defined by a multivariate normal distribution. The probability of ith driver error can be written as:

$$Prob[Choice_{in}] = Prob[U_{in} > U_{jn}, j = 1, 2, 3, \ldots\ldots, I; i \neq j] \tag{2}$$

Using above formulation, the probability of occurrence of ith driver error can be specified as:

$$\begin{aligned} &Prob[Choice_{in}|X_n] > Prob[(\varepsilon_{in} - \varepsilon_{1n})\rangle X_n'(\beta_1 - \beta_i), \\ &(\varepsilon_{in} - \varepsilon_{2n}) > X_n'(\beta_2 - \beta_i), \ldots., \left(\varepsilon_{in} - \varepsilon_{(i-1)n}\right) \\ &> X_n'\left(\beta_{(i-1)} - \beta_i\right), \left(\varepsilon_{in} - \varepsilon_{(i+1)n}\right) \\ &> X_n'\left(\beta_{(i+1)} - \beta_i\right), \ldots.(\varepsilon_{in} - \varepsilon_{In}) > X_n'(\beta_1 - \beta_I)] \end{aligned} \tag{3}$$

The estimated coefficient β_i can be interpreted as the marginal effect of X_i on the log odds ratio of ith alternative to the baseline alternative. A "margin" is a statistic computed from predictions from a model while manipulating the values of the covariates. The marginal effect of X_i on the probability of ith alternative can be expressed as follows:

$$\frac{\partial P_r(Y = Y_n|X_{in})}{\partial X_i} = \frac{\partial E(Y_n|X_i)}{\partial X_i} = \emptyset(\beta_i X_{in})\beta_i \tag{4}$$

where, $\emptyset$ represents cumulative normal density function.

Data Description

Data for segment-related crashes that occurred on the Wisconsin state trunk network system between 2013 and 2015 were collected, excluding deer-related crashes *(26)*. After removing all crashes without good location information, 48,441 rural crashes and 46,221 urban crashes were available. Specific driver errors were extracted from the Wisconsin Motor Vehicle Accident Reporting Form 4000 (MV4000), in which the investigating police officers documented detailed accident information *(26, 27)*. There is a list of fourteen driver-related factors. When a crash is associated with multiple driver factors, the most severe driver factor is noted based on the police investigation.

Modeling fourteen choices may not be effective because of the sample size, strong correlation between some error types, and difficulties of interpretation. Based on the similarities in driver errors, the NMVCCS study classified driver-related critical reasons into recognition errors, decision errors, performance errors, and non-performance errors (*1*). Recognition error includes driver inattention, internal and external distraction, inadequate surveillance, and so forth; aggressive driving behavior, driving too fast, and so on, are categorized as decision error; overcompensation and poor directional controls are categorized as performance error; sleep and physical impairment are considered as non-performance error. This categorization combines narrative errors with similar traits. The driver factors in Wisconsin crash data were grouped into four NMVCCS driver error categories based on the NMVCCS criteria and definition of each category. Table 1 shows the NMVCCS driver error types, and corresponding Wisconsin driver factors along with summary statistics for each category.

The broad categorization of driver errors follows a sequence of information processing. When driving, a driver needs to detect and identify a hazard, decide what to do, and react accordingly. Driver errors leading to a crash are also categorized following the driving tasks. A driver's recognition efficiency can be affected by any internal or external distraction or by any form of inattention. Recognition error refers to all the driver factors that may lead to lack of awareness or failure in recognition of hazardous situations. In Wisconsin, 18% and 20% of total crashes that occurred between 2013 and 2015 were attributable to inattentive driving in rural and urban areas, respectively.

A driver's decision on what to do directly leads to what happens next, whether it is a decision after detecting a hazard or a decision while driving. A bad maneuver decision after recognizing a hazardous situation may cause a crash. A reckless decision such as "exceeding the speed limit" may go wrong even without an imminent hazard. In Wisconsin, 35% and 38% of crashes occurred as a result of decision error in rural and urban areas, respectively.

If a maneuver is not properly performed, a crash may happen. Poorly performed driving tasks are categorized as performance error, which is dependent on the driver's experience and skills. Although non-performance error is not related to driver behavior, it represents a driver's health conditions, fatigue, level of impairment, or other non-performance issues.

The crash dataset does not contain detailed roadway geometric information at the crash location. Roadway geometry, pavement characteristics, mobility, safety, and other roadway-related data tables stored in Meta-Manager at the Wisconsin Department of Transportation (WisDOT) were linked with crash data using spatial join in ArcGIS. The joined dataset contains the data collected by the crash investigating police officer, roadway geometry, and traffic information for each crash. Table 2 provides the summary statistics.

Analysis of Results

The coefficient estimates of the final MNP models for rural and urban crashes are presented in Tables 3 and 4, respectively. The STATA command "mprobit" was used to estimate the coefficient of the MNL model (*28*). In

Table 1. Categorization and Distribution of Driver Error

Error type	NMVCCS criteria	Wisconsin criteria	Rural	Urban
Recognition error	• Inadequate surveillance • Internal distraction • External distraction • Inattention	• Inattentive driving	8,659 (17.88%)	9,044 (19.57%)
Decision error	• Too fast for conditions • Too fast for curve • False assumption of other's action • Illegal maneuver • Misjudgment of gap or other's action • Following too closely • Aggressive driving behavior	• Too fast for condition • Exceed speed limit • Disregard traffic control • Following too close • Improper overtake • Improper turn	17,139 (35.38%)	17,662 (38.21%)
Performance error	• Overcompensation • Poor directional control • Panic/Freezing • Other performance error	• Failure to keep vehicle under control • Left of center • Unsafe backing • Failure to yield	10,288 (21.24%)	9,867 (21.35%)
Non-performance error	• Sleep • Heart attack • Other non-performance error	• Disability • Driver condition • Others	2,402 (4.96%)	3,030 (6.55%)
No error			9,953 (20.55%)	6,620 (14.32%)

Table 2. Summary Statistics of Explanatory Variables

Variable	Description	Type	Rural Mean	Rural Std. dev.	Urban Mean	Urban Std. dev.
AADT	Annual average daily traffic	Continuous	21,610.37	26,554.21	55,450.19	48,566.43
Truck	Truck percentage (%)	Continuous	11.44	4.59	7.735	2.86
Speed	Posted speed limit (MPH)	Continuous	57.49	11.31	46.95	14.17
Lane	Number of lanes (Count)	Continuous	2.13	0.43	2.59	0.715
LW	Lane width (feet)	Continuous	12.10	0.83	12.34	1.08
SW	Shoulder width (feet)	Continuous	8.60	3.87	5.58	5.49
Rut	Pavement rutting (inch)	Continuous	0.088	0.08	0.07	0.07
Percent passing	Passing percentage (%)	Continuous	26	31.90	3.25	15.85
Highway type	Interstate	Categorical with 3 levels	9,840 (20.31%)		12,012 (25.99%)	
	State highway		37,377 (77.16%)		30,062 (65%)	
	Other state roadway		1,224 (2.53%)		4,147 (9%)	
Roadway type	Undivided	Categorical with 3 levels	24,177 (49.91%)		8,225 (17.79%)	
	Divided		23,889 (49.32%)		36,324 (78.59%)	
	One-way		375 (0.77%)		1,672 (3.62%)	
Presence of median	No	Categorical with 2 levels	31,530 (65.09%)		20,170 (43.64%)	
	Yes		16,911 (34.91%)		26,051 (56.36%)	
Roadway condition	Dry	Categorical with 4 levels	27,830 (57.45%)		31,740 (68.67%)	
	Wet		5,255 (10.85%)		7,517 (16.26%)	
	Snow		10,281 (21.22%)		5,307 (11.48%)	
	Ice		5,075 (10.48%)		1,657 (3.58%)	
Weather condition	Clear	Categorical with 5 levels	20,591 (42.51%)		22,378 (48.42%)	
	Fog/cloudy		13,619 (28.11%)		14,671 (31.74%)	
	Wind		1,041 (2.15%)		140 (0.3%)	
	Rain		3,057 (6.31%)		4,157 (8.99%)	
	Snow/sleet		10,133 (20.92%)		4,875 (10.55%)	
Lighting condition	Day	Categorical with 3 levels	33,065 (68.26%)		30,046 (73.66%)	
	Night-unlit		13,477 (27.82%)		3,026 (6.55%)	
	Night-lit		1,899 (3.92%)		9,149 (19.79%)	
Horizontal curve	No	Categorical with 2 levels	39,390 (81.32%)		41,750 (90.33%)	
	Yes		9,051 (18.68%)		4,471 (9.67%)	
Vertical curve	No	Categorical with 2 levels	38,865 (80.23%)		40,542 (87.71%)	
	Yes		9,576 (19.77%)		5,679 (12.29%)	
Age group	Adolescent (<18 years)	Categorical with 5 levels	2,363 (4.88%)		1,789 (3.87%)	
	Young adults (18–25 years)		11,206 (23.13%)		11,271 (24.39%)	
	Adults (26–35 years)		10,309 (21.28%)		11,406 (24.68%)	
	Middle age (36–65 years)		20,223 (41.47%)		18,294 (39.58%)	
	Old (>65 years)		4,340 (8.96%)		3,461 (7.49%)	
Gender	Male	Categorical with 2 levels	30,090 (62.12%)		26,989 (58.39%)	
	Female		18,351 (37.88%)		19,232 (41.61%)	
Vehicle	Passenger car	Categorical with 4 levels	35,498 (73.28%)		38,051 (82.32%)	
	Motorcycle		888 (1.83%)		545 (1.18%)	
	Light truck		8,096 (16.71%)		4,898 (10.6%)	
	Heavy truck		3,959 (8.17%)		1,727 (5.9%)	
Alcohol	No	Categorical with 2 levels	45,725 (94.39%)		44,470 (96.21%)	
	Yes		2,716 (5.61%)		1,751 (3.79%)	
Drug	No	Categorical with 2 levels	47,920 (98.92%)		45,881 (99.26%)	
	Yes		521 (1.08%)		340 (0.74%)	
Visibility obscured	No	Categorical with 2 levels	48,078 (99.25%)		46,013 (99.55%)	
	Yes		363 (0.75%)		208 (0.45%)	
Work zone	No	Categorical with 2 levels	47,625 (98.32%)		45,339 (98.09%)	
	Yes		816 (1.68%)		882 (1.91%)	
Debris on road	No	Categorical with 2 levels	47,695 (98.46%)		45,860 (99.22%)	
	Yes		746 (1.54%)		361 (0.78%)	

Table 3. Coefficient Estimates for MNP Model for Driver Errors in Rural Crashes

Variable	Recognition error		Decision error		Performance error	
	Coef.	Std. err.	Coef.	Std. err.	Coef.	Std. err.
AADT (in thousands)	**0.002**	**0.000**	**0.010**	**0.000**	**0.004**	**0.000**
Truck	**−0.005**	**0.003**	**−0.006**	**0.002**	**−0.011**	**0.002**
Speed	**−0.02**	**0.002**	**−0.019**	**0.002**	**−0.01**	**0.002**
Lanes	−0.046	0.032	**−0.084**	**0.03**	**−0.12**	**0.032**
Shoulder width	**0.012**	**0.004**	0.001	0.004	**−0.014**	**0.004**
Pavement rutting	**−0.462**	**0.166**	**−0.515**	**0.154**	**−0.518**	**0.158**
Highway type						
Interstate	Base condition					
State highway	−0.033	0.037	**−0.102**	**0.031**	**0.08**	**0.034**
Other state roadway	−0.114	0.077	**−0.186**	**0.072**	**0.338**	**0.072**
Roadway type						
Undivided	0.07	0.047	**−0.146**	**0.043**	4E-5	0.044
Divided	Base condition					
One-way	**−0.36**	**0.143**	0.036	0.123	−0.026	0.13
Horizontal curve						
No	Base condition					
Yes	**0.153**	**0.031**	**0.255**	**0.027**	**0.356**	**0.028**
Vertical curve						
No	Base condition					
Yes	-0.029	0.029	**0.065**	**0.025**	**0.054**	**0.026**
Roadway condition						
Dry	Base condition					
Wet	−0.026	0.048	**0.413**	**0.045**	**0.208**	**0.046**
Snow	**−0.892**	**0.057**	**1.072**	**0.039**	**0.284**	**0.042**
Ice	**−1.56**	**0.076**	**0.947**	**0.038**	**0.18**	**0.041**
Weather condition						
Clear	Base condition					
Fog/cloudy	**0.157**	**0.026**	**0.16**	**0.025**	**0.19**	**0.026**
Wind	**−0.895**	**0.169**	0.054	0.070	−0.068	0.077
Rain	**−0.163**	**0.065**	**0.269**	**0.057**	0.019	0.06
Snow/sleet	**−0.335**	**0.063**	**0.17**	**0.041**	0.085	0.044
Lighting condition						
Day	Base condition					
Night-unlit	**−0.207**	**0.026**	**−0.341**	**0.023**	**−0.116**	**0.024**
Night-lit	−0.020	0.059	**−0.281**	**0.056**	**−0.194**	**0.058**
Visibility						
No	Base condition					
Yes	**−0.364**	**0.132**	−0.046	0.108	−0.105	0.114
Work zone						
No	Base condition					
Yes	**0.210**	**0.082**	**0.565**	**0.076**	−0.074	0.089
Debris on road						
No	Base condition					
Yes	**−2.136**	**0.117**	**−1.863**	**0.094**	**−1.762**	**0.101**
Age group						
Adolescent	Base condition					
Young adults	**−0.181**	**0.056**	**−0.174**	**0.052**	**−0.178**	**0.054**
Adults	**−0.420**	**0.056**	**−0.341**	**0.052**	**−0.278**	**0.054**
Middle age	**−0.561**	**0.053**	**−0.527**	**0.050**	**−0.44**	**0.051**
Old	**−0.339**	**0.061**	**−0.582**	**0.058**	**−0.211**	**0.059**
Gender						
Male	0.022	0.024	0.006	0.021	**−0.056**	**0.022**
Female	Base condition					
Vehicle						
Passenger car	**0.194**	**0.040**	**0.332**	**0.037**	**0.216**	**0.039**
Motorcycle	**−0.564**	**0.091**	**0.225**	**0.081**	**0.488**	**0.078**
Light truck	**0.193**	**0.046**	**0.325**	**0.041**	**0.237**	**0.044**
Heavy truck	Base condition					

(continued)

Table 3. (continued)

Variable	Recognition error		Decision error		Performance error	
	Coef.	Std. err.	Coef.	Std. err.	Coef.	Std. err.
Alcohol						
No	Base condition					
Yes	**1.120**	**0.067**	**1.282**	**0.066**	**1.459**	**0.065**
Drug						
No	Base condition					
Yes	**0.754**	**0.129**	**0.740**	**0.129**	**0.875**	**0.128**
Intercept	**1.277**	**0.255**	**1.324**	**0.244**	**1.138**	**0.252**

Note: Variables that are statistically significant at 90% confidence interval are presented in bold font.

Table 4. Coefficient Estimates for MNP Model for Driver Errors in Urban Crashes

Variable	Recognition error		Decision error		Performance error	
	Coef.	Std. err.	Coef.	Std. err.	Coef.	Std. err.
AADT (in thousands)	–3E-05	0.000	**0.002**	**0.000**	–0.0002	0.000
Truck	**–0.015**	**0.004**	**–0.015**	**0.004**	**–0.01**	**0.004**
Speed	**–0.006**	**0.001**	–0.001	0.001	**–0.011**	**0.001**
Lanes	**0.057**	**0.019**	**0.049**	**0.018**	**0.095**	**0.019**
Lane width	**0.047**	**0.011**	**0.045**	**0.011**	**0.036**	**0.011**
Shoulder width	0.003	0.003	**0.006**	**0.003**	**0.017**	**0.003**
Pavement rutting	**–0.482**	**0.162**	**–0.393**	**0.152**	**0.387**	**0.159**
Percent passing	**0.005**	**0.001**	**0.003**	**0.001**	**0.002**	**0.001**
Highway type						
Interstate	Base condition					
State highway	0.001	0.033	–0.017	0.029	**–0.244**	**0.032**
Other state roadway	**–0.12**	**0.049**	**–0.335**	**0.045**	**–0.146**	**0.047**
Roadway type						
Undivided	0.043	0.043	**–0.105**	**0.04**	**–0.069**	**0.042**
Divided	Base condition					
One-way	–0.056	0.063	**–0.171**	**0.06**	**–0.248**	**0.064**
Horizontal curve						
No	Base condition					
Yes	**–0.196**	**0.043**	0.053	0.037	**0.283**	**0.038**
Vertical curve						
No	Base condition					
Yes	**–0.082**	**0.036**	–0.029	0.032	**–0.12**	**0.034**
Roadway condition						
Dry	Base condition					
Wet	–0.069	0.045	**0.157**	**0.042**	**0.203**	**0.044**
Snow	**–0.802**	**0.063**	**0.323**	**0.049**	**0.104**	**0.052**
Ice	**–1.991**	**0.123**	0.061	0.055	**–0.408**	**0.063**
Weather condition						
Clear	Base condition					
Fog/cloudy	**0.145**	**0.026**	**0.168**	**0.025**	**0.198**	**0.026**
Wind	–0.456	0.303	0.119	0.173	–0.074	0.2
Rain	**–0.153**	**0.06**	**0.219**	**0.054**	0.077	0.057
Snow/sleet	**–0.296**	**0.07**	**0.181**	**0.052**	**0.12**	**0.056**
Lighting condition						
Day	Base condition					
Night-unlit	**–0.19**	**0.05**	**–0.321**	**0.043**	**–0.159**	**0.047**
Night-lit	**–0.243**	**0.029**	**–0.37**	**0.027**	**–0.225**	**0.028**
Visibility						
No	Base condition					
Yes	**–0.588**	**0.166**	**–0.682**	**0.153**	0.22	0.142

(continued)

Table 4. (continued)

Variable	Recognition error		Decision error		Performance error	
	Coef.	Std. err.	Coef.	Std. err.	Coef.	Std. err.
Construction zone						
No	Base condition					
Yes	–0.043	0.083	**0.289**	**0.075**	–0.052	0.084
Debris on road						
No	Base condition					
Yes	**–2.04**	**0.149**	**–1.895**	**0.113**	**–1.925**	**0.14**
Age group						
Adolescent	Base condition					
Young adults	**–0.202**	**0.065**	**–0.213**	**0.061**	**–0.188**	**0.065**
Adults	**–0.376**	**0.065**	**–0.431**	**0.061**	**–0.296**	**0.065**
Middle age	**–0.393**	**0.064**	**–0.5**	**0.059**	**–0.307**	**0.063**
Old	**–0.271**	**0.073**	**–0.531**	**0.069**	**0.001**	**0.072**
Gender						
Male	**–0.069**	**0.024**	–0.011	0.022	–0.035	0.023
Female	Base condition					
Vehicle						
Passenger car	**0.413**	**0.046**	**0.512**	**0.042**	**0.44**	**0.046**
Motorcycle	**–0.536**	**0.118**	0.02	0.099	**0.436**	**0.099**
Light truck	**0.44**	**0.056**	**0.512**	**0.051**	**0.434**	**0.054**
Heavy truck	Base condition					
Alcohol						
No	Base condition					
Yes	**0.917**	**0.086**	**0.849**	**0.083**	**1.226**	**0.083**
Drug						
No	Base condition					
Yes	**0.667**	**0.18**	**0.527**	**0.176**	**0.783**	**0.176**
Intercept	0.035	0.195	–0.011	0.184	0.036	0.194

Note: Variables that are statistically significant at 90% confidence interval are presented in bold font.

both tables, the coefficient estimates represent the log odds ratio between the probability of a defined driver error type and no error type, with a positive sign for increase and a negative sign for decrease. "No error" category was considered as the base outcome in the MNP model.

The modeling results for "Non-performance error" were excluded as this error category does not include behavioral driver factors. For a quick summary, middle age and old age groups are more prone to non-performance error. Alcohol and drug consumption also increases the probability of non-performance error compared with no error.

In Table 3, it can be noted that both traffic variables annual average daily traffic (AADT) and truck percentage are significantly related to all driver error categories. A thousand-unit change in AADT results in increased probability 1.002 ($e^{0.002}$) times, 1.01 ($e^{0.010}$) times and, 1.004 ($e^{0.004}$) times in recognition error, decision error, and performance error compared with no error, respectively. The signs of estimated coefficients of truck percentage, speed, number of lanes, shoulder width, and pavement rutting suggest the reduction in the probability of an error compared with no error.

Interesting results found in roadway classification show that the highway type is significantly related to both decision and performance error types, but not recognition error. This suggests that a driver's recognition/inattentive driving error may not depend on highway type. Decision error mostly occurs on Interstate highway, whereas performance error occurs least on the Interstate. The change in the probability of performance error is the highest in other highways which include rural city or town roads. One-ways reduce recognition error, and undivided highways lead to fewer decision errors. Horizontal and vertical curves significantly increase the probability of all error categories, with a maximum increase in performance error for horizontal and decision error for vertical curves.

Roadway events have a significant effect on driver errors. A comparison between roadway and weather condition variables illustrates a few important observations. For example, snowy pavement increases decision error from no error by 4.13 times, whereas snow precipitation only increases this by 1.24 times. Another important observation is that snowy pavement has a higher increase in probability than icy pavement. Drivers tend to be more cautious during adverse weather events because of the negative impact of recognition error. A construction zone increases the probability of decision and recognition error but is not statistically significant for performance error. The negative impact of roadway debris on all types of errors suggests drivers may be more vigilant toward unusual objects on the roadway.

Driver age, gender, vehicle type, alcohol, and drug impairment were found to be statistically significant in predicting all driver error categories. Adolescents are more prone to driver errors compared with all other age groups. For decision error, the probability gradually reduces with the increase in age. But for performance and recognition error types, old drivers are more prone to error compared with young and middle-aged drivers. Decision and recognition error types do not vary by gender, whereas female drivers were found to have a higher probability of performance error. Motorcycle drivers are least likely to have a recognition error, but they are most likely to commit a performance error. Alcohol or drug impairment increases the probability of all error categories, with a maximum increase in performance error.

Table 4 provides the coefficient estimates for the MNP model with urban crash data. Except for median variable, all explanatory variables were found statistically significant at 10% significance level to predict driver error categories.

There are dissimilarities found in the urban crash analysis compared with rural crashes. AADT is only significant in predicting decision error. This means the probability of making a performance or recognition error in an urban setting does not vary by AADT. The posted speed limit does not affect decision error, which is counterintuitive because one of the major driver errors in this category is "Exceeding Speed Limit." Plausibly, speed violation-related crashes may occur at any posted speed limit. The number of lanes, lane width, shoulder width, and passing percent have a positive effect on driver errors.

For the highway type variable, both decision and performance errors mostly occur on Interstate highways in urban areas. For recognition error, other highway types have the highest increase in probability compared with no error. The roadway type variable is not significant in predicting recognition error but is significant for both decision and performance error types at all levels. Divided highways increase the probability of both decision and performance error types compared with no error. In urban areas, drivers are least likely to make performance mistakes with ice on the roadway. For other explanatory variables, similar trends as discussed for rural crashes are observed.

Discussion of Factors Contributing to Driver Errors

With numerous factors contributing to a variety of driver error types, it is challenging to summarize their individual effects. Thus, a review of contributing factors by error type is necessary. The marginal effect of a variable in the driver error model for rural crashes can be seen in Table 5. The marginal effect has varying definitions based upon the variable type. For a continuous independent variable, the marginal effect is the difference in the probability at each level following a one-unit change; for a categorical independent variable, the marginal effect is calculated as the changes in the probability for each level caused by a change in the value from its base condition.

Traffic and roadway variables significantly affect the probability of decision error. For example, a unit increase in AADT increases the probability of decision error, whereas an increase in truck percentage decreases the probability. In the other error types, the marginal effect of truck percentage and AADT are not statistically significant. However, the posted speed limit decreases the probability of recognition error but increases the probability of performance error.

Compared with no errors, the higher probability of recognition error is likely to happen on undivided highways and/or at places where vertical and horizontal curves are present. On foggy/cloudy and/or windy days drivers are more likely to make recognition mistakes. Furthermore, nighttime with (street) light has a positive impact on recognition error. On the other hand, drivers are less likely to commit a recognition error on other highways and one-way streets than on the Interstate and state highways. In addition, recognition error is low when the pavement is either wet or covered in snow, or weather type is snow/sleet/rain, or nighttime without light. This suggests that drivers may exercise caution when traveling in adverse weather or dark conditions. Similarly, when visibility is low or roadway debris is present, the probability of making a recognition error is low. Another source of low recognition error is people that are older than 18, with middle-aged drivers having the lowest probability of recognition error. For different vehicle types, motorcyclists have the lowest probability of recognition error.

Despite the fact that recognition error shares many similar circumstances leading to decision error, the latter is more likely to take place on Interstate highways or one-way streets, but less likely to take place on undivided highways, whereas the opposite pattern is observed for recognition error. A deterrent for decision error seems to be poor pavement condition (i.e., large rutting value). However, the probability of decision error is higher under adverse weather (e.g., fog/cloudy, snow/sleet, rain) and/or on slippery pavement (snow, ice, and wet) as well as the night condition irrespective of the availability of street lighting. Heavy trucks have the lowest probability of decision error among all vehicle types, perhaps because of the safety regulations imposed on drivers. Work zones may

Table 5. Review of Marginal Effects for Rural Crashes

Variable	Recognition error	Decision error	Performance error
Traffic variables	–	Truck (–0.0001)	Truck (–0.0001)
	AADT (0.0002)	AADT (0.0001)	–
Roadway geometry	–	–	Lanes (-0.015)
	Speed (–0.002)	Speed (–0.003)	Speed (0.001)
Highway type (base: interstate)	Other highways (–0.028)	Other highways (–0.076)	Other highways (0.014)
	–	State highways (–0.032)	State highways (0.006)
Roadway type (base: divided)	Undivided (0.026)	Undivided (–0.042)	Undivided (0.015)
	One-way (–0.043)	One-way (0.041)	–
Alignment	Ver.Curve = Yes (0.010)	Hor.Curve = Yes (0.0177)	Hor.Curve = Yes (0.048)
	Hor.Curve = Yes (0.009)	Ver.Curve = Yes (0.0129)	Ver.Curve = Yes (0.010)
Pavement	–	Rutting (–0.047)	–
Roadway condition (base: dry)	Snow (-0.197)	Snow (0.342)	
	Wet (-0.048)	Wet (0.094)	Wet (0.015)
		Ice (0.340)	
Weather condition (base: clear)	Fog/cloudy (0.007)	Fog/cloudy (0.011)	Fog/cloudy (0.017)
	Snow/sleet (–0.085)	Snow/sleet (0.061)	Snow/sleet (0.025)
	Rain (–0.045)	Rain (0.082)	–
	Wind (0.137)	–	–
Lighting condition (base: day)	Night-unlit (–0.012)	Night-unlit (0.061)	Night-unlit (0.026)
	Night-lit (0.021)	Night-lit (0.056)	Night-lit (0.020)
Events	Debris = Yes (–0.145)	Debris = Yes (–0.226)	Debris = Yes (–0.131)
	–	Work zone = Yes (0.145)	Work zone = Yes (0.084)
	Visibility: Yes (–0.058)	–	–
Impairment	Alcohol (–0.028)	Alcohol (0.022)	Alcohol (0.073)
	–	Drug (0.021)	Drug (0.060)
Age (base: adolescent)	Old (–0.022)	Old (–0.112)	Old (0.023)
	Adult (–0.040)	Adult (–0.033)	–
	Middle age (–0.046)	Middle age (–0.058)	Middle age (–0.016)
Gender (base: female)	–	–	Male (–0.003)
Vehicle type (base: heavy truck)	Motorcycle (–0.104)	Motorcycle (0.032)	Motorcycle (0.133)
	–	Passenger car (0.053)	Passenger car (0.014)
	–	Light truck (0.047)	–

Note: Marginal effect presented with "–" is not significant at 90% confidence interval.

see a higher probability of decision error. Finally, decision error is increased by the use of alcohol and/or drugs.

Compared with the other error types, performance error is the most probable with the change in roadway geometry and traffic configuration. When making comparisons with no error, the probability of performance error is high for all highway types, horizontal or vertical alignments, adverse weather, wet pavement surfaces, and during the night. Similar to decision error, performance error is more likely to occur when a work zone is present and is aggravated by the use of alcohol and drugs. Finally, both motorcycles and passenger cars are associated with a higher probability of performance error than truck drivers. On the other hand, middle-aged male drivers have the lowest probability of committing performance error.

Table 6 provides the estimates of marginal effects of covariates for urban crashes. Only the statistically significant variables at 90% confidence interval are shown in the table.

Similar to rural crashes, traffic variables significantly affect the probability of decision error. An increase in AADT increases the probability of decision error, and an increase in truck percentage decreases the probability of decision error. In urban crashes, changes in roadway geometric configuration affect performance error more compared with other error categories. The effect of highway type, roadway type, and existence of horizontal and vertical curve on recognition error in urban crashes is similar to rural crashes. The weather and roadway conditions are more likely to affect decision error, whereas only wet or snowy roadway surface is responsible for performance error.

The gender variable is only statistically significant in predicting the probability of recognition error. Males are less likely to commit recognition error in urban crashes. Having alcohol in the blood while driving contributes to the probability of all driver error, but the presence of drug only affects the probability of decision error. This coincides with practical knowledge that having taken a

Table 6. Review of Marginal Effects for Urban Crashes

Variable	Recognition error	Decision error	Performance error
Traffic variables	–	Truck (–0.002)	Truck (0.0001)
	AADT (–1.45E–7)	AADT (4.89E–7)	–
Roadway geometry	–	Speed (0.002)	Speed (–0.002)
	–	–	Lanes (0.013)
	Lane width (0.004)	Lane width (0.006)	–
	–	–	Shoulder width (0.003)
	Percent passing (0.001)	–	–
Highway type (base: interstate)	Other highways (0.011)	Other highways (–0.070)	–
	State highways (–0.015)	State highways (–0.019)	State highways (-0.057)
Roadway type (base: divided)	Undivided (0.022)	Undivided (–0.029)	–
	–	One-way (-0.026)	One-way (–0.039)
Alignment	Hor.Curve = Yes (–0.060)	–	Hor.Curve = Yes (0.075)
	Ver.Curve = Yes (–0.006)	Ver.Curve = Yes (0.014)	Ver.Curve = Yes (–0.017)
Pavement	Rutting (–0.084)	Rutting (–0.105)	Rutting (0.164)
Roadway condition (base: dry)	Snow (–0.159)	Snow (0.157)	Snow (0.031)
	Wet (–0.040)	Wet (0.037)	Wet (0.041)
	Ice (–0.210)	Ice (0.200)	–
Weather condition (base: clear)	–	Fog/cloudy (0.011)	Fog/cloudy (0.016)
	Snow/sleet (–0.074)	Snow/sleet (0.073)	Snow/sleet (0.030)
	Rain (–0.056)	Rain (0.067)	–
	–	Wind (0.091)	–
Lighting condition (base: day)	Night-unlit (–0.004)	Night-unlit (0.061)	Night-unlit (0.008)
	Night-lit (–0.002)	Night-lit (0.059)	Night-lit (0.002)
Events	Debris = Yes (–0.142)	Debris = Yes (–0.230)	Debris = Yes (–0.142)
	Work zone = Yes (–0.003)	Work zone = Yes (0.096)	Work zone = Yes (–0.040)
	Visibility: Yes (–0.075)	Visibility: Yes (–0.162)	Visibility: Yes (0.183)
Impairment	Alcohol (–0.033)	Alcohol (–0.074)	Alcohol (0.064)
		Drug (–0.051)	–
Age (base: adolescent)	Young Adult (–0.006)	–	–
	Adult (–0.052)	Adult (–0.053)	–
	Middle age (–0.015)	Middle age (–0.075)	Middle age (–0.011)
	–	Old (–0.134)	Old (0.075)
Gender (base: female)	Male (–0.013)	–	–
Vehicle type (base: heavy truck)	Passenger car (0.023)	Passenger car (0.079)	Passenger car (0.031)
	Motorcycle (–0.095)	–	Motorcycle (0.151)
	Light truck (0.028)	Light truck (0.074)	Light truck (0.027)

Note: Marginal effect presented with "–" is not significant at 90% confidence interval.

drug will decrease the attentiveness of the driver and eventually increase the probability of making a decision error. The interpretation of the marginal effect for all other variables can be expressed in a similar way as discussed for rural crashes.

Conclusion

More than 90% of crashes that occurred on a roadway segment involve driver error. Driver error can be categorized as recognition, decision, performance, and nonperformance, based on the definition of each error category investigated in NMVCCS. The reasons behind these errors can be complicated, including highway and traffic characteristics, environmental factors, roadway events, driver characteristics, and vehicle types.

This study established a statistical relationship between driver errors with a series of factors including roadway, traffic, and crash data elements. MNP models were applied to quantify the effect of each explanatory variable. The model results suggest that many of the roadway geometry, highway classification, traffic characteristics, roadway event, and driver-related variables are statistically correlated with different driver error categories in both rural and urban areas. Dissimilarities were found by comparing results between rural and urban crashes, which suggests a possible influence of safety culture.

To better understand the impact of various factors contributing to driver error, a review was conducted using marginal effects from the MNP models. The marginal effect of each explanatory variable represents the quantity of increase or decrease in the probability of a specific driver error type. Thus, each error category can be characterized by a combination of unique variables that help to differentiate future safety treatments. These findings provide evidence-based information to support safety professionals in the development of cost-effective engineering countermeasures, safety enforcement, or driver training programs focused on specific driver errors.

Author Contributions

The authors confirm contribution to the paper as follows: study conception and design: Qin and Shaon; data collection: Shaon; literature review: Shaon, Qin and Chen; Analysis and interpretation of results: Shaon, Qin and Chen; draft manuscript preparation: Shaon, Qin, Chen and Zhang. All authors reviewed the results and approved the final version of the manuscript.

Reference

1. National Highway Traffic Safety Administration. National Motor Vehicle Crash Causation Survey: Report to Congress. *National Highway Traffic Safety Administration Technical Report DOT HS*, Vol. 811, 2008, p. 59.
2. USDOT. *MMUCC Guideline: Model Minimum Uniform Crash Criteria*. US Department of Transportation, Washington, D.C., 2014.
3. Hauer, E. Traffic Safety Toolbox: A Primer on Traffic Safety. *Institute of Transportation Engineers*, Washington, D.C., 1999.
4. Campbell, J. L. *NCHRP Report 600: Human Factors Guidelines for Road Systems*, 2nd ed. Transportation Research Board of the National Academies, Washington, D.C., 2012.
5. Hauer, E. Speed and Safety. *Transportation Research Record: Journal of the Transportation Research Board*, 2009. 2103: 10–17.
6. Tate, F., and S. Turner. Road Geometry and Drivers' Speed Choice. *Road & Transport Research: A Journal of Australian and New Zealand Research and Practice*, Vol. 16, No. 4, 2007, p. 53.
7. Liu, C., and C.-L. Chen. *An Analysis of Speeding-Related Crashes: Definitions and the Effects of Road Environments*. Report No. HS-811 090. NHTSA, Washington, D.C., 2009.
8. NHTSA. Traffic Safety Facts: Driver Electronic Device Use Observation Protocol. *DOT HS*, Vol. 811, 2010, p. 361.
9. Box, S. New Data from VTTI Provides Insight Into Cell Phone Use and Driving Distraction. *Virginia Tech Transportation Institute*, Blacksburg, VA, 2009. Available at: https://vtnews.vt.edu/articles/2009/07/2009-571.html. Accessed July 15, 2017.
10. Pickrell, T. M., and C. Liu. *Occupant Restraint Use in 2014: Results from the NOPUS Controlled Intersection Study*. Report No. DOT HS 812 244. NHTSA, Washington, D.C., 2016.
11. Juergens, W. *Construction Zone, Detour and Temporary Connection Accidents*. Report No. 1263-38103. Bridge Department, California Division of Highways, Sacramento, CA, 1962.
12. Graham, J. L., R. J. Paulsen, and J. C. Glennon. *Accident and Speed Studies in Construction Zones*. Report No. FHWA-RD-77-80. FHWA, U.S. Department of Transportation, Washington, D.C., 1977.
13. Rouphail, N. M., Z. S. Yang, and J. Fazio. *Comparative Study of Short-And Long-Term Urban Freeway Work Zones. Transportation Research Record: Journal of the Transportation Research Board*, 1988. 1163: 4–14.
14. Blomberg, R. D., R. C. Peck, H. Moskowitz, M. Burns, and D. Fiorentino. Fiorentino. *Crash Risk of Alcohol Involved Driving: A Case-Control Study*. Final Report to the National Highway Traffic Safety Administration, Dunlap and Associates, Inc., Stamford, CT, 2005.
15. NHTSA. 2015 Motor Vehicle Crashes: Overview. *Traffic Safety Facts Research Note*, Vol. 2016, 2016, pp. 1–9.
16. Compton, R., M. Vegega, and D. Smither. *Drug-Impaired Driving: Understanding the Problem and Ways to Reduce It: A Report to Congress*. Report No. DOT HS 811 268. NHTSA, U.S. Department of Transportation, Washington, D.C., 2009.
17. Asbridge, M., J. A. Hayden, and J. L. Cartwright. Acute Cannabis Consumption and Motor Vehicle Collision Risk: Systematic Review of Observational Studies and Meta-Analysis. *Bmj*, Vol. 344, 2012, p. e536.
18. Walsh, J. M., A. G. Verstraete, M. A. Huestis, and J. Mørland. Guidelines for Research on Drugged Driving. *Addiction*, Vol. 103, No. 8, 2008, pp. 1258–1268.
19. Hedlund, J. *Drug Impaired Driving: A Guide for States*. Governors Highway Safety Association (GHSA), Washington, D.C., 2017.
20. Devlin, A., N. Candappa, B. Corben, and D. Logan. Designing Safer Roads to Accommodate Driver Error. *Psychopharmacology*, Vol. 10, 2011, pp. 193–212.
21. Bao, S., and L. N. Boyle. Age-Related Differences in Visual Scanning at Median-Divided Highway Intersections in Rural Areas. *Accident Analysis & Prevention*, Vol. 41, No. 1, 2009, pp. 146–152.
22. Dingus, T. A., F. Guo, S. Lee, J. F. Antin, M. Perez, M. Buchanan-King, and J. Hankey. Driver Crash Risk Factors and Prevalence Evaluation Using Naturalistic Driving Data. *Proceedings of the National Academy of Sciences*, Vol. 113, No. 10, 2016, pp. 2636–2641.
23. Wang, K., and X. Qin. Exploring Driver Error at Intersections: Key Contributors and Solutions. *Transportation Research Record: Journal of the Transportation Research Board*, 2015. 2514: 1–9.
24. Borooah, V. K. *Logit and Probit: Ordered and Multinomial Models*. Thousand Oaks, CA: Sage, 2002.
25. Greene, W. H. *Econometric Analysis*. Prentice Hall, Upper Saddle River, NJ, 2000.
26. Parker, S. T., and Y. Tao. WisTransPortal: A Wisconsin Traffic Operations Data Hub. In K. C. P. Wang, B. L.

Smith, D. R. Uzarski, and S. C. Wong (eds.), *Applications of Advanced Technology in Transportation.* American Society of Civil Engineers, Reston, VA, 2006, pp. 611–616.

27. WisDOT. *Wiscosnin Crash Data User Guide.* http://transportal.cee.wisc.edu/documents/database/crash/crash-data-user-guide.pdf. Accessed June 20, 2017.

28. Long, J. S., and J. Freese. *Regression Models for Categorical Dependent Variables Using Stata*, 2nd ed. Stata press, College Station, TX, 2006.

The Safety Section (ANB00) peer-reviewed this paper (18-03131).

Article

TRR
JOURNAL OF THE TRANSPORTATION RESEARCH BOARD

Transportation Research Record
2018, Vol. 2672(38) 35–45

DOI: 10.1177/0361198118793499
journals.sagepub.com/home/trr
SAGE

Safety Evaluation of Change in Posted Speed Limit from 65 to 70 mph on Rural Virginia Interstate System

Scott Himes[1], Frank Gross[1], Michael Nichols[2], and Mena Lockwood[2]

Abstract

Effective July 1, 2010, the Virginia Department of Transportation (VDOT) increased the maximum posted speed limit on interstates and similar facilities from 65 to 70 mph, if recommended following an engineering study. As a result, VDOT performed engineering studies on selected rural interstates posted at 65 mph. By November 2010, VDOT had increased the speed limit from 65 to 70 mph for approximately 670 centerline miles of select rural interstates. This paper presents the results of an empirical Bayes before–after study into the safety and operational effects of the speed limit increase. The analysis focused on total, injury, run-off-road, and truck-related crashes. Comparison segments were used to develop annual adjustment factors, account for regional differences, and identify underlying crash trends in the period before the speed limit increase. At the aggregate level, the results indicated no increase in any of the focus crash types after the increase. Focusing on sites without other changes, the increased speed limit did not increase or decrease any of the crash types. The disaggregate analysis provided further insight into the circumstances in which the change in posted speed limit had more and less pronounced impacts; specifically, that segment type (base or interchange) influenced safety: interchange segments observed statistically significant increases in total, run-off-road, and truck-related crashes. The disaggregate analysis also showed that roadway improvements may help to offset the safety impact of increasing the posted speed limit.

Effective July 1, 2010, the Virginia Department of Transportation (VDOT) increased the maximum allowable posted speed limit on interstates and similar facilities from 65 to 70 mph, if recommended following an engineering study. As a result, VDOT performed engineering studies on selected rural interstates already posted at 65 mph. By November 2010, VDOT had increased the speed limit from 65 to 70 mph for approximately 670 centerline miles of rural interstates statewide, where justified by the engineering studies.

Consideration for an increase in the speed limit was limited to corridors where the speed limit was already 65 mph and where the characteristics were rural in nature (e.g., interchange spacing over 3 mi and free of recurring congestion). Corridors were further distinguished by a relatively low density of crashes. Detailed engineering studies were initiated for those corridors. The studies further constrained speed limit increase recommendations to sections where roadway departure crash densities did not generally exceed the critical rate determined for that corridor. VDOT further scrutinized miscellaneous "hot-spot" roadway departure crash locations along corridors and identified mitigation strategies for implementation in conjunction with any recommended speed limit increase. Where the studies recommended an increase in the speed limit, input and concurrence was then obtained by law enforcement.

The primary objective of this study was to quantify the relationship between safety, as measured by crash frequency, and the increase in the posted speed limit. The primary question to be answered by the crash-based analysis was "what is the effect of changing the posted speed limit from 65 to 70 mph on crash frequency and severity?" Target crash types were identified based on common industry practice and those crashes that are most likely to be affected by changes in the posted speed limit. The following target crash types were defined and analyzed:

[1]Vanasse Hangen Brustlin (VHB), Raleigh, NC
[2]Virginia Department of Transportation, Richmond, VA

Corresponding Author:
Address correspondence to Scott Himes: shimes@vhb.com

- Total crashes: All crashes reported during the study period. Injury crashes: A subset of total crashes in which the crash resulted in one or more K, A, B, or C injuries as defined by the KABCO scale (an injury scale in which K = fatal (killed), A = incapacitating injury, B = nonincapacitating injury, C = possible injury, and O = property damage only).
- Due to changes in crash severity coding in Virginia, aggregate and disaggregate analyses focused on KABC crashes combined.
- Run-off-road crashes: A subset of total crashes in which the crash involved a single vehicle or multiple vehicles leaving the travel lane to the left or right.
- Truck-related crashes: A subset of total crashes in which the crash involved a heavy vehicle.

A secondary question of interest is, "how do roadway and traffic characteristics affect the relationship between safety (i.e., target crashes) and change in posted speed?" Specific characteristics of interest include:

- Roadway geometry (e.g., number of lanes)
- Segment type (e.g., base freeway segment or interchange area)
- Traffic operations (e.g., mean speed, pace speed the 10-mph increment in speed in which the highest percentage of drivers are observed, and speed variance

Literature Review

Based on a review of studies of posted speed limit change on rural interstates or expressways (excluding most studies focusing on the repeal of the national maximum 55-mph speed limit), several studies investigated the safety impact of changing the posted speed limit from 65 to 70 mph. Three of the studies concluded that changes in posted speed limit had relatively little impact (no statistically significant change) on safety performance. Malyshkina and Mannering focused on the impact of speed limit increases on crash-injury severities in Indiana (*1*). They examined roadway sections using multinomial logit models and concluded that the increase from 65 to 70 mph did not significantly affect the severity of crashes. Renski et al. focused on the safety impacts with respect to single vehicle crashes in North Carolina and concluded that no statistically significant changes in injury severity were associated with the change in posted speed limit from 65 to 70 mph (*2*). Souleyrette et al. analyzed crashes over a 17-year period and indicated that none of the changes were statistically significant at the 5% level after the increase in posted speed limit (*3*). However, two studies indicated a statistically significant increase in crashes after increasing the posted speed from 65 to 70 mph (*4, 5*). Haselton et al. focused on the safety impacts of speed limit changes in California and concluded that there was a statistically significant increase in total and fatal crashes after the increase in posted speed limit (*4*). They did note, however, that there was no statistically significant change in fatal and injury crashes (combined) in any of the analyses. Souleyrette and Cook found a statistically significant (at the 10% level) 25% increase in total crashes for 4 years of data in Iowa (*5*). The researchers also found increases of 52% and 24% for nighttime fatal crashes and in serious cross median crashes, respectively, neither of which were statistically significant. Donnell et al. compared the before and after crash rates on pilot Pennsylvania Interstate sections where the posted speed limit was increased from 65 to 70 mph to reference sections where the posted speed limit remained unchanged (*6*). Based on a limited sample size, the relative risk of a crash in the 70-mph posted speed limit zone increased for all pilot sections for total crashes. The relative risk decreased on two interstate sections for fatal and injury crashes and for speeding-related crashes.

Another recent study investigated the impacts of speed limit policies from 49 states on traffic fatalities on rural interstates using Fatality Analysis Reporting System (FARS) data (*7*). FARS data were selected because uniform crash data are available for all states, reducing differences due to crash reporting practices. The authors used random parameter negative binomial models to conduct a cross-sectional analysis using 13 years of data (1999 to 2011). The results indicated that a 70-mph speed limit was associated with 22.2% more fatal crashes and 31.7% more truck or bus-involved crashes compared with a baseline of a 60 or 65-mph maximum speed limit.

Several studies were identified that investigated the safety impacts of changes in posted speed limits after the implementation, increase, and repeal of the national maximum speed limit in 1974, 1987, and 1995, respectively (*8–15*). The primary finding is that there is considerable variability in the results among states, even within groups of states that implement similar changes in posted speed limits on similar types of roads. One of the studies, specific to Virginia (*8*), concluded that the repeal of the national maximum speed limit in 1995 resulted in an initial increase in speed, fatalities, and fatal crashes, followed by a plateau or reduction in both speed and crashes. Although past studies may be criticized for various reasons (e.g., not comparing the change in safety on adjacent, parallel roadways), the relevance to this study is that impacts may vary by state and over time, so it is important to develop Virginia-specific estimates of the effects on safety.

Previous work was conducted to establish the safety impacts of raising the maximum speed limit from 55 to

65 mph for passenger cars in 1988, and commercial buses in 1989 within the state of Virginia. The data showed that fatal crashes and fatalities increased on Virginia's rural interstates after the implementation of the 65-mph speed limit. However, the increases plateaued after 2 years. It was unclear whether the general increase in fatal crashes and fatalities were the result of the change in speed limit or other factors, as the study did not account for other changes. There have been substantial advances in the state of the practice for estimating safety impacts since that research was completed. These advances allow for more rigorous control of other factors such as regression-to-the-mean (RTM) and changes in traffic volume.

Methodology

An empirical Bayes (EB) before–after study was deemed appropriate for meeting the objectives of this study. The EB method can account for RTM and changes in traffic volume.

In the EB method, the change in safety for a given crash type at a treated segment is given by:

$$\pi - \lambda \tag{1}$$

where π is the expected number of crashes that would have occurred in the after period without the treatment, and λ is the number of reported crashes in the after period. Due to changes in safety that may result from changes in traffic volume, RTM, and temporal trends, the count of crashes before a treatment by itself is not a good estimate of π (*16*). Instead, π is estimated from an EB procedure in which a safety performance function (SPF) is used to first predict the number of crashes in each year of the before period based on locations with traffic volumes and other characteristics similar to the treatment sites (*16*). The sum of these annual SPF estimates (P) is then combined with the count of crashes (x) in the before period at the treatment site to obtain an estimate of the expected number of crashes (m) before the treatment. This estimate of m is denoted by:

$$m = (w)(P) + (1-w)(x) \tag{2}$$

The weight w is estimated as:

$$w = 1/(1 + \alpha \times P) \tag{3}$$

where α is the dispersion parameter of the negative binomial distribution that is assumed for the crash counts used in estimating the SPF. The value of α is estimated during the development of the SPF using a maximum likelihood procedure, and smaller values of α indicate a better fit (i.e., less dispersion).

A factor is then applied to m to account for the length of the after period and differences in traffic volumes between the before and after periods. This factor is typically the sum of the annual SPF predictions for the after period divided by P, the sum of these predictions for the before period. In this case, the factor also includes information from the comparison group (adjacent untreated segments) to help account for general temporal trends. The result, after applying this factor, is an estimate of π.

The estimate of π is then summed over all sites in a treatment group of interest (to obtain π_{sum}) and compared with the count of crashes during the after period in that group (λ_{sum}). The variance of π is also summed over all sections in the group of interest.

The index of safety effectiveness (θ) represents the crash modification factor (CMF) and is estimated as:

$$\theta = (\lambda_{sum}/\pi_{sum})/\{1 + [Var(\pi_{sum})/\pi_{sum}^2]\} \tag{4}$$

The standard error of θ is given by:

$$Stderr(\theta) = [\theta^2\{[Var(\lambda_{sum})/\lambda_{sum}^2] + [Var(\pi_{sum})/\pi_{sum}^2]\}/[1 + Var(\pi_{sum})/\pi_{sum}^2]^2]^{0.5} \tag{5}$$

The percent change in crashes is $100(1-\theta)$; thus, a value of $\theta = 0.70$ with a standard error of 0.12 indicates a 30% reduction in crashes with a standard error of 12%.

Data Collection

Figure 1 provides an overview of the interstate sections that have a 70-mph posted speed limit and segments that have not been recommended for a 70-mph posted speed limit. The blue segments are those that have received the 70-mph posted speed limit and the red segments are those that have not. Table 1 provides the extents of segments that were included in this study as treatment sections, encompassing most of the sections shown in Figure 1. Notably, Interstate 85 was excluded from the analysis as the increase in posted speed occurred prior to 2010. Only segments treated in 2010 were included in the analysis. Table 1 also provides the extents of comparison sites used to account for time-related trends. The comparison segments are those that were considered for an increase in posted speed, but remained unchanged. Note the posted speed limit increased to 70 mph in December of 2013 on a large portion of the Charlottesville segment along Interstate 64 between milepost 107.54 and 125.28. A portion of this section was utilized as a comparison site and, as such, the 2014 data for MP 107.54 to 117.66 were excluded from the analysis.

VDOT provided segment-level data from their roadway inventory database, which included information on annual average daily traffic volume (AADT), number of lanes, segment type, area type, and segment boundaries.

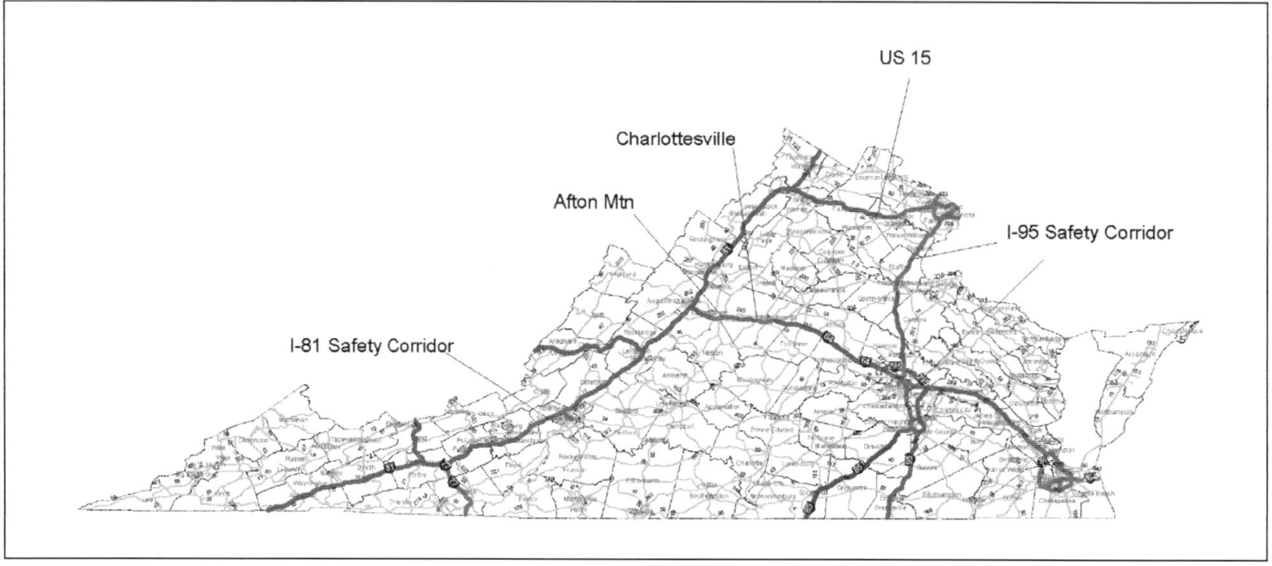

Figure 1. Overview of Virginia 70-mph segments.

New segments were defined in the database by changes in these features as well as changes in shoulder width or median width. The minimum and maximum segment lengths were 0.1 mi and 6.97 mi for base segments, and 0.1 mi and 6.19 mi for interchange segments, respectively. VDOT provided data for the before period from 2006 to 2009 and the after period from 2011 to 2014. VDOT also provided crash data for the same period, indicating crash-injury severity, a run-off-road indicator, and an indicator for involvement of heavy trucks.

Table 1 shows that despite the area-type designation, defined by the VDOT functional classification, urban roadways included in the treatment group often had a similar mean AADT to their rural counterparts. In all cases, the rural segments had a higher mean AADT than their urban counterparts. It is likely that the urban segments included in the 70-mph postings have certain characteristics consistent with freeways coded as rural by VDOT, and are not representative of urban segments for the facility type if all statewide segments are considered. For the current study sites, urban segments tend to be shorter than rural segments, especially for base (i.e., non-interchange) segments.

Table 2 presents a summary of the data elements for treatment site indicator variables. The majority of sites have two lanes in one direction (81%) whereas the remainder has three and four lanes in the travel direction (15% and 4%, respectively). Approximately two-thirds of the segments are base freeway segments whereas one-third has interchanges. Some districts have fewer segment miles of freeway segments (e.g., District 9), and therefore districts were combined into regions for analysis.

SPF Development

Generalized linear modeling techniques were applied to estimate SPFs for each target crash type. A log-linear relationship was specified using a negative binomial error structure, which is consistent with the state of research in developing SPFs. Model coefficients were estimated using the software package Stata.

Several functional forms were considered for developing the SPFs, but the following were considered most applicable:

$$Crashes/year = Constant{*}AADT^{\beta_1} {*}L{*}exp(C_1{*}X_1 + \ldots + C_n{*}X_n) \quad (6)$$

$$Crashes/year = Constant{*}AADT^{\beta_1}{*}L^{\beta_2} {*}exp(C_1{*}X_1 + \ldots + C_n{*}X_n) \quad (7)$$

where

AADT = annual average daily traffic on roadway

L = length of road segment in miles

X_i = vector of independent variables related to the roadway. In this case, X_i corresponds to regions for regional variations

Constant, β_i, and C_i are parameters estimated from data in the SPF development process.

The selection of the final functional form was based on the log-likelihood where a greater (less negative) value indicates a better model fit. In all cases, Equation 7 provided a better fit. The difference between Equation 6 and Equation 7 is that Equation 6 includes segment length (L) as an offset (i.e., restricts the coefficient to 1.0) while

Table 1. Data Summary for Treatment and Comparison Sites

Facility type	Segment length (mi)		Before period							After period						
			AADT							AADT						
	Total	Mean	Min	Mean	Max	Total	Injury	ROR	Truck	Min	Mean	Max	Total	Injury	ROR	Truck
Two-lane rural base	823.97	1.63	3,279	17,246	36,638	3.94	1.33	2.13	0.80	3,583	17,479	35,795	3.96	1.22	2.11	0.82
Two-lane urban base	103.97	0.82	4,902	17,158	31,808	2.71	0.91	1.34	0.59	5,125	17,363	32,703	2.30	0.72	1.18	0.52
Two-lane rural interchange	92.74	0.36	3,279	17,839	36,638	1.15	0.43	0.61	0.23	3,583	18,025	35,795	1.55	0.47	0.80	0.31
Two-lane urban interchange	16.93	0.24	5,196	16,702	36,825	0.92	0.29	0.38	0.29	5,494	16,882	30,737	1.27	0.42	0.65	0.32
Three-lane rural base	71.43	2.04	15,841	37,729	58,887	9.14	2.89	4.44	2.09	13,892	38,785	58,263	9.05	2.58	4.06	1.87
Three-lane urban base	66.06	0.80	9,612	24,182	58,887	2.11	0.74	1.03	0.40	10,663	24,416	58,263	2.40	0.74	1.12	0.43
Three-lane rural interchange	4.73	0.28	20,869	40,093	50,408	1.71	0.75	0.51	0.56	23,234	41,115	52,591	3.00	0.63	1.09	0.81
Three-lane urban interchange	16.26	0.38	10,118	20,025	55,427	1.43	0.51	0.65	0.29	10,931	20,485	54,671	1.80	0.59	0.89	0.33
Four-lane urban base	10.84	0.45	23,937	38,839	50,234	2.67	1.02	1.15	0.55	30,645	42,331	52,440	2.47	0.96	1.06	0.51
Four-lane urban interchange	10.62	0.53	23,937	40,516	50,234	2.90	0.96	1.55	0.43	30,645	43,599	52,440	2.54	0.95	1.31	0.41
Comparison sites	253.8	0.76	8,129	33,784	84,726	5.73	1.53	1.90	0.88	9,880	34,076	73,149	5.81	1.54	1.47	0.92

Note: Crash statistics are presented as crashes/segment/year. AADT = annual average daily traffic; Min = minimum; Max = maximum; ROR = run-off-road.

Equation 7 allows the coefficient to be estimated for segment length.

Due to variation in sample size within VDOT districts, the districts were combined into regions with similar characteristics to account for variation between regions across the state. Three regions were used, consisting of the following districts:

- Region A: Bristol, Salem, and Staunton (Districts 1, 2, and 8, respectively). These are all rural areas with mountainous terrain and a similar mix of truck and driver population. Region A consists of the west and southwest portions of Virginia. This area is more rural and mountainous, following the Shenandoah Mountains and Valley.
- Region B: Culpeper and Northern Virginia (Districts 7 and 9, respectively). These areas have similar traffic levels, vehicle mix, and (level) terrain with relatively straight geometrics. Region B consists of north and northeast Virginia. It is the suburban and rolling areas surrounding suburban Washington DC.
- Region C: Richmond, Hampton Roads, and Fredericksburg (Districts 4, 5, and 6, respectively). These areas have similar levels of congestion, geometrics, and driver population. Region C consists of east and southeast Virginia. This area is flat and mostly makes up the tidewater portion of Virginia.

In specifying a negative binomial error structure, the dispersion parameter (α) was iteratively estimated from the data. The dispersion parameter (α) relates the mean and variance of the SPF and is used in EB methodology. The value of the dispersion parameter (α) is such that the larger its value, the greater the variance relative to the mean for a given set of data. For the models presented in this paper, a smaller value is preferred.

SPFs were estimated for each of the crash types considering a variable dispersion parameter that is a function of the segment length. Hauer identified the possibility that dispersion can logically be related to the segment length; shorter segments will tend of have a higher dispersion than longer segments (*17*). Further research by Cafiso et al. used the logarithm of segment length in the log-dispersion model, resulting in the functional form (*18*):

$$Dispersion = Constant^{*}L^{\beta_1} \quad (8)$$

The value of the constant in Equation 8 is the exponent of the coefficient from the model of the logarithm of dispersion. If the coefficient of the segment length is negative, the model shows that dispersion is inversely related

Table 2. Data Summary for Treatment Site Indicator Variables

Variable	Category	Number of segments	Percent of segments	Length (mi)	AADT		
					Min	Mean	Max
Lanes	2	961	81.23	1,037.61	3,279	17,354	36,825
	3	178	15.05	158.48	9,612	27,361	58,887
	4	44	3.72	21.46	23,937	39,602	50,234
Area type	Rural	816	68.98	992.87	3,279	18,788	58,887
	Urban	367	31.02	224.68	4,902	21,686	58,887
Segment type	Base	776	65.60	1,076.27	3,279	19,565	58,887
	Interchange	407	34.40	141.28	3,279	19,918	55,427
District (region)	1 (A)	249	21.05	223.11	10,660	16,734	31,285
	2 (A)	89	7.52	109.32	15,790	19,847	24,373
	4 (C)	251	21.22	231.32	8,312	24,718	58,887
	5 (C)	118	9.97	103.07	11,375	20,901	32,439
	6 (C)	29	2.45	46.94	37,074	44,360	49,291
	7 (B)	65	5.50	88.93	11,440	17,837	23,014
	8 (A)	373	31.53	409.25	3,279	16,292	36,825
	9 (B)	9	0.76	5.61	17,290	18,121	19,481

Note: AADT = annual average daily traffic; Min = minimum; Max = maximum.

to the length of the segment, which is consistent with the logic established by Hauer (*17*).

SPFs were estimated using the before data from the treatment sites, excluding segment-years with roadway improvement activities, such as paving or installation of safety strategies. The average daily traffic and segment length were considered in the model to account for exposure. Factors were also included to account for differences across years and among VDOT regions. Separate SPFs were developed by number of lanes, area type, and segment type as needed and as sample size allowed. In some cases, separating the SPFs resulted in sample sizes fewer than 100 crashes per year, which the *Highway Safety Manual* suggests is the minimum sample size. In these cases, SPFs were combined, using an indicator variable and an interaction term between the characteristic (e.g., rural versus urban) and traffic volume to account for the impact of the characteristic and the differential impacts of traffic volume. To be consistent with the *Highway Safety Manual*, a minimum of 100 crashes per year are recommended to develop separate SPFs for rural and urban segments. Final SPFs are provided in Table 3.

Results

Aggregate Analysis

The aggregate safety analysis results are presented in Table 4, including the EB estimate of the crashes expected in the after period had the speed limit not been changed, the observed crashes with the increase in speed limit, the CMF, and the standard error of the estimate. CMFs shown in bold font are statistically significant at the five-percent significance level. Table 4 provides results for the following analysis levels:

- All sites. The analysis results indicate the effect of the change in posted speed limit for all sites, including those that received roadway improvements along with the increase in posted speed and those that did not. It reflects the combined effects of the increased posted speed limit and improvement activities; however, improvement activities only occurred on a subset of these segments whereas all segments had an increase in the posted speed limit.
- Sites with improvements excluding improvement years. The analysis results indicate the effect of the change in posted speed limit in combination with all roadway improvement activities. This analysis reflects the combined effects of the increased posted speed limit and improvement activities based on the subset of segments with improvement activities. This analysis excludes segment-years in which improvements occurred, and all sites for which there were no improvements.
- Sites without improvements. The analysis results indicate the effect of the change in speed limit only, removing any time-related effects of roadway improvement activities. Removing years during and after improvement activities eliminates the potential time-effects of improvements (e.g., rumble strips have a lasting impact on run-off-road crashes). This set of results is most reflective of the effects of the change in posted speed limits. The disaggregate analyses is based on this group

Table 3. SPF Coefficients (Standard Errors) by Crash and Facility Type

				Region							Dispersion		
Model	Constant	AADT	Length	B	C	Urban	Urban-AADT	Int	Int-AADT	Four-lane	α	Constant	Length
Total crashes													
Two-lane base	−6.089	0.735	0.798	0.072	−0.097	0.167	NA	NA	NA	NA	NA	−1.422	−0.514
	(0.361)	(0.037)	(0.016)	(0.049)	(0.036)	(0.043)						(0.075)	(0.068)
Two-lane interchange	−5.897	0.667	0.607	0.638	0.425	−4.030	0.430	NA	NA	NA	0.393	NA	NA
	(1.007)	(0.104)	(0.048)	(0.124)	(0.079)	(2.463)	(0.252)						
Three-lane rural	−10.869	1.141	1.038	NA	0.194	NA	NA	14.232	−1.298	NA	0.155	NA	NA
	(3.230)	(0.321)	(0.089)		(0.232)			(4.850)	(0.460)				
Three- and four-lane urban	−9.657	1.062	0.689	NA	−0.023	NA	NA	0.190	NA	NA	NA	−1.583	−0.764
	(0.914)	(0.091)	(0.041)		(0.093)			(0.084)				(0.216)	(0.168)
KABC crashes													
Two-lane	−6.413	0.654	0.798	0.018	0.089	−3.078	0.326	NA	NA	NA	NA	−1.453	−0.526
	(0.537)	(0.055)	(0.018)	(0.068)	(0.046)	(1.552)	(0.158)					(0.141)	(0.122)
Three- and four-lane	−9.864	0.979	0.780	NA	0.022	NA	NA	3.263	−0.293	NA	NA	−1.293	−0.359
	(1.240)	(0.122)	(0.045)		(0.114)			(2.356)	(0.229)			(0.230)	(0.221)
Run-off road crashes													
Two-lane	−5.572	0.615	0.799	0.111	0.090	NA	NA	−0.108	NA	NA	NA	−1.018	−0.642
	(0.430)	(0.044)	(0.019)	(0.059)	(0.041)			(0.058)				(0.077)	(0.067)
Three- and four-lane urban	−8.347	0.893	0.717	NA	−0.447	NA	NA	NA	NA	0.299	NA	−1.315	−0.711
	(1.414)	(0.141)	(0.052)		(0.121)					(0.140)		(0.278)	(0.240)
Three-lane rural	−7.872	0.763	1.057	NA	0.511	NA	NA	NA	NA	NA	0.236	NA	NA
	(4.191)	(0.416)	(0.074)		(0.314)								
Truck crashes													
Two-lane	−9.924	0.982	0.793	−0.676	−0.728	0.196	NA	NA	NA	NA	NA	−1.135	−0.573
	(0.678)	(0.069)	(0.023)	(0.102)	(0.066)	(0.067)						(0.167)	(0.141)
Three- and four-lane	−12.632	1.217	0.825	NA	−0.209	NA	NA	4.359	−0.404	NA	0.173	NA	NA
	(1.528)	(0.150)	(0.055)		(0.136)			(2.942)	(0.285)				

Note: AADT = annual average daily traffic; Int = interchange.

Table 4. Aggregate Safety Results for Change in Posted Speed Limit

Analysis group	Statistic	Total	KABC	ROR	Truck
All sites	EB estimate of crashes expected	14,058.0	4,778.4	7,235.5	2,913.9
	Count of crashes observed	14,163	4,363	7,132	2,929
	CMF	1.007	0.913*	0.986	1.005
	Standard error	0.011	0.017	0.015	0.023
Sites with improvements (excluding improvement years)	Expected	4,138.0	1,418.9	2,131.3	879.2
	Observed	4,168	1,326	2,168	884
	CMF	1.007	0.934*	1.017	0.027
	SE	0.019	0.030	0.027	0.039
Sites without improvements	Expected	4,540.3	1,593.9	2,387.3	899.8
	Observed	4,712	1,521	2,331	969
	CMF	1.038	0.954	0.976	1.076
	SE	0.020	0.030	0.026	0.042

Note: $*p < .05$. EB = empirical Bayes; CMF = crash modification factor; SE = standard error; ROR = run-off-road.

(i.e., removing years of data during and after improvement activities). It is important to note that although the intent is to remove the effects of roadway improvement activities, the results may still be influenced by the safety effects of other changes in the roadway environment. For example, the data may include sites with undocumented roadway maintenance activities. Further, there is potential for spillover effects from upstream segments with roadway improvements.

The following is a summary of the results based on various groupings of sites with and without other roadway improvements.

Results for all sites:

- Total crashes, run-off-road crashes, and truck-related crashes did not change after the change in posted speed limit.
- Total injury crashes (K, A, B, and C) decreased by 8.7% after the change in posted speed limit.

Results for sites with improvements (excluding improvement years):

- Total crashes, run-off-road crashes, and truck-related crashes did not change after the change in posted speed limit.
- Total injury crashes (K, A, B, and C) decreased by 6.6% after the change in posted speed limit.

Results for sites without improvements:

- There was no statistically significant change in any of the crash types after the change in posted speed limit. This result is consistent with analyses of observed speeds. The 2008 and 2014 speeds were measured at 114 locations within segments with posted speed increases. Across all sites, there was a 1.1-mph increase in mean speed and a 1.79-mph increase in 85th percentile speeds, which is less than the 5-mph increase in posted speed limit.

In regard to the decrease in injury crashes that occurred in two of the three scenarios after the change in the speed limit, the 2006 and 2007 injury crash data (i.e., K, A, B, and C injuries on the KABCO scale) indicate KABC crashes in 2006 and 2007 were 15% larger than the average KABC crashes from 2008 to 2014 at treatment sites (1,274 crashes per year in 2006 and 2007 versus 1,081 from 2008 to 2014). Although 2006 and 2007 marked the end of an upward trend in KABC crashes statewide, the trend at the comparison sites was less pronounced than the trend at the treatment sites. It is unknown why there was a difference in the 2006 and 2007 KABC crash trends at the treatment sites and comparison sites. Excluding 2006 and 2007 data from the analysis, the results indicate no change in injury crashes. As such, the results for injury crashes should be considered with caution.

Disaggregate Analysis

A disaggregate analysis was conducted for each of the target crash types to determine whether safety effects are more or less pronounced for specific conditions. Specific conditions explored include: years after implementation, number of lanes, segment type, area type, facility subtype, AADT, mean speed before implementation, change in mean speed, pace speed before implementation, change in speed standard deviation, expected crash frequency before the speed limit change, district, and roadway improvements.

Table 5. Disaggregate Safety Results for Change in Posted Speed Limit

Crash type	Measure	EB estimate of crashes without change	Observed crashes with treatment	CMF (standard error)
Segment type				
Total	Base	3,773.7	3,763	0.997 (0.021)
	Interchange	766.5	949	1.237 (0.053)*
Injury	Base	1,311.7	1,205	0.918 (0.032)*
	Interchange	282.2	316	1.118 (0.076)
Run-off-road	Base	2,003.6	1,886	0.941 (0.027)*
	Interchange	383.8	445	1.158 (0.070)*
Truck-related	Base	756.9	748	0.988 (0.043)
	Interchange	142.9	221	1.544 (0.125)*
Years after speed limit increase				
Total	Year 1	1,658.1	1,669	1.006 (0.027)
	Year 2	1,224.3	1,221	0.997 (0.031)
	Year 3	821.9	903	1.098 (0.040)*
	Year 4	836.0	919	1.099 (0.040)*
Injury	Year 1	568.9	561	0.986 (0.044)
	Year 2	433.3	406	0.937 (0.050)
	Year 3	293.4	268	0.913 (0.059)
	Year 4	298.3	286	0.958 (0.061)
Run-off-road	Year 1	872.0	833	0.955 (0.036)
	Year 2	647.8	627	0.968 (0.042)
	Year 3	430.6	438	1.017 (0.053)
	Year 4	437.0	433	0.990 (0.052)
Truck-related	Year 1	316.0	320	1.012 (0.060)
	Year 2	254.3	278	1.093 (0.070)
	Year 3	162.7	165	1.013 (0.084)
	Year 4	166.9	206	1.234 (0.093)*
Before period expected crash frequency				
Total	<3 crashes/year	1,266.5	1,570	1.239 (0.040)*
	3 + crashes/year	3,273.7	3,142	0.960 (0.022)
Injury	<2 crashes/year	790.0	818	1.035 (0.043)
	2 + crashes/year	803.9	703	0.874 (0.041)*
Run-off-road	<2 crashes/year	772.3	882	1.141 (0.049)*
	2 + crashes/year	1,615.1	1,449	0.897 (0.030)*
Truck-related	<1 crashes/year	375.7	454	1.207 (0.066)*
	1 + crashes/year	524.1	415	0.982 (0.053)

Note: *$p < .05$. EB = empirical Bayes; CMF = crash modification factor.

The disaggregate analysis focuses on segments with no improvement activities (i.e., excluding all sites with improvements). The results of the disaggregate analysis are provided in Table 5.

For base segments, there was no change in total or truck-related crashes, an 8.2% reduction in injury crashes, and a 5.9% reduction in run-off-road crashes. For interchange segments, there was a 23.7% increase in total crashes, no change in injury crashes, a 15.8% increase in run-off-road crashes, and a 54.4% increase in truck-related crashes. Similar to the aggregate analysis, the results for injury crashes were influenced by the trend in KABC crashes in 2006 and 2007. Excluding the data for 2006 and 2007, the results indicated no change in KABC crashes for base segments and an 18.7% increase in KABC crashes for interchange segments.

The results are consistent across time (by year) after the speed limit increase. There was no change in any crash type considering the cumulative after periods from 1 to 4 years (e.g., years 1–2, years 1–3, and years 1–4). Considering each individual year after implementation (e.g., Year 1, Year2, Year 3, Year 4), there were no changes in crashes from year to year, except for years 3 and 4 for total crashes and Year 4 for truck-related crashes. Total crashes increased by 9.8% and 9.9% for years 3 and 4, respectively. Truck-related crashes increased by 23.4% for Year 4.

The disaggregate analysis indicated a notable difference related to the expected crashes before increasing the posted speed limit. Specifically, there was an increase or no change in all crash types for sites with less than 3.0 expected crashes per year before increasing the posted speed limit. For sites with 3.0 or more expected crashes per year before increasing the posted speed limit, there was a reduction or no change in all crash types. This is at least partially explained by the correlation between segment type and the expected number of crashes in the before period. Interchange segments are often short and have a smaller expected crash frequency than base segments, and interchange segments had a general increase in crashes compared with base segments.

Additionally, the study explored the effects of roadway improvements in the disaggregate analysis. The findings related to the safety improvement benefits are reflected in the CMF values displayed in Table 4. VDOT provided a list of roadway improvements and the specific locations where they were implemented during and after the increase in posted speed limit. The improvements include those targeting run-off-road crashes as well as general maintenance and construction activities. Improvement activities include rumble strip installation/reinstallation, pavement resurfacing, guardrail, shoulder activities, pavement markers and markings, and various warning signage. It is likely that these improvements affect the safety performance of the treated segments. As such, they may offset or mask the safety effect of the speed limit increase.

As a general note, the safety results were disaggregated by various speed-related measures; however, there were only 109 segments with before and after speed data out of 1,183 total segments. The 109 sites with speed data were not found to be representative of all study segments due to several factors. Specifically, study segments with speed data tended to be located on base freeway segments (248.2 miles of base segments compared with 13.1 miles of interchange segments). Due to the location of speed collection sites (e.g., tangent sections with limited horizontal and vertical curvature), it is intuitive that the safety performance at these sites would be better than the aggregate safety performance across all study sites. The results indicated an aggregate increase in mean speeds of 1.1 mph, and provided no consistent relationships between speed measures and changes in crash frequency.

Conclusions

The objective of this study was to evaluate the impacts on safety after changing the posted speed limit from 65 to 70 mph on rural interstates in Virginia. Safety effects were estimated using the EB before–after method with a focus on total, injury, run-off-road, and truck-related crashes. SPFs were estimated and used to account for changes in traffic volume. Comparison segments were used to develop annual adjustment factors, account for regional differences, and identify underlying crash trends in the period before the increase in speed limit.

The study considered the effects of the 70 mph speed limits on crashes at both aggregate and disaggregate levels. The aggregate analysis considered the effect of the increased speed limit at all treatment sites as well as two subgroups: sites with improvements and sites without improvements during the study period. Sites without improvements are considered most indicative of the impacts of the increased speed limit as they exclude the influence of improvements or changes other than the increased speed limit. The disaggregate analysis focused on sites without improvements during the study period.

At the aggregate level, the raised speed limit did not increase crash frequency for any crash type across any of the groups (i.e., all sites, sites with improvement, or sites without improvements). Focusing on the results for sites without improvements, which again are most indicative of the impacts of the increased speed limit, the increased speed limit did not change (i.e., increase or decrease) any of the crash types. Focusing on the other two groups, all sites and sites with improvements, the results are consistent for total, run-off-road, and truck crashes (i.e., no change in crashes after the increased speed limit). Although injury crashes decreased for these two groups, all sites and sites with improvements, this is explained by two factors. First, the other improvements at these sites may have contributed to a reduction in injury crashes. Second, the results were influenced by a statewide upward trend in injury crashes in the "before" period, which was unequally reflected at the comparison sites compared with the treatment sites. It is unknown why there was a difference in the 2006 and 2007 injury crash trends at the treatment and comparison sites. There was no change in injury crashes at all sites and sites with improvements after accounting for this trend. As such, the results for injury crashes should be considered with caution. In summary, the results suggest that the increased speed limit did not increase any crash types, and the other improvements may help to reduce injury crashes.

The disaggregate analysis provides further insight into the circumstances where the change in posted speed limit had more and less pronounced impacts. The disaggregate analysis showed that segment type (base or interchange) influenced the safety impact. For base segments, the results indicated a reduction in injury crashes and run-off-road crashes of 8.2% and 5.9%, respectively. There was no change in total crashes and truck crashes. For interchange segments, the results indicated an increase in all crash types except injury crashes. Specifically, there was a 23.7% increase in total crashes, 15.8% increase in run-off-road crashes, and 54.4% increase in truck crashes. There was no change in injury crashes. Again, the results for injury crashes were influenced by the trend in KABC crashes in 2006 and 2007. Excluding 2006 and

2007 data from the analysis, the results indicated no change in injury crashes at base segments and an 18.7% increase in injury crashes at interchange segments.

The disaggregate analysis showed that roadway improvements may help to offset the safety impact of increasing the posted speed limit. In general, there was a relative reduction in total and truck crashes at sites with roadway improvements compared with sites without roadway improvements; however, the differences were not statistically significant. Further, there was a reduction in injury crashes for sites with roadway improvements and no change in injury crashes for sites without roadway improvements.

As the study data are limited to Virginia, caution should be exercised in generalizing the results to other states. There are many factors affecting the transferability of results among jurisdictions, including differences in geometric, traffic, driver, and temporal characteristics, as well as the level of safety before the increase in posted speed limit.

Author Contributions

The authors confirm contribution to the paper as follows: study conception and design: SH, FG, MN, ML; data collection: SH; analysis and interpretation of results: SH, FG, MN, ML; draft manuscript preparation: SH, FG. All authors reviewed the results and approved the final version of the manuscript.

References

1. Malyshkina, N., and F. Mannering. Analysis of the Effect of Speed Limit Increases on Accident-Injury Severities. *Transportation Research Record: Journal of the Transportation Research Board*, 2008. 2083: 122-127.
2. Renski, H., A. J. Khattak, and F. M. Council. Effect of Speed Limit Increases on Crash Injury Severity: Analysis of Single-Vehicle Crashes on North Carolina Interstate Highways. *Transportation Research Record: Journal of the Transportation Research Board*, 1999. 1665: 100–108.
3. Souleyrette, R. R., T. B. Stout, and A. Carriquiry. *Evaluation of Iowa's 70 mph Speed Limit – 2.5 Year Update*. Iowa Department of Transportation, Center for Transportation Research and Education, Ames, IA, 2009.
4. Haselton, C. B., A. R. Gibby, and T. C. Ferrara. *Collision Experience with Speed Limit Changes on Selected California Highways*. FHWA/CA-TO-2001/04. California Department of Transportation, Sacramento, CA, 2001.
5. Souleyrette, R., and D. Cook. *Evaluation of Iowa's 70 mph Speed Limit – 4 Years Later*. Iowa Department of Transportation Office of Traffic and Safety, Project 06-247. Ames, IA, December 2010.
6. Donnell, E. D., B. Hamadeh, L. Li, and J. Wood. *70 mph Study*. FHWA-PA-2016-009-PSU WO 13. Pennsylvania Department of Transportation, Harrisburg, PA, 2016.
7. Davis, A., E. Hacker, P. Savolainen, and T. Gates. Longitudinal Analysis of Rural Interstate Fatalities in Relation to Speed Limit Policies. *Transportation Research Record: Journal of the Transportation Research Board*, 2015. 2514: 21–31.
8. Jernigan, J. D., S. E. Strong, and C. W. Lynn. *Impact of the 65 mph Speed Limit on Virginia's Rural Interstate Highways: 1989-1992*. Final Report. VTRC 95-R7. Virginia Transportation Research Council, Charlottesville, VA, 1994.
9. Farmer, C. M., R. A. Retting, and A. K. Lund. *Effect of 1996 Speed Limit Changes on Motor Vehicle Occupant Fatalities*. Insurance Institute for Highway Safety, Washington, D.C., 1997.
10. National Highway and Traffic Safety Administration. *The Effects of the Increased Speed Limits in the Post-NMSL Era 1990: A Report to Congress*. U.S. Department of Transportation, Washington, D.C., 1998.
11. Khan, N. M., K. C. Sinha, and P. McCarthy. *An Analysis of Speed Limit Policies for Indiana*. FHWA/IN/JTRP-99/14. Joint Transportation Research Program, Indiana Department of Transportation and Purdue University, West Lafayette, IN, 2000.
12. Balkin, S., and J. K. Ord. Assessing the Impact of Speed-Limit Increases on Fatal Interstate Crashes. *Journal of Transportation and Statistics*, Vol. 4, No. 1, 2001, pp. 1–26.
13. Kockelman, K. M. *NCHRP Report 90: Safety Impacts and Other Implications of Raised Speed Limits on High-Speed Roads*. Transportation Research Board of the National Academies, Washington, D.C., 2006.
14. Retting, R. A., and E. R. Teoh. Traffic Speeds on Interstates and Freeways 10 Years after Repeal of National Maximum Speed Limit. *Traffic Injury Prevention*, Vol. 9, No. 2, 2008, pp. 119–124.
15. Friedman, L. S., D. Hedeker, and E. D. Richter. Long-Term Effects of Repealing the National Maximum Speed Limit in the United States. *American Journal of Public Health*, Vol. 99, No. 9, 2009, pp. 1626–1631.
16. Hauer, E. *Observational Before-After Studies in Road Safety: Estimating the Effect of Highway and Traffic Engineering Measures on Road Safety*. Pergamon Press, Elsevier Science Ltd., Oxford, UK, 1997.
17. Hauer, E. Overdispersion in Modelling Accidents on Road Sections and in Empirical Bayes Estimation. *Accident Analysis and Prevention*, Vol. 33, No. 6, 2001, pp. 799–808.
18. Cafiso, S., G. Silvestro, B. Persaud, and M. Begum. Revisiting Variability of Dispersion of Safety Performance for Two-Lane Rural Roads. *Transportation Research Record: Journal of the Transportation Research Board*, 2010. 2148: 38–46.

The Standing Committee on Safety Data, Analysis and Evaluation (ANB20) peer-reviewed this paper (18-01486).

Article

TRR
JOURNAL OF THE TRANSPORTATION RESEARCH BOARD

Transportation Research Record
2018, Vol. 2672(38) 46–57

DOI: 10.1177/0361198118776782
journals.sagepub.com/home/trr
SAGE

Developing a Comprehensive Road Safety Index: A Revisit of the SUNflowerNext Study

Yongjun Shen[1,2]

Abstract

Given that more and more indicators are used to describe the complex character of the road safety phenomenon, the development of a road safety index is increasingly recognized as a useful tool to reduce the size of the topic without dropping underlying information and to perform meaningful comparison and monitoring of overall road safety performance. In this respect, a scientifically sound methodology for index construction is crucial. In the SUNflowerNext study, a comprehensive road safety index that aggregated 21 sub-indicators belonging to four different types of road safety indicators was created for a set of European countries. As a revisit of this study, this paper focuses on two methodological challenges when developing such an index. One is to reflect the hierarchical structure of the indicators, and the other is to distinguish between quantitative and qualitative data. To this end, a mathematical programming model based on the data envelopment analysis approach is proposed. By solving a constrained optimization problem, the concept of layered hierarchy is embodied in the model and the presence of both quantitative and qualitative data is properly integrated. Based on the country-specific models, the optimal road safety index score is computed for each country, and a more acceptable country ranking is obtained compared with the one from the SUNflowerNext study. In addition, by deducing the weights allocated in each layer of the hierarchy for each country separately, valuable information on prioritizing policy action per country is derived.

Crash data such as the number of fatalities and casualties, normally viewed as road safety final outcomes, are traditionally used to assess the road safety performance of a country. However, they are imperfect indicators to represent the entire level of road safety because they are only the "worst case scenario" in the insecure operational conditions of road traffic (*1*). Today, having recognized the complex character of the road safety phenomenon, countries all over the world are encouraged to develop data collection procedures for road safety indicators to cover not only the final outcomes, but also intermediate outcomes (e.g., levels of mean traffic speeds, drink driving, seat belt wearing, and vehicle safety ratings), exposure measures (for relating outcomes to population levels, licensed driver numbers, or distances traveled), institutional delivery outputs (including different categories of enforcement effort), and socio-economic costs associated with road trauma as well (*2*). Moreover, several initiatives and research studies have been implemented which look at the development of a road safety index, which combines individual indicator values into a single score, with the purpose of capturing the multi-dimensional concept of road safety and of performing meaningful comparisons (*3–5*). Among others, the SUNflowerNext study (*6*) is a representative one, in which four different types of road safety indicators were distinguished corresponding to the road safety target hierarchy originated in New Zealand (see Section 2). They were final outcome indicators, safety performance indicators, policy performance indicators, and structure and culture indicators. For each type of indicator, a set of sub-indicators with a layered hierarchy was developed. An overall road safety index combining all the sub-indicators was then created using two statistical weighting approaches, principal component analysis (PCA) and factor analysis (FA), based on the data collected for 27 European countries.

The SUNflowerNext study clearly demonstrated the feasibility of creating a comprehensive index for road

[1]School of Transportation, Southeast University, Nanjing, China
[2]Transportation Research Institute (IMOB), Hasselt University, Diepenbeek, Belgium

Corresponding Author:
Address correspondence to Yongjun Shen: shenyongjun@seu.edu.cn

safety evaluation and comparison among different countries, and the results revealed that such an index gave an enriched picture of road safety (*6*). However, when the detail of the index construction is examined, some research attention is still needed. First of all, obtainment of measurable and quantitative indicators is commonly the prerequisite of any index research. However, this becomes more and more difficult because the natural uncertainty of reality often leads to imprecision and vagueness inherent in information that can only be represented by means of qualitative data, such as the policy performance indicators developed in the SUNflowerNext study (see Section 2). Simply treating them as quantitative ones may result in unreasonable conclusions. Second, in the SUNflowerNext study, a set of sub-indicators was developed for different types of road safety indicators constituting a multilayer hierarchical structure. However, all these indicators were actually treated as belonging to the same layer when deducing the indicator weights and the final index score as well. In fact, the structure of the indicators contains valuable information worth considering in the index construction. In addition, the weighting methods applied in the SUNflowerNext study, like most of the common weighting schemes, assigned the same indicator weights for all the countries under study and the final index score was calculated as the weighted sum of indicator values. In this way, full use is not made of country-specific characteristics. In other words, the importance level of each indicator in different countries is ignored, which makes the examination of root causes of poor performance in each country difficult.

In this paper, an alternative approach is proposed to take all the above methodological challenges into account when developing a comprehensive road safety index. More specifically, a mathematical programming model based on the data envelopment analysis (DEA) approach is developed to deal with the nature of quantitative and qualitative data, to reflect the hierarchical structure of the indicators, and to generate country-specific indicator weights. By solving a constrained optimization problem, the concept of layered hierarchy is embodied in the model and the presence of both quantitative and qualitative data is properly integrated. Moreover, each country obtains its own best possible indicator weights, based on which the optimal road safety index score of each country, relative to the other countries in the data set, is determined. The results are further compared with the ones from the SUNflowerNext study.

The remainder of this paper is structured as follows. In Section 2, the road safety indicators developed in the SUNflowerNext study and the data used for this study are briefly introduced. Section 3 focuses on the development of the new model for road safety index construction, and the results are presented and compared in Section 4. The paper ends with discussion and concluding remarks in Section 5.

Road Safety Indicators and Data in the SUNflowerNext Study

In Europe, the need for road safety indicators and data was stated in the 2001 Transport White Paper of the European Commission (*7*), and further specified in the 3rd Road Safety Action Plan in 2003 (*8*). In doing so, a road safety target hierarchy was adopted. The concept originated in New Zealand (*9*), and was further used in the SUNflower study (*10*), the SafetyNet study (*11*), and the DaCoTA study (*12*). Now, it has also become the theoretical basis for the creation of the European Road Safety Observatory with the purpose of bringing together all Community activities in relation to safety data and knowledge. In general, the target hierarchy describes road safety as a pyramid consisting of five layers as presented in Figure 1.

From bottom to top they are: structure and culture (describing the background conditions of a country or its policy context); safety measures and programs (or the road safety policy performance); safety performance indicators (also known as intermediate outcomes); the number of fatalities and casualties (as the final outcomes); and the social costs of crashes and injuries at the very top.

Based on this framework, the first SUNflower study (*10*) was carried out in 2002, which compared road safety in three countries—Sweden, the United Kingdom, and the Netherlands—considering not only the numbers of crashes and victims, but also the factors, circumstances, and development that have an influence on the risk of a crash and the severity of its outcome. Given the long list of interesting findings and valuable recommendations, the study was quickly enlarged by the following SUNflower + 6 study (*13*), and later on the SUNflowerNext study, with the purpose of developing a knowledge-based framework for comprehensive benchmarking of road safety performance and development for a country or other sub-national jurisdictions (*6*).

The Indicator Structure

In the SUNflowerNext study, four types of road safety related indicators, that is, final outcome indicators, safety performance indicators, policy performance indicators, and structure and culture indicators, were defined to reflect the different layers of the road safety pyramid as shown in Figure 1. Moreover, for each type of the

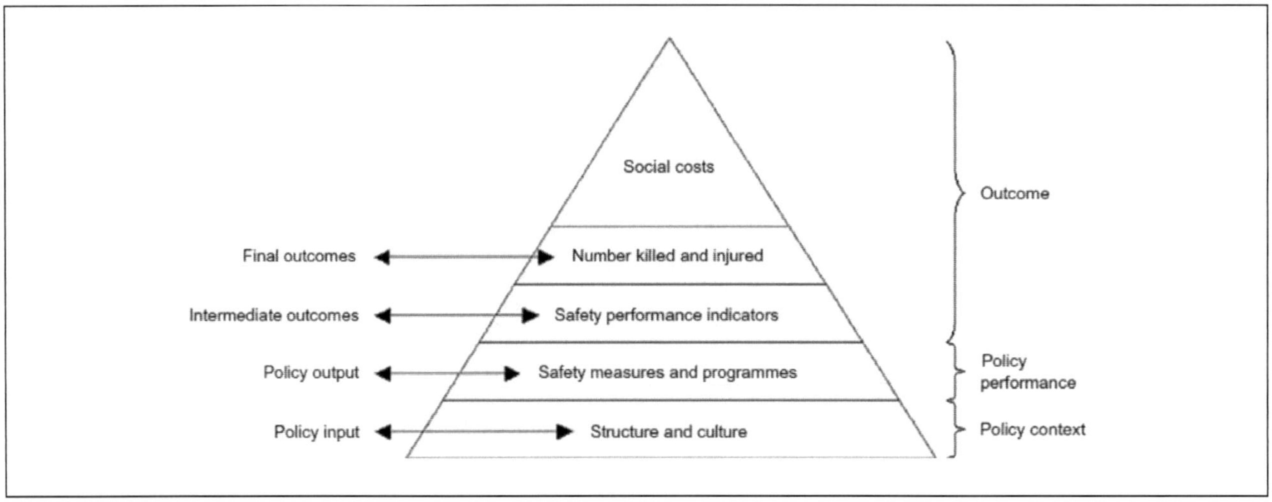

Figure 1. A target hierarchy for road safety (*13*).

Table 1. The Hierarchical Structure of the Indicators Developed in the SUNflowerNext Study: An Overall Road Safety Index

Indicator types			Indicators
Final outcome indicators	Personal safety		Fatalities per million inhabitants (F1)
	Traffic safety		Fatalities per million passenger cars (F2)
			Fatalities per 10 billion passenger-km travelled (F3)
	Traffic injury		Injury accidents per fatality (F4)
	Vulnerable road users		Share of pedestrian fatalities out of the total fatalities (F5)
			Share of bicyclist fatalities out of the total fatalities (F6)
			Share of motorcyclist fatalities out of the total fatalities (F7)
Safety performance indicators	Alcohol		Share of the total fatalities in drink-driving accidents (S1)
	Protective systems		Daytime wearing rates of seat belts in the front seats (S2)
			Daytime wearing rates of seat belts in the rear seats (S3)
	Vehicles	Crashworthiness of the passenger car fleet	Average EuroNCAP score of the passenger car fleet (S4)
			Median age of the passenger car fleet (S5)
		Vehicle fleet composition	Share of motorcycles in the vehicle fleet (S6)
			Share of heavy goods vehicles in the vehicle fleet (S7)
Policy performance indicators			The availability and ambition of national safety targets (P1)
			Selection of interventions (P2)
			Economic evaluation (P3)
			Monitoring the programme's performance (P4)
			Programme's stakeholders (P5)
Structure and culture indicators			Number of passenger cars per 1000 inhabitants (C1)
			Population per square kilometer of country's territory (C2)

indicator, a set of sub-indicators with a multilayer hierarchical structure was developed, which is illustrated in Table 1.

Specifically, the final outcome indicators, used to assess the output level of road safety in a country, were classified into four aspects: personal safety, traffic safety, traffic injury, and the issue of vulnerable road users. For each aspect, one or several sub-indicators were defined and denoted as F1 to F7. For the safety performance indicators, which are used to characterize the quality of implemented road safety policies, three safety areas were considered at this point because of data availability. They were alcohol, protective systems, and the crashworthiness and composition of the vehicle fleet. For each area, several sub-indicators were formulated (S1–S7). For the third type of the indicator, or the policy performance indicators, a set of five sub-indicators was developed describing the level of national road safety strategies (P1–P5). Finally, with regard to the bottom level of the pyramid, two structure and culture indicators (C1–C2) were selected to represent the background characteristics of countries, which were motorization level and population density.

Data Description

Taking a look at these 21 road safety indicators developed in the SUNflowerNext study, most of them can be measured by quantitative data, except for the five policy performance indicators, which are qualitative in nature. In other words, they should take the form of ordered classes from best to worst or some other formulation, rather than numerical data. Indicator values related to 2006 (except for S1–S3 with data for 2005, and S4–S5 for 2003) were collected (or imputed) for the 27 European countries in the SUNflowerNext study. 24 of them are considered in this paper (see Table 2) because Malta had very distinctive data and Cyprus and Luxembourg provided no data on the policy performance indicators (these were given as imputed numbers in the SUNflowerNext study). So these three countries have been removed from this study.

For the three types of quantitative indicators, that is, the final outcome indicators, the safety performance indicators, and the structure and culture indicators, to eliminate the scale differences of the indicators and to ensure that all of them are expressed in the same direction as their expected road safety impact, that is, a high indicator value corresponds to a better road safety level, the raw data are normalized using the distance to a reference approach (*14*), in which the maximum or minimum value of each indicator in the data set is treated as the reference when a benefit respective to a cost indicator is taken into account. As a result, the country with the highest safety performance receives a normalized value of 1.00 whereas the others are expressed as percentage share of that country's value. Taking the fatalities per million inhabitants (F1) as an example, the Netherlands performs the best (1.00), Lithuania the worst (0.20), and all other countries' values lie within this interval.

For the policy performance indicators, ordinal data rated on either a 3-point or a 4-point scale are used and shown in Table 2 as well, in which rank 1 indicates a better performance than 2, which is better than 3, and so on. However, they cannot be simply treated as quantitative data for which a score of 4 is twice as large as a score of 2. Therefore, when combining all the indicator values into an overall road safety index score, the numerical and ordinal ones need to be treated differently.

Model Development

Different from the models applied in the SUNflowerNext study, in this paper an alternative approach is proposed for the development of a comprehensive road safety index based on the same set of indicators. More specifically, a mathematical programming model based on the DEA approach is proposed to 1) generate country-specific indicator weights; 2) reflect the hierarchical structure of the indicators; and 3) deal with the nature of quantitative and qualitative data.

DEA was originally developed as a non-parametric frontier technique to measure the so called relative efficiency of a homogeneous set of decision making units (DMUs) (*15*). Since its first introduction in 1978, DEA has been quickly recognized as a powerful analytical research tool for modeling operational processes in terms of performance evaluation, benchmarking, and decision making, and it has been successfully applied to a wide range of activities in many contexts (*16*, *17*). Lately, this technique has also been introduced to the field of index construction, and various DEA-based index models have been developed (*18*, *19*). The most attractive feature of DEA, relative to other methods in developing an index is that each DMU, or country in this case, obtains its own best possible indicator weights, that is, the weights resulting in the highest index score for a country. This implies that dimensions on which the country performs relatively well get a higher weight. In this way, main strengths and key problems can be identified for each country separately.

Mathematically, consider a set of n countries, each having m different indicators. The DEA-based index model for a particular country c can be formulated as

$$I_c = \max \sum_{i=1}^{m} w_i x_{ic}$$
$$s.t.(\text{subject to}) = \sum_{i=1}^{m} w_i x_{ij} \leq 1,\ j = 1, \cdots, n \qquad (1)$$
$$w_i \geq \varepsilon,\ i = 1, \cdots, m$$

where x_{ij} is the ith indicator of the jth country, w_i is the weight given to indicator i, and ε is a small non-negative number for restricting the model to assign a weight of 0 to unfavorable indicators.

This linear program is computed separately for each country to determine its best possible indicator weights.

Table 2. Ordinal and Normalized Numerical Data of the 21 Road Safety Indicators for 24 European Countries

Country		F1	F2	F3	F4	F5	F6	F7	S1	S2	S3	S4	S5	S6	S7	P1	P2	P3	P4	P5	C1	C2
Austria	AT	0.51	0.54	0.42	0.85	0.62	0.20	0.21	0.40	0.86	0.58	0.91	0.67	0.11	0.73	1	3	2	2	1	0.49	0.15
Belgium	BE	0.45	0.44	0.44	0.60	0.82	0.15	0.25	0.55	0.73	0.27	0.94	0.75	0.23	0.46	2	2	2	2	2	0.53	0.04
Czech Republic	CZ	0.43	0.36	0.29	0.32	0.49	0.13	0.36	0.44	0.74	0.15	0.91	0.45	0.09	0.57	2	3	2	2	2	0.62	0.11
Denmark	DK	0.80	0.62	0.75	0.27	0.47	0.13	0.27	0.10	0.88	0.71	0.91	0.68	0.21	0.27	1	2	2	2	2	0.67	0.11
Estonia	EE	0.30	0.24	0.24	0.20	0.34	0.32	0.95	0.08	0.76	0.34	0.96	0.39	0.74	0.36	2	3	2	2	1	0.60	0.48
Finland	FI	0.70	0.70	0.79	0.31	0.78	0.12	0.41	0.10	0.91	0.88	0.90	0.55	0.13	0.44	1	2	2	2	2	0.52	0.92
France	FR	0.58	0.62	0.66	0.27	0.82	0.34	0.17	0.08	1.00	0.79	0.92	0.60	0.22	0.37	1	1	2	1	1	0.49	0.13
Germany	DE	0.73	0.86	0.74	1.00	0.66	0.14	0.22	0.47	0.99	1.00	0.89	0.83	0.13	1.00	1	2	2	1	2	0.44	0.06
Greece	EL	0.30	0.25	0.24	0.15	0.58	1.00	0.13	0.22	0.81	0.31	0.91	0.63	0.08	0.29	1	2	2	2	2	0.61	0.17
Hungary	HU	0.35	0.21	0.15	0.25	0.41	0.11	0.39	0.27	0.69	0.38	0.91	0.63	0.38	0.41	1	3	2	2	2	0.84	0.13
Ireland	IE	0.52	0.45	0.32	0.29	0.49	0.43	0.24	0.16	0.89	0.52	0.91	0.98	0.88	0.35	2	2	2	1	2	0.59	0.23
Italy	IT	0.47	0.59	0.57	0.65	0.74	0.27	0.16	1.00	0.73	0.24	0.93	0.63	0.07	0.59	3	4	3	3	4	0.41	0.07
Latvia	LV	0.25	0.18	0.16	0.16	0.24	0.22	0.98	0.11	0.79	0.26	0.94	0.50	0.38	0.42	2	3	2	2	2	0.69	0.41
Lithuania	LT	0.20	0.19	0.22	0.14	0.27	0.25	0.34	0.20	0.62	0.15	1.00	0.34	1.00	0.66	2	3	2	2	2	0.53	0.28
the Netherlands	NL	1.00	0.93	0.88	0.52	0.99	0.07	0.21	0.16	0.93	0.72	0.89	0.83	0.22	0.45	1	1	2	1	1	0.56	0.04
Norway	NO	0.87	0.82	0.67	0.51	0.64	0.39	0.25	0.08	0.94	0.93	0.93	0.63	0.15	0.30	2	1	1	2	2	0.56	1.00
Poland	PL	0.33	0.23	0.18	0.14	0.29	0.12	1.00	0.24	0.80	0.26	0.91	0.42	0.30	0.35	2	3	2	2	2	0.70	0.12
Portugal	PT	0.49	0.42	0.32	0.57	0.58	0.32	0.16	0.21	0.89	0.51	0.92	0.52	0.16	0.24	2	3	2	2	2	0.61	0.12
Slovakia	SK	0.42	0.22	0.20	0.21	1.00	0.24	0.53	0.20	0.78	0.27	0.93	0.71	0.38	0.43	2	3	2	2	2	1.00	0.13
Slovenia	SI	0.34	0.35	0.37	0.69	0.68	0.25	0.19	0.07	0.90	0.34	0.93	1.00	0.29	0.81	2	3	2	2	2	0.51	0.15
Spain	ES	0.48	0.47	0.36	0.38	0.61	0.72	0.22	0.20	0.76	0.57	0.91	0.66	0.10	0.31	1	3	2	2	2	0.53	0.16
Sweden	SE	0.92	0.89	0.93	0.63	0.75	0.22	0.25	0.07	0.95	0.82	1.00	0.71	0.15	0.55	1	1	1	1	1	0.54	0.71
Switzerland	CH	0.90	1.00	1.00	0.90	0.45	0.14	0.18	0.12	0.85	0.60	0.91	0.68	0.11	0.80	1	1	1	2	1	0.48	0.08
Great Britain	GB	0.83	0.82	0.88	0.91	0.44	0.28	0.22	0.14	0.93	0.94	0.91	1.00	0.38	0.48	1	2	2	1	1	0.52	0.06

In other words, the weights in the objective function are chosen to maximize country c's index score I_c and meanwhile respect the less than unity constraint for all the countries including c itself. Consequently, a country is considered to be best-performing if it obtains a score of 1.00 whereas a score less than 1.00 implies that it is an underperforming country.

To apply the DEA-based index model for this study, two methodological challenges have to be properly handled. One is to embody the hierarchical structure of indicators, and the other is to distinguish between quantitative and qualitative data. As a consequence, Model (1) has to be modified accordingly. Suppose that a set of n countries is to be evaluated in terms of m_1 numerical indicators and m_2 ordinal indicators with an L layers' hierarchical structure. The index score of each country can now be computed by solving the following mathematical programming problem:

$$\begin{aligned} I_c &= \max \sum_{i=1}^{m_1} v_i x_{ic} + \sum_{r=1}^{m_2} \sum_{k=1}^{K} \mu_{rkc} \\ s.t. &= \sum_{i=1}^{m_1} v_i x_{ij} + \sum_{r=1}^{m_2} \sum_{k=1}^{K} \mu_{rkj} \leq 1, \; j = 1, \cdots, n \\ \underset{i_l \in A_{i_{l+1}}^{(l+1)}}{p_{i_l}^{(l)}} &= \frac{\sum\limits_{i \in A_{i_l}^{(l)}} v_{ic}}{\sum\limits_{i \in A_{i_{l+1}}^{(l+1)}} v_{ic}} \in \Theta, \; i_l = 1, \cdots, m_1^{(l)}, \; l = 1, \cdots, L-1 \\ & v_i \geq \varepsilon, \; i = 1, \cdots, m_1 \\ & \mu_{rk} - \mu_{rk+1} \geq \varepsilon^2, \; k = 1, \cdots, K-1, \; r = 1, \cdots, m_2 \\ & \mu_{rK} \geq \varepsilon^2, \; r = 1, \cdots, m_2 \end{aligned} \tag{2}$$

where

v_i denotes the new weights assigned to the numerical indicator i,

$\sum_{k=1}^{K} \mu_{rkc}$ denotes the weighted virtual value of each ordinal indicator,

$A_{i_l}^{(l)}$ denotes the set of the indicators of the ith category in the lth layer, $i_l = 1, \cdots, m_1^{(l)}, l = 1, \cdots, L-1$,

$\underset{i_l \in A_{i_{l+1}}^{(l+1)}}{p_{i_l}^{(l)}}$ denotes the internal weights of the indicators in each category of each layer, and

Θ denotes the weight restrictions.

The main idea of incorporating the layered hierarchy into Model (1) is to first aggregate the values of the indicators within a particular category of a particular layer by the weighted sum approach in which the sum of the internal weights (p) equals 1.0. With respect to the final layer, the four types of road safety indicators representing the different levels of the road safety pyramid, the weights are determined using the standard DEA approach. For the detailed deduction process, refer to (*20*). In particular, the internal weights of the indicators in each category of each layer can be deduced as $\underset{i_l \in A_{i_{l+1}}^{(l+1)}}{p_{i_l}^{(l)}} = \sum_{i \in A_{i_l}^{(l)}} v_i / \sum_{i \in A_{i_{l+1}}^{(l+1)}} v_i$. In such a way, the value judgment from decision makers can be incorporated into the model by restricting the weight flexibility in a category, which is denoted as Θ. In this study, the sub-indicators in the same category of each layer are considered to be of similar importance. Thus, absolute weight restrictions are applied, in which the weights are required to vary within a range from 0.8 to 1.2 of their average weights. Taking the structure and culture indicators as an example, the weights of the two sub-indicators C1 and C2 (the average weight is 0.5) are required to lie between 0.4 and 0.6.

With respect to the last layer, the four types of road safety indicators, an ordinal relation in terms of share (i.e., $Share_F > Share_S > Share_P > Share_C$) is utilized in accordance with the target hierarchy for road safety presented in Figure 1, which implies the importance ordering of these four types of indicators in assessing the road safety performance of a country. Here, the "share" is the sum of the products of the indicator values and the corresponding weights, divided by the final index score.

Moreover, to model the m_2 ordinal indicators, a latter part of the objective function in Model (2) is added. Specifically, consider a set of ordinal data ($n \times 1$) with a

Table 3. Conversion of Ordinal Data

PI		AI		
1		1	0	0
2		0	1	0
2		0	1	0
1		1	0	0
2		0	1	0
1		1	0	0
1		1	0	0
1		1	0	0
1	⇨	1	0	0
1		1	0	0
2		0	1	0
3		0	0	1
2		0	1	0
2		0	1	0
1		1	0	0
2		0	1	0
2		0	1	0
2		0	1	0
2		0	1	0
2		0	1	0
1		1	0	0
1		1	0	0
1		1	0	0
1		1	0	0

K-point scale that is related to indicator r ($r = 1, \cdots, m_2$). First convert it into an $n \times K$ matrix A_r in which the value of the rank is translated into its position. Taking the first policy performance sub-indicator, that is, Safety targets (P1), as an example, the 24 ordinal values can be translated as shown in Table 3.

Subsequently, the matrix can be multiplied by a vector $[x_{r1}, x_{r2}, \cdots x_{rK}]^T$ to obtain the virtual value of each ordinal indicator, that is, $x_r = A_r[x_{r1}, x_{r2}, \cdots x_{rK}]^T$, which should satisfy the minimal requirement that $x_{r1} > x_{r2} > \cdots > x_{rK} \geq \varepsilon$. Thus, the aggregation of all the m_2 ordinal indicators for country c can be expressed as: $\sum_{r=1}^{m_2} v_r x_{rc}$, which can be further linearized by $\sum_{r=1}^{m_2} \sum_{k=1}^{K} \mu_{rkc}$ as presented in Model (2). Moreover, to realize the above ordinal restrictions on x_{rk}, set $x_{rk} - x_{rk+1} \geq \varepsilon$ ($k = 1, \cdots, K-1$), and $x_{rK} \geq \varepsilon$. Then $\mu_{rk} - \mu_{rk+1}$ which equals to $v_r x_{rk} - v_r x_{rk+1}$ should at least satisfy the minimum constraint that $\mu_{rk} - \mu_{rk+1} \geq \varepsilon^2$, and $\mu_{rK} \geq \varepsilon^2$. Both of the inequalities are included as the last two constraints of Model (2).

Results

Now the proposed model is applied to combine the 21 hierarchically structured road safety indicators (with both numerical and ordinal data) into an optimal index score for each country by selecting its best possible indicator weights that respect all the imposed constraints. As the model leads up to country-specific outcomes, the index scores resulting from all 24 models are shown in Table 4.

This is a 24 $\times$ 24 matrix. Each column represents the index score of the 24 European countries with the best possible weights of the country under study. For instance, the values in the second column are obtained from the model for Austria, in which Austria obtains its optimal index score of 0.84, while Germany receives the benchmark value of 1.00. Moreover, the elements along the major diagonal represent the optimal road safety index score of each country. It can be seen that the optimal value of 1.00 is obtained by three countries—Germany, Norway, and Sweden—followed by Switzerland, Great Britain, and the Netherlands, for which the index score is also quite close to 1.00. It means that they are better-performing countries under study with respect to the overall level of road safety. Conversely, the remaining 18 countries obtaining a value much less than 1.00 are considered to be underperforming.

To fully rank all these 24 countries, besides their optimal index score, the results from the other 23 models are employed as well. In particular, the average value of a certain row in Table 4 is calculated as the final road safety index score of that country. Because all the countries are now evaluated based on the same weighting set, their comparisons can be made directly, with a higher index score indicating better overall performance. The results are shown in the second column of Table 5 in decreasing value of the index score. As can be seen, Germany is the top country for overall road safety performance, followed by Sweden and Norway, whereas Lithuania is the worst. To further make comparison with the SUNflowerNext study, the results based on the PCA approach are also listed in the fourth column of Table 5. In addition, as a basis for judging the results of other rankings, the number of road fatalities per million inhabitants, which is used as the traditional approach for countries' comparison, is considered as well and is presented in the sixth column of Table 5.

It should be noted that in the SUNflowerNext study, both PCA and FA were utilized to combine the individual indicators, and five different trials were performed, named as PCA-all, PCA-group, FA-4factors, FA-2factors-noS4, and FA-2factors, respectively. Detailed description of these trials were provided in (*21*). Here, the results from the PCA-all trial are used for comparison, because in this trial all the indicators were analyzed together, whereas in the FA-related trials, some indicators were excluded to apply this approach for index construction.

By comparing the three rankings shown in Table 5, it can be seen that the ranking from either the current study or the SUNflowerNext study is different from the traditional ranking of countries based only on the fatality rate. It is mainly because apart from the final outcome indicators, three other types of road safety indicators, that is, the safety performance indicators, the policy performance indicators, and the structure and culture indicators, are also taken into account in both the current study and the SUNflowerNext study. The combination of these indicators therefore gives a more enriched picture of road safety and provides different countries' rankings as well.

Nevertheless, the ranking from the current study is found to be more consistent with the traditional ranking compared with that from the SUNflowerNext study given that the correlation coefficient between the former two rankings is relatively higher (0.91 versus 0.70). It can be interpreted by understanding that the importance ordering of these four types of indicators in the development of a comprehensive road safety index is considered in the proposed model, that is, $Share_F > Share_S > Share_P > Share_C$, according to the road safety pyramid shown in Figure 1. As a consequence, the final outcome indicators receive the highest share in the final index score relative to the others. In other words, the ranking of countries based on the proposed model relies more on the final outcome indicators than the others. Such a consideration, however, cannot be realized in

Table 4. Overall Results for the 24 European Countries

	AT	BE	CZ	DK	EE	FI	FR	DE	EL	HU	IE	IT	LV	LT	NL	NO	PL	PT	SK	SI	ES	SE	CH	GB
AT	**0.84**	0.83	0.83	0.82	0.83	0.80	0.81	0.81	0.83	0.83	0.83	0.82	0.83	0.82	0.82	0.80	0.83	0.83	0.83	0.83	0.83	0.81	0.79	0.80
BE	0.77	**0.80**	0.80	0.78	0.77	0.75	0.77	0.75	0.80	0.79	0.79	0.79	0.79	0.78	0.78	0.74	0.79	0.77	0.79	0.78	0.79	0.76	0.74	0.75
CZ	0.67	0.70	**0.70**	0.69	0.68	0.65	0.66	0.61	0.69	0.70	0.69	0.68	0.69	0.68	0.67	0.64	0.70	0.67	0.69	0.66	0.69	0.65	0.63	0.64
DK	0.77	0.82	0.82	**0.83**	0.81	0.79	0.81	0.70	0.82	0.82	0.82	0.80	0.82	0.81	0.82	0.80	0.82	0.81	0.82	0.77	0.82	0.81	0.81	0.82
EE	0.66	0.68	0.67	0.68	**0.69**	0.65	0.64	0.57	0.66	0.68	0.67	0.65	0.69	0.68	0.65	0.66	0.69	0.66	0.67	0.64	0.66	0.65	0.59	0.61
FI	0.86	0.90	0.90	0.91	0.89	**0.93**	0.90	0.75	0.90	0.90	0.89	0.90	0.90	0.90	0.89	0.91	0.90	0.89	0.90	0.86	0.90	0.88	0.87	0.88
FR	0.74	0.77	0.76	0.78	0.76	0.77	**0.79**	0.68	0.77	0.77	0.77	0.77	0.76	0.77	0.78	0.76	0.76	0.78	0.77	0.74	0.77	0.76	0.76	0.76
DE	1.00	1.00	1.00	1.00	1.00	1.00	1.00	**1.00**	1.00	1.00	1.00	1.00	1.00	1.00	1.00	0.98	1.00	1.00	1.00	1.00	1.00	1.00	1.00	1.00
EL	0.66	0.66	0.65	0.66	0.65	0.62	0.66	0.57	**0.68**	0.65	0.68	0.67	0.66	0.66	0.64	0.63	0.67	0.68	0.65	0.66	0.68	0.63	0.59	0.61
HU	0.64	0.66	0.66	0.66	0.66	0.61	0.62	0.55	0.66	**0.67**	0.66	0.63	0.66	0.65	0.63	0.62	0.66	0.65	0.66	0.63	0.66	0.60	0.57	0.60
IE	0.75	0.76	0.75	0.76	0.77	0.72	0.74	0.67	0.76	0.76	**0.78**	0.75	0.77	0.75	0.75	0.74	0.77	0.76	0.77	0.74	0.77	0.74	0.72	0.74
IT	0.83	0.87	0.88	0.85	0.83	0.84	0.84	0.83	0.88	0.86	0.86	**0.89**	0.85	0.86	0.83	0.80	0.86	0.83	0.85	0.83	0.87	0.82	0.81	0.82
LV	0.64	0.65	0.65	0.66	0.66	0.62	0.61	0.54	0.63	0.66	0.65	0.62	**0.66**	0.65	0.61	0.63	0.66	0.63	0.65	0.62	0.63	0.62	0.56	0.58
LT	0.57	0.58	0.57	0.57	0.58	0.54	0.55	0.50	0.57	0.58	0.58	0.57	0.58	**0.59**	0.55	0.55	0.58	0.56	0.57	0.55	0.57	0.54	0.52	0.53
NL	0.91	0.96	0.95	0.95	0.95	0.92	0.96	0.87	0.96	0.96	0.96	0.94	0.95	0.94	**0.98**	0.93	0.95	0.95	0.97	0.91	0.96	0.96	0.96	0.97
NO	0.96	0.97	0.97	0.99	0.98	0.98	0.96	0.82	0.98	0.98	0.99	0.97	0.98	0.97	0.96	**1.00**	0.98	0.99	0.98	0.94	0.99	0.96	0.95	0.95
PL	0.63	0.66	0.66	0.66	0.66	0.61	0.61	0.56	0.64	0.66	0.65	0.62	0.66	0.65	0.62	0.62	**0.67**	0.63	0.65	0.61	0.64	0.62	0.56	0.59
PT	0.75	0.75	0.74	0.75	0.75	0.71	0.74	0.68	0.75	0.75	0.75	0.73	0.75	0.74	0.74	0.72	0.75	**0.76**	0.75	0.74	0.76	0.73	0.70	0.71
SK	0.72	0.76	0.76	0.75	0.75	0.69	0.72	0.62	0.75	0.76	0.75	0.72	0.75	0.74	0.73	0.69	0.76	0.74	**0.76**	0.72	0.75	0.69	0.63	0.67
SI	0.76	0.73	0.73	0.73	0.74	0.70	0.74	0.71	0.73	0.73	0.74	0.73	0.74	0.74	0.74	0.70	0.73	0.75	0.74	**0.77**	0.73	0.72	0.70	0.72
ES	0.75	0.75	0.74	0.75	0.75	0.72	0.75	0.68	0.76	0.75	0.77	0.75	0.76	0.75	0.74	0.74	0.76	0.76	0.75	0.74	**0.77**	0.74	0.71	0.72
SE	0.97	1.00	0.99	1.00	1.00	1.00	1.00	0.88	1.00	1.00	1.00	1.00	1.00	1.00	1.00	1.00	1.00	1.00	1.00	0.97	1.00	**1.00**	1.00	1.00
CH	0.95	0.95	0.95	0.94	0.96	0.93	0.96	0.92	0.95	0.95	0.95	0.95	0.96	0.96	0.96	0.93	0.95	0.95	0.95	0.94	0.95	0.97	**0.99**	0.99
GB	0.97	0.96	0.95	0.97	0.97	0.95	0.97	0.93	0.96	0.96	0.97	0.96	0.97	0.97	0.97	0.95	0.96	0.97	0.96	0.96	0.96	0.98	0.99	**0.99**

Table 5. Road Safety Rankings of the 24 European Countries

Country	Index score from this study	Ranking	Index score from the SUNflowernext study	Ranking	Number of road fatalities per million inhabitants	Ranking
DE	0.999	1	−0.516	4	62	7
SE	0.992	2	−0.999	1	49	2
NO	0.967	3	−0.500	5	52	4
GB	0.964	4	−0.750	2	54	5
CH	0.952	5	−0.112	10	50	3
NL	0.946	6	−0.198	9	45	1
FI	0.887	7	−0.245	8	64	8
IT	0.846	8	0.761	24	96	14
AT	0.820	9	−0.032	12	88	11
DK	0.806	10	−0.047	11	56	6
BE	0.775	11	0.259	15	101	15
FR	0.762	12	−0.697	3	77	9
IE	0.749	13	−0.425	6	86	10
ES	0.744	14	0.166	14	93	13
PT	0.738	15	0.274	17	92	12
SI	0.732	16	−0.294	7	131	19
SK	0.725	17	0.264	16	107	17
CZ	0.672	18	0.686	23	104	16
EE	0.657	19	0.029	13	152	22
EL	0.649	20	0.576	20	149	21
HU	0.636	21	0.645	22	129	18
PL	0.631	22	0.644	21	137	20
LV	0.628	23	0.360	18	178	23
LT	0.561	24	0.500	19	224	24

the PCA approach. Thus, a more consistent result with the traditional one is observed when applying the proposed model.

Looking into the detail of the rankings of each country, it appears that some better-performing countries like Sweden and Norway and some worse-performing countries like Greece and Poland are relatively consistent in their rankings across the different approaches. The widest deviation in rankings happens to the countries like Italy and Slovenia. Taking Italy as an example, it receives a rank of 14 based on the traditional final outcome indicator, but is ranked last out of the 24 countries in the SUNflowerNext study. To explain the reason for such a big difference in the ranking, it is necessary to look back on the indicator data collected for this country. It can be found that Italy has the worst scores for all the five policy performance indicators among the 24 European countries. They are ordinal data rated on either a 3-point or 4-point scale but were treated as numerical ones in the SUNflowerNext study. In other words, a score of 4 was considered as four times larger than a score of 1 in the index calculation using the PCA approach, which thereby led up to an unrealistically bad rank for Italy. In this study, however, the model distinguishes these ordinal indicators from those numerical ones, and also groups indicators and assigns importance ordering via different types. As a result, a better rank is achieved for Italy given that it has relatively good indicator values on some other final outcome indicators (e.g., F4) and some safety performance indicators (e.g., S1) as well.

In addition to the index score, because of the consideration of hierarchical structure of indicators in the proposed model, the weights allocated in each layer of the hierarchy can be deduced for each country separately. Based on the principle of DEA, an indicator is assigned a high weight if the country performs relatively well on that aspect. On the contrary, low weights provide policymakers with valuable information about the kind of measures most urgently needed to improve the road safety performance of a country. In Figure 2, the assigned weights (the values in brackets are shares) in each layer are presented for the case of Austria. It can be seen that the assigned weights (and shares) are in accordance with the imposed restrictions described in Section 3. For example, the indicators located in a particular category of a layer are of similar importance (with a maximum of 20% variability of their average weights), such as the two indicators related to the vehicle fleet composition (i.e., S6 and S7). Meanwhile, the share of final outcome indicators (52%) is larger than that of safety performance indicators (23%), which is also larger than that of policy performance indicators (16%) and that of structure and culture indicators (9%) as well. Furthermore, because the importance level of each indicator belonging to a particular category of a layer can be

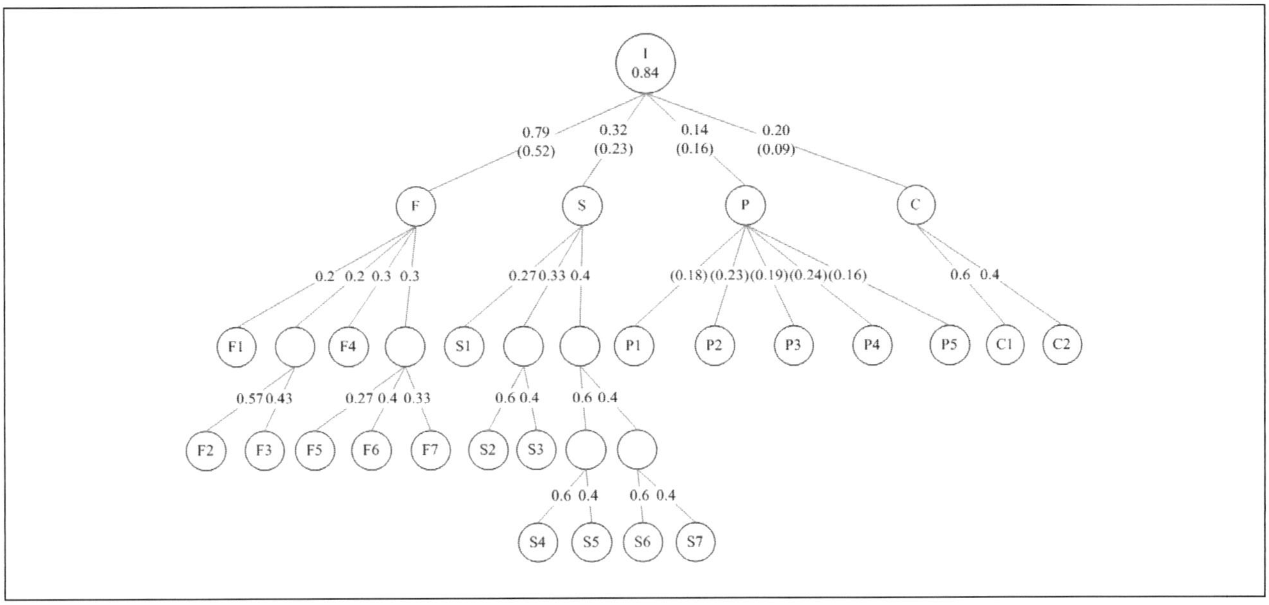

Figure 2. Assigned weights (and shares) in each layer of the hierarchy for Austria.

reflected by the allocated internal weight, more detailed insight with respect to road safety priorities can be gained accordingly. For instance, within the category of safety performance indicators, more policy attention should be paid to the alcohol aspect (S1) in Austria, and improving rear seat belt wearing rates (S3) is also urgently needed.

Discussion and Concluding Remarks

Over the last decade, several studies have been implemented looking at the development of road safety indicators and indices to monitor the road safety progress in a country and to compare the level of road safety between different countries. In the SUNflowerNext study, it was highlighted that designing a comprehensive road safety index that captures relevant information from the different components of the road safety pyramid is realistic and meaningful. As a revisit of this study, this paper proposes an alternative approach to take two methodological challenges into account when developing such an index. One is to reflect the hierarchical structure of the indicators in the index construction, and the other is to distinguish between quantitative and qualitative data. To this end, a mathematical programming model based on the mechanism of DEA is developed. Different from most of the common weighting schemes that assign the same indicator weights for all the countries under study, such as the PCA and FA approaches used in the SUNflowerNext study, the DEA-based index model provides a new way of combining multiple indicators. By solving a constrained optimization problem, the model assesses the relative performance of a particular country by taking the performance of all other countries into account, and each country obtains its own best possible indicator weights. In this way, the importance level of each indicator in different countries can be reflected, which makes it possible to identify the key problems on road safety in each country separately. More importantly, by incorporating some extra restrictions into the model, the aforementioned two methodological challenges linked to the structure of indicators and the type of data, which are usually ignored or difficult to solve in traditional weighting approaches, are properly handled.

In the application, the proposed model is used to combine the same set of indicators developed in the SUNflowerNext study into an overall road safety index for 24 European countries. By using country-specific models, the optimal road safety index score is calculated for each country, based on which the best-performing and underperforming countries are distinguished, and all the countries are further ranked by computing their cross index score. Subsequently, the results are compared with the ones from the SUNflowerNext study, and some new insights are gained. First of all, because of the consideration of four different types of road safety indicators, the overall index-based country ranking is different from the traditional one based only on the fatality rate. However, by taking the importance ordering of these four types of indicators into account, which cannot be realized in the PCA and FA approaches, a more consistent ranking with the traditional one is derived based on the proposed model. Moreover, because of the different treatment of numerical and ordinal data in the index score calculation, the inappropriate ranks obtained by some countries, such as Italy, which is ranked last out of all the countries in the SUNflowerNext study, can be clearly explained and further adjusted. In addition, by reflecting the layered hierarchy of the indicators and further incorporating weight restrictions at different layers in the proposed model, the weights allocated in each layer of the hierarchy can be deduced for each country separately, and they provide valuable information on the relative importance of the corresponding indicator and lead to priorities for policy action. All this, however, is not attainable by applying the PCA and FA approaches.

In spite of its effectiveness, it should be noted here that as a weighting approach, DEA itself is not superior to any other techniques in combining multiple indicators, including PCA and FA utilized in the SUNflowerNext study. This is because none of them is a priori the best one to use in all circumstances. Moreover, in the road safety context, there is no single criterion that can be used to judge the correctness of the results generated from different approaches. However, the ability to extend the basic model to take various methodological challenges related to indicators and data into account, which in this study means reflecting the hierarchical structure of indicators on the one hand, and dealing with the nature of quantitative and qualitative data on the other, makes DEA a powerful analytical research tool for developing a comprehensive road safety index. Nevertheless, attention is still needed when applying this approach. First, as an extreme point technique, the model proposed in this study is susceptible to outliers especially when the number of countries under study increases dramatically. Moreover, the results obtained from the proposed model are also sensitive to indicator specification and chosen weight restrictions. Therefore, comparable countries should be considered, appropriate indicators selected, and accepted views from experts adopted to ensure the robustness of the model to the utmost extent. In addition, given that road safety research toward a thorough understanding of this complex problem is still an ongoing process, more and more indicators are to be developed and data collected in the future. New methodological challenges may appear accordingly (such as the treatment of negative indicator values), which have to be tackled carefully and the solutions should be able to be integrated with the existing model appropriately.

Acknowledgments

This research was supported by the National Natural Science Foundation of China (Grant No. 71701045), and the Fundamental Research Funds for the Central Universities (Grant No. 2242018K40002).

Author Contributions

The authors confirm contribution to the paper as follows: study conception and design: Yongjun Shen; data collection: Yongjun Shen; analysis and interpretation of results: Yongjun Shen; draft manuscript preparation: Yongjun Shen. All authors reviewed the results and approved the final version of the manuscript.

References

1. Vis, M. A. (ed.). *State-of-the-Art Report on Road Safety Performance Indicators*. Deliverable D3.1 of the EU FP6 Project SafetyNet, 2005.
2. Organization for Economic Co-operation and Development/International Transport Forum. *Towards Zero: Ambitious Road Safety Targets and the Safe System Approach*. Joint Transport Research Centre of the OECD and the ITF, Paris, 2008.
3. Bao, Q., D. Ruan, Y. Shen, E. Hermans, and D. Janssens. Improved Hierarchical Fuzzy TOPSIS for Road Safety Performance Evaluation. *Knowledge-Based Systems*, Vol. 32, 2012, pp. 84–90.
4. Shen, Y., E. Hermans, Q. Bao, T. Brijs, G. Wets, and W. Wang. Inter-National Benchmarking of Road Safety: State of the Art. *Transportation Research Part C: Emerging Technologies*, Vol. 50, 2015, pp. 37–50.
5. Rosić, M., D. Pešić, D. Kukić, B. Antić, and M. Božović. Method for Selection of Optimal Road Safety Composite Index with Examples from DEA and TOPSIS Method. *Accident Analysis & Prevention*, Vol. 98, 2017, pp. 277–286.
6. Wegman, F., J. Commandeur, E. Doveh, V. Eksler, V. Gitelman, S. Hakkert, D. Lynam, and S. Oppe. *SUNflowerNext: Towards a Composite Road Safety Performance Index*. Deliverable D6.16 of the EU FP6 Project SafetyNet, 2008.
7. European Commission. *White Paper: European Transport Policy for 2010—Time to Decide*. Commission of the European Communities, Brussels, 2001.
8. European Commission. *European Road Safety Action Programme: Halving the Number of Road Accident Victims in the European Union by 2010—A Shared Responsibility*. Commission of the European Communities, Brussels, 2003.
9. National Road Safety Committee. *Road Safety Strategy 2010*. Land Traffic Safety Authority, Wellington, 2000.
10. Koornstra, M., D. Lynam, G. Nilsson, P. Noordzij, H. E. Pettersson, F. Wegman, and P. Wouters. *SUNflower: A Comparative Study of the Development of Road Safety in Sweden, the United Kingdom, and the Netherlands*. SWOV Institute for Road Safety Research, Leidschendam, 2002.
11. Thomas, P., G. Yannis, G. Duchamp, M. Vis, H. Jahi, G. Vallet, and E. Dupont. *SafetyNet Final Activity Report*. The EU FP6 Project SafetyNet, 2009.
12. Bax, C., P. Wesemann, V. Gitelman, Y. Shen, C. Goldenbeld, E. Hermans, E. Doveh, S. Hakkert, F. Wegman, and L. Aarts. *Developing a Road Safety Index*. Deliverable 4.9 of the EC FP7 Project DaCoTA, 2012.

13. Wegman, F., V. Eksler, S. Hayes, D. Lynam, P. Morsink, and S. Oppe. *SUNflower + 6: A Comparative Study of the Development of Road Safety in the SUNflower + 6 Countries—Final Report*. SWOV Institute for Road Safety Research, Leidschendam, 2005.
14. Shen, Y., E. Hermans, T. Brijs, and G. Wets. Data Envelopment Analysis for Composite Indicators: A Multiple Layer Model. *Social Indicators Research*, Vol. 114, No. 2, 2013, pp. 739–756.
15. Charnes, A., W. W. Cooper, and E. Rhodes. Measuring the Efficiency of Decision Making Units. *European Journal of Operational Research*, Vol. 2, No. 6, 1978, pp. 429–444.
16. Liu, J. S., L. Y. Y. Lu, W. Lu, and B. J. Y. Lin. A Survey of DEA Applications. *Omega*, Vol. 41, No. 5, 2013, pp. 893–902.
17. Emrouznejad, A., and G. Yang. A Survey and Analysis of the First 40 Years of Scholarly Literature in DEA: 1978–2016. *Socio-Economic Planning Sciences*, Vol. 61, 2018, pp. 4–8.
18. Cherchye, L., W. Moesen, N. Rogge, T. van Puyenbroeck, M. Saisana, A. Saltelli, R. Liska, and S. Tarantola. Creating Composite Indicators with DEA and Robustness Analysis: The Case of the Technology Achievement Index. *Journal of the Operational Research Society*, Vol. 59, No. 2, 2008, pp. 239–251.
19. Shen, Y., E. Hermans, T. Brijs, and G. Wets. Fuzzy Data Envelopment Analysis in Composite Indicator Construction. In *Performance Measurement with Fuzzy Data Envelopment Analysis, Studies in Fuzziness and Soft Computing 309* (Emrouznejad, A., and M. Tavana, eds.), Springer-Verlag, Berlin Heidelberg, 2014, pp. 89–100.
20. Shen, Y., E. Hermans, D. Ruan, G. Wets, T. Brijs, and K. Vanhoof. A Generalized Multiple Layer Data Envelopment Analysis Model for Hierarchical Structure Assessment: A Case Study in Road Safety Performance Evaluation. *Expert Systems with Applications*, Vol. 38, No. 12, 2011, pp. 15262–15272.
21. Gitelman, V., E. Doveh, and S. Hakkert. Designing a Composite Indicator for Road Safety. *Safety Science*, Vol. 48, No. 9, 2010, pp. 1212–1224.

The Standing Committee on Safety Data, Analysis (ANB20) and Evaluation peer-reviewed this paper (18-01835).

Article

TRR
JOURNAL OF THE TRANSPORTATION RESEARCH BOARD

Transportation Research Record
2018, Vol. 2672(38) 58–68

DOI: 10.1177/0361198118797802
journals.sagepub.com/home/trr
SAGE

Enhancing the Prediction Performance of Real-Time Crash Prediction Models: A Cell Transmission-Dynamic Bayesian Network Approach

Ananya Roy[1], Moinul Hossain[2], and Yasunori Muromachi[1]

Abstract

Predicting crash probability in real time is a concept that has inspired studies into complex modeling methods, exploring more sophisticated data collection methods, reducing specification errors by introducing new traffic variables, etc. Most existing real-time crash prediction models (RTCPMs) are based on loop detector data and the model architecture, and their prediction performances are sensitive to the location and layout of detectors with respect to crashes, that is, the spacing of detectors on the road section. Different expressways/freeways have different detector layouts and these vary substantially even within the same expressway/freeway. This limitation is a major obstacle in developing a universal RTCPM. To address this, this paper takes one-minute aggregated loop detector data along with detector layouts as input and employs a cell transmission model (CTM) to interpolate states of traffic flow variables for a pre-defined hypothetical detector layout. Next, several RTCPMs are constructed using Bayesian networks (BNs) and dynamic Bayesian networks (DBNs) for existing and CTM-based detector layouts. It is observed that the CTM can generate traffic flow parameters with a mean percentage error of 12.96%. The results further suggest that the model constructed with the CTM-based uniformly spaced simulated detector data coupled with the DBN method performed better than the model based on the BN and existing non-uniform detector layout.

The concept of building a real-time crash prediction model (RTCPM) is based on the hypothesis that the probability of a crash on a specific road section can be predicted for a very short time window using instantaneous traffic flow data (*1*, *2*). Most of the studies in the literature have made use of traffic flow data from dual loop detectors placed at fixed locations beneath the road surface, but where the inter-detector spacing varied substantially, with an average of around 0.80 km and a high standard deviation (*3*–*5*). One study (*3*) showed a standard deviation ranging between 0.88 and 3.60 km for an AVI system. Another study (*5*) reported a range of spacing from a minimum spacing of 0.16 km to a maximum of 5.90 km, with a standard deviation ranging between 0.16 and 1.56 km. Hong and Fuduka (*6*) used a cell transmission model (CTM) with ensemble Kalman filter to analyze the effect of various configurations of detector location on estimation of travel speed, and concluded that sensors located at large distances from each other without location optimization lead to an overestimation of travel speed, whereas sensor numbers can be reduced if their locations are optimal to achieve a better estimation of travel speed. Many studies have attempted to minimize the number of detectors (*7*–*9*), whereas Morrison and Martonosi (*10*) examined whether there are sufficient conditions for optimal solutions for the detector locations. This non-uniformity in existing detector layouts along with high variability in detector spacing among expressways/freeways raises questions about the universality as well as the transferability of such models. The doubt is exacerbated by two studies (*3*, *11*) which examined the performance of RTCPMs built with traffic data collected from six different combinations of detector layouts and found different results for each combination. Therefore, there is sufficient evidence that variation

[1]Department of Civil and Environmental Engineering, School of Environment and Society, Tokyo Institute of Technology, Yokohamashi, Japan
[2]Civil and Environmental Engineering (CEE), Islamic University of Technology (IUT), Dhaka, Bangladesh

Corresponding Author:
Address correspondence to Ananya Roy: roy.a.aa@m.titech.ac.jp

in detector layout has an impact on the states of traffic flow variables.

The existing RTCPMs are developed depending on the detector layout that exists in the study area. Hence, variation in detector layout raises the issue of spatial transferability of the RTCPMs, as altering the locations of detectors or installing new ones to replicate the detector layout of the model is neither practical nor cost effective. Moreover, even for new roads, it may not always be feasible to follow a specific detector spacing. To address these issues, there is a need to devise a mechanism to simulate traffic sensor data of desired spacing from any detector layout that can be fed into an RTCPM. This study employed CTM, a macroscopic dynamic freeway traffic model introduced by Daganzo (*12, 13*), which is consistent with the hydrodynamic theory of traffic flow (*14*), to transform the traffic states obtained from non-uniform detector layouts into a pre-defined detector layout. CTM was chosen for its analytical simplicity and ability to reproduce congestion propagation dynamics.

Crashes occurring on freeways/expressways are considered to relate closely to preceding traffic states occurring before the crash, which are time-varying (*15*). Thus, it is essential to establish a single model that can address such time series data and the evolving process of traffic flow to predict crash risk in real time. Substantial effort has been put into improving the RTCPMs by employing sophisticated modeling methods such as artificial neural networks, Bayesian structural equations, Bayesian networks (BNs), dynamic Bayesian networks (DBNs), Bayesian classifier, support vector machines, genetic programming, etc. (*4, 16–19*). Among these, the modeling architecture of a DBN conforms to this requirement (*15, 16, 20*). Therefore, this study develops RTCPMs by simulating uniformly spaced detector data from fixed detector layouts and uses DBN as the modeling method to distinguish between crash prone and normal traffic conditions. Finally, the model performance is compared with the models constructed with fixed detector layouts and for which BNs are employed as the modeling method.

Study Area and Methodology

Tokyo Metropolitan expressway's 11.9-km long Route 3 Shibuya Line was selected for this study. It sustains a substantial number of crashes throughout the year. The expressway has two lanes in each direction with 48 detectors in the inbound direction and 46 in the outbound direction (roughly 240 m apart). Figure 1 shows the entire Shibuya Route 3 along with the enlarged view of a crash prone segment as a sample to illustrate the distribution of detectors (yellow lines). The fixed loop detectors collect data on flow, average speed, and occupancy for each minute round the clock. For this study, traffic flow data from March 2014 to August 2014 were used for model building.

Cell Transmission Model

The workflow followed in this study is illustrated by Figure 2. According to the CTM (*12, 17*), if the relationship between traffic flow (q) and density (k) can be expressed in a triangular form, as in Figure 3, then the Lighthill, Whitham, and Richards (LWR) equations (*14*) for a single highway link can be approximated by a set of difference equations where current conditions (the state of the system) are updated with every clock interval as:

$$q = \min\{vk, q_{max}, w(k_j - k)\}, \text{for } 0 \leq k \leq k_j \quad (1)$$

where v = the free flow speed;
q_{max} = the maximum flow (or capacity);
w = the back wave or wave speed; and
k_j = the jam density.

In the CTM, a road segment is divided into several homogeneous cells, i, whose length is equal to the free flow speed times one clock interval. The state of the system at instant t is then given by the number of vehicles contained in each cell, $n_i(t)$. The following parameters are defined for each cell:

$N_i(t)$ = maximum number of vehicles that can be present in cell i at time t;

$Q_i(t)$ = maximum number of vehicles that can flow into cell i when the clock advances from t to $t + 1$.

If cells are numbered consecutively starting with the upstream end of the road from i = 1 to I, the recursive relationship of the CTM can be expressed as:

$$n_i(t+1) = n_i(t) + y_i(t) - y_{i+1}(t) \quad (2)$$

where $y_i(t)$ is the inflow to cell i in the time interval (t, $t+$ 1), given by:

$$y_i(t) = min\{n_{i-1}(t), Q_i(t), \mathrm{d}[N_i(t) - n_i(t)]\} \quad (3)$$

where $d = w/v$.

The general procedure for networks involves two steps for each clock interval:

(i) Determine the flow on each link with the equivalent of Equation 3.
(ii) Update the cell occupancies by transferring the flows of step (i) from the beginning cell to the end cell of each link.

In CTM, a road segment is described by nodes and directed arcs where each arc possesses some physical data that includes its length and the parameters defining a flow–density relation for a steady state of traffic of the

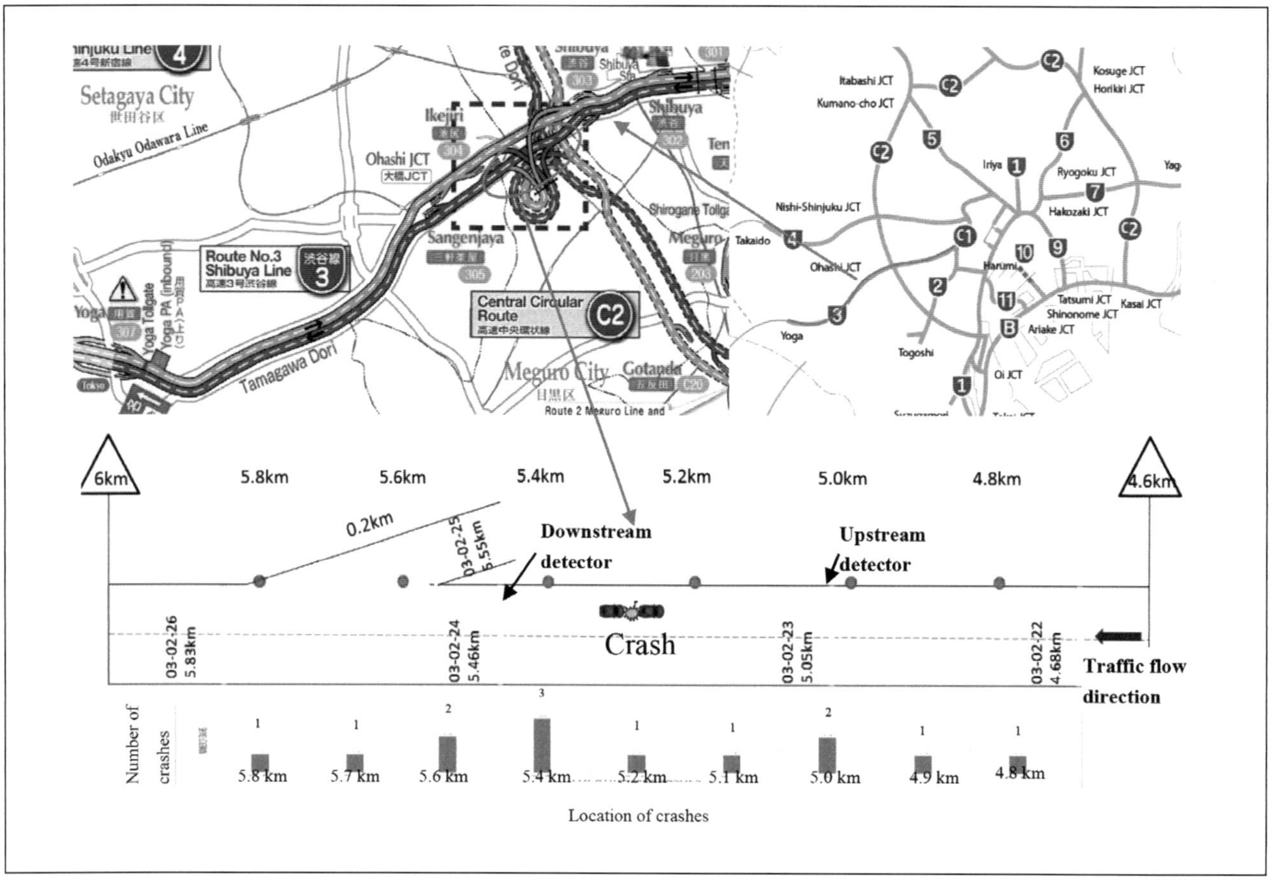

Figure 1. Route 3 Shibuya line of Tokyo Metropolitan expressway.
Source: Official website of Tokyo Metropolitan Expressway Company Limited.

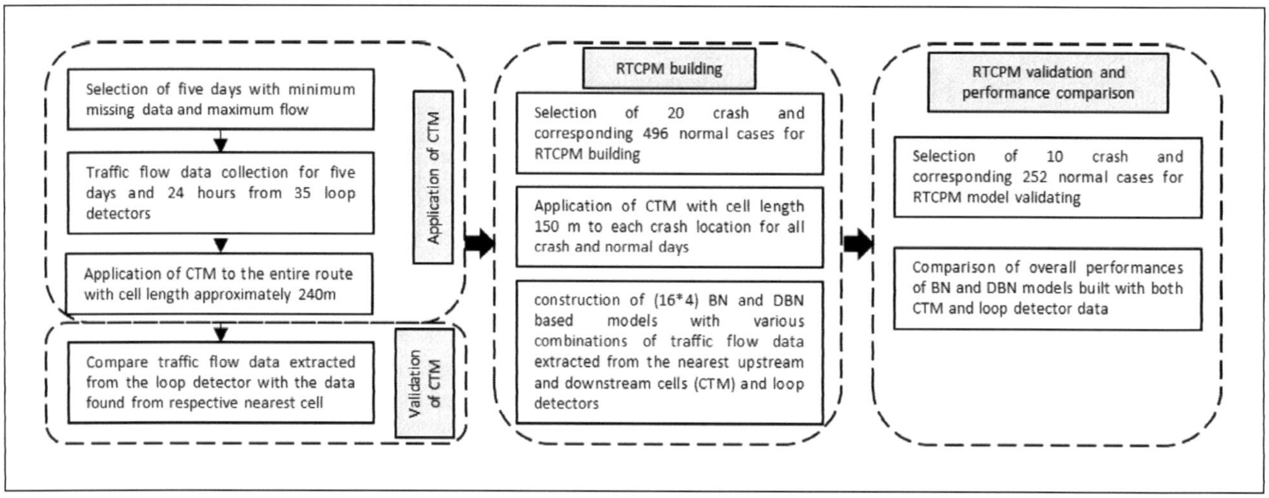

Figure 2. Workflow diagram.

type shown in Figure 3. There are three kinds of cells: (i) ordinary, (ii) diverge, and (iii) merge cells, as shown in Figure 3.

For this study, five days (April 1, August 21, June 26, March 26, and May 2, 2014) were selected as on these days most of the detectors in Shibuya Route 3 line

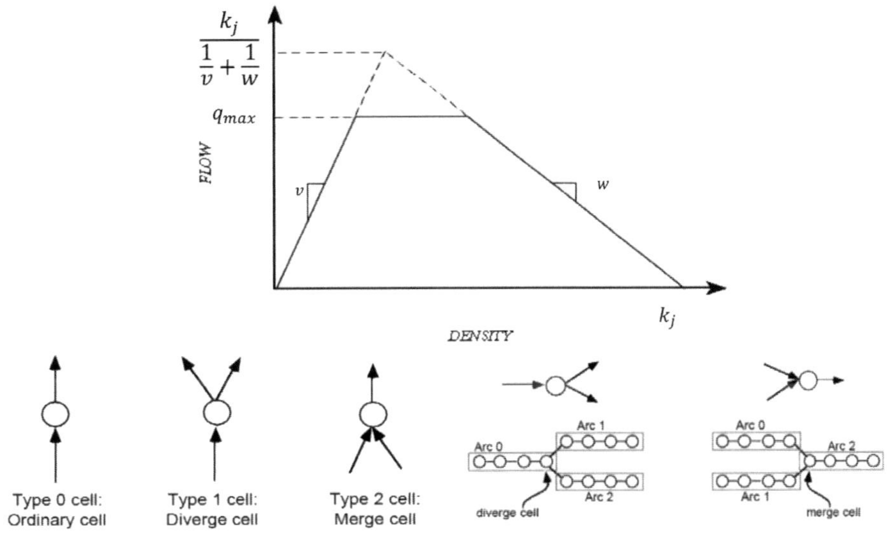

Figure 3. The fundamental diagram and ordinary, diverge, and merge cells of a cell transmission method.

inbound direction were in sound condition (less missing data) and there was enough traffic flow to ensure the fundamental diagram (FD) to fulfill the conditions to form the CTM. The FD was generated in the following way:

a. For each sound detector on the freeway, the flow and density data were extracted.
b. The maximum value of the flow was considered as the capacity, q_{max}.
c. According to Dervisoglu et al. and Kurzhanskiy and Varaiya (*21*, *22*), the density–flow pair provides a good fit for free flow speed, v. The least-squares method was used to estimate the free flow speed, v.
d. The critical density was found using $k_c = \frac{q_{max}}{v}$.
e. The constrained least-squares method was employed to determine the back-wave speed, w.

Afterwards, CTM was applied to the entire Shibuya Route 3 (inbound direction) for 24 hours on the aforementioned five days using traffic flow data for each minute (total of 1440 minutes per day). Since the cell length is the product of free flow speed and time step, to keep the cell length approximately uniform, the time step was chosen accordingly. To build the CTM model, the AuroraRNM (*23*) simulator was used and it has an option that enables the user to choose different time steps for different links. The traffic flow data generated using CTM was validated using the traffic data extracted from the fixed loop detectors.

BN and DBN

Crashes occurring on freeways/expressways are considered to be closely related to the traffic conditions prevalent before the crash, which are time variant. It has been indicated that, for different combinations of upstream and downstream traffic states, the crash involvement rates and crash risk ratios (ratio of crash cases and non-crash cases) are inconsistent. Establishing a single model that can address such time series data and the evolving process of the traffic flow is essential. Therefore, in this study, a DBN model of the time sequence traffic data has been proposed to investigate the relationship between the crash occurrence and dynamic traffic flow data with a one-minute interval.

There are two types of variables in the BN: information variable and hypothesis variable. Information variables are those whose values are expected to be obtained to calculate the probability of the hypothesis variable. DBN is an extension of BN, allowing the explicit modeling of variations over time. The assumption behind the model is that an event can cause another event in the future, but not vice versa. The BN can be extended to a DBN model by introducing relevant temporal dependencies that capture the dynamic behaviors of the domain variables between the representations of the static network at different times. Two types of dependencies can be distinguished in the DBN: contemporaneous dependencies (relations within a time slice) and non-contemporaneous dependencies (relations between the time slices).

A DBN consists of the probability distribution function (PDF) for the sequence of T hidden state variables $X = \{x_0,\ldots,x_{(T-1)}\}$ and the sequence of T observable variables $Y = \{y_0,\ldots,y_{(T-1)}\}$, where T is the time boundary for the given event under investigation.

This can be expressed as:

$$\Pr(X, Y) = \prod_{t=1}^{T-1} \Pr(x_t|x_{t-1}) \prod_{t=0}^{T-1} \Pr(y_t|x_t)\Pr(x_0) \quad (4)$$

To completely specify a DBN three sets of parameters must be defined:

- state transition PDFs $\Pr(X_T|X_{T-1})$ which specify time dependencies between the states;
- observation PDFs $\Pr(Y_T|X_T)$ which specify dependencies of observation nodes regarding other nodes at time t; and
- initial state distribution $\Pr(X_0)$ which brings initial probability distribution in the beginning of the process.

In this study, the hidden state variables are the crash likelihood and the observation variables are the traffic flow variables (i.e., flow, speed, occupancy, difference between upstream and downstream flow, speed, occupancy, etc.). Thus, the state transition could be denoted as Pr(crash|previous crash).

Traffic data was collected from the nearest upstream and downstream loop detectors of the crash location (Figure 4). In the case of CTM-based RTCPMs, for each crash case, one segment of the road was selected where both upstream and downstream detectors were present and traffic flow data was extracted from the closest upstream and downstream cells from the crash location. RTCPMs are highly dependent on the information variables chosen. In previous studies (*15, 16, 20*) BN and DBN models were generated using various combinations of traffic flow variables, that is, flow, speed, occupancy, etc., along with their descriptive statistics. Existing literature also suggests that the overall performance of DBN models built with a combination of traffic parameters is better than the models with only one traffic parameter (*17, 20*). From these studies, it was established that the information variables such as downstream speed, occupancy, difference of upstream and downstream speed occupancy, and upstream flow are suitable predictors to develop BN and DBN models. In this study, 16 different combinations of six base variables and three relative variables were produced to construct BN and DBN models to compare their performances in terms of sensitivity, specificity, etc. to identify the best possible combination of the variables.

Models 1, 3, and 5 consist of all downstream variables and one relative variable, assuming that the downstream traffic condition is more reliable to demonstrate a crash prone situation. Models 2, 4, and 6 include the upstream variables along with a relative variable that deals with the changes in traffic upstream of the incident. Models 7 to 11 basically account for all the relative traffic parameters intending to detect the crash prone situation from only the longitudinal changes of traffic. Models 12 and 13 consist of only the base variables whereas models 14 to 16 are developed with different combinations of base and relative traffic variables to investigate the performance of models in the case of missing predictors.

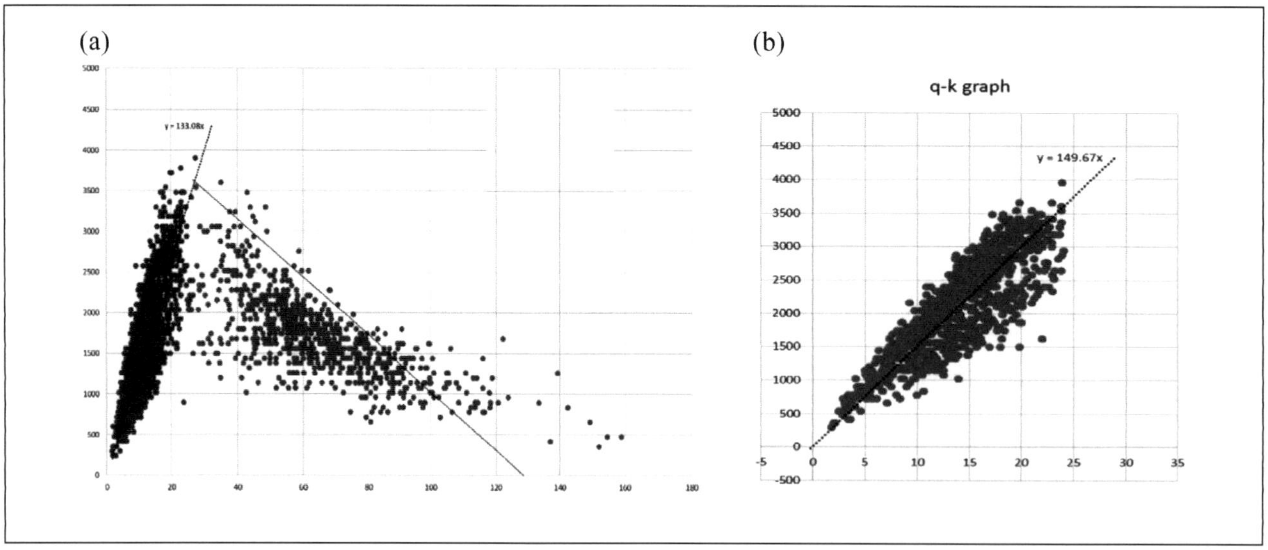

Figure 4. Estimating the fundamental diagram: (*a*) good data; (*b*) poor data.

Table 1. List of BN and DBN Models

Description of RTCPMs (both BN and DBN)	
Model 1	Downstream flow, speed, occupancy, difference of upstream and downstream flow
Model 2	Upstream flow, speed, occupancy, difference of upstream and downstream flow
Model 3	Downstream flow, speed, occupancy, difference of upstream and downstream occupancy
Model 4	Upstream flow, speed, occupancy, difference of upstream and downstream occupancy
Model 5	Downstream flow, speed, occupancy, difference of upstream and downstream speed
Model 6	Upstream flow, speed, occupancy, difference of upstream and downstream speed
Model 7	Difference of upstream and downstream flow, speed, occupancy, downstream occupancy
Model 8	Difference of upstream and downstream flow, speed, occupancy, upstream occupancy
Model 9	Difference of upstream and downstream flow, speed, occupancy, downstream flow
Model 10	Difference of upstream and downstream flow, speed, occupancy, upstream flow
Model 11	Difference of upstream and downstream flow, speed, occupancy, downstream speed
Model 12	Downstream flow, speed, occupancy
Model 13	Upstream flow, speed, occupancy
Model 14	Downstream speed, occupancy, difference of upstream and downstream flow
Model 15	Upstream speed, occupancy, difference of upstream and downstream flow
Model 16	Difference of upstream and downstream flow, speed, occupancy, upstream speed

In the analysis conducted for this paper, 30 crash cases and a corresponding 748 normal cases were extracted to construct BN- and DBN-based RTCPMs. The objective of such models is to maintain a high detection rate for the detection of crash prone traffic conditions but at the same time maintain low false alarm. In order to construct the model, first, the raw data was screened for faulty detectors and the accurate crash time was estimated from the reported crash time by investigating the upstream and downstream detectors' traffic data for each crash. Since the traffic data is recorded every minute, the variation in traffic flow variables was evident and it was possible to locate the actual time of each crash. Then, for crash cases, data was collected for one minute just before the incident occurred. For the corresponding normal data, traffic flow data was collected from the same detectors and cells during the same time as crash cases, but for those days when no incident took place. For example, if a crash had occurred on Wednesday March 3 at 15:00, then traffic flow data on March 3 at 14:59 was counted as crash data and traffic flow data from all other Wednesdays at 14:59 was considered normal data. In addition, for a DBN model with three time slices, traffic data was extracted at 14:59, 14:58, and 14:57 timestamps as one, two, and three minutes before crash, respectively. Traffic data was collected during a six-month period. Thus, for each crash case approximately (5 months $\times$ 4 days + 1 month $\times$ 3 days = 23 days $\times$ 33 crash cases =) 759 days of normal data was found; of which 748 were selected after removing erroneous data. This data was divided into training (20 crash and corresponding 496 normal) and test (10 crash and corresponding 252 normal) data for RTCPM building and validation purposes, respectively. Several studies conducted by Abdel-Aty et al. (*3*), Sun and Sun (*16*), and Lee et al. (*1*) used both crash and normal traffic data to construct RTCPMs in a similar manner. A total of 16 $\times$ 2 $\times$ 2 = 64 BN- and DBN-based RTCPMs were generated by employing existing loop detectors and the simulated CTM data (Table 1). Performances based on accuracy of crash detection and false alarm rate were calculated using validation data.

Analysis, Results, and Discussion

To employ the CTM for the entire route, a FD of traffic was generated using a flow–density graph. Figure 4*a* and *b* show examples of good and poor datasets, respectively.

If a detector with poor data was found, the next closest detector data was used for analysis. The values of FD parameters ranged as follows: capacity (q_{max}) 1800 to 3900 vehicles/h; free flow speed (v) 66.92 to 139.2 km/h; critical density (k_c) 22.6 to 31.52 vehicles/km; jam density (k_j) 75 to 130 vehicles/km; and back wave (w) 33.6 to 40.3 km/h. The cell length was kept at 240 m, the average inter-detector spacing for this study route, to generate uniformly distributed detectors.

Validation of the CTM was done by comparing traffic flow parameters of the cells located closest to each of the 35 detectors. For example, on June 26, 2014 between detectors 03-01-24 and 03-01-27, the free flow speed was found to be 139.2 km/h with the distance between these two detectors being 1.14 km. To maintain the cell length at approximately 240 m, the average inter-detector spacing for this route, a time interval of six seconds was chosen (cell length = free flow speed times time interval). Three cells in the link between detectors 03-01-24 and 03-01-27 were generated by the CTM (Figure 5). Between 07:00 and 09:00, the flow in this segment was uninterrupted and, therefore, a plot of the flow, speed,

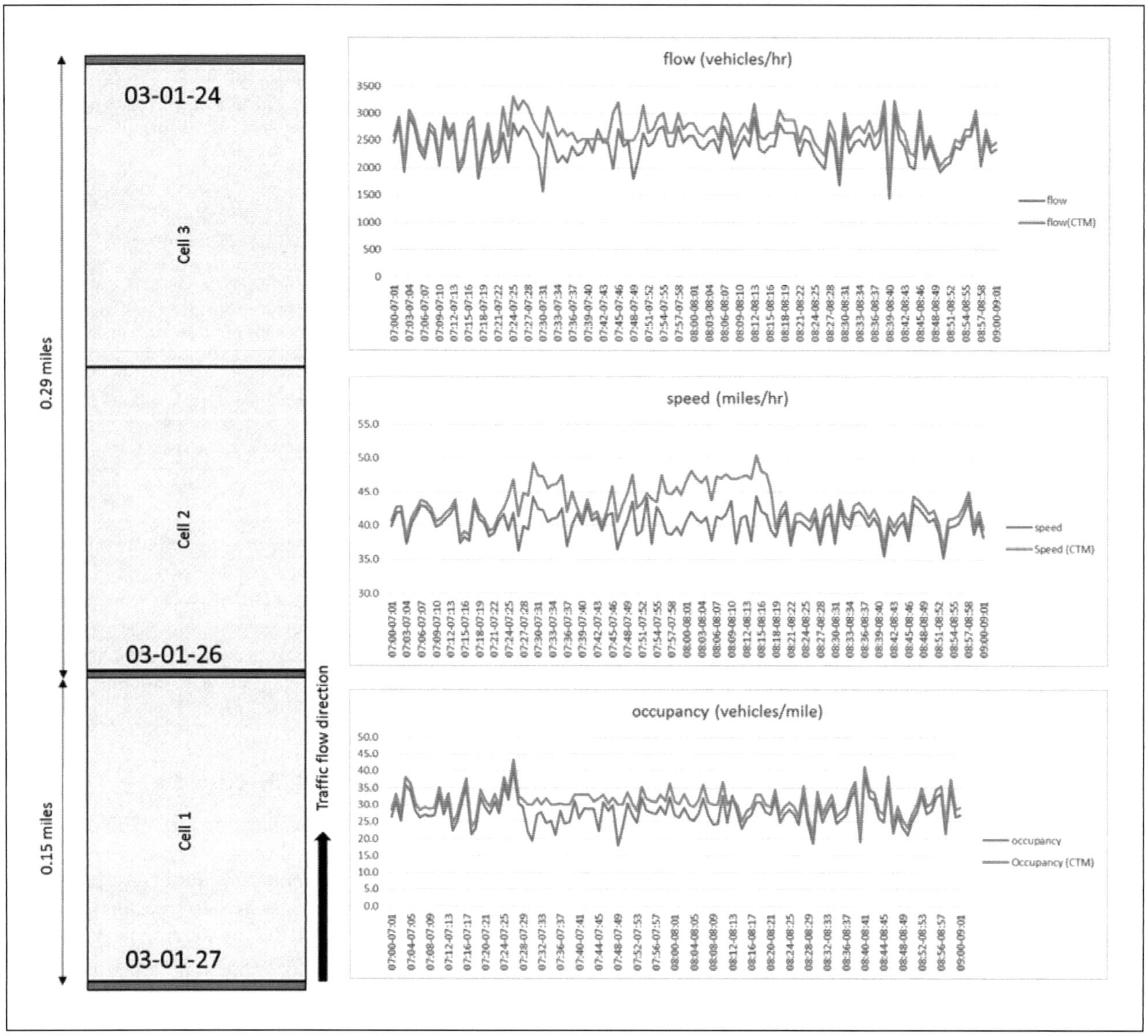

Figure 5. Measured and simulated flow, speed, and occupancy between detectors 03-01-27 and 03-01-24 on June 26, 2014 (07:00–09:00).

and occupancy data during this time stretch was generated to allow a comparison between the CTM and loop detector data. For comparative purposes cell 2, which is closest to detector 03-01-26, was selected. It can be observed from Figure 5 that the simulated value follows the measured value well when it comes to flow and occupancy data; however, the speed data differ from the original data to some extent. Such a comparison was conducted for all 35 detectors with their nearest cells for 24 hours for every minute of five days as suggested by a previous study (*24*). The mean percentage error (MPE) defined by $\frac{1}{N}\sum_{1}^{N}\left|\frac{Value_{Original}-Value_{CTM}}{Value_{original}}\right|$ was calculated to validate the data, where N = total number of data for a day = number of detectors × time (min) = 35 × 1440 (Table 2).

The next objective was to generate traffic flow data for crash prediction analysis. While generating BN and DBN models, CTMs were reproduced separately for each road segment corresponding to a crash for each day, thus the FDs for separate detectors were different based on their locations and days of data collection. A BN and corresponding DBN model structure of model 13 is shown in Figure 6.

Separate test data was used to validate all (16 × 4) BN and DBN models built with both loop detectors and

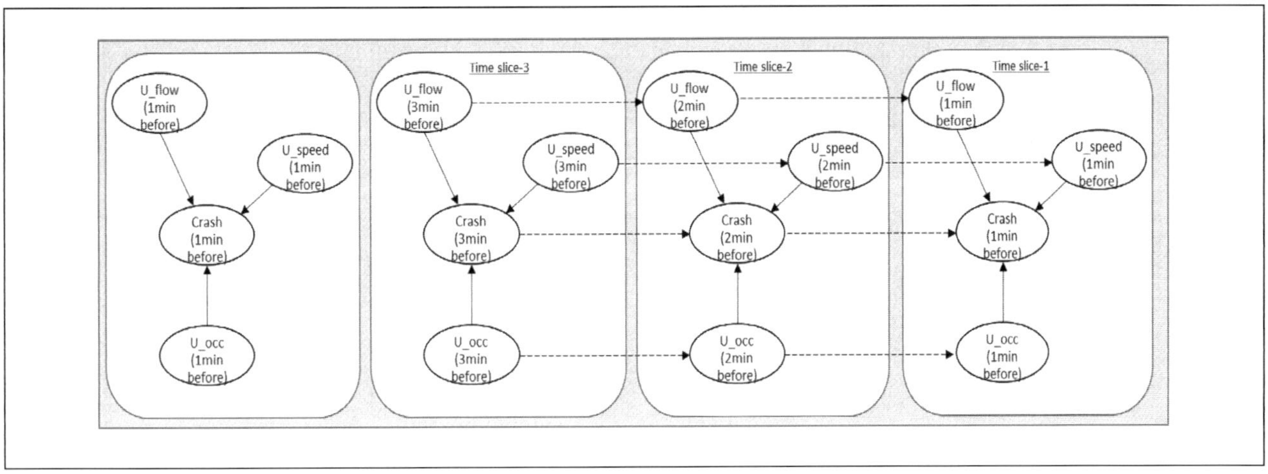

Figure 6. BN model structure (*left*) and DBN model structure with three time slices (*right*).

Table 2. Mean Percentage Error for Flow, Speed, and Occupancy Estimates

	Mean percentage error (%)		
Date	Flow	Speed	Occupancy
April 1	0.177	0.403	0.080
August 21	0.040	0.180	0.057
June 26	0.090	0.314	0.108
March 26	0.084	0.111	0.064
May 2	0.025	0.142	0.071

the CTM data for different threshold values. Crash prediction accuracy and sensitivity–specificity were calculated for each model (Table 3). Figure 7 shows the comparison of overall accuracy of prediction of crash likelihood of DBN- and BN-based models built with the loop detectors and the CTM data. All the models can predict with an overall accuracy over 65%. In the case of loop detectors, both BN- and DBN-based models perform well and 10 BN models showed better prediction accuracy than the DBN models. Similar results are seen in the case of models with the CTM data. Since the cells are shorter in length (150 m) than the average loop detector spacing, this result shows that the cells can generate traffic flow data that can be used for RTCPMs. In addition, in situations of faulty detectors, the CTM can be used to produce the traffic data from the vicinity of that incident.

From the overall accuracy of prediction of crash likelihood, loop-detector-based model 12 has the highest performance at 93% (BN) and 92% (DBN) with false alarm rates of 4% and 6%, respectively. In the case of CTM-based models, the overall accuracy of prediction of crash likelihood of model 14 was 84% (BN) and 86% (DBN) and with the corresponding false alarm rates at 16% and 14%, respectively. The information variables of models 12 and 14 are: downstream flow, speed, occupancy, and difference of upstream and downstream flow. In previous studies (*1*, *2*, *20*, *25*), downstream flow and speed and the relative flow values were found to play a significant role in RTCPM. Model 12 consists of downstream flow, speed, and occupancy, which implies that the change in traffic situation downstream of a crash location can be used to detect an impending hazardous situation with great accuracy. On the other hand, information variables of model 14 explain hazardous situations with not only downstream speed and occupancy, but also with the sudden change of traffic flow. Compared with loop detectors, 11 out of 16 BN models using the CTM data showed on average 8% more overall accuracy of prediction of crash likelihood, whereas 9 out of 16 DBN models showed on average 10% more overall accuracy.

Conclusion

In this paper, a framework has been proposed for building a universal RTCPM that will incorporate traffic flow data from detectors at various spacings and from that generate uniformly distributed simulated detector data which will eventually be fed into the DBN-based RTCPMs to predict crash risk with low false alarm. This way, the method expects to overcome the transferability issue of the existing RTCPMs and thereby approach closer to attaining universality. For this, a simple method of the CTM was employed in the study area to generate the simulated detector data. It was found that the CTM-based method could reproduce values of the traffic flow variables with an average error of 12.96% where the speed data showed higher MPE as compared with flow and occupancy data. One of the reasons is that the basic

Table 3. Performance Evaluation of BN- and DBN-Based Real-Time Crash Prediction Models with Loop Detectors and CTM Data

Modeling method		BN				DBN			
Data type	Model number	Sensitivity (%)	Specificity (%)	Accuracy (%)	False alarm (%)	Sensitivity (%)	Specificity (%)	Accuracy (%)	False alarm (%)
Loop detector data (LP)	1	0.50	0.82	0.81	0.18	0.30	0.73	0.71	0.27
	2	0.60	0.85	0.84	0.15	0.20	0.88	0.85	0.12
	3	0.20	0.85	0.82	0.15	0.40	0.86	0.84	0.14
	4	0.40	0.85	0.83	0.15	0.40	0.84	0.83	0.16
	5	0.20	0.82	0.79	0.18	0.20	0.82	0.80	0.18
	6	0.50	0.78	0.77	0.22	0.40	0.85	0.84	0.15
	7	0.60	0.73	0.73	0.27	0.20	0.74	0.72	0.26
	8	0.70	0.65	0.65	0.35	0.20	0.68	0.66	0.32
	9	0.60	0.77	0.76	0.23	0.40	0.77	0.76	0.23
	10	0.70	0.69	0.69	0.31	0.30	0.70	0.68	0.30
	11	0.60	0.71	0.71	0.29	0.20	0.74	0.72	0.26
	12	0.10	0.96	0.93	0.04	0.20	0.94	0.92	0.06
	13	0.20	0.74	0.72	0.26	0.10	0.76	0.73	0.24
	14	0.60	0.92	0.91	0.08	0.20	0.93	0.90	0.07
	15	0.70	0.85	0.84	0.15	0.30	0.86	0.83	0.14
	16	0.70	0.71	0.71	0.29	0.20	0.73	0.71	0.27
Cell transmission model data (CTM)	1	0.70	0.85	0.84	0.15	0.60	0.84	0.83	0.16
	2	0.60	0.83	0.82	0.17	0.60	0.83	0.82	0.17
	3	0.80	0.82	0.82	0.18	0.70	0.83	0.82	0.17
	4	0.80	0.85	0.84	0.15	0.60	0.83	0.82	0.17
	5	0.80	0.85	0.84	0.15	0.80	0.84	0.84	0.16
	6	0.60	0.81	0.81	0.19	0.60	0.83	0.82	0.17
	7	0.80	0.83	0.82	0.17	0.80	0.83	0.83	0.17
	8	0.80	0.83	0.82	0.17	0.80	0.82	0.82	0.18
	9	0.70	0.81	0.81	0.19	0.80	0.82	0.82	0.18
	10	0.40	0.82	0.80	0.18	0.60	0.81	0.80	0.19
	11	0.80	0.83	0.82	0.17	0.80	0.81	0.81	0.19
	12	0.20	0.81	0.79	0.19	1.00	0.82	0.83	0.18
	13	0.60	0.82	0.81	0.18	0.60	0.85	0.84	0.15
	14	0.80	0.84	0.84	0.16	1.00	0.86	0.86	0.14
	15	0.80	0.81	0.81	0.19	0.60	0.82	0.81	0.18
	16	0.44	0.84	0.81	0.16	0.44	0.84	0.81	0.16

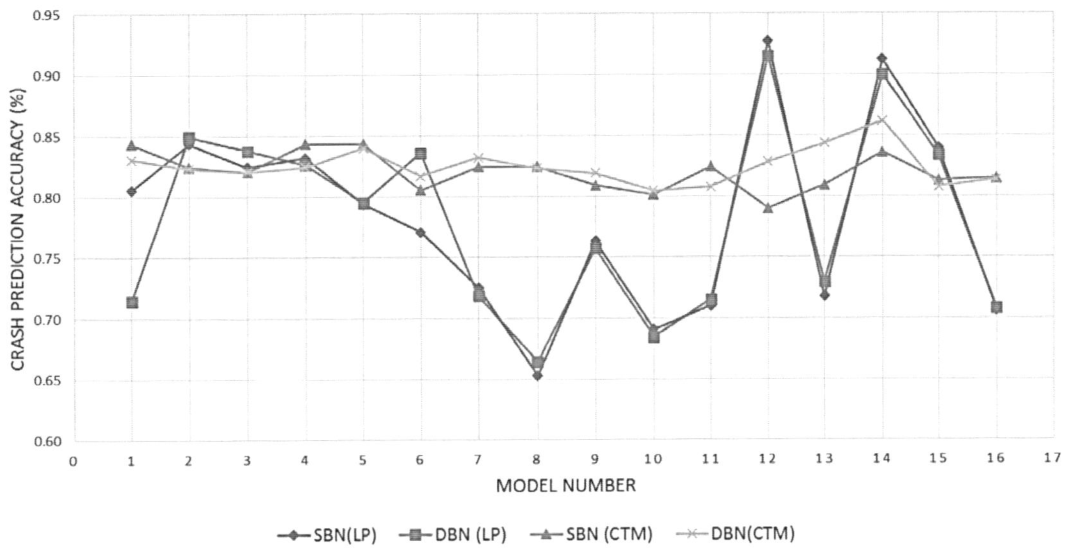

Figure 7. Comparison of overall crash prediction performances (%).

FD does not allow for control of speed, resulting in simulated speed values unresponsive to the traffic situation. However, with a flexible FD the speed can be controlled which could generate speed data with greater accuracy. Moreover, recent studies (*26*, *27*) identified speed data to be the most vulnerable data considering the detector type, traffic state, and the quality of other traffic variables.

The performance of BN- and DBN-based RTCPMs built with traffic data from both fixed detectors and CTM has been also investigated. A series of models with different information variables suggested the most influential variables to be downstream flow, speed, occupancy, and difference of upstream and downstream flow. The results indicate that there is not much performance boost in the RTCPMs constructed with CTM-based simulated uniformly distributed detector data as compared with the fixed detector based RTCPMs. This may be due to the nature of the study area which harbored detectors roughly 240 m apart with low standard deviation. However, the contribution of this paper is as a proof of concept and it is expected that if a similar study is repeated in a study area where the inter-detector spacing varies substantially, the CTM-based RTCPMs may outperform the fixed detector layout based RTCPMs and thereby address the issue of transferability of these models. One area of future work will be to investigate the optimum inter-detector spacing and detector location as well as cells by experimenting with different cell lengths at different locations (50, 100, and 150 m upstream and downstream). Moreover, other than freeway stretches, the model can be expanded in future to incorporate ramps rather than considering only the basic freeway segments. The RTCPM is highly dependent on the quantity as well as the quality of the traffic flow data. Therefore, it is recommended that more case studies and different time steps should be incorporated to ensure an efficient model that can be implemented in real time.

Acknowledgments

The authors thank Tokyo Metropolitan Expressway Company Limited for providing necessary data to conduct this study. This study was also supported by JSPS KAKENHI Grant Number 25420535.

Author Contributions

The authors confirm contribution to the paper as follows: study conception and design: YM, MH; data collection: YM, AR; analysis and interpretation of results: YM, MH, AR; draft manuscript preparation: AR. All authors reviewed the results and approved the final version of the manuscript.

References

1. Lee, C., B. Hellinga, and F. Saccomanno. Real-time Crash Prediction Model for the Application to Crash Prevention in Freeway Traffic. *Transportation Research Record: Journal of the Transportation Research Board*, 2003. 1840: 67–77.
2. Pande, A., and M. Abdel-Aty. A Freeway Safety Strategy for Advanced Proactive Traffic Management. *Journal of*

Intelligent Transportation Systems, Vol. 9, No. 3, 2005, pp. 145–158.
3. Abdel-Aty, M., H. M. Hassan, M. Ahmed, and A. S. Al-Ghamdi. Real-Time Prediction of Visibility Related Crashes. *Transportation Research Part C*, Vol. 24, 2012, pp. 288–298.
4. Hossain, M., and Y. Muromachi. A Bayesian Network Based Framework for Real-Time Crash Prediction on the Basic Freeway Segments of Urban Expressways. *Accident Analysis and Prevention*, Vol. 45, 2011, pp. 373–381.
5. Shi, Q., and M. Abdel-Aty. Big Data Applications in Real-Time Traffic Operation and Safety Monitoring and Improvement on Urban Expressways. *Transportation Research Part C*, Vol. 58, 2015, pp. 380–394.
6. Hong, Z., and D. Fukuda. Effects of Traffic Sensor Location on Traffic State Estimation. *Proceedings of the 15th Meeting of the EURO Working Group on Transportation, Procedia - Social and Behavioral Sciences*, Vol. 54, 2012, pp. 1186–1196.
7. Bianco, L., G. Confessore, and P. Reverberi. A Network Based Model for Traffic Sensor Location with Implications on O/D Matrix Estimates. *Transportation Science*, Vol. 35, No. 1, 2001, pp. 50–60.
8. Bianco, L., G. Confessore, and M. Gentili. Combinatorial Aspects of the Sensor Location Problem. *Annals of Operations Research*, Vol. 144, No. 1, 2006, pp. 201–234.
9. Bianco, L., R. Cerulli, and M. Gentili. New Resolution Approaches for the Sensor Location Problem. *Proc., Tristan VI Symposium*, Phuket Island, Thailand, 2007.
10. Morrison, D., and S. Martonosi. Characteristics of Optimal Solutions to the Sensor Location Problem. *Annals of Operations Research*, Vol. 226, No. 1, 2014, pp. 463–478.
11. Hossain, M., and Y. Muromachi. Optimum Detector Spacing for Real-time Monitoring of Hazardous Locations on Urban Expressways. *Japanese Society of Civil Engineers*, Vol. 27, No. 5, 2010, pp. 1045–1054.
12. Daganzo, C. F. The Cell Transmission Model, Part II: Network Traffic. *Transportation Research Part B*, Vol. 29, 1995, pp. 79–93.
13. Daganzo, C. F. The Cell Transmission Model: A Dynamic Representation of Highway Traffic Consistent with the Hydrodynamic Theory. *Transportation Research Part B*, Vol. 28, 1994, pp. 269–287.
14. Lighthill, M., and G. Whitham. On Kinematic Waves II. A Theory of Traffic Flow on Long Crowded Roads. *Proceedings Royal Society of London, Part A*, Vol. 229, No. 1178, 1955.
15. Mihajlovic, V., and M. Petkovic. *Dynamic Bayesian Network: A State of the Art*. CTIT Technical Report Series, Vol. 01-34, 2001.
16. Sun, J., and J. Sun. Dynamic Bayesian Network Model for Real-Time Crash Prediction using Traffic Speed Conditions Data. *Transportation Research Part C*, Vol. 54, 2015, pp. 176–186.
17. Lee, C., M. Abdel-Aty, and L. Hsia. Potential Real-Time Indicators of Sideswipe Crashes on Freeways. *Transportation Research Record: Journal of the Transportation Research Board*, 2006. 1953: 41–49.
18. Sun, J., and J. Sun. Real-Time Crash Prediction on Urban Expressways: Identification of Key Variables and a Hybrid Support Vector Machine Model. *IET Intelligent Transport Systems*, Vol. 10, No. 5, 2016, pp. 331–337.
19. Park, H., and A. Haghani. Real-Time Prediction of Secondary Incident Occurrences using Vehicle Probe Data. *Transportation Research Part C*, Vol. 70, 2015, pp. 69–85.
20. Roy, A., R. Kobayashi, M. Hossain, and Y. Muromachi. Real-Time Crash Prediction Model for Urban Expressway using Dynamic Bayesian Network. *Journal of Japan Society of Civil Engineers, Series D3*, Vol. 72, No. 5, 2016, pp. 1331–1338.
21. Dervisoglu, G., G. Gomes, J. Kwon, R. Horowitz, and P. Varaiya. Automatic Calibration of the Fundamental Diagram and Empirical Observations on Capacity. Presented at 88th Annual Meeting of the Transportation Research Board, Washington, D.C., 2009.
22. Kurzhanskiy, A. A., and P. Varaiya. Active Traffic Management on Road Networks: A Macroscopic Approach. *Philosophical Transactions of the Royal Society A*, Vol. 368, 2010, pp. 4607–4626.
23. Chow, A. H. F., G. Gomes, A. A. Kurzhanskiy, and P. Varaiya. *AURORA RNM – A Macroscopic Simulation Tool for Arterial Traffic Modeling and Control*. PATH Technical Note, Institute of Transportation Studies, University of California, Berkeley, CA. 2009.
24. Muñoz, L., X. Sun, D. Sun, R. Horowitz, and L. Alvarez. Traffic Density Estimation with Cell Transmission Model. *Proc., American Control Conference*. Denver, CO, 2003.
25. Lu, X., Z. Kim, M. Cao, P. Varaiya, and R. Horowitz. Deliver a Set of Tools for Resolving Bad Inductive Loops and Correcting Bad Data. *California PATH Research Report*, 2010.
26. Hossain, M., and Y. Muromachi. A Real-Time Crash Prediction Model for the Ramp Vicinities of Urban Expressway. *IATSS Research*, Vol. 37, No. 1, 2013, pp. 68–79.
27. Coifman, B. Using Dual Loop Speed Traps to Identify Detector Errors. *Transportation Research Record: Journal of the Transportation Research Board*, 2014. 1683: 47–58.

The Standing Committee on Safety Data, Analysis and Evaluation (ANB20) peer-reviewed this paper (18-05257).

Article

TRR
JOURNAL OF THE TRANSPORTATION RESEARCH BOARD

Transportation Research Record
2018, Vol. 2672(38) 69–80

DOI: 10.1177/0361198118780709
journals.sagepub.com/home/trr
SAGE

Identification of Site Characteristics for Proactive High-Friction Surface Treatment Site Selection using Sensor-Based, Detailed, Location-Referenced Curve Characteristics Data

Yichang (James) Tsai[1], Yi-Ching Wu[1], Cibi Pranav P. S.[1], and Chengbo Ai[1,2]

Abstract

The Georgia Department of Transportation (GDOT) has developed a proactive high-friction surface treatment (HFST) program for curve sites prone to run-off-road (ROR) crashes. Using crash data and a single-criterion, ball bank indicator (BBI) value, GDOT seeks to maximize the return on its HFST investment. GDOT has partnered with Georgia Tech to identify additional factors for its HFST site-selection (HFST-SS) decision-making process by leveraging high-resolution, full-coverage sensor data (e.g., GPS and LiDAR). This paper proposes a methodology to identify site characteristics that can be used in GDOT's HFST-SS process by leveraging the sensor data and automatically extracting roadway curve features as follows: (a) roadway data collection using state-of-the-art sensing technologies, (b) automatic extraction of detailed site characteristics data and curve information, (c) curved-based roadway segmentation using the extracted curve information; (d) spatial integration of curve-site characteristics data (CSCD); (e) analysis of CSCD and ROR crashes to identify additional factors for HFST site selection. A case study using CSCD extracted from Georgia State Route 2 demonstrates the proposed methodology. Results show that on sharp curves having comparable site characteristics, vertical grades greater than 3% play an important role in ROR crashes. Therefore, a vertical grade greater than 3% could be considered as an additional HFST-SS factor along with the current BBI criterion.

High-friction surface treatment (HFST) is an effective countermeasure that can effectively restore pavement friction to improve the safety of horizontal curves (*1*). HFST is usually composed of a layer of thin aggregates (i.e., basalt, calcine bauxite, emery, granite, steel slag, or taconite) on top of a special binder (i.e., epoxy-resin, rosin ester, polyurethane-resin, or acrylic resin). HFST aggregates resist polish and abrasion, and reduce hydroplaning on wet pavement surfaces (*2*, *3*). However, HFST is a relatively expensive countermeasure compared with other conventional curve countermeasures, such as advanced curve warning signs, chevrons, rumble strips, raised pavement markers, etc. (*4*). Thus it is important to install HFST at the right places to achieve the highest level of effectiveness. Most transportation agencies, such as those in Kentucky (*5*), Thurston County, Washington (*6*), and Placer and Nevada Counties, California (*7*), primarily rely on crash data to identify HFST candidate sites with high ROR crashes, using different crash attributes (e.g., crash number, crash rate, etc.). Some transportation agencies, such as Georgia Department of Transportation (GDOT), prioritize the sites by the severity of crashes, giving a higher weight to fatal crashes; some transportation agencies, such as that of Kentucky, place more weight on pavement conditions during crashes, such as wet pavements (*5*).

GDOT has developed a proactive HFST site-selection process using a two-stage procedure. GDOT first uses crash data to prioritize and select the corridors with high ROR crash rates that could benefit from HFST. Then, within the selected corridors, GDOT uses single-criterion, ball bank indicator (BBI) values to proactively select curve sites for HFST installation. A BBI value combines the impact of the radius of curvature, superelevation, driving speed, and pavement friction. GDOT collects the BBI values along the selected corridors, and the maximum BBI value at each curve is recorded. The curves with a maximum BBI value greater than or equal to 12 are then selected for HFST installation, regardless of the crash history.

[1]Georgia Institute of Technology, Atlanta, GA
[2]University of Massachusetts, Amherst, MA

Corresponding Author:
Address correspondence to Cibi Pranav P. S.: cibipranav@gatech.edu

While GDOT has a developed HFST-SS decision-making process, the process is still evolving, and GDOT is seeking to improve its HFST-SS criteria to maximize the return on its investment in HFST. GDOT has partnered with Georgia Tech to leverage currently available, high-resolution, full-coverage sensor data (e.g., GPS, LiDAR, etc.), and identify additional curve characteristics that contribute to ROR crashes and could be used, in addition to BBI values, in its HFST-SS decision process.

Horizontal curve safety has been a prominent topic of research in the last two decades and several approaches have focused on curve-level site characteristics (number of curves, curve radii, grade, length, etc.) to identify the contributing factors related to horizontal curve crashes (*8–12*). With advances in sensing technologies (including 2D imaging, 3D laser, LiDAR, IMU, and GPS/GIS technologies), high-resolution, full-coverage sensor data that captures pavement and roadway features can be collected at highway speeds at dense intervals for extracting curve characteristics (e.g., curve radius, the point of curve (PC), the point of tangent (PT), vertical grade, superelevation, etc.). Such data, in conjunction with crash data, can support safety analyses on curves. Mraz and Nazef (*13*) used a multi-purpose survey vehicle to measure cross-slope, grade, smoothness, and rut depth to map roadways prone to hydroplaning. Brown and Brennan (*14*) used sensing data to recreate the roadway digitally and ran vehicle simulation tests to identify potential locations for improvement based on vehicle responses (e.g., lateral offset). Ambros et al. (*15*) collected GPS data along a road section in the Czech Republic, extracted the horizontal alignment, and segmented the road into curves and tangents to study the effect of curve speed differential on predicted number of crashes. There are studies focusing on extracting information on road characteristics, such as road geometry (*16*), curves (*17–19*), signs (*20*), superelevation (*21*, *22*), etc. using sensing data for inventory purposes. There have been limited (*19*) attempts to perform comprehensive curve safety analyses using detailed-level, comprehensive curve-site characteristics data (CSCD) (including curve, grade, superelevation, etc.) extracted from state-of-the-art sensing technologies. Thus there is a need to leverage this data to support comprehensive analysis of curve safety, especially to support HFST-SS.

The objective of this paper is to develop a methodology to extract, integrate, and analyze effectively the detailed data on location-referenced curve characteristics collected using state-of-the-art sensing technologies to identify additional curve-site characteristic factors (besides BBI values) that could be incorporated into GDOT's HFST-SS decision-making criteria for its proactive HFST program and for maximizing the return on HFST investment.

The paper is organized as follows. First, the background and the objective of this paper are presented. Second, the proposed methodology and its steps are described. Third, a case study is conducted to demonstrate the curve safety analysis performed with the data extracted using the proposed methodology on Georgia State Route 2. Finally, conclusions and recommendations are made.

Proposed Methodology

To identify the additional curve-site characteristic factors that, along with the use of BBI, could be incorporated into GDOT's HFST-SS process, a methodology has been proposed and presented below. This methodology effectively extracts, spatially integrates, and analyzes CSCD values with ROR crashes to identify the factor(s) that can be incorporated for an effective HFST-SS decision-making process.

The proposed methodology consists of the following steps: (a) collection of sensor data on curves, using state-of-the-art sensing technologies, including 2D imaging, 3D LiDAR, IMUs, and GPS/GIS technologies; (b) automatic feature extraction of the detailed, location-referenced curve information, including curve radius, PC and PT, etc. and continuous site characteristics data such as superelevation, grade, BBI, etc.; (c) curve-based segmentation using the extracted PC and PT; (d) spatial integration of CSCD; (e) analysis of correlation between CSCD and ROR crashes to identify additional factors for HFST-SS. Figure 1 illustrates the steps of the proposed methodology.

Sensor Data Collection

In this study, the Georgia Tech sensing vehicle (GTSV), equipped with high-resolution cameras, light detection and ranging (LiDAR), a 3D pavement laser system, a global positing system (GPS), an inertial measurement unit (IMU), and a distance measuring instrument (DMI), was used to collect the sensor data for extracting detailed, location-referenced CSCD. The GTSV, as shown in Figure 2*a*, was sponsored by the US DOT and GDOT. It consists of the following: (a) four high-resolution video cameras at 2448 × 2048 pixels for generating a panoramic view of the roadway and a detailed, downward view of the pavement; (b) two line-scanning LiDARs at 10 kHz–15 kHz with 3-cm ranging error for generating full 3D coverage of the roadway and roadside; (c) a pair of high-frequency pavement profiling lasers at 5.6 Hz with 0.5 mm resolution of elevation measurement for generating a detailed 3D scan of the pavement; and (d) GPS, IMU, and DMI for location-referencing and synchronizing the data. While various data were collected using multiple sensors, in this study, the GPS and LiDAR data are used for extracting curve information, cross-slope, and vertical grade. Figure 2*b* shows examples of the collected data, including video log images, a LiDAR point cloud, a GPS trajectory, and a 3D pavement scan. In addition, the BBI measurements were collected using a Rieker Digital BBI device fixed on the dashboard of the GTSV.

Automatic Extraction of Continuous CSCD

In this step, the raw sensor data are processed using various roadway feature extraction algorithms to automatically

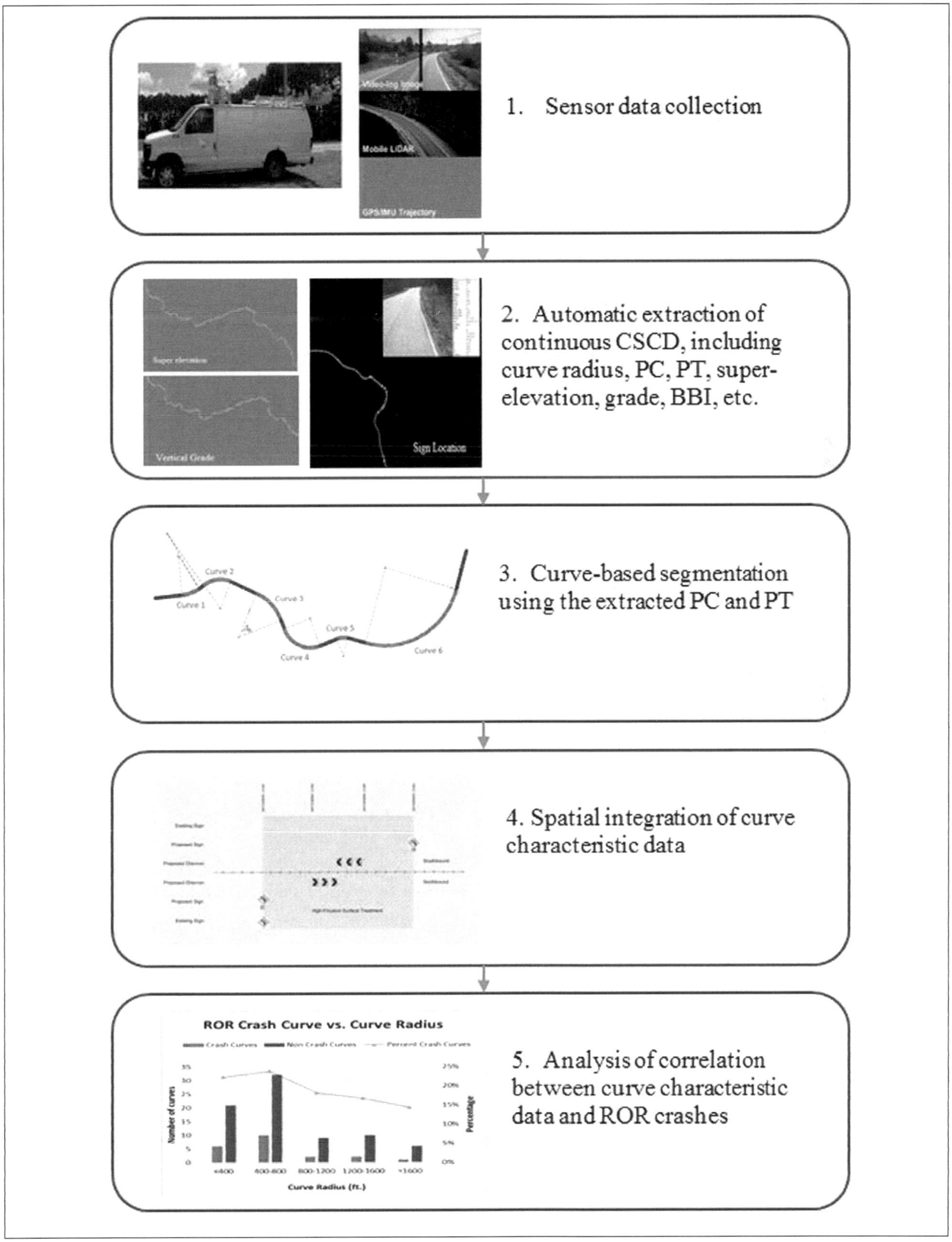

Figure 1. Illustration of the steps in the proposed methodology.

Figure 2. (*a*) Georgia Tech sensing vehicle equipped with sensors and (*b*) sample of data collected using the sensors.

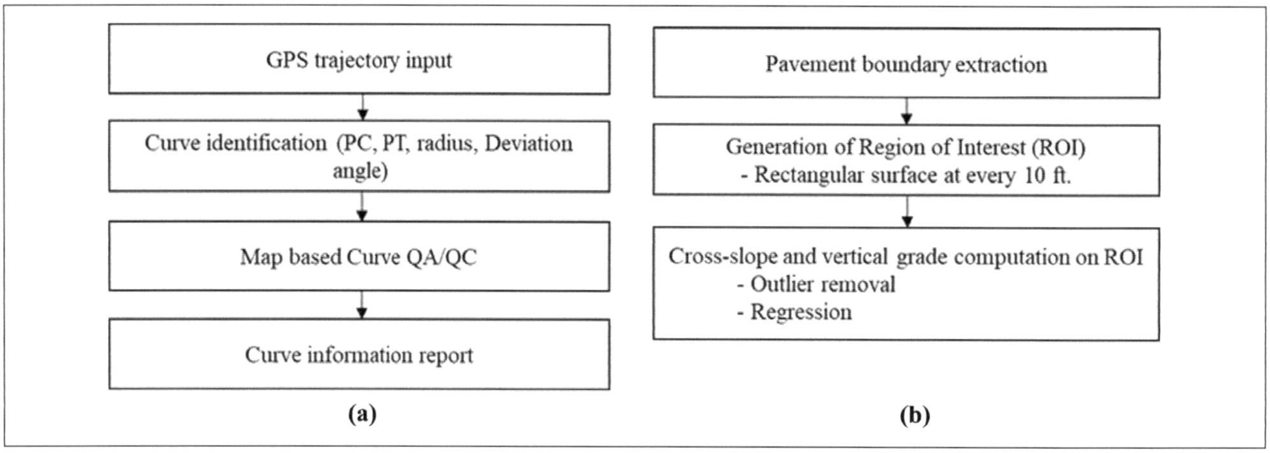

Figure 3. Flowchart showing (*a*) GPS-based curve features extraction and (*b*) LiDAR-based superelevation and grade extraction.

extract the detailed, location-referenced, continuous roadway site characteristics data. First, the GPS trajectory is used to extract the curve information (foundation for CSCD), including PC and PT of the curve, radius of curvature, deviation angle, length, and direction of curve, using the Smart-CIE (smart curve information extraction) tool developed by Tsai et al. (*18*) featuring a curve-identification algorithm developed by Ai and Tsai (*17*) and map-based QA/QC. Second, the LiDAR data is processed using an algorithm develop by Tsai et al. (*22*) to measure the superelevation and vertical grade along the road. This involves three steps: (a) extraction of pavement boundary, (b) extraction of the region of interest (ROI), i.e., the rectangular region within a single lane between the identified pavement boundary at a fixed interval (e.g., 10 feet), and (c) performing regression-based superelevation estimation at each ROI. Similarly, regression is performed in the longitudinal direction along the direction of travel to determine the vertical grade. Figure 3 shows in brief the methodology for extraction of curve geometrical information using GPS and continuous roadway information extraction using LiDAR.

The final output of processing the sensor data contains the curve characteristics data including PC and PT of the curve, the radius of curvature, deviation angle, length, and direction of curve; and continuous data including superelevation, grade, pavement width, and BBI.

Curve-Based Segmentation using PC and PT

In this step, the PC and PT locations extracted from the GPS point are georeferenced on the roadway map in the GIS platform and then the roadway is segmented into individual curves and tangents using the PC and PT. This curve-based segmentation allows transportation agencies to identify the exact curve locations, analyze crashes on curves easily, and further identify the issues with regard to curve safety. Further,

Table 1. An Example of Integrated CSCD at Curve and Site Levels

Site No.	1	2	3		4	5				6
AADT (truck percentage)	2500 (10%)	2500 (10%)	2500 (10%)		2500 (10%)	2500 (10%)				2500 (10%)
Type of curve site	Single curve	Single curve	Reverse curve*		Single curve	Continuous curves**				Single curve
Curve no.	C1	C2	C3	C4	C5	C6	C7	C8	C9	C10
Point of curve mile point	14.25	14.68	15.77	15.95	16.20	17.11	17.23	17.33	17.41	17.89
Point of tangent mile point	14.31	14.80	15.95	16.11	16.34	17.22	17.32	17.41	17.66	18.03
Radius (ft)	877	1158	1206	718	571	410	404	934	1448	1808
Deviation angle (°)	11	17	28	34	37	42	37	14	26	12
Length (ft)	325.9	667.3	598.9	858.3	743.1	582.3	508.5	442.1	1309.2	737.4
Direction (Left = −1 Right = 1)	1	−1	−1	1	−1	1	−1	−1	1	1
Avg. lane width (ft)	3.8	4.0	3.9	3.7	3.7	4.0	4.1	3.9	3.8	3.7
Super elevation (%) – 90 percentile	7	5.5	9	11	12	14	12	10	7	7
Avg. vertical grade (%)	0.6	4.8	−5.0	−0.2	7.3	4.4	5.1	1.6	−1.7	−5.7
Maximum BBI (°)	9	5	5	6	8	9	10	9	5	5

Note: * Site 5 contains reverse curves (C3 and C4).

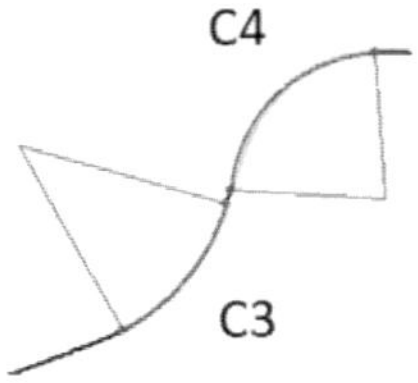

**Site 5 contains continuous curves (C6, C7, C8, and C9).

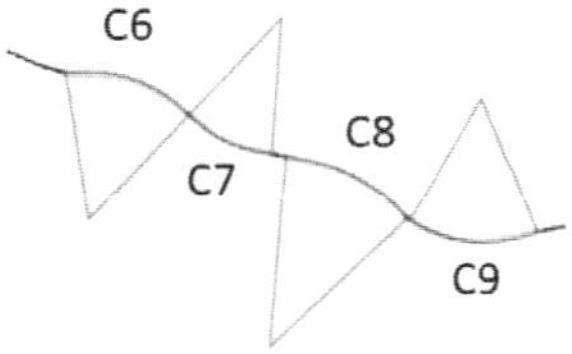

the curve segments are clustered into sites to allow transportation agencies to identify non-single-curve sites that may require attention from a safety point of view. A site, which consists of one or more than one curve, can be categorized into a single curve, reversed curve, compound curve, or continuous curve (one with more than two curves). Table 1 shows the sites and curves on a section of road. Adjacent curves with a distance less than 80 feet are clustered into a single site; the underlying assumption of using 80 feet considers the fact that drivers have to respond to the changing alignment within 1.5 seconds when traveling at the typical curve speed (35 mph). Accordingly, if the adjacent curves are within 80 feet of each other, drivers do not have sufficient time for steering properly while entering the next curve.

Spatial Integration of CSCD

In this step, the extracted detail-level continuous data (e.g., grade and superelevation measured at every 3-meter interval, BBI) and crash data are spatially integrated with curve segments to support further analyses. A multi-level data model was designed to support the analyses at various levels. First, all the data extracted from sensor and crash data are linear referenced. All extracted continuous data falling within the limits of PC and PT of a particular curve are assigned to that curve, enhancing the curve characteristics data. At this level, there is no aggregation of data. Such CSCD allows detailed analysis of individual curves in conjunction with crash location. Next, this data is aggregated into each individual curve to support the curve-level analysis, such as curve-to-curve comparisons and statistical analysis of site characteristics across multiple curves. In this case, each curve is assigned a single value for each of the continuous data points, such as the maximum absolute BBI falling within each curve is assigned as the curve's BBI. Similarly, the average value of the vertical grade points and the 90-percentile value for the superelevation points that fall within a curve is assigned as the curve's grade and superelevation, respectively. At third level, the curves are aggregated into the sites, such as continuous curves, to support analysis at the site level. Table 1 shows an example of CSCD integration at multiple levels. Site 3 in Table 1 is an example of curve aggregation in which curves C3 and C4 are grouped together, and so is Site 5.

Analysis of Correlation between CSCD and ROR Crashes

In this step, a curve-level analysis is conducted to identify factors that correlate to ROR crashes so that they may be considered in HFST-SS criteria. An over-representation analysis is conducted to analyze qualitatively each site characteristic that may be related to ROR crashes at horizontal curves; it studies the number of curves with and without crashes based on different sets of a site characteristic. A particular set is said to be over-represented if the crashes in that set are more frequent than in its complement set. Over-representation is mathematically represented using the over-representation factor,

$$\mathrm{ORF} = \frac{R_{set}}{R_{comp}} \tag{1}$$

where R_{set} is the percentage of curves with crashes in the set of interest (say BBI < 4) and R_{comp} is the percentage of curves with crashes in the complementary set (say BBI 6–8, 8–12, etc.)

An ORF value higher than one means that the ROR crashes occur more frequently in that set of a site characteristic than the others. Accordingly, site characteristics, including curve radius, cross-slope (superelevation), and vertical grade, are analyzed in this study. In addition, BBI, the HFST-SS criterion used by GDOT, is analyzed to verify the criterion. However, this analysis is not limited to the identified site characteristics alone. It may extend to other site characteristics, such as roadside hazards and pavement conditions (*7*), type of site (single curve, reversed curve, compound curve, and continuous curve), etc.

Case Study

A case study using the detailed, location-referenced CSCD collected on a test section (31 miles of State Route 2), was conducted to demonstrate the proposed methodology to identify additional CSCD factors that impact ROR crashes, besides BBI. The State Route 2 sections in Rabun and Towns Counties are located in northern Georgia, a mountainous terrain. They cover many curves with various radii and a wide range of grades and superelevations. Sensor data, including GPS, LiDAR, video, and acceleration data, were collected by the GTSV along eastbound direction for extracting detailed, location-referenced CSCD.

CSCD Data

After processing the sensor data collected on the test section, 140 horizontal curves were extracted from 31 miles of the road. The map in Figure 4*a* shows the 140 curves (with circle size indicating curve radii) automatically extracted using GPS data. In addition, the geometric properties of the curves, including curve radii, PC and PT, grade, and superelevation, were extracted and spatially integrated using linear referencing. Results are shown in Figure 4. Figure 4*a* collectively represents the mile-point location of the curves along the test section, direction of turn, radius, and length of the curve. The height of the bar in Figure 4*a* represents the radius of the curve; a positive radius implies that the curve turns in the right-hand direction, and a negative radius implies the curve is turning towards the left. The width of the bar represents the curve length. Figure 4*b, c,* and *d* represent the detailed elevation, superelevation, and BBI of a 0.16-mile subsection containing a continuous curve site. The elevation and superelevation are measured every 10 feet, while BBI is collected at 1 Hz frequency.

Before further analysis, a few CSCD were plotted against each other to verify whether the data follow the expected trends. A review of BBI values and curve radii shows a negative correlation (as shown in Figure 5*a*); that is, as radius decreases, BBI values increase. The superelevation and curve radii also have a negative correlation, as shown in Figure 5*b*. As the radius decreases, superelevation increases. The trends shown in Figure 5*a* and *b* are reasonable.

In addition to CSCD, ROR crash data between 2010 and 2015 were extracted from the crash database (i.e., Georgia Electronic Accident Reporting System - GEARS) and spatially integrated with the curves to facilitate the subsequent analysis. There were 55 ROR crashes in the eastbound direction, 34 of which were located within curves. ROR crashes occurring within 50 feet from the PC of the curve were also assigned to the curve since the crash site in the database may not represent the true start of crash events. In most cases, one or no crash per curve occurred; only five curves had more than one crash. After spatial integration, CSCD and crash data were correlated and analyzed to identify additional CSCD factors that affect ROR crashes.

Over-representation Analysis

An over-representation analysis was conducted for BBI, radius, grade, and superelevation. Each of these four characteristics was quantitatively categorized into sets; the

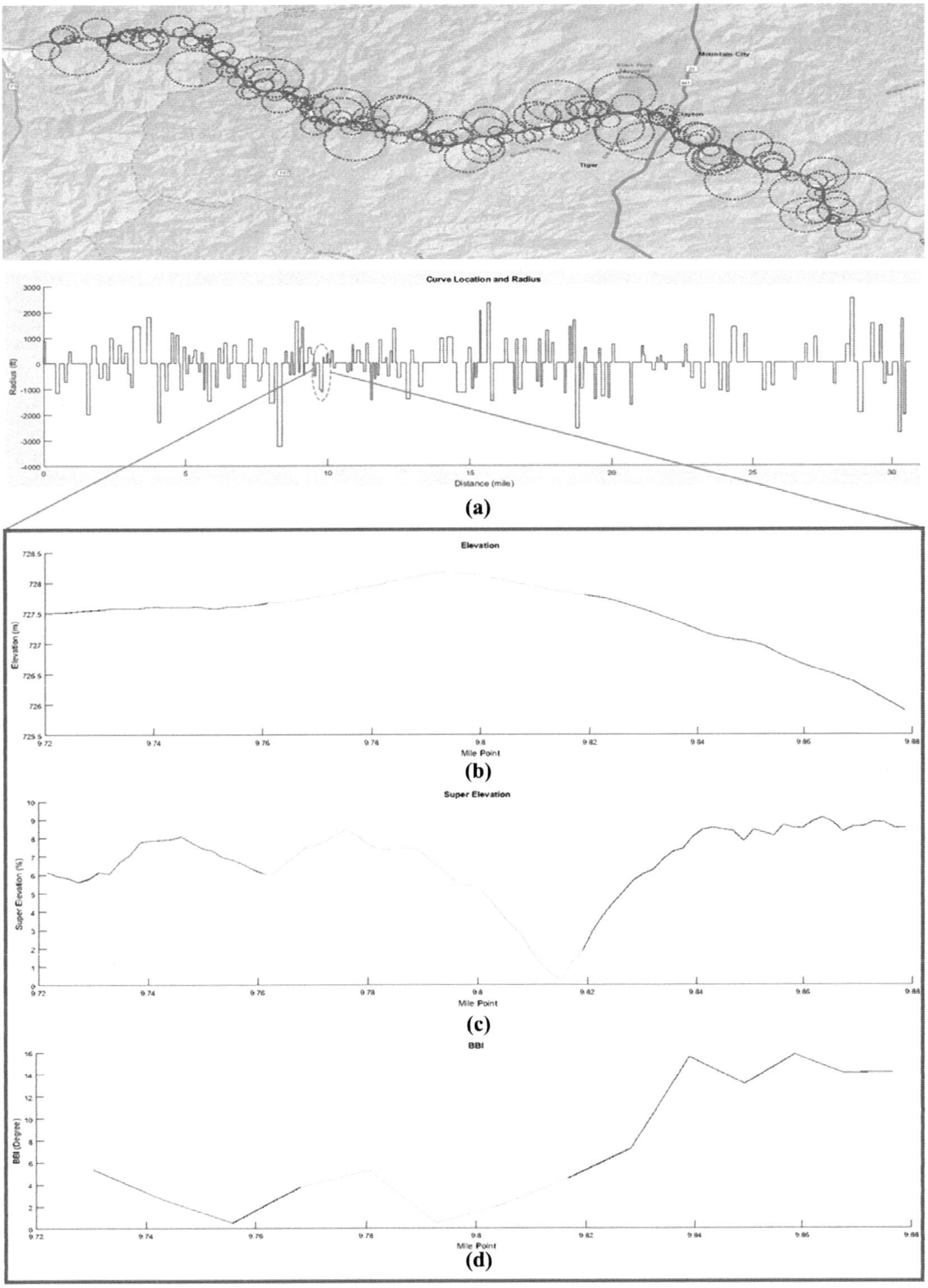

Figure 4. Visualization of (*a*) Curve location and radius of all the curves, and example of detailed, location-referenced CSCD including, (*b*) Elevation, (*c*) Superelevation distribution, and (*d*) BBI values for a subsection on State Route 2 consisting of three continuous curves highlighted by red, green, and blue.

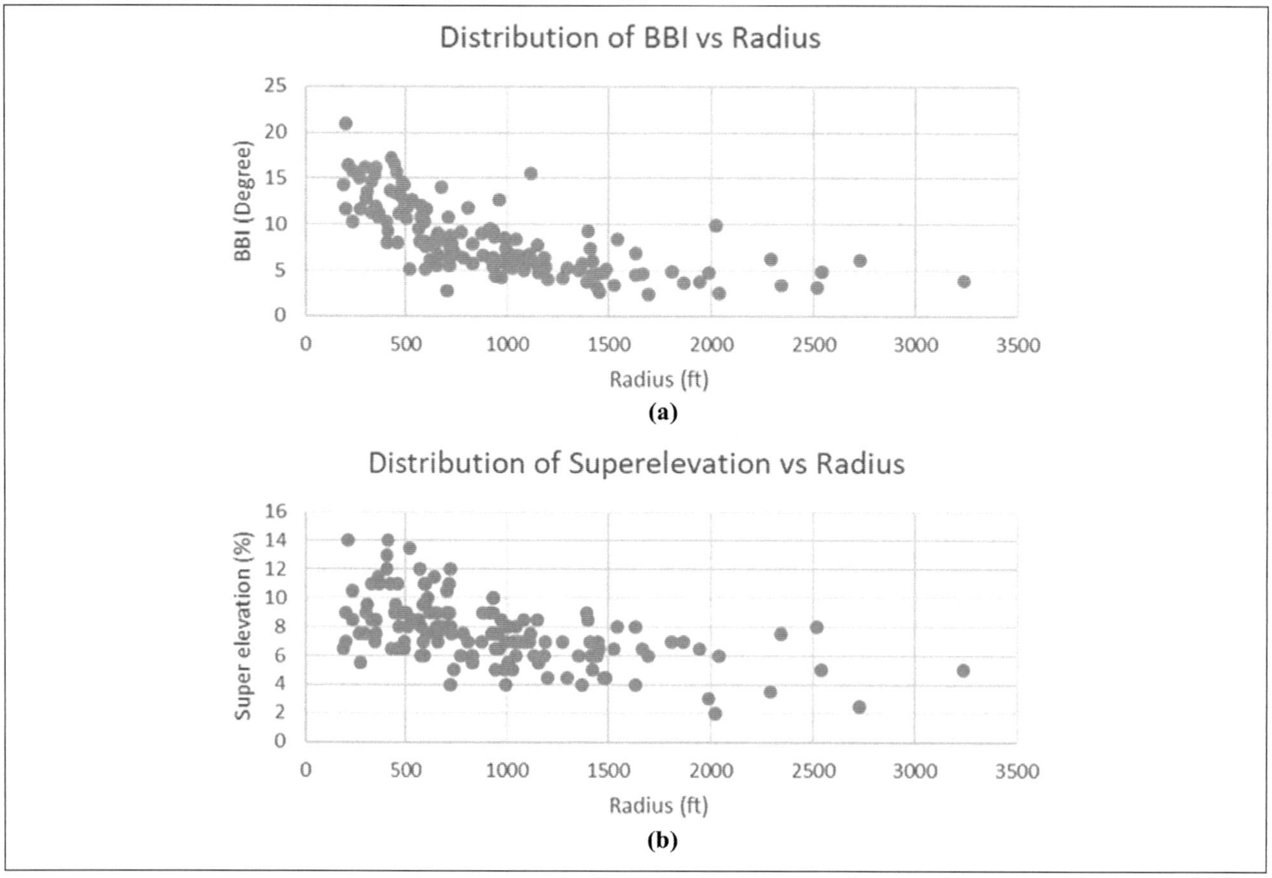

Figure 5. Relationship between (*a*) the radius and the BBI and (*b*) the radius and the superelevation.

numbers of curves with and without ROR in each set were then computed to assess whether or not a site characteristic with a certain group(s) is over-represented. Results are shown in Figure 6 and discussed as follows.

BBI. BBI values are currently used by GDOT as the single criterion to identify curves proactively for application of HFST, regardless of the crash history. BBI "measures the overturning force (side friction), measured in degrees on a vehicle negotiating a horizontal curve" (*23*), AASHTO's Geometric Design of Highways and Streets (*24*) correlates the side friction demand with the BBI values. Accordingly, a high value of BBI indicates low friction and a higher possibility of ROR crash. Figure 6*a* compares the number of curves, with and without crashes, and the ORF in each BBI set. It is observed that the ORF for BBI greater than 16 is very high compared with other categories. This result confirms and supports GDOT's current HFST-SS using BBI as the decision-making factor.

Curve Radius. Elvik (*25*) analyzed the effect of horizontal curve radius on highway crashes for 10 countries and concluded that the number of crashes typically increases as the radius decreases; the increase is significant when the radius of curvature is below 200 m (~656 feet). Although previous studies for crash-contributing factors considered curve radius for analysis, only one study confirmed that curve radius is a significant contributing factor (*12*).

In this study, the curve radius was categorized into five sets and the numbers of curves with and without ROR crashes were computed and ORF compared, as shown in Figure 6*b*. It is observed that the ORF for radius less than 400 feet is just over one and it tends to decrease mildly with increasing radius except for a peak at the 1600–2000 feet set. This may be because of the presence of other site characteristics, such as superelevation and vertical grade. On closer analysis of the curves in this set, it is found that the respective curves lie on downward slopes of various percentages and among the curves with crashes the downward slope was 3% for one curve and 6% for two curves. Besides, there is no significant pattern suggesting that as curve radius decreases, the number of ROR crashes increases.

Superelevation. The superelevation was grouped into six sets (each set varied by 2% increase). Figure 6*c* summarizes the crashes based on superelevation category. The ORF for

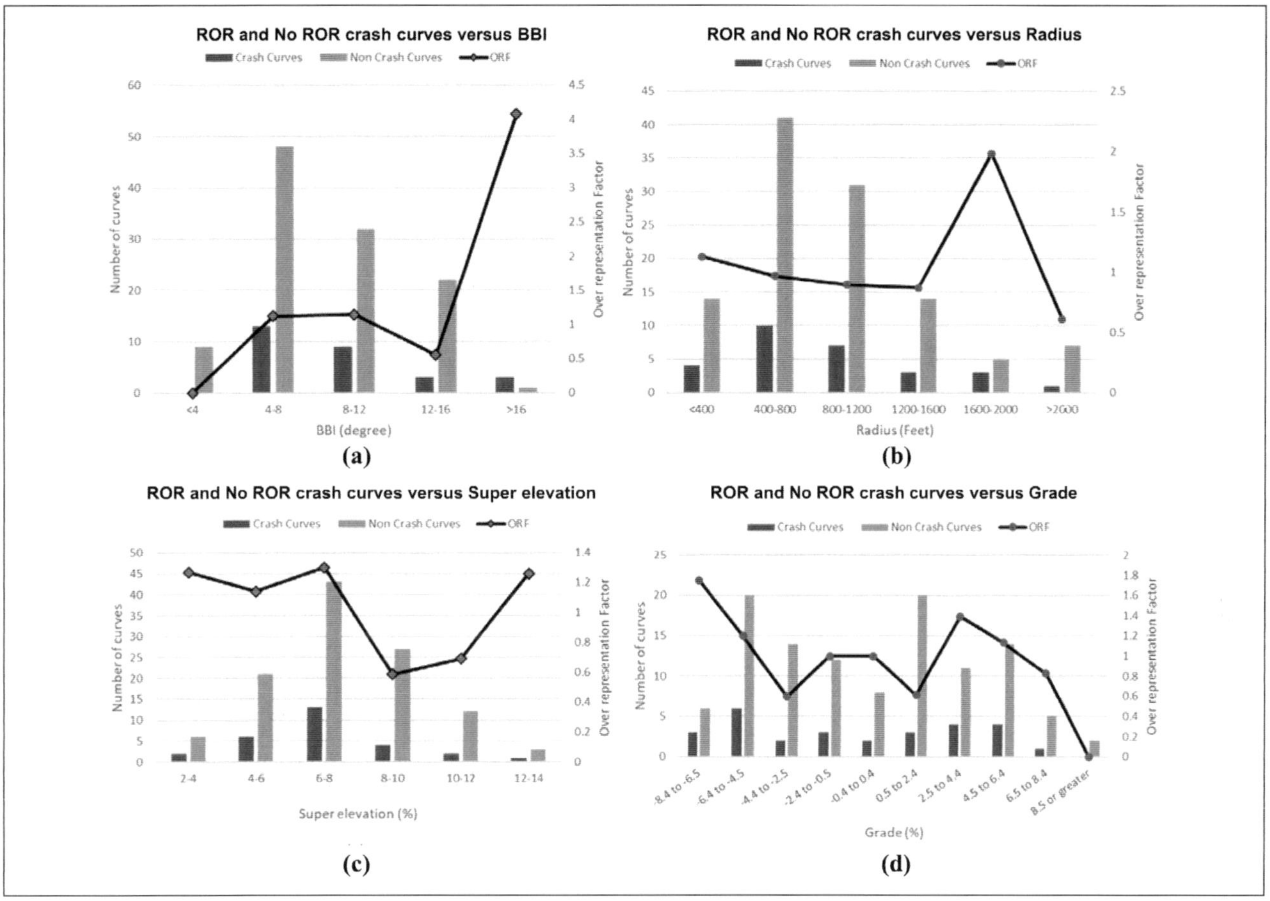

Figure 6. Over-representation for (*a*) BBI, (*b*) radius, (*c*) superelevation, and (*d*) grade.

superelevation less than 8 and greater than 12 show values slightly more than one. However, no significant trend can be observed from Figure 6*c*, which implies superelevation cannot be used in its absolute sense as a contributory factor. It is noted that the superelevation may be correlated with other HFST-SS data, such as advisory speed and curve radius.

Vertical Grade. The vertical grade was categorized into ten sets based on the code for grade in the highway performance monitoring system (HPMS) field manual (*26*). While absolute grade is reported in the HPMS, positive and negative grades were analyzed in this study. Strong positive or strong negative values (greater than 3%) indicate that the curve is located on an uphill or downhill slope. Figure 6*d* shows the ORF values for vertical grades. It is observed that there is a noticeable trend in ORF which indicates that the percentage of ROR crashes on curves with steep downhill grades increases with each 2% increase in the grade. It is also seen that the percentage of curves with ROR crashes decreases noticeably on uphill slopes greater than 4.5%. Such a trend could be explained based on acceleration. On an uphill steep grade, drivers tend to have a lower speed due to the uphill deceleration and drivers having to apply throttle to maintain the speed; on a downhill steep grade, drivers tend to have a higher speed due to the downhill acceleration and drivers having to brake to decelerate.

The four qualitative results based on the over-representation of the curves with ROR crashes show that the vertical grade is one potential indicator, besides BBI, that indicates an ROR crash-prone curve. However, a full-scale quantitative analysis of data using confidence intervals is still needed to validate that a downhill steep grade is an additional indicator.

Analysis of Selected Curves

It is observed from the over-representation analysis that the vertical grade is a significant site characteristic that may be closely related to ROR crashes on horizontal curves. Therefore, the authors further explored the CSCD of all the curves in the test section to identify comparable curves that contain different vertical grades, but which have site characteristics that are similar, such as curve radius, superelevation, BBI values, etc. Two cases were identified among the 140 curves. Both cases contain two curves whose curve radii, curve lengths, cross

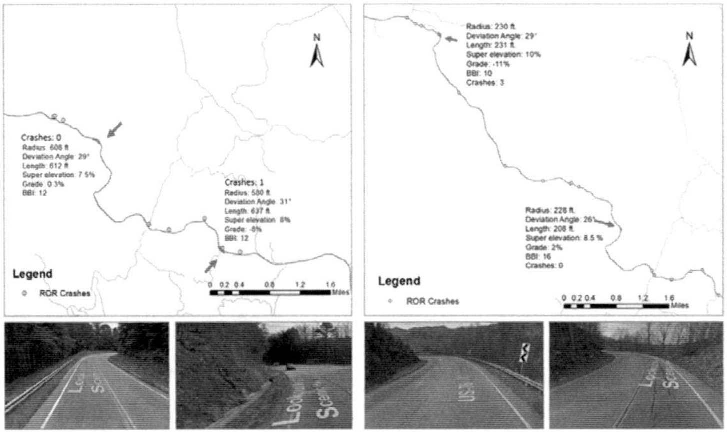

Figure 7. Sample curve pairs with similar site characteristics except for vertical grade.

slopes, and BBI values are comparable; however, the vertical grades are different (flat versus steep downhill). Figure 7 shows the two identified cases with the respective locations of the pair of curves and the perceptions of the curves from Google Street View. The green dot represents ROR crashes. It can be observed that no ROR crash occurred on the curve with a flat vertical grade, but ROR crashes occurred on the steep downhill vertical grade when approaching the curve. Drivers on a steep, downhill grade tend to have higher speeds due to additional acceleration on the downward slope and they must apply brakes constantly to decelerate. Downhill is constantly in an acceleration condition, leading to the higher likelihood of increasing and excessive speed unless a driver deliberately tries to reduce speed. If the driver does not brake and decrease speed properly, a dangerous situation occurs.

It is also observed that both cases show a small curve radius, which implies that the presence of a steep, downhill vertical grade is likely to synergize with sharp curves to create high friction demand, which eventually leads to potentially hazardous situations.

This preliminary study of the selected curves shows that the vertical grade (with a steep downgrade slope greater than 3%) is one of the factors that should be considered and further evaluated during HFST-SS, especially when a sharp curve radius (less than 800 feet) is present. The analysis shows that vertical grade (greater than 3%) plays an important role in ROR crashes on sharp curves when their site characteristics are similar. Thus, a vertical grade greater than 3% could be used as an additional indicator for HFST-SS. Additional study with a large data set is recommended to support these preliminary findings.

Conclusions and Recommendations

HFST is an effective countermeasure for restoring pavement friction to improve the safety of horizontal curves. Because it is a relatively expensive countermeasure, compared with other conventional curve countermeasures, such as advanced curve warning signs, chevrons, rumble strips, raised pavement markers, etc., HFST is often applied at selected sites based on crash data. GDOT has adopted a proactive approach to identify curves for HFST application using BBI collected on selected segments/corridors. GDOT is partnering with Georgia Tech to identify additional site characteristics data (besides BBI) that can be incorporated into its HFST-SS decision-making process to maximize its return on HFST investment by leveraging data from state-of-the-art sensing technologies

This paper makes the following contributions: (a) it proposes a methodology to extract, integrate and analyze curve site characteristics data obtained from state-of-the-art sensing technologies to identify additional site characteristics for HFST-SS, and (b) it offers preliminary findings on the additional site characteristics for HFST-SS using the proposed methodology by analyzing the data collected on 31 miles of State Route 2.

The proposed methodology consists of five steps: (a) collection of sensor data using state-of-the-art sensing technologies, (b) automatic feature extraction of the detailed, location-referenced, continuous CSCD, (c) curve-based segmentation with the extracted curve information; (d) spatial integration of CSCD; (e) analysis of correlation between CSCD and ROR crashes to identify additional factors for HFST-SS.

The preliminary findings show that a vertical grade greater than 3% plays an important role in ROR crashes on sharp curves when their site characteristics are comparable. Therefore, a vertical grade greater than 3% could be considered as an additional criterion for HFST-SS decision-making along with the current criterion (BBI value equal to or greater than 12).

The following are recommendations for future research:

1) Studying a larger data set on other roadways using the proposed methodology is recommended to confirm

the preliminary findings. Also, it is recommended to further incorporate combinations of factors such as radius and downgrade while performing the analysis.

2) It is further recommended to extract the sign information to help study driver response to the curves signs and their speed behavior using the detailed, location-referenced continuous CSCD.
3) It is recommended to compare the accuracy and efficiency of different data sources such as GPS, LiDAR, and IMU for curve, superelevation, and grade extraction to generate the CSCD.
4) The detailed, location-referenced continuous CSCD collected at different times can be used to track detailed changes in CSCD. For example, countermeasures, such as HFST and curve warning signs, installed on each curve can be identified and this allows safety engineers to evaluate a countermeasure's effectiveness objectively (i.e., crash reduction rate) by factoring other roadway conditions (such as superelevation and grade) to which the combined countermeasures have been applied.
5) With the detailed, location-referenced, continuous CSCD, transportation agencies can review and analyze safety issues on both individual curves and adjacent curves to gain an in-depth understanding of the combined CSCD effects on curve safety. Since the design of the curves is based on the appropriate balancing of multiple factors such as designed speed, curve radius, superelevation, and pavement friction to provide sufficient resistance to safely and effectively keep a vehicle on a curve, engineers can simultaneously assess the multiple factors proactively and systematically to identify curves with safety issues.

Acknowledgments

This research project is sponsored by GDOT RES PROJ 15-04. The authors thank Binh Bui from the GDOT Office of Research and Andrew Heath, Michael Turpeau, and David Adams from the Office of Traffic Operations for their support and valuable inputs. In addition, the authors thank area engineers, Chris D. York and Jerry Thomson, and field engineers for their support in data collection.

Author Contributions

The authors confirm contribution to the paper as follows: study conception and design: Tsai, Wu, Pranav, Ai; data collection: Wu, Pranav, Ai; analysis and interpretation of results: Tsai, Wu, Pranav, Ai; draft manuscript preparation: Wu, Pranav. All authors reviewed the results and approved the final version of the manuscript.

References

1. McGee, Hugh W., and R. Fred. Hanscom. *Low-Cost Treatments for Horizontal Curve Safety*. FHWA-SA-07-002. Federal Highway Administration, U.S. Department of Transportation, 2006.
2. Roa, J. A. *Evaluation of International Friction Index and High Friction Surfaces*. Virginia Polytechnic Institute, and State University, 2008. https://vtechworks.lib.vt.edu/handle/10919/36277. Accessed May 23, 2016.
3. Brimley, B., and P. Carlson. *Using High Friction Surface Treatments to Improve Safety at Horizontal Curves*. Texas Transportation Institute, College Station, 2012.
4. Albin, R., V. Brinkly, J. Cheung, F. Julian, C. Satterfield, W. Stein, E. Donnell, H. McGee, A. Holzem, M. Albee, J. Wood, and F. Hanscom. *Low-cost Treatments for Horizontal Curve Safety 2016*. FHWA-SA-15-084. Federal Highway Administration, U.S. Department of Transportation, 2016.
5. Quintus, H. V., and A. Mergenmeier. *Case Study: Kentucky Transportation Cabinet's High Friction Surface Treatment and Field Installation Program*. Federal Highway Administration, Washington, D.C., 2005.
6. Holloway, S., S. Castleberry, T. Davini, and K. Kochevar. *High Friction Surface Treatment (HFST) Applications for Cycle 6 Call-for-Projects*. Statewide Local Safety Training Webinar, Washington, D.C., 2013. http://www.dot.ca.gov/hq/LocalPrograms/HSIP/Combined-%20HFST-webinar-052213.pdf Accessed May 20, 2016.
7. Davis, S. *High Friction Surface Site Selection Analysis*. EDC Exchange for Local and Tribal Agencies, Washington, D.C., 2014. http://prltap.org/eng/wp-content/uploads/2014/06/EDC_June2014_2+revised+Scott.pdf Accessed May 20, 2016.
8. Shankar, V., F. Mannering, and W. Barfield. Effect of Roadway Geometrics and Environmental Factors on Rural Freeway Accident Frequencies. *Accident Analysis & Prevention*, Vol. 27, No. 3, 1995, pp. 371–389.
9. Khan, G., A. Bill, M. Chitturi, and D. Noyce. Safety Evaluation of Horizontal Curves on Rural Undivided Roads. *Transportation Research Record: Journal of the Transportation Research Board*, 2013. 2386: 147–157.
10. Hummer, J. E., W. Rasdorf, D. J. Findley, C. V. Zegeer, and C. A. Sundstrom. Curve Collisions: Road and Collision Characteristics and Countermeasures. *Journal of Transportation Safety & Security*, Vol. 2, No. 3, 2010, pp. 203–220.
11. Bauer, K., and D. Harwood. Safety Effects of Horizontal Curve and Grade Combinations on Rural Two-Lane Highways. *Transportation Research Record: Journal of the Transportation Research Board*, 2013. 2398: 37–49.
12. Khan, G., A. Bill, M. Chitturi, and D. Noyce. Safety Evaluation of Horizontal Curves on Rural Undivided Roads. *Transportation Research Record: Journal of the Transportation Research Board*, 2013. 2386: 147–157.
13. Mraz, A., and A. Nazef. Development of a Road Deficiency GIS Using Data from Automated Multi-Sensor Systems. In *Progress in Location-Based Services 2016*, Springer International Publishing, 2017, pp. 233–251.
14. Brown, A., and S. Brennan. Simulating Vehicle Dynamics on Both Design Plans and Laser-Scanned Road Geometry to Guide Highway Design Policy. *Transportation Research Part C: Emerging Technologies*, Vol. 50, 2015, pp. 28–36.
15. Ambros, J., V. Valentová, and P. Oríšek. Safety Ranking of Rural Curves Based on Design Consistency Measures. *Proc., 5th International Symposium on Highway Geometric Design*, Vancouver, Canada. 2015.
16. Wu, J., and Y. Tsai. Enhanced Roadway Geometry Data Collection Using an Effective Video Log Image Processing

Algorithm, *Transportation Research Record: Journal of the Transportation Research Board*, 2016. 1972: 133–140.
17. Ai, C., and Y. Tsai. Automatic Horizontal Curve Identification and Measurement Methodology Using GPS Data. *Journal of Transportation Engineering*, Vol. 141. No. 2, 2014,p. 04014078. https://doi.org/10.1061/(ASCE)TE.1943-5436.0000740.
18. Tsai, Y., C. Ai, and Y. Wu. *Curve Identification for High Friction Surface (HFST) Installation Recommendation*. Final Report, FHWA STIC and GDOT. RP 15-05, 2017. Georgia Dept. of Transportation, Office of Research. Accessed from https://rosap.ntl.bts.gov/view/dot/32777
19. Findley, D. J., C. V. Zegeer, C. A. Sundstrom, J. E. Hummer, W. Rasdorf, and T. J. Fowler. Finding and Measuring Horizontal Curves in a Large Highway Network: A GIS Approach. *Public Works Management & Policy*, Vol. 17, No. 2, 2012, pp. 189–211.
20. Tsai, Y., P. Kim, and Z. Wang. A Generalized Image Detection Model for Developing a Sign Inventory. *Journal of Computing in Civil Engineering*, Vol. 23, No. 5, 2009, pp. 266–276.
21. Tsai, Y., and C. Ai. An Automated Superelevation Measurement Methodology for Horizontal Curve Safety Assessment Using Low-Cost Mobile Device. *Transportation Research Record: Journal of the Transportation Research Board*, 2017. 2621: 62–70.
22. Tsai, Y., C. Ai, Z. Wang, and E. Pitts. Mobile Cross-Slope Measurement Methodology Using Lidar Technology. *Transportation Research Record: Journal of the Transportation Research Board*, 2013. 2367: 53–59.
23. Rieker Incorporated. *Ball Bank Indicator: Mechanical and Digital*. http://www.riekerinc.com/ballbankindicators.htm. Accessed November 10, 2016.
24. American Association of State Highway and Transportation Officials. *A Policy on Geometric Design of Highways and Streets*. American Association of State Highway and Transportation Officials, Washington, D.C., 2004.
25. Elvik, R. International Transferability of Accident Modification Functions for Horizontal Curves. *Accident Analysis & Prevention*, Vol. 59, 2013, pp. 487–496.
26. FHWA. *Highway Performance Monitoring System - Field Manual*. Office of Highway Policy Information, U.S. Department of Transportation, 2016, pp. 4.89–4.101.

The Standing Committee on Safety Data, Analysis and Evaluation (ANB20) peer-reviewed this paper (18-01446).

Article

TRR
JOURNAL OF THE TRANSPORTATION RESEARCH BOARD

Transportation Research Record
2018, Vol. 2672(38) 81–92

DOI: 10.1177/0361198118773889
journals.sagepub.com/home/trr
SAGE

Revisiting Hit-and-Run Crashes: A Geo-Spatial Modeling Method

Jun Liu[1], Asad J. Khattak[1], Cong Chen[2], Dan Wan[3], Jiaqi Ma[4], and Jia Hu[5]

Abstract

Hit-and-run crashes often delay emergency response and may result in increasing/secondary harms/damages to the victims in the crash. This study revisited hit-and-run crashes using a geo-spatial modeling approach, specifically, Geographically Weighted Regression (GWR), to explore geo-referenced crash data. The data cover motor vehicle crashes (N = 138,529) in Southeast Michigan including 20,813 hit-and-run crashes in 2015. This study presented the results from both traditional regression and GWR models. GWR model results can be mapped in space, and the maps offer visual insights about the spatially varying correlates of hit-and-run crashes that are not available from previous studies. Results from traditional binary logit model are generally consistent with findings in previous studies. For example, hit-and-run is more likely to occur on weekends or during nighttime (especially without street lights on). Driving under impairment (DUI) seems to increase the likelihood of hit-and-run. GWR models also uncovered spatially varying correlates of hit-and-run. For example, DUI crashes in the northwest of the Detroit metropolitan area are associated with an even greater hit-and-run likelihood than those in other parts in this area. In addition, the local socio-economic factors are included in the analysis. Results show that hit-and-run is more likely to occur in census tracts with a higher unemployment rate, a lower household income, a smaller portion of college-educated population, and a greater population density. The study demonstrates a way of making sense of geo-referenced traffic safety data. The geo-spatial modeling method is useful for prioritizing specific geographic regions/corridors for safety improvement countermeasures, and outperforms traditional modeling techniques.

Hit-and-run is an action of causing or contributing to a traffic crash (either with another vehicle, a pedestrian/biker, or a fixed object) and leaving the crash scene without stopping afterwards. Almost all states in the United States have laws regarding hit-and-run crashes, and committing hit-and-run is generally considered as a serious law violation/offense/crime regardless of the at-fault party in the crash (*1*). Hit-and-run often delays emergency response and may result in increasing/secondary harms/damages to the victims (*2*, *3*). According to National Highway Traffic Safety Administration (NHTSA), there are nationwide around 1,400–1,500 fatalities in hit-and-run crashes annually (*4*). Hit-and-run is also a major contributing factor leading to the fatalities of pedestrians. In 2015, 19% of pedestrian fatalities were caused by hit-rand-run crashes (*5*).

Researchers have extensively discussed hit-and-run-associated factors from crash attributes, roadway features, environmental contexts, and time of day/year, to characteristics of involved vehicles and persons (*2*, *3*, *6*–*9*). A commonly used method by these researchers was to build regression models based on crash data. Models presented in these studies generally assumed that relationships between factors and hit-and-run are stationary. However, hit-and-run crashes involve complex mechanisms of driver behavior that are potentially influenced by diverse social contexts. In addition, hit-and-run crashes (and all traffic crashes) are distributed in space, interacting with spatially varying geographic environments including the roadway features and environmental contexts. It remains unclear whether hit-and-run crashes and their correlates are likely to vary in space.

[1]Department of Civil, Construction and Environmental Engineering, University of Alabama, Tuscaloosa, AL
[2]Center for Urban Transportation Research, University of South Florida, FL
[3]Department of Civil Engineering, City College of New York/CUNY, NY
[4]University of Cincinnati, Cincinnati, OH
[5]Tongji University, Shanghai, China

Corresponding Author:
Address correspondence to Jun Liu: jliu@eng.ua.edu

The objective of this study is to investigate the spatial correlates of hit-and-run crashes through a geo-spatial modeling approach. If the correlates substantially vary in space, hit-and-run prevention strategies/recommendations may need to be localized. The contributions of this study are clear. Methodologically, the study applies the geo-spatial modeling to investigate hit-and-run crashes. The method takes advantage of the increasingly available geo-referenced data and powerful computational tools. Empirically, this study offers new insights into correlates of hit-and-run crashes that are likely to vary in the spatial domain.

Literature Review

Previous studies have identified a long list of factors related to hit-and-run crashes, and the results are used to help develop hit-and-run prevention strategies. An early study by Solnick and Hemenway investigated hit-and-run crashes that involved pedestrian fatalities (*6*). The study reported that 20% of the drivers left the scene after collisions. Regarding the crash context and time, drivers were more likely to hit-and-run in urban areas, at night, and on weekends. Tay et al. built a logistic model to analyze hit-and-run crashes (*7*), and some findings were consistent with the study by Solnick and Hemenway (*6*). They both found that hit-and-run crashes were more likely to occur at night and when the involved driver was a male. Kim et al. built regression models to find that males, tourists, driving under impairment (DUI) drivers, or those driving a stolen car were more likely to leave the scene after a crash (*10*).

Studies from recent years continued to scrutinize hit-and-run crashes, and most of their findings are consistent with earlier studies. MacLeod et al. investigated crashes that involved pedestrian fatalities through a logistic model (*11*). Results implied an increased likelihood of hit-and-run on weekends, at night, or in the early morning. This study also found that drivers with traffic violation histories (e.g., DUI convictions) were more likely to hit-and-run. Regarding DUI, Frencha and Gumus particularly discussed the effect of blood alcohol concentration (BAC) limits on driver behavior after collisions, and concluded that decreasing the BAC limit caused an increase in hit-and-run fatality count (*12*). In addition, researchers also added insights from developing countries (*3*, *8*). Aidoo et al. analyzed pedestrian hit-and-run crashes in Ghana through a binary logit model. Hit-and-run was found to be associated with nighttime, unclear weather, and straight and flat roads without medians and junctions (*8*). Zhang et al. examined hit-and-run crashes in Guangdong, China (*3*). They related the hit-and-run crashes to factors from crash attributes and human/vehicle characteristics to road and environmental factors.

In a different approach to studies based on data analysis, Hopkins et al. completed a recent study based on interviews from 52 drivers and industry stakeholders who were convicted for hit-and-run offenses (*13*). This study offered new insights from the angle of hit-and-run drivers. Four major reasons for leaving the crash scene were identified, including self-preservation, alcohol use, no knowledge of the crash, and too trivial to report.

Most previous data-driven studies applied logistic regression to examine the correlates of hit-and-run crashes and reported one set of stationary correlates of hit-and-run crashes. The safety improvement recommendations from these studies are often suggested for the entire study region without differentiating the localities. Researchers focusing on other types of traffic crashes have applied geo-spatial modeling techniques to explore the possible spatial patterns of crashes. Traffic crashes do have certain spatial patterns that reflect diverse geographic conditions (*14–21*). Geographically Weighted Regression (GWR) is one type of spatial analytic technique that has gained extensive attention from researchers aiming to understand the spatially varying correlations between factors across geographic regions (*18*, *22–26*). The results from GWR models offer great insights that help prioritize certain parts of the study region for localized safety countermeasures.

Methodology

Data

This study used geo-referenced crash data from Southeast Michigan (including the counties of Livingston, Macomb, Monroe, Oakland, St. Clair, Washtenaw, and Wayne) (*27*). It is a pre-cleaned database. In the 2015 database, there are 138,529 crash records including 20,813 hit-and-run crashes. The variables in the database include crash type, crash severity, road condition, weather and time, the involvement of pedestrian/biker, young/senior driver, DUI driver, and so on. Some key information about involved vehicles and persons is limited by the authors, including driver gender, vehicle type, and so forth.

To relate crashes to local contexts, Census data at tract level were downloaded and linked to crashes according to their geo-coded locations. This study used the 2015 American Community Survey (ACS) estimates (*28*) to couple with 2015 crash records. ACS is huge database with over 35,000 variables. This study selected and summarized five important socio-economic variables, including population density and the rates/percentages of unemployment, household income under 25k, persons (>25 years old) without college education, and persons uninsured.

Table 1 lists variables of interest for this study and their descriptive statistics, as well as comparison between hit-and-run crashes and other crashes. The hit-and-run crashes account for 15% of total sampled crashes. Most crashes did not cause any injuries, and 0.3% of sampled crashes were fatal. As relative to other crashes, hit-and-run crashes appear to be less severe. Regarding crash type, rear-end was the most frequent type and accounted for total sample crashes. For sideswipe crashes, hit-and-run held a much greater share than other crashes (28.6% vs. 14.7%). In total 34.3% of crashes involved young drivers aged 24 and younger, and 15.5% of crashes involved senior drivers aged 65 and older, but such types had lower shares in hit-and-run crashes. Note that all variables related to drivers are for the at-fault drivers.

Statistics also show that 1.6% of crashes involved pedestrians or bikers, and 4.2% occurred owing to collisions with animals. Among hit-and-run crashes, 3.1% involved pedestrians or bikers, which is 133.3% greater than the share in non-hit-and-run crashes. In 3.4% of crashes, the at-fault driver was found to be impaired by drug or alcohol, termed DUI. DUI accounted for 4.8% of hit-and-run crashes, which is 53.7% greater than other crashes. Most crashes occurred during daytime, in clear/cloudy weather, and on dry roads. There were 12.4% of crashes that occurred on roads with icy, snowy, muddy, slushy surface, 10.3% in darkness without street lighting, and 16.2% under foggy, smoky, raining, snowing, or severely windy weather conditions. Noticeably, 25.5% of hit-and-run crashes occurred in darkness with street lighting; this number is 93.5% higher than in other crashes. About 44% of crashes occurred in winter time (from November to March), and 22.5% occurred during weekends (Saturdays and Sundays). Hit-and-run crashes seemed to be more likely to occur during weekends. Crashes in rural areas accounted for 8.2% of total crashes, and interstate crashes for 8.2%.

The socio-economic variables indicate the social contexts of crashes. Comparison between hit-and-run and other crashes shows that hit-and-run crashes are related to more populous communities and higher percentages of unemployed population, low household incomes (below 25K), uninsured population, and persons (>25 years old) without college education.

The geo-referenced data offer an additional angle for researchers to observe the traffic crashes from the space. Figure 1 presents the spatial distributions of (*a*) all sampled crashes and (*b*) hit-and-run crashes. As expected, crashes are clustered in urbanized areas (Detroit and Ann Arbor) with higher population densities. In some rural regions, for example, north and west parts in the study region, few crashes involved hit-and-run behavior. The clusters of all crashes and hit-and-run crashes generally overlap in urbanized regions, but their clusters pose different shapes. This implies their spatial distributions are not homogeneous.

Model

Following previous studies, this study first builds a logistic regression model to estimate stationary correlations of hit-and-run crashes. As the hit-and-run is a yes/no action (whether a driver left the scene of a crash), a binary logistic model is appropriate, as follows (*29*):

$$P(Y_i = Whether\, hit^-and^-run) = \frac{\exp(\alpha + \beta X_i)}{1 + \exp(\alpha + \beta X_i)} \quad (1)$$

Equation 1 could also be expressed with a logit transformation,

$$\text{Logit}(P|Y_i = Whether\, hit^-and^-run) = \alpha + \beta X_i \quad (2)$$

where Y_i is the *i*th crash's driver action (whether he or she left the scene after the crash) in the database, $Y_i = 1$ indicates it was a hit-and-run crash and 0 otherwise; X_i is a vector of independent variables found for *i*th crash, summarized in Table 1; α is the intercept of the binary logit model; β is a set of coefficients estimated for independent variables X_i.

The binary logit model in Equation 2 takes all valid observations in estimation, and often outputs one set of coefficients which represent the global relationship between hit-and-run crashes and associated factors. The model assumes that all observations share the same set of coefficients, and therefore the relationship is stationary. Traditional models, such as the binary logit model above, work well when the data do not have a spatial nature, for example, data from laboratory tests. For data that contain spatial characteristics, such as traffic crashes, such models may still be used to estimate the overall relationships between factors, but are unable to reveal the detailed local relationships which may vary substantially in space. With the advances of geo-referenced data, this study examines the spatial patterns of hit-and-run crashes and compares the geo-spatial modeling results with the outputs of the traditional binary logit model. If the spatial variations are substantial, then it is of practical necessity and importance to consider different strategies for different regions to prevent/reduce the occurrence of hit-and-run.

The geo-spatial modeling approach in this study is GWR. GWR has two core features which show the differences from traditional approaches: (1) GWR takes only a sub-group of observations into the estimation. The estimated results represent only the local relationships between factors; (2) observations in the sub-group are unequal but weighted according to the distance

Table 1. Descriptive Statistics of Selected Variables and Comparison between Hit-and-Run and Other Crashes

Variable	Total N = 138,529	Hit-&-run N = 20,813	Other N = 117,716	Difference in %
	Percent or mean	Percent or mean	Percent or mean	
Hit-and-run[a] (1 = yes; 0 = otherwise)	15.0%	–	–	–
Crash severity				
1 = Property damage only (PDO)	79.3%	85.0%	78.3%	8.5%
2 = Possible injury	14.5%	11.2%	15.1%	–25.8%
3 = Non-incapacitating injury	4.8%	2.9%	5.1%	–43.4%
4 = Incapacitating injury	1.1%	0.7%	1.2%	–44.8%
5 = Fatal injury	0.3%	0.2%	0.3%	–7.3%
Crash type				
1 = Single	17.5%	7.9%	19.2%	–58.8%
2 = Head on	3.8%	3.2%	3.9%	–17.7%
3 = Angle	17.5%	15.4%	17.9%	–13.9%
4 = Rear-end	37.2%	30.8%	38.3%	–19.6%
5 = Sideswipe	16.8%	28.6%	14.7%	93.9%
6 = Other	7.3%	14.1%	6.1%	133.3%
Young driver[a] (<25 years old. 1 = yes; 0 = otherwise)	34.3%	19.1%	37.0%	–48.5%
Senior driver[a] (>64 years old. 1 = yes; 0 = otherwise)	15.5%	8.0%	16.8%	–52.4%
Pedestrian/biker involved (1 = yes; 0 = otherwise)	1.6%	3.1%	1.3%	133.3%
Driving under impairment[a] (1 = yes; 0 = otherwise)	3.4%	4.8%	3.1%	53.7%
Animal involved (1 = yes; 0 = otherwise)	4.2%	0.0%	5.0%	–99.1%
Road surface condition				
1 = Dry	73.0%	71.5%	73.3%	–2.4%
2 = Wet	14.5%	13.6%	14.7%	–7.3%
3 = Other (icy, snowy, muddy, slushy, etc.)	12.4%	14.8%	12.0%	23.9%
Lighting condition				
1 = Daylight	69.1%	55.5%	71.5%	–22.4%
2 = Dawn	2.2%	1.8%	2.2%	–19.8%
3 = Dusk	2.2%	2.2%	2.2%	0.0%
4 = Dark - with street lights on	15.0%	25.5%	13.2%	93.5%
5 = Dark- without street lights on	10.3%	10.2%	10.3%	–0.4%
6 = Unknown	1.2%	4.8%	0.6%	664.9%
Weather				
1 = Clear or cloudy	83.8%	81.6%	84.2%	–3.1%
2 = Other (fog, smoke, rain, snow, severe wind, etc.)	16.2%	18.4%	15.8%	16.5%
Winter (from Nov. to Mar. 1 = yes; 0 = otherwise)	44.3%	45.2%	44.1%	2.4%
Weekend (1 = yes; 0 = otherwise)	22.5%	28.2%	21.5%	31.2%
Speed limit				
1 = "≤ 30 mph"	22.9%	42.5%	19.5%	118.1%
2 = "35~45 mph"	47.1%	37.0%	48.9%	–24.4%
3 = "50~60 mph"	17.9%	11.8%	19.0%	–37.8%
4 = ">65 mph"	12.1%	8.7%	12.7%	–31.0%
Rural (1 = yes; 0 = otherwise)	8.2%	2.8%	9.2%	–69.3%
Interstate (1 = yes; 0 = otherwise)	14.5%	13.2%	14.7%	–9.8%
Percent of unemployed population in labor force at tract level	11.2%	15.1%	10.5%	44.1%
Percent of household income below 25K at tract level	20.1%	27.0%	18.8%	43.2%
Percent of uninsured population at tract level	10.1%	12.4%	9.7%	27.2%
Percent of population (>25 years old) without college education at tract level	61.5%	68.7%	60.2%	14.2%
Population density at tract level (persons per square mile)	3,304	4,291	3,130	37.1%

Note: Difference in % = (Hit-and-run statistics – other crash statistics)/other crash statistics × 100%.
[a]Variables are for the at-fault driver.

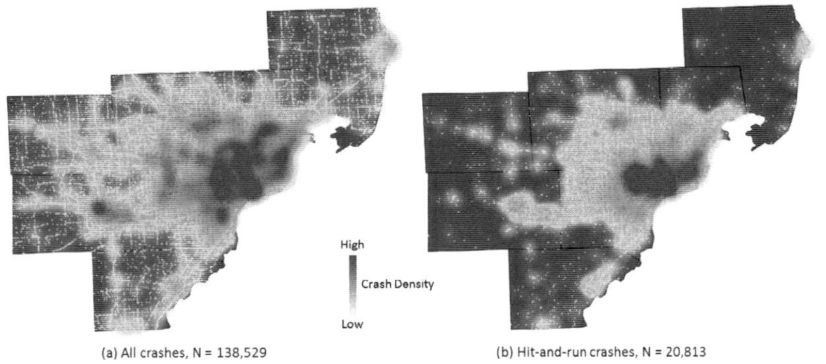

Figure 1. Spatial distributions of (*a*) all sampled crashes and (*b*) hit-and-run crashes.

between their locations and the sub-group's geographic center. The sub-group's center moves from one observation to another. For each observation, GWR estimates a local set of coefficients based on the weighted observations in the sub-group. GWR samples observations into a sub-group based on their geographical distance from the sub-group's center; if an observation's distance is within a given bandwidth, it is sampled into the local GWR modeling. The bandwidth is a critical configuration parameter in GWR, determining the spatial extent of sub-groups (*30*). It is equal to the geographical distance of the farthest observation within the sub-group from the center. To bridge the GWR with traditional regression approaches, GWR is often written in an extended form of the traditional models. Extended from Equation 2, the form can be written as (*30*):

$$\begin{aligned}\text{Logit}(\text{P}|Y_i &= \textit{Whether hit}-\textit{and}-\textit{run})\\ &= \alpha(u_i, v_i) + \beta(u_i, v_i)X_i\end{aligned} \tag{3}$$

where $\alpha(u_i, v_i)$ and $\beta(u_i, v_i)$ are the local coefficients for *i*th crash located at (u_i, v_i). Y_i is the response, that is, hit-and-run behavior. X is a vector of explanatory variables. The observations in sub-groups are weighted based on their spatial distances to the sub-group's center. Adaptive bi-square function is often used to determine weights in GWR modeling (*30*):

$$w_j(u_i, v_i) = \left(1 - \left(\frac{d_{\text{ij}}}{d_{\text{bandwidth}}}\right)^2\right)^2 \tag{4}$$

where $w_j(u_i, v_i)$ is the weight for *j*th observation in the sub-group which is centered at *i*th observation in the overall sample; d_{ij} is the geographical distance of *j*th observation from the center; $d_{\text{bandwidth}}$ is bandwidth determining the spatial extent of sub-groups. As traffic crashes are not uniformly distributed in space, non-fixed bandwidth was used in this study to ensure that each sub-group has the same number of observations for local GWR modeling.

For local coefficients from GWR, the z-value is calculated to verify the significance of estimates: $z = \frac{\beta(u_i, v_i)}{SE(u_i, v_i)}$, where $SE(u_i, v_i)$ the standard error of local coefficients. The model uses a value of ± 1.96 which approximately corresponds to a two-sided hypothesis test with a significance level of p-value ≤ 0.05. Maximum absolute z-values, $|z|$, are used to determine whether there is at least one significant local estimate. If the maximum |z-value| for an independent variable is less than 1.96, it means this variable is not associated with the dependent variable anywhere in the study area. This variable may be removed from the final model.

Random-parameter models are also able to consider the varying correlates across observations caused by the unobserved heterogeneity (*31*). GWR assumes that the unobserved heterogeneity is related to the space, while random-parameter models do not have such an assumption. In GWR modeling, the extent of heterogeneity between observations is associated with their spatial separations. The parameters or coefficients from GWR models often present clear and smooth spatial patterns, whereas the random-parameter models do not. If the heterogeneity is strongly related to the space, then the random-parameter models would also output spatial patterns which may not be as smooth as the patterns from GWR models. This study is focused on the driver behavior (i.e., hit-and-run) and motor vehicle crashes which have a spatial nature. The driver behavior also interacts with the socio-economics, which vary from different land uses. The use of GWR is to link the hit-and-run behavior and motor crashes with the space. Though both random-parameter and GWR models may result in similar local estimates, from a practical aspect, GWR outperforms random-parameter models by giving clear spatial patterns that can be used to develop local safety improvement countermeasures.

Model Comparison

Though GWR can reveal the spatial variations of relationships between responses and explanatory factors, the variations may be insubstantial, indicating that observations from different places have similar relationships. In that case, the coefficients from GWR are approximately equal to the stationary coefficients from traditional models. To verify whether the local estimates from GWR model vary significantly in space, this study performed non-stationarity tests. GWR outputs a set of coefficients for each observation, and the coefficients for the same variable have a distribution with minimum, lower quartile, median, upper quartile, and maximum value. Considering that the minimum and maximum coefficients may be too extreme, the non-stationarity test takes a conservative range *Delta* from the lower quartile and upper quartile and compares this range with the corresponding coefficient's standard error in the traditional model. The non-stationarity test is performed for each variable by comparing the estimates from GWR and traditional models (*32*):

$$Delta = \beta^{\text{upper}} - \beta^{\text{lower}} \begin{cases} >1.96(S.E.) \text{ and } Max.|z|>1.96, & \text{Pass the Non} - \text{stationarity Test} \\ \text{otherwise}, & \text{Fail the Non} - \text{stationarity Test} \end{cases} \quad (5)$$

where $S.E.$ is the corresponding coefficient's standard error in the traditional model, and $|z_i|$ is the z-value of a local GWR coefficient $\beta(u_i, v_i)$. If *Delta* is greater than $1.96(S.E.)$ and $\text{Max.}|z|$ is greater than 1.96, the non-stationarity test is passed, indicating significant spatial variations of GWR estimates.

In addition, the overall model's goodness-of-fit was compared by calculating the Deviance and Akaike Information Criterion (AIC). Deviance at a regressed model can be calculated:

$$D = \sum_{i=1}^{N} \left(s_i \sqrt{-2(y_i \log \hat{p}_i + (1 - y_i) log(1 - \hat{p}_i)} \right)^2 \quad (6)$$

where y_i is the observed response, that is, hit-and-run behavior in the *i*th crash (1 = yes, 0 = otherwise); $\hat{p}_i$ is the predicted probability of hit-and-run for the *i*th crash; $s_i = 1$, if $y_i = 1$ and $s_i = -1$, if $y_i = 0$. A better goodness-of-fit is associated with a smaller deviance. The deviance reduces if a newly added variable/information in the model helps explain the data. The extent of information explained by the model can be reflected by the explained deviance, which is also known as Pseudo R^2 for logistic models:

$$Pseudo\ R^2 = 1 - \frac{D}{D(null)} \quad (7)$$

where $D(null) = \sum_{i=1}^{N} \left(s_i \sqrt{-2(y_i \log P + (1 - y_i) log(1 - P)} \right)^2$, and P is the overall probability of hit-and-run among all crashes.

AIC has been often used for model comparison, and a smaller AIC value often points to a better goodness-of-fit (*33*). A three-point decrease of AIC in model comparison indicates a significantly better goodness-of-fit (*30*). The AIC can be calculated based on the deviance derived from the model:

$$AIC = D + 2k \quad (8)$$

where k is the number of parameters in the model.

Modeling Results

Table 2 reports the estimation results of traditional binary logit and GWR models. As the GWR outputs a set of coefficients for each observation, only the estimates' mean, minimum, and maximum are summarized in the Table. The modeling results imply that the GWR approach outperformed the traditional binary logit model, according to its greater Pseudo R^2 and significantly lower AIC value. The estimates from the traditional logit model are in general reasonable and consistent with previous studies (*2, 3, 6–8, 11, 12*). Most factors examined were found to be significantly associated with hit-and-run. As expected, estimates from the traditional model are within the ranges of GWR estimates, and traditional estimates are in general close to the mean of GWR estimates. The estimated coefficients represent the log of odds of hit-and-run given a set of conditions as relative to the base conditions. Figure 2 presents GWR outputs mapped in space and the coefficients of selected variables are shown in maps. Only eight variables are shown in maps, mainly because of the space limitation. Some variables (e.g., fatal crash, DUI, and socio-economic factors) are of special interest, and other variables (pedestrian/biker, street light and socio-economic factors) are reported because they are likely related to actionable recommendations (compared with variables such as weekend and weather). In these maps, colors from blue to red represent coefficients from small to large, and te relative magnitude of these coefficients are indicated by color tones.

The relationships between crash severity and hit-and-run could be bi-directional: (1) knowing the severity of a

Table 2. Traditional Binary Logistic Model and GWR Model Estimates

Model (1 = hit-and-run, 0 = otherwise)	Traditional binary logit Std. error Pr(>|z|)			GWR model Min β Max β Max |Z|					
Variable	*β*	*Std. error*	*P-value*	*Mean β*	*Min β*	*Max β*	*Max |Z|*	*Delta*	*Test*
Intercept	−3.374	0.049	0.000	−3.061	−4.055	−2.193	19.075	0.431	YES
Crash severity (base = property damage only)									
Possible injury	−0.410	0.026	0.000	−0.480	−0.833	−0.161	5.223	0.185	YES
Non-incapacitating injury	−0.733	0.049	0.000	−0.777	−1.491	−0.200	6.504	0.325	YES
Incapacitating injury	−1.121	0.098	0.000	−1.207	−3.104	−0.190	4.215	0.526	YES
Fatal injury	−0.897	0.171	0.000	−1.061	−6.328	0.441	2.303	0.982	YES
Crash type (base = single)									
Head on	0.815	0.054	0.000	0.773	−0.284	1.841	10.158	0.567	YES
Angle	0.863	0.037	0.000	0.800	0.066	1.665	12.910	0.408	YES
Rear-end	0.970	0.033	0.000	0.947	0.208	1.658	15.011	0.303	YES
Sideswipe	1.554	0.034	0.000	1.558	1.107	2.080	18.530	0.368	YES
Other	1.387	0.038	0.000	1.383	0.641	2.057	15.361	0.452	YES
Young drivers involved (<25 years old. 1 = yes; 0 = otherwise)	−0.911	0.020	0.000	−0.883	−1.175	−0.595	16.154	0.139	YES
Senior drivers involved (>64 years old. 1 = yes; 0 = otherwise)	−0.862	0.028	0.000	−0.868	−1.339	−0.640	10.352	0.240	YES
Pedestrian/biker involved (1 = yes; 0 = otherwise)	1.113	0.060	0.000	1.170	−0.082	2.088	9.165	0.576	YES
Driving under impairment (1 = yes; 0 = otherwise)	0.381	0.042	0.000	0.394	−0.705	1.400	6.163	0.874	YES
Animal involved (1 = yes; 0 = otherwise)	−4.078	0.334	0.000	−1.274	−10.040	2.460	5.227	2.242	YES
Road surface condition (base = dry)									
Wet	−0.062	0.030	0.039	−0.074	−0.403	0.397	2.156	0.198	YES
Other (icy, snowy, muddy, slushy, etc.)	−0.103	0.033	0.002	−0.127	−0.736	0.288	2.653	0.195	YES
Lighting condition (base = daylight)									
Dawn	0.242	0.061	0.000	0.218	−0.338	0.751	2.657	0.265	YES
Dusk	0.287	0.055	0.000	0.241	−0.350	0.735	3.566	0.282	YES
Dark - with street lights on	0.800	0.021	0.000	0.799	0.586	1.203	12.277	0.134	YES
Dark- without street lights on	0.872	0.030	0.000	0.892	0.553	1.151	8.556	0.136	YES
Unknown	2.051	0.062	0.000	2.053	0.971	4.131	9.589	0.623	YES
Not clear or cloudy weather (fog, smoke, rain, snow, severe wind, etc.)	−0.004	0.031	0.896	0.023	−0.680	0.342	3.148	0.152	YES
Winter (from Nov. to Mar. 1 = yes; 0 = otherwise)	−0.040	0.018	0.028	−0.046	−0.371	0.096	2.229	0.081	YES
Weekend (1 = yes; 0 = otherwise)	0.262	0.019	0.000	0.245	0.010	0.456	5.582	0.119	YES
Speed limit (base = "≤ 30 mph)									
35–45 mph	−0.520	0.020	0.000	−0.519	−0.818	−0.070	10.496	0.226	YES
50–60 mph	−0.575	0.029	0.000	−0.543	−1.411	−0.032	8.274	0.381	YES
>65 mph	−0.580	0.042	0.000	−0.511	−1.182	0.100	5.825	0.372	YES
Interstate (1 = yes; 0 = otherwise)	0.049	0.034	0.140	−0.063	−0.742	0.401	2.656	0.264	YES
Percent of unemployed population in labor force at tract level	2.394	0.136	0.000	1.065	−3.392	3.615	5.603	1.543	YES
Percent of household income below 25K at tract level	0.173	0.078	0.026	0.268	−2.753	3.190	3.786	0.537	YES
Percent of uninsured population at tract level	−0.099	0.171	0.562	−0.566	−4.609	2.161	3.317	0.972	YES
Percent of population (>25 years old) without college education at tract level	1.004	0.059	0.000	0.909	−0.367	2.134	5.878	0.719	YES
Population density at tract level (1000 persons per square mile)	0.051	0.003	0.000	0.040	−0.034	0.114	4.087	0.039	YES
Summary statistics									
Number of observations	138529			138529					
Deviance at null model	117231.6			117231.6					
Deviance at regression model	97167.74			94558.58					
Pseudo R^2 (percent of explained deviance)	0.171			0.193					
AIC	97235.74			94626.58					

Note: If maximum $|z| >= 1.96$, a local estimate is significant at 95% confidence level, i.e., p-value ≤ 0.05; *Delta* = $\beta^{\text{upper quantile}} - \beta^{\text{lower quantile}}$; Test = Non-Stationarity Test; YES = Passed the Test; NO = Failed the Test.

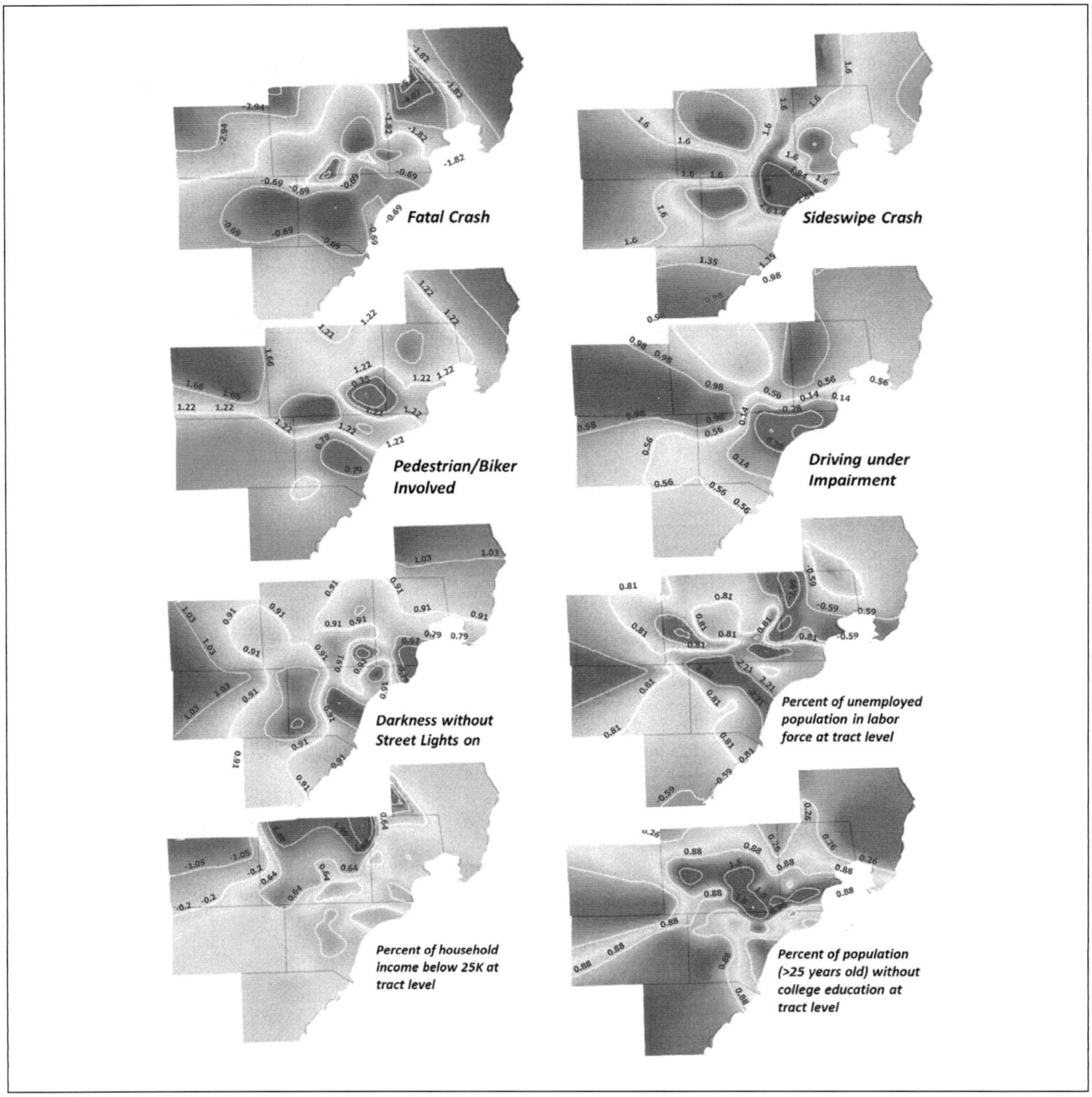

Figure 2. GWR estimates of selected variables mapped in space (blue to red colors representing small to large coefficients).

crash, the at-fault driver chose to leave or not to leave the scene after the crash; and (2) because of hit-and-run, crash severity increased because of delayed emergency services. Therefore, it might be perilous to claim either that crash severity affects hit-and-run decision or vice versa. That being said, this study revealed that crash severity is significantly correlated with hit-and-run. From the traditional logit model, crashes that caused higher severities are associated with a lower hit-and-run likelihood. As Hopkins et al. (*13*) reported, one of the motivations for a driver leaving the scene was based on his or her judgment that the crash was too trivial to report. GWR models uncovered spatially varying associations between crash severity and hit-and-run. The non-stationarity test showed that their spatial variations are significant. Figure 2 shows the spatially varying relationships between fatal crash and hit-and-run. Within the entire spatial domain, fatal crashes are associated with a smaller hit-and-run likelihood, as relative to the base—property-damage-only crashes—that is, all coefficients for fatal crashes are negative. In addition, if only fatal crashes are considered in the discussion, those in red regions are bonded with a higher likelihood of hit-and-run. It shows that red regions are located in the northeastern (Marysville and Port Huron) and mid-southern (Ann Arbor, Ypsilanti, Romulus) regions, as well as the City of Pontiac in the mid-north. Single-vehicle crashes were less likely to involve hit-and-run, as relative to other crashes. Sideswipe crashes (among all other types of crashes) seemed to have the highest likelihood of hit-and-run. GWR results show that sideswipe crashes in downtown of Detroit were associated with an even higher chance of hit-and-run than those in other areas.

The traditional model revealed that young (<25 years old) or senior (>64 years old) drivers were less likely to hit-and-run as relative to middle-aged drivers (25–64 years old). Crashes that involved pedestrians/bikers were associated with a greater likelihood of hit-and-run than other crashes without pedestrians/bikers. The GWR models uncovered that pedestrians/biker-involved crashes were bonded with an even greater likelihood of hit-and-run in the northwestern part (City of Howell) of the study region and the City of Farmington Hills (located at northeast of Detroit). DUI is a societal concern, and this study found that DUI drivers were more likely to hit-and-run, as also revealed by Hopkins et al. (*13*). GWR models showed that DUI crashes were associated with an even greater hit-and-run likelihood in northwestern part of the study region (from the City of Farmington Hills, Brighton, to Howell).

In addition, this study found that hit-and-run likelihood is likely to increase when crashes occurred on dry road surface, at nighttime, or on weekends. Crashes that occurred in darkness without street lighting had a slightly greater chance of turning into hit-and-run, as relative to those that occurred in places with street lights on. GWR revealed an even larger likelihood of hit-and-run in the City of Taylor and Wyandotte (located at the south of Detroit downtown) as well as the western (City of Chelsea) and northern (Imlay City) parts of the study region, if a crash occurred in darkness without street lighting, as shown in Figure 2. Modeling results also showed that hit-and-run is associated with the speed limits. Higher speed limits appear to relate to a lower hit-and-run likelihood. Higher speed limits are often associated with roadway of a higher classification and imply better driving conditions. Therefore a lower-class roadway seems to be related to higher rates of hit-and-run. Two reasons may explain this possibility: (1) lower-class roads are often coupled with lower speed limits and therefore crashes are less likely to be severe. Studies have found that small crashes are more likely to be unreported and involve hit-and-run (*34*, *35*); and (2) lower-class roads may be under-enforced and lack surveillance systems, and therefore drivers may feel less pressured to hit-and-run.

The modeling also showed that socio-economic contexts are significantly associated with hit-and-run crashes. Crashes in census tracts with a higher unemployment rate, lower household income, or more population without college education were more likely to involve hit-and-run. Crashes in more populous areas were related to a greater chance of hit-and-run. GWR revealed that with an increase in unemployed population in the labor force, some areas are associated with a greater increase in the chance of hit-and-run given a crash. As shown in Figure 2, these areas include the western part of the study region (City of Chelsea) and areas from the City of Farmington Hills, Taylor to Wyandotte. Regarding household income, the northern areas (from the City of Flint to Lapeer) are found to have a greater increase in hit-and-run likelihood if more households' income is under 25K. In most of the southern area and north side of Detroit, the GWR model revealed a greater increase in hit-and-run likelihood with an increase in the undereducated population. The socio-economic variables are included in the analysis because driving is also regarded as a type of culture and it interacts with the socio-economics of drivers and land use or circumstances. Based on the modeling results, reducing unemployment rates may reduce the hit-and-run rate, and relevant educational efforts may be needed for areas with low household income and under-education in order to reduce the occurrence of hit-and-run crashes.

Limitations

Results in this study depend heavily on data accuracy, especially the geo-referenced information. The data used in this study were summarized from police crash reports and pre-processed by SEMCOG (*27*) or an organization associated with SEMCOG. The extent of errors in the data remains unknown.

In addition, small crashes without casualties or severe property damage are likely to be underreported. As noted in Michigan's Traffic Crash Report Instruction Manual (*36*), all crashes that involved a motor vehicle in transport and on the roadway, and that resulted in death, injury, or property damage of \$1,000 or more should be reported. Ye and Lord suggested a method of Weighted Exogenous Sample Maximum Likelihood Estimator (WESMLE) to treat the reported crashes as sample data (*34*). When the underreporting rates are known for each crash severity level, this method can improve the estimation of crash models. A recent NHTSA report (*35*) indicated that roughly 3 in 10 of all crashes are unreported, and property-damage-only crashes are more likely to be unreported. However, there is no information about the reporting rates related to the spatial contexts. For this study, the effects of the underreporting issue may include that a higher rate of underreported crashes may be associated with rural or small-town areas where law enforcement may be understaffed; (*37*, *38*) and underreported hit-and-run crashes are likely to be property-damage-only crashes.

All crashes are linked to census tracts according to the crash geo-coded locations and the tract polygon edges. Some crashes occurred on road segments that are the boundaries of census tracts. Boundary crashes are randomly linked to the tracts on two sides of the boundary. Though the socio-economic statistics can be similar in two neighboring tracts under most circumstances, there are still some cases in which the socio-economics can be substantially different across a boundary. Therefore, the information about associated socio-economics may be inaccurate.

This study examined a limited number of variables and their relationships with hit-and-run crashes, for two reasons: (1) some variables that were investigated by other researchers are not available from the database used in this study; (2) the geo-spatial modeling is computationally intensive, and adding one variable in the model structure may exponentially increase the computation time. Therefore, the results presented in this study may be influenced by unobserved variables. Further research is needed to address this issue focusing on increasing model complexity while maintaining acceptable computational efficiency.

Conclusion

This study revisited hit-and-run crashes through a geo-spatial modeling approach, specifically, GWR. The study also demonstrated a way of scientific use of geo-referenced traffic safety data. The geo-spatial modeling method is useful for prioritizing specific regions/corridors for traffic safety improvements that outperforms traditional modeling techniques. This study compared the results from traditional regression and GWR models. Results from the traditional binary logit model are generally consistent with previous findings (*2*, *3*, *6–8*, *11*, *12*), whereas the key finding from GWR modeling is that the correlates of hit-and-run crashes significantly vary in space.

The results showed that hit-and-run is significantly associated with the crash severity, and more likely to occur when a crash is not severe. It is worthy to mention that the hit-and-run behavior and crash severity may interact as both cause and effect: (1) a driver leaves the scene based on his or her judgment that the crash was too trivial to report; and (2) crash severity may increase because of the delayed the emergency service resulting from hit-and-run, which was not found in this study but was reported in previous studies (*2*, *3*). In addition, hit-and-run was more likely to happen in two-vehicle or multi-vehicle sideswipe crashes, pedestrian or biker-involved crashes, crashes on dry roads, crashes on road segments with low speed limits, and weekend or nighttime crashes. Senior drivers were less likely to leave the sense of a crash, whereas drivers under drug or alcohol influence were bonded with a higher chance of leaving the scene immediately. Moreover, hit-and-run was more likely to occur in census tracts with a higher unemployment rates, a lower household income, a smaller portion of college-educated population, and a greater population density.

The spatially varying correlates of hit-and-run from GWR modeling could be visualized in maps. In addition to the stationary associations between factors and hit-and-run within the entire region, the GWR result maps also show the variations of the associations across different parts of the study region, which provides critical reference regarding prioritization of countermeasure implementation to mitigate hit-and-run crashes. For example, GWR results showed that DUI crashes are more likely to involve hit-and-run in the northwest of the study region (from the City of Farmington Hills, Brighton, to Howell) than those in other places. Therefore, the DUI enforcement or relevant education efforts may be recommended for this region. Further, GWR results revealed that in several areas, including the Cities of Taylor and Wyandotte (located at the south of Detroit downtown), the City of Chelsea and Imlay City, crashes that occurred in darkness without street lighting are associated with a higher likelihood of hit-and-run than such crashes in other places. Accordingly, adding street lighting may be an effective solution to reduce the occurrence of hit-and-run in these areas, if financially feasible. More local countermeasures can be developed based on the GWR results.

Although limited to the investigation of hit-and-run crashes, the methodology of this research has extended applicability as well as expanded values in other traffic safety research. The geo-spatial modeling process takes advantage of geo-referenced safety data and powerful computational tools, which nowadays are increasingly accessible by researchers and practitioners who are concerned about traffic safety. The modeling process and results could help researchers and practitioners to target traffic safety improvement strategies specifically for specific regions or corridors.

Author Contributions

The authors confirm contribution to the paper as follows: study conception and design: Liu, Khattak; data collection: Liu;

analysis and interpretation of results: Liu, Chen, Wan, Ma, Hu; draft manuscript preparation: Liu, Khattak, Chen. All authors reviewed the results and approved the final version of the manuscript.

References

1. FindLaw. *Leaving the Scene of an Accident/Hit and Run: State Laws*. http://traffic.findlaw.com/traffic-tickets/leaving-the-scene-of-an-accident-hit-and-run-state-laws.html. Accessed April 16, 2017.
2. Tay, R., U. Barua, and L. Kattan. Factors Contributing to Hit-and-Run in Fatal Crashes. *Accident Analysis & Prevention*, Vol. 41, No. 2, 2009, pp. 227–233.
3. Zhang, G., G. Li, T. Cai, D. M. Bishai, C. Wu, and Z. Chan. Factors Contributing to Hit-and-Run Crashes in China. *Transportation Research Part F: Traffic Psychology and Behavior*, Vol. 23, 2014, pp. 113–124.
4. National Highway Traffic Safety Administration. *2015 Motor Vehicle Crashes: Overview*. DOT HS 812 318. U.S. Department of Transportation, Washington, D.C., 2016.
5. National Highway Traffic Safety Administration. *2015 Traffic Safety Facts, Pedestrians*. DOT HS 812 375, U.S. Department of Transportation, Washington, D.C., 2017. https://crashstats.nhtsa.dot.gov/Api/Public/Publication/812375. Accessed February 4, 2017.
6. Solnick, S. J., and D. Hemenway. The Hit-and-Run in Fatal Pedestrian Accidents: Victims, Circumstances and Drivers. *Accident Analysis & Prevention*, Vol. 27, No. 5, 1995, pp. 643–649.
7. Tay, R., S. M. Rifaat, and H. C. Chin. A Logistic Model of the Effects of Roadway, Environmental, Vehicle, Crash and Driver Characteristics on Hit-and-Run Crashes. *Accident Analysis & Prevention*, Vol. 40, No. 4, 2008, pp. 1330–1336.
8. Aidoo, E. N., R. Amoh-Gyimah, and W. Ackaah. The Effect of Road and Environmental Characteristics on Pedestrian Hit-and-Run Accidents in Ghana. *Accident Analysis & Prevention*, Vol. 53, 2013, pp. 23–27.
9. Jiang, C., L. Lu, S. Chen, and J. J. Lu. Hit-and-Run Crashes in Urban River-Crossing Road Tunnels. *Accident Analysis & Prevention*, Vol. 95, 2016, pp. 373–380.
10. Kim, K., P. Pant, and E. Yamashita. Hit-and-Run Crashes: Use of Rough Set Analysis with Logistic Regression to Capture Critical Attributes and Determinants. *Transportation Research Record: Journal of the Transportation Research Board*, 2008. 2083: 114–121.
11. MacLeod, K. E., J. B. Griswold, L. S. Arnold, and D. R. Ragland. Factors Associated with Hit-and-Run Pedestrian Fatalities and Driver Identification. *Accident Analysis & Prevention*, Vol. 45, 2012, pp. 366–372.
12. Frencha, M. T., and G. Gumus. Hit-and-Run or Hit-and-Stay: Do Stricter BAC Limits Encourage Drivers to Flee the Crash Scene? *Proc., 2015 Fall Conference: The Golden Age of Evidence-Based Policy*, Miami, FL, 2015.
13. Hopkins, M., S. Chivers, and G. Stevenson-Freer. *Hit-and-Run: Why Do Drivers Fail to Stop after an Accident? Contexts of Incidents, Driver Motivations and Preventative Strategies*. https://www.mib.org.uk/media/350114/hit-and-run-why-do-drivers-fail-to-stop-after-an-accident.pdf. Accessed May 23, 2017.
14. Graham, D. J., and S. Glaister. Spatial Variation in Road Pedestrian Casualties: The Role of Urban Scale, Density and Land-Use Mix. *Urban Studies*, Vol. 40, No. 8, 2003, pp. 1591–1607.
15. Meliker, J. R., R. F. Maio, M. A. Zimmerman, H. M. Kim, S. C. Smith, and M. L. Wilson. Spatial Analysis of Alcohol-Related Motor Vehicle Crash Injuries in Southeastern Michigan. *Accident Analysis & Prevention*, Vol. 36, No. 6, 2004, pp. 1129–1135.
16. Noland, R. B., and M. A. Quddus. Congestion and Safety: A Spatial Analysis of London. *Transportation Research Part A: Policy and Practice*, Vol. 39, No. 7, 2005, pp. 737–754.
17. Quddus, M. A. Modelling Area-Wide Count Outcomes with Spatial Correlation and Heterogeneity: An Analysis of London Crash Data. *Accident Analysis & Prevention*, Vol. 40, No. 4, 2008, pp. 1486–1497.
18. Wang, X., J. Liu, A. J. Khattak, and D. Clarke. Non-Crossing Rail-Trespassing Crashes in the Past Decade: A Spatial Approach to Analyzing Injury Severity. *Safety Science*, Vol. 82, 2016, pp. 44–55.
19. Liu, J., A. J. Khattak, and B. Wali. Do Safety Performance Functions Used for Predicting Crash Frequency Vary Across Space? Applying Geographically Weighted Regressions to Account for Spatial Heterogeneity. *Accident Analysis & Prevention*, Vol. 109, 2017, pp. 132–142.
20. Zhang, M., A. J. Khattak, J. Liu, and D. Clarke. A Comparative Study of Rail-Pedestrian Trespassing Crash Injury Severity Between Highway-Rail Grade Crossings and Non-Crossings. *Accident Analysis & Prevention*, Vol. 117, 2018, pp. 427–438.
21. Liu, J., A. Khattak and X. Wang. A Comparative Study of Driving Performance in Metropolitan Regions Using Large-scale Vehicle Trajectory Data: Implications for Sustainable Cities. *International Journal of Sustainable Transportation*, Vol. 11, 2017, pp. 170–185.
22. Hadayeghi, A., A. S. Shalaby, and B. N. Persaud. Development of Planning Level Transportation Safety Tools Using Geographically Weighted Poisson Regression. *Accident Analysis & Prevention*, Vol. 42, No. 2, 2010, pp. 676–688.
23. Wang, Y., K. Kockelman, and X. Wang. Anticipation of Land Use Change through Use of Geographically Weighted Regression Models for Discrete Response. *Transportation Research Record: Journal of the Transportation Research Board*, 2011. 2245: 111–123.
24. Li, Z., W. Wang, P. Liu, J. M. Bigham, and D. R. Ragland. Using Geographically Weighted Poisson Regression for County-Level Crash Modeling in California. *Safety Science*, Vol. 58, 2013, pp. 89–97.
25. Pirdavani, A., T. Bellemans, T. Brijs, and G. Wets. Application of Geographically Weighted Regression Technique in Spatial Analysis of Fatal and Injury Crashes. *Journal of Transportation Engineering*, Vol. 140, No. 8, 2014, pp. 14–32.
26. Liu, J., X. Wang, A. J. Khattak, J. Hu, J. Cui, and J. Ma. How Big Data Serves for Freight Safety Management at

Highway-Rail Grade Crossings? A Spatial Approach Fused with Path Analysis. *Neurocomputing*, Vol. 181, 2016, pp. 38–52.

27. Southeast Michigan Council of Governments. *Open Data Portal*. http://www.semcog.org/. Accessed May 15, 2017.
28. Census. *TIGER/Line with Selected Demographic and Economic Data*. 2017. https://www.census.gov/geo/maps-data/data/tiger-data.html. Accessed June 20, 2017.
29. StataCorp. *Stata User Guide*. 2013. http://www.stata.com/manuals13/u.pdf. Accessed April 5, 2015.
30. Fotheringham, A. S., M. E. Charlton, and C. Brunsdon. Geographically Weighted Regression: A Natural Evolution of the Expansion Method for Spatial Data Analysis. *Environment and Planning A*, Vol. 30, No. 11, 1998, pp. 1905–1927.
31. Brunsdon, C., M. Aitkin, S. Fotheringham, and M. Charlton. A Comparison of Random Coefficient Modelling and Geographically Weighted Regression for Spatially Non-Stationary Regression Problems. *Geographical and Environmental Modelling*, Vol. 3, No. 1, 1999, pp. 47–62.
32. Liu, J., and A. J. Khattak. Gate-Violation Behavior at Highway-Rail Grade Crossings and the Consequences: Using Geo-Spatial Modeling Integrated with Path Analysis. *Accident Analysis & Prevention*, Vol. 109, 2017, pp. 99–112.
33. Bozdogan, H. Model Selection and Akaike's Information Criterion (AIC): The General Theory and Its Analytical Extensions. *Psychometrika*, Vol. 52, 1987, pp. 345–370.
34. Ye, F., and D. Lord. Investigation of Effects of Underreporting Crash Data on Three Commonly Used Traffic Crash Severity Models: Multinomial Logit, Ordered Probit, and Mixed Logit. *Transportation Research Record: Journal of the Transportation Research Board*, 2011. 2241: 51–58.
35. National Highway Traffic Safety Administration. *National Telephone Survey of Reported and Unreported Motor Vehicle Crashes*. DOT HS 812 183. U.S. Department of Transportation, Washington, D.C., 2015.
36. Michigan Department of State Police. *Traffic Crash Report Instruction Manual*. 2014. https://www.michigan.gov/documents/msp/UD10_Manual_Plus_Electronic_2014_Revision_441848_7.pdf. Accessed June 30, 2015.
37. Liu, J., A. Khattak and M. Zhang. What Role do Precrash Driver Actions Play in Work one Crashes? Application of Hierarchical Models to Crash Data. *Transportation Research Record: Journal of Transportation Research Board*, 2016. 2555: 1–11.
38. Liu, J., A. Khattak, S. Richards and S. Nambisan. What are the Differences in Driver Injury Outcomes at Highway-Rail Grade Crossings? Untangling the Role of Pre-Crash Behaviors. *Accident Analysis & Prevention*, Vol. 85, 2015, pp. 157–169.

The Standing Committee on Safety Data, Analysis and Evaluation (ANB20) peer-reviewed this paper (18-03009).

Article

TRR
JOURNAL OF THE TRANSPORTATION RESEARCH BOARD

Transportation Research Record
2018, Vol. 2672(38) 93–104

DOI: 10.1177/0361198118782763
journals.sagepub.com/home/trr
SAGE

Comprehensive Assessment of Temporal Treatments in Crash Prediction Models

Gurdiljot Singh Gill[1], Wen Cheng[1], Jiao Zhou[1], and Xudong Jia[1]

Abstract

This study conducted a comprehensive comparison of temporal treatments employed in crash prediction models. Nine groups of methodological approaches based on different ways of addressing temporal correlations, including the newly proposed time adjacency matrix, were developed. Moreover, three types of models were developed for each group in terms of spatial dependency. Finally, ten different assessment criteria were utilized for the evaluation purpose. All models and performance-checking criteria applied to 8 years of county-level crash counts in California. The modeling results illustrated that the space–time models consistently enhanced the precision associated with the intercepts. The serial and spatial correlations also appeared to be statistically significant. In terms of model complexity, the models with spatial correlations outperformed the ones without considering spatially structured heterogeneity, and the models accounting for the temporal dependency revealed more benefits compared with those without temporal treatments. The opposite trends were found by prediction-pertinent criteria based on the aggregation results, even though the first-order autoregressive process space–time models with spatiotemporal interaction claimed the first place of prediction in most cases. The correlation analysis among all ten criteria illustrated that the efficiency in reducing the effective number of parameters tended to have larger impacts on the value of deviance information criterion than did the mean deviance, which demonstrated the statistically significant correlations with all other prediction-related measures.

Traditional crash prediction models often disaggregated the crash counts into small time intervals for different reasons such as enlarging the dataset to circumvent the issue of small sample size or incorporating the information of time-varying factors. Such division introduced a potential bias, as the multiple crash observations of an entity may be correlated over time (*1*). The bias may be addressed by the accommodation of serial correlations, which were noted to enhance performance at model fit and precision of estimates in roadway crash studies (*2, 3*).

The safety literature is replete with studies pertaining to different treatments of the temporal aspect of crash data. Wang and Abdel-Aty developed four models with different types of temporal treatment, namely: independent, exchangeable, autoregressive (AR), and unstructured (*4*). The comparison of different correlation structures demonstrated the superior fit associated with the autoregressive structure. Huang et al. proposed the full Bayesian hierarchical approach for identification of hot spots by employing the same AR-1 model, along with five others (*5*). The best fit of the AR-1 model demonstrated the benefit of serial correlation to capture a time-dependent safety effect. The segment-level study by Aguero-Valverde compared the fixed- and independent-over-time random effects (*6*). The notable advantage at model fit, more precise estimates, and consistent site-ranking performance associated with fixed-over-time indicated that such temporal treatment allowed more flexibility for model parameters. The subsequent study by El-Basyouny et al. developed four models which included with and without linear time trend, yearly varying intercept, and yearly varying coefficients. The last one demonstrated the best performance at model fit (*7*).

Apart from the above studies focused on temporal correlation, another dimension of crash prediction models incorporated spatial correlation structures to account for the spatial dependency of crash counts among neighboring sites (*8, 9*). Such correlation structures allow the flexibility to "pool strength" from neighboring entities and incorporate the unmeasured spatial confounding factors (*10*). Many studies observed a significant improvement of model fit, crash prediction, and site ranking

[1]Department of Civil Engineering, California State Polytechnic University, Pomona, CA

Corresponding Author:
Address correspondence to Wen Cheng: wcheng@cpp.edu

performance with the inclusion of spatial correlation (*4, 11–13*).

Some studies developed spatiotemporal models to further extended the benefits of accommodating space and time. The study by Wang and Abdel-Aty incorporated independent spatial and temporal correlations for rear-end crashes among intersections for 3-year longitudinal crash data (*4*). The modeling results revealed the presence of high correlations between the longitudinal and spatially correlated rear-end crashes. A recent study by Cheng et al. developed two spatiotemporal models, one with a linear time trend and another with time-varying coefficients (*14*). Statistically significant spatial correlation was observed among crash types of intersections. Concerning the temporal treatment, the time-varying coefficient model performed better at model fit, but the relatively less complex linear trend model was superior at crash prediction.

Some studies allowed the interaction of space and time, as roadway crashes may cluster both spatially and temporally. Based on the formulation proposed by Bernardinelli et al. (*15*), Aguero-Valverde and Jovanis applied a spatiotemporal model to county-level fatal and injury crashes with the interaction of linear time trend and spatial correlation (*11*). The random effect terms for spatial correlation, time trend, and space–time interactions were observed to be statistically significant for the injury crashes. Another spatiotemporal study by Dong et al. for hot spot identification at traffic analysis zone level analyzed area-specific crash trends (*16*). The space–time interaction model (linear trend with spatial) significantly outperformed the independent space–time model (AR-1 with spatial). A more recent study by Ma et al. explored a finer temporal scale of daily crash data for estimation of injury severity levels of segment crashes (*17*). The proposed spatiotemporal models were observed to be superior to other models which accounted only for heterogeneity among crash types and spatial correlations.

As is evident from the literature review, although some studies have attempted the comparison among few temporal models, a comprehensive comparison of temporal treatments of crash data is not well documented. To this end, the authors developed different models which represent various ways of addressing the serial correlations. Specifically, the models were classified into nine groups: (1) independent-over-time (serving as the base); (2) linear time trend; (3) quadratic time trend; (4) yearly varying intercept; (5) autoregressive-1 (AR-1); (6) AR-2; (7) moving average-1 (MA-1); (8) MA-2; and (9) time adjacency. These models (except the last one) were selected as representative of the temporal models commonly employed in the safety literature. The time adjacency model is the first proposed in the traffic safety field and is borrowed from a previous disease mapping study (*18*). This model marks the innovative contribution of this study toward the field of traffic safety and is proposed with the intention to compare its performance relative to the prevailing temporal models. In addition, the research developed three types of models for each temporal treatment group which contains: (1) temporal correlation only; (2) independent temporal and spatial correlations; (3) temporal and spatial correlations with spatiotemporal interactions. Furthermore, ten different evaluation criteria were employed to observe the modeling performance of 27 models from different perspectives.

Methodology

Bayesian Hierarchical Model Specification

At the first level, the authors defined a Poisson model for the within-county variability of crash counts:

$$y_{it} \sim Poisson(e_{it}\lambda_{it}) \tag{1}$$

Where y_{it} is the observed crash count for county i and year t, λ_{it} is associated mean expected crash rate, and e_{it} is the corresponding exposure, which is the total daily vehicle-miles (DVMT) by county in this study.

At the second level, λ_{it} can be split into portions explained by overall risk (β_0), main temporal effects (ξ_t) and spatial effects (π_i, applies merely to model types 2 & 3), space–time interactions (υ_{it} applies only to model type 3) and random errors (ε_{it}), which can be assumed with the noninformative normal priors. As a result, λ_{it} can be expressed as follows:

$$\begin{cases} \ln(\lambda_{it}) = \beta_0 + \xi_t + \pi_i + \upsilon_{it} + \varepsilon_{it} \\ \varepsilon_{it} \sim Normal(0, \tau^2) \end{cases} \tag{2}$$

It should be noted that generally the model development incorporates some potential influential factors. However, this study did not incorporate any covariates for multiple reasons. First, the inclusion of various factors might blur the comparison between the model performance of temporal effects. Second, some spatial-only covariates would have different impacts on the models with and without spatially structured random effects and thus potentially yield biased comparative results. Third, the exclusion of covariates would allow enough heterogeneity to support the estimation of various random effects, especially the space–time interaction one.

At the third level, the various parameters that are involved in Equation 2 are generally treated with unknown and hyper-prior distributions. As with many other Bayesian analyses (*19*), the authors chose inverse gamma for τ^2 with parameters 0.5 and 0.0005.

Nine Groups of Temporal Treatments

Group 1: Independent-Over-Time Random Effect (Base). In this group, all crash counts of different years are assumed to be independent of each other, without any applicable serial trends or correlations (*20*). Equation 2 is reformed using the following expression:

$$\ln(\lambda_{it}) = \beta_0 + \pi_i + \upsilon_{it} + \varepsilon_{it} \tag{3}$$

Group 2: Linear Trend. Under this group, a global linear temporal trend is applied to crash counts of all counties. Equation 2 is modified to the following form:

$$\ln(\lambda_{it}) = \beta_0 + \beta_{t1}t + \pi_i + \upsilon_{it} + \varepsilon_{it} \tag{4}$$

Where β_{t1} is the scalar parameter for linear yearly trend, and t represents various years. The crash counts of counties are anticipated to be decreased or increased at the same rate with time going.

Group 3: Quadratic Trend. Within this group, an additional quadratic term is applied to crash counts in addition to the linear trend. The underlying premise is that the quadratic term can capture some heterogeneity which escapes from the linear term (*21*). The resulting expression is shown as follows:

$$\ln(\lambda_{it}) = \beta_0 + \beta_{t1}t + \beta_{t2}t^2 + \pi_i + \upsilon_{it} + \varepsilon_{it} \tag{5}$$

Where β_{t2} is the scalar parameter for quadratic yearly trend.

Group 4: Time-Varying Intercepts. This group assumes that the crashes of differing years have a time-changing overall risk (intercepts). The corresponding formulation is expressed in Equation 6:

$$\ln(\lambda_{it}) = \beta_{0t} + \pi_i + \upsilon_{it} + \varepsilon_{it} \tag{6}$$

Where β_{0t} is the vector of yearly varying intercepts. Essentially, it can be regarded as a "random parameter" model in terms of intercepts) (*7*). It is worth mentioning that other covariates are not utilized. Otherwise, the associated variable coefficients might be time-varying as well.

Group 5: First-Order Autoregressive Process (AR-1). This group takes into the consideration the AR safety effect by relating the crashes of current year (t) to the ones from previous year (t − 1) with one certain coefficient (*5*, *20*). The study chose AR-1 based on the assumption of stationarity restriction. Under this formulation, the distributions of ε_{it} in Equation 2 are given by:

$$\begin{cases} \varepsilon_{it} \sim normal\left(0, \frac{\sigma_{it}^2}{(1-\varnothing_1^2)}\right) \\ \varepsilon_{it} \sim normal\left(\varnothing_1 \varepsilon_{i,t-1}, \sigma_{it}^2\right), t>1 \end{cases} \tag{7}$$

Where $\varnothing_1$ is the autocorrelation coefficient with the range of $-1 < \varnothing_1 < 1$ (*22*).

Group 6: Second-Order Autoregressive Process (AR-2). Different from AR-1, AR-2 assumes that the crashes of the present year are dependent on the ones of two previous years. Correspondingly, the distributions of ε_{it} are shown in the following expression:

$$\begin{cases} \varepsilon_{it} \sim normal\left(0, \frac{\sigma_{it}^2}{(1-\varnothing_2)^2-\varnothing_1^2} \cdot \frac{1-\varnothing_2}{1+\varnothing_2}\right) \\ \varepsilon_{it} \sim normal\left(\varnothing_1 \varepsilon_{i,t-1} + \varnothing_2 \varepsilon_{i,t-2}, \sigma_{it}^2\right), t>2 \end{cases} \tag{8}$$

Where $\varnothing_1$ and $\varnothing_2$ are the autocorrelation coefficients. An AR-2 process is stationary provided $|\varnothing_2| < 1, and |\varnothing_1| + \varnothing_2 < 1$ (*22*).

Group 7: First-Order Moving Average Process (MA-1). Under this group, the crashes of the present year are a linear function of errors of past year. Specifically, MA-1 takes the form as follows:

$$\begin{cases} \ln(\lambda_{it}) = \beta_0 + \pi_i + \upsilon_{it} + \varepsilon_{it} + \theta_1 \cdot \varepsilon_{i,t-1} \\ \varepsilon_{it}, \varepsilon_{i,t-1} \sim Normal(0, \sigma^2) \end{cases} \tag{9}$$

Where θ_1 is the coefficient associated with error of past year. The error terms of current and past years are assumed to be independently distributed with a normal distribution with mean zero and a constant variance σ^2. MA-1 process is invertible provided $|\theta_1| < 1$ (*23*).

Group 8: Second-Order Moving Average Process (MA-2). An MA-1 process can be easily extended to MA-2 where crashes of present year are related to errors of the past two years. The resulting expression is shown as below:

$$\begin{cases} \ln(\lambda_{it}) = \beta_0 + \pi_i + \upsilon_{it} + \varepsilon_{it} + \theta_1 \cdot \varepsilon_{i,t-1} + \theta_2 \cdot \varepsilon_{i,t-2} \\ \varepsilon_{it}, \varepsilon_{i,t-1}, \varepsilon_{i,t-2} \sim Normal(0, \sigma^2) \end{cases} \tag{10}$$

Where θ_1 and θ_2 are the coefficients associated with errors of the past two years. Again, the error terms of current and past two years are assumed to be independently distributed with a normal distribution with mean zero and a constant variance σ^2. MA-2 process is invertible provided $|\theta_j| < 1, j = 1, 2$ (*23*).

Group 9: Time Adjacency. As indicated by the name, this group assumes strong influence of the adjacent years on crash risk, similar to the dependency of adjacent roadway entities in the case of spatial correlation. Compared with the AR or MA models which capture the impact of crash risk of the year(s) prior to the concerned time, this model also accounts for the potential impact of the subsequent year(s). To the best knowledge of the authors, such formulation has not yet been applied in the traffic safety field. This treatment is borrowed from a previous study regarding public health issues (*18*) with the intention to explore its capabilities in analyzing safety data.

Similar to many other spatial correlation studies, the authors estimated temporal effects (ξ_t) in Equation 2 by applying conditional autoregressive (CAR) formulation, which incorporates the temporal correlation among crashes occurring at neighboring years. The details of CAR can be found in the following subsection. It is important to note that the relatively small sample size ($t = 8$) might harm the performance of this group to some degree.

Three Types of Models

Aside from the temporal treatment, the 27 models can also be divided into three types in terms of different ways of handling the spatial correlations among counties. Each of the model types is applied to all groups of temporal treatments. The detailed description is provided below in order.

Type 1: Temporal Correlation Only. This type of model assumes that the spatial feature of the crash risk at other counties has ignorable impacts on the crashes of concerned county. In response, Equation 2 is modified by removing the terms of π_i and υ_{it}:

$$\ln(\lambda_{it}) = \beta_0 + \xi_t + \varepsilon_{it} \tag{11}$$

Type 2: Space–Time Model without Spatiotemporal Interactions. Different from Type 1, this model type also accounts for the spatial correlations among the crash risks of counties, but the associated spatial and temporal interactions are not considered. Therefore, the updated Equation 2 is shown as below for this type:

$$\ln(\lambda_{it}) = \beta_0 + \xi_t + \pi_i + \varepsilon_{it} \tag{12}$$

In this study, the authors selected the commonly used CAR in the safety field with the distance-based weight structure (*6*, *10–13*, *21*). The specific formulation is shown in Equation 13:

$$\pi_i \sim N\left(\frac{\overline{\pi_i}, 1}{\tau_i}\right) \quad \overline{\pi_i} = \frac{\sum_{i \neq k} \pi_k w_{ik}}{\sum_{i \neq k} w_{ik}} \quad \tau_i = \frac{\tau_c}{\sum_{i \neq k} w_{ik}} \quad and\ w_{ij} = \frac{1}{d_{ij}} \tag{13}$$

Where w_{ij} is the proximity weight matrix, τ_c is the precision parameter in the CAR prior, and d_{ij} is the distance between counties. As shown in the above equation, estimation of the risk in any county is conditional on risks in other locations. Subscripts i and k refer to a county and its neighbor, respectively, and $k \varepsilon N_i$, where N_i represents the set of neighbors of county i. The current study employs the distance-based structure to explore the spatial correlation, where all other counties are considered neighboring counties and relative influence is inversely related with distances in between. It is noteworthy that, for the serial correlation in the time adjacency group, the authors chose adjacency-based weight structure where the weights are given by $w_{ij} = 1$ if counties i, j are adjacent, and 0 otherwise. Therefore, the proximity of immediate neighbors establishes a temporal dependency while more distant years are considered not to have a significant impact.

Once π_i is estimated, it is also interesting to calculate the percentage of extra-Poisson variability that is caused by spatial clustering:

$$\alpha = \frac{sd(\pi)}{sd(\pi) + sd(\varepsilon)} \tag{14}$$

Where α is defined as fraction, and sd is the marginal standard deviation function. The larger the fraction value, the more variability explained by the spatially structured random effects.

Type 3: Space–Time Model with Spatiotemporal Interactions. Type 3 is distinct from Type 2 with the inclusion of interactions between temporal and spatial correlations. This research chose the linear interaction that has been used in a previous safety study (*11*). In this case, Equation 2 is changed with the following form:

$$\ln(\lambda_{it}) = \beta_0 + \xi_t + \pi_i + \delta_i \cdot T + \varepsilon_{it} \tag{15}$$

Where T is the vector representing various time periods and δ_i represents the interacted spatial correlations which are also estimated with CAR formulation.

Analogously, the updated formula for the fraction is shown in Equation 16:

$$\alpha = \frac{\mathrm{sd}(\pi) + \mathrm{sd}(\delta)}{\mathrm{sd}(\pi) + \mathrm{sd}(\delta) + \mathrm{sd}(\varepsilon)} \tag{16}$$

The fraction, in this case, indicates the proportion of crash variability explained by both spatial effects: π_i and δ_i.

Evaluation Criteria

The brief background and analytics of each criterion are provided as follows.

Criteria 1–3: Deviance Information Criterion and Its Components. Deviance information criterion (DIC) is a penalized fit criterion developed by Spiegelhalter et al. (*24*). Usually, the estimation of DIC can be conducted through the following expression:

$$\begin{cases} DIC = \bar{D} + P_d \\ D(\theta) = -2 \cdot log(p(y|\theta)) \\ \bar{D} = E^{\theta}[D(\theta)], P_d = \bar{D} - D(\bar{\theta}) \end{cases} \quad (17)$$

Where y are the data; θ are the unknown parameters of the model; $p(y|\theta)$ is the likelihood function; $\bar{\theta}$ is the expectation of θ; $\bar{D}$ is the posterior mean deviance, or the expectation of the deviance, $D(\theta)$; $\bar{D}$ measures how well the model fits the data; the smaller the $\bar{D}$, the better the fit. P_d represents the effective number of parameters. Usually, smaller values of DIC are preferred. Some general guidelines are also suggested by Lunn et al. (*25*): models with DIC values within 5 of the "best" model are comparable and also strongly supported, values within 5 and 10, weakly supported, and models with a DIC greater than 10 are substantially inferior.

Criterion 4: Log Pseudo Marginal Likelihoods (LPML). This criterion checks the predictive capability of models based on a special case of cross-validation, leave-one-out cross-validation, where the number of folds equals the number of instances in the dataset. It has the benefit of avoiding selection bias in the sense that it treats each data observation as a single-item test set while using all other instances as a training set. Calculation of LPML can be done through the following expression (*26*, *27*):

$$\begin{cases} CPO = \left(\frac{1}{T}\sum_{t=1}^{T}\frac{1}{p(Y_i|\beta^{(t)})}\right)^{-1} \\ LPML = \sum_{i=1}^{n}\log(CPO_i) \end{cases} \quad (18)$$

Where CPO is conditional predictive ordinate; Y_i is the ith observation ($i = 1, 2, 3, \ldots, n$) for all counties; β is the vector of estimated model parameters; $p(Y_i|\beta^{(t)})$ is the likelihood; t represents interaction. The larger the LPML value, the better the predictive performance tends to be.

Criterion 5: Mean Absolute Deviation (MAD). MAD is also frequently used by researchers to check the fitness of data irrespective of the data distribution. It is merely based on the model deviation or residue. The corresponding expression is shown in Equation 19:

$$MAD = \frac{1}{n}\sum_{i=1}^{n}|Y_i - O_i| \quad (19)$$

Where Y_i is the Bayesian-estimated crash frequency and O_i is the observed crash count for county i by a model during the same time period. The smaller the MAD value, the better fitness to the data.

Criterion 6: Mean Square Predictive Error (MSPE). MSPE differs from MAD by calculating the square of deviation, rather than the corresponding absolute value. Specifically, MSPE assumes the form shown below:

$$MSPE = \frac{1}{n}\sum_{i=1}^{n}(Y_i - O_i)^2 \quad (20)$$

Where terms are as defined previously. It is expected that MSPE assigns more penalties to the counties whose deviations are larger as a result of the squared deviation. Again, the larger MSPE indicates an inferior performance.

Criterion 7: The R_p^2 Statistic. A Poisson regression model is usually associated with the nonlinearity of conditional mean (E[y|**X**]) and heteroscedasticity (*28*). Therefore, R-square in ordinary linear regression cannot be directly applied to the crash frequency model. Instead, the present research chose an equivalent measure, R_p^2, which is based on standardized residuals and computed as follows:

$$R_p^2 = 1 - \frac{\sum_{i=1}^{n}\left[\frac{O_i - Y_i}{\sqrt{Y_i}}\right]^2}{\sum_{i=1}^{n}\left[\frac{O_i - \bar{O}}{\sqrt{\bar{O}}}\right]^2} \quad (21)$$

Where Y_i and O_i are as defined previously, and $\bar{O}$ represents the mean value of the observed counts. Similar to R-square, a larger R_p^2 value is preferred.

Criterion 8: The G^2 Statistic. The sum of model deviances, G^2, gives a test of whether the model gives an adequate explanation of the data relative to the saturated model (*29*). The G^2 statistic is given as:

$$G^2 = 2\sum_{i=1}^{n}O_i LN\left(\frac{O_i}{Y_i}\right) \quad (22)$$

A large G^2 indicates that the model fits poorly as compared with the saturated model.

Criterion 9: Chi-Squared Residual Sum of Squares (RSS). RSS is defined as:

$$RSS = \sum_{i=1}^{n}\frac{(O_i - Y_i)^2}{Y_i} \quad (23)$$

Under MAD and MSPE, the larger counties are expected to subject to more penalties as a result of greater counts and residues. RSS (*30*) tends to remove such bias by

Table 1. Descriptive Statistics of Variables for Various Counties in California

Variables	Description	Year	Min	Max	Median	Mean	S.D.
Collision	Total annual fatal and injury vehicle-involved collisions	2006	23	48,107	758	2,791	6,767
		2007	10	46,558	699	2,672	6,515
		2008	15	41,794	631	2,389	5,841
		2009	20	40,197	611	2,290	5,612
		2010	18	39,560	538	2,249	5,531
		2011	14	38,933	576	2,185	5,430
		2012	16	38,477	560	2,171	5,389
		2013	21	38,855	544	2,141	5,437
DVMT	Daily vehicle-miles traveled (in thousand)	2006	180	217,444	5,092	15,577	32,178
		2007	162	218,027	5,146	15,608	32,261
		2008	168	214,971	5,005	15,387	31,894
		2009	171	214,237	4,837	15,318	31,744
		2010	169	211,877	5,449	15,483	31,421
		2011	165	214,458	4,762	15,353	31,871
		2012	167	214,482	4,789	15,438	31,927
		2013	165	215,818	4,791	15,549	32,162
Distance	Distance among centroids of counties (unit: miles)	N/A	25	962	227	273	176

Note: S.D. = standard deviation; min = minimum; max = maximum; N/A = not applicable.

calculating the squared residual relative to the estimated number of crashes. A particular model is considered more reliable if smaller RSS value is observed.

Criterion 10: Total Rank Difference (TRD). Finally, rather than use the magnitude of residue value, TRD accounts for the rank deviations based on the observed and estimated crash counts (*14*). The rank difference is calculated by using the following equation:

$$TRD = \sum_{i=1}^{n} |R(i_o) - R(i_Y)| \quad (24)$$

Where $R(i_o)$ is the observed data rank at county i and $R(i_Y)$ is the rank based on estimated crash counts for the same time period. A model is considered superior if smaller TRD value is revealed, which signifies that the specific model assigns rankings close to the observed crash counts.

Data Preparation

For ease of comparison of various temporal treatments, only the number of fatal and injury vehicle-involved collisions and a main exposure-related yearly varying factor of county safety performance, that is, DVMT, were collected for analysis. The first one was obtained from the website of TIMS, which provides tools for accessing and mapping collision data from the California Statewide Integrated Traffic Records System. DVMT was collected from Highway Performance Monitoring System. Eight years of data, which span from 2006 to 2013, were used for both variables. In addition, for implementation of distance-based CAR priors for spatial correlations, a matrix containing distances among centroids of all counties in California was developed based on the information provided by Southern California of Association of Governments (SCAG). The descriptive statistics of variables are illustrated in Table 1.

Results

The three sets and nine groups of models were implemented in freeware WinBUGS package (*24*). For model calibration, two Markov chains were utilized to visually inspect trace plots of posterior estimates for convergence.

Modeling Results

The posterior mean and standard deviation for the different model coefficient estimates are illustrated in Table 2. The intercept term (β_0) is observed to be statistically significant for all the groups, and the impact of inclusion of spatial correlations may be noted on closer scrutiny with the standard deviation for β_0 consistently decreasing across all groups for Type 2 and Type 3, compared with Type 1. The fraction α is observed to be statistically significant for all spatiotemporal models, which reflects the presence of spatial clustering of crashes. This also bolsters the loss of precision noticed for intercept of Type 1 across all groups. Apart from the spatial correlation, the time dependency also appears to be statistically significant, as indicated by the different temporal treatment model terms such as β_{t1}, β_{t2}, φ_1, φ_2, θ_1, θ_2, etc.

Table 2. Parameter Estimates and Associated Standard Deviation for Various Models

Parameters	Type 1	Type 2	Type 3
Group 1: independent-over-time random effect (base)			
β_0	−2.068 (0.0140)	−2.072 (0.0068)	−2.114 (0.0119)
α		0.678 (0.0082)	0.654 (0.0095)
Group 2: linear trend			
β_0	−1.893 (0.0304)	−1.902 (0.0111)	−1.901 (0.0109)
β_{t1}	−0.0379 (0.0061)	−0.0379 (0.0022)	−0.0382 (0.0020)
α		0.754 (0.0075)	0.754 (0.0076)
Group 3: quadratic trend			
β_0	−1.871 (0.0368)	−1.825 (0.0207)	−1.838 (0.0186)
β_{t1}	−0.053 (0.0167)	−0.084 (0.0107)	−0.076 (0.0097)
β_{t2}	0.002 (0.0018)	0.005 (0.0012)	0.004 (0.0011)
α		0.759 (0.0075)	0.781 (0.0106)
Group 4: time-varying intercept			
β_0^1	−1.907 (0.0383)	−1.914 (0.0132)	−1.913 (0.0127)
β_0^2	−1.941 (0.0386)	−1.950 (0.0129)	−1.950 (0.0124)
β_0^3	−1.941 (0.0386)	−1.950 (0.0129)	−1.950 (0.0124)
β_0^4	−2.053 (0.0384)	−2.069 (0.0137)	−2.070 (0.0132)
β_0^5	−2.115 (0.0410)	−2.153 (0.0144)	−2.152 (0.0132)
β_0^6	−2.113 (0.0374)	−2.123 (0.0134)	−2.123 (0.0131)
β_0^7	−2.134 (0.0384)	−2.143 (0.0143)	−2.144 (0.0128)
β_0^8	−2.169 (0.0422)	−2.187 (0.0137)	−2.189 (0.0129)
α		0.763 (0.0075)	0.787 (0.0109)
Group 5: AR-1			
β_0	−2.028 (0.0147)	−2.057 (0.0064)	−2.052 (0.0132)
φ_1	−0.401 (0.0221)	−0.491 (0.0384)	−0.456 (0.0392)
α		0.677 (0.0078)	0.618 (0.0161)
Group 6: AR-2			
β_0	−2.093 (0.0203)	−2.051 (0.0118)	−2.075 (0.0116)
φ_1	1.013 (0.0519)	1.103 (0.0438)	1.191 (0.0239)
φ_2	−0.527 (0.0242)	−0.418 (0.0438)	−0.540 (0.0294)
α		0.631 (0.0438)	0.621 (0.0219)
Group 7: MA-1			
β_0	−2.065 (0.0171)	−2.071 (0.0089)	−2.122 (0.0146)
θ_1	0.924 (0.0530)	0.700 (0.0915)	0.849 (0.0765)
α		0.721 (0.0122)	0.728 (0.0119)
Group 8: MA-2			
β_0	−2.049 (0.0182)	−2.067 (0.0106)	−2.135 (0.0189)
θ_1	0.976 (0.0219)	0.830 (0.0787)	0.955 (0.0331)
θ_2	0.627 (0.0449)	0.428 (0.0700)	0.514 (0.0394)
α		0.728 (0.0126)	0.753 (0.0126)
Group 9: time adjacency			
β_0	−2.068 (0.0160)	−2.073 (0.0093)	−2.084 (0.0062)
α		0.762 (0.0075)	0.786 (0.0101)

Note: The shaded cell indicates non-significant coefficient at 95% confidence level. The standard deviations are shown in the parentheses. Refer to the Methodology section for details of various parameters and groups/types of models.

This illustrates the strong serial correlation among crashes of successive years and the tendency of crashes to cluster both spatially and temporally, as reflected by the space–time interaction of Type 3. The importance of accounting for temporal changes is also evident from the time-varying intercept models (Group 4), which demonstrate the different magnitude of intercepts for 8 years. The parameter values in Table 2 can also be used to predict the annual crash counts of various counties. For example, the 8-year DVMT (in thousand) for Amador County are (1193.12, 1223.70, 1204.87, 1201.17, 1067.05, 1058.22, 1027.92, 1035.65), and the associated observed crash counts are (265, 233, 203, 204, 212, 180, 161, 142). Under the Type 1, Group 1 Model, $\beta_0 = -2.068$, the predicted crash counts ($DVMT{*}e^{\beta_0}$) would be (150.9, 154.8, 152.4, 152.0, 135.0, 133.9, 130.0, 131.0). However, if the estimated independent-over-time random effect, ε_{it} = (0.5384, 0.3885, 0.2696, 0.2767, 0.4273, 0.2772, 0.1978, 0.0716), is also included for calculation, the predicted values ($DVMT{*}e^{(\beta_0 + \varepsilon_{it})}$) would be (258.5, 228.7, 200.1, 200.9, 207.4, 177.1, 158.9, 141.2), which are much closer to the observed counts. Such phenomenon

Table 3. Evaluation Results of Various Models under Different Criteria

Group	Type	$\bar{D}$	P_d	DIC	LPML	MAD	MSPE	R_p^2	G^2	RSS	TRD
1	1	4328	443.5	4771	−2090.0	2.611	**10.814**	0.994	27.2	37.1	382
	2	4342.3	426.4	4768	−2094.4	4.755	42.192	0.989	35.4	68.5	571
	3	4328.5	465.3	4793	−2088.6	2.284	18.359	0.998	22.2	12.3	216
2	1	4331.3	442.2	4773	−2091.1	2.684	11.780	0.993	11.3	41.7	399
	2	4360.1	392.4	4752	−2102.2	5.974	76.870	0.982	102	116.7	862
	3	4369.6	393.4	4763	−2106.2	5.992	76.448	0.982	111	116.2	862
3	1	4331.5	442.3	4773	−2091.2	2.685	11.763	0.993	26.8	41.8	402
	2	4358.9	389.3	4748	−2102.0	6.119	80.481	0.982	99.8	118.5	864
	3	4373.0	390.1	4763	−2106.0	6.202	87.479	0.98	182.1	130.5	918
4	1	4331.2	442.2	4773	−2091.0	2.697	11.855	0.993	61.1	41.1	401
	2	4358.3	388.6	4746	−2101.9	6.263	83.228	0.981	112.7	119.0	871
	3	4370.4	387.7	4758	−2105.4	6.327	90.072	0.98	70.9	131.3	908
5	1	4350.7	450.3	4801	−2092.0	2.685	14.752	0.994	9.3	39.5	383
	2	4355.1	420.4	4775	−2094.2	5.464	57.852	0.989	102	72.6	611
	3	**4322.8**	451.2	4774	**−2084.2**	**1.858**	15.378	**0.998**	53.1	**6.9**	**163**
6	1	4491.1	**332.7**	4823	−2142.7	16.41	2727.5	0.956	274.1	286.6	1097
	2	4397.9	337.3	**4735**	−2110.2	9.429	215.03	0.971	131.6	188.4	1079
	3	4388.4	346.9	**4735**	−2106.8	8.865	193.64	0.974	155.9	167.8	1033
7	1	4346.1	413.4	4759	−2091.7	3.897	29.842	0.99	26.2	65.3	586
	2	4384.1	394.3	4778	−2103.2	6.242	82.645	0.982	117.8	117.9	821
	3	4365.8	429.9	4795	−2095.1	4.370	44.389	0.991	55.7	58.2	577
8	1	4361.6	398.4	4760	−2092.4	4.342	41.117	0.987	**−8.4**	80.6	676
	2	4405.6	386.7	4792	−2105.4	6.405	93.739	0.979	108	132.3	861
	3	4377.7	434.5	4812	−2094.3	3.967	39.525	0.991	27	58.7	563
9	1	4330.7	442.3	4773	−2090.9	2.690	11.878	0.993	96.3	40.6	394
	2	4354.1	385.7	**4739**	−2100.9	6.161	81.223	0.982	62	118.1	867
	3	4364.8	382.2	4747	−2104.1	6.277	88.442	0.98	130.1	130	902

Note: The shaded cells indicate the worst performance of models under different evaluation criteria. The bold values indicate the best performance of models under different evaluation criteria. There are three models sharing the best performance in terms of DIC since the associated difference in DIC is < 5. 4. Refer to Table 2 for group names.

indicates the importance of random effect for crash count estimation.

Evaluation Results

Generally speaking, $\bar{D}$, LPML, MAD, MSPE, R_P^2, G^2, RSS, and TRD are akin to each other as they are centered on the predictive accuracies of models, while P_d is mainly focused on the model complexity. DIC is the penalized criterion which considers the combined effects of both perspectives. Table 3 demonstrates the evaluation results of each of the 27 individual models. For ease of description, G#T# will be used hereafter to denote the model of certain group and type. For example, G3T2 represents model of Group 3 and Type 2. Inspection of Table 3 reveals some interesting findings: (1) In terms of the prediction-pertinent criteria, G6T1 appears to have the worst performance while G5T3 tends to be ranked first in most cases; (2) In terms of model complexity (P_d), G6T1 performs the best while G1T3 is placed last; (3) In terms of the penalized criterion (DIC), G6T2, G6T3, and G9T2 are observed to be superior, whereas G6T1 has the disadvantage compared with others. Such phenomena indicate that (a) the various measures for prediction accuracies tend to be highly correlated with each other; (b) the models' prediction capabilities may be inconsistent with their efficiency in reducing the effective number of parameters; (c) under the DIC, the models with spatial random effects (either independent or interactive ones) seem to have better performance than the ones with temporal random effects only under most conditions.

Table 4 demonstrates the aggregated model performance based on two perspectives: across different temporal treatments and the three model types. In the case of aggregation by types, the proposed G9 models record the lowest DIC value (4753.36) while G8 models exhibit the worst fit. The base models may be expected to demonstrate the overall best predictions with the best rankings in most prediction-related measures, but the model complexity turns out to be the governing factor, which is the highest, resulting in overall inferior performance. This finding may indicate another perspective to quantify the impact of incorporating the temporal treatment, as compared with the base group, in the sense that the other model groups exhibit considerably lower complexity. The best performance of G9 in terms of DIC suggests that this kind of temporal treatment is worth further exploration in the safety field in future. The possible explanation

Table 4. Aggregated Evaluation Results under Different Criteria

Comparison unit	$\bar{D}$	P_d	DIC	LPML	MAD	MSPE	R_p^2	G^2	RSS	TRD
Models				Aggregation by nine groups of temporal treatments						
1	**4355.8**	423.0	4778	**−2097**	**4.52**	319.04	**0.989**	**58.259**	**74.96**	**524**
2	4368.5	**391.2**	**4759**	−2101.6	6.31	90.36	0.982	96.854	116.94	823
3	4362.3	409.0	4771	−2099	5.12	**72.63**	0.986	89.835	90.27	682
Groups				Aggregation by three types of models						
1	**4332.9**	445.1	4778	−2091	**3.21**	**23.78**	**0.994**	**28.325**	**39.35**	389
2	4353.7	409.3	4763	−2099.8	4.88	55.03	0.986	74.817	91.59	707
3	4354.5	407.2	4761	−2099.8	5	59.90	0.985	102.935	97.01	728
4	4353.3	406.2	4759	−2099.4	5.09	61.71	0.985	81.618	97.19	726
5	4342.8	440.6	4783	**−2090.1**	3.33	29.32	0.994	54.858	39.7	**385**
6	4425.8	**338.9**	4764	−2119.9	11.57	1045.41	0.967	187.263	214.3	1069
7	4365.4	412.5	4777	−2096.6	4.83	52.29	0.988	66.625	80.49	661
8	4381.6	406.5	4788	−2097.4	4.90	58.12	0.986	42.231	90.59	700
9	4349.8	403.4	**4753**	−2098.6	5.04	60.51	0.985	96.175	96.31	721

Note: The shaded cells indicate the worst performance of models under different evaluation criteria. The bold values indicate the best performance of models under different evaluation criteria.

for best predictions associated with the base models might be because of the myriad of factors (e.g., policy, population, weather, and so on) which affect the safety of large spatial unit, county, which makes it difficult to predict the county-level crashes by using a universal temporal trend or correlation.

In the case of aggregation by temporal treatments, Type 2 models are observed to have best overall fit as reflected by the lowest DIC (4759.82), followed by Type 3, whereas Type 1 records the highest DIC. Once again, the governing factor is the model complexity, which significantly decreases with the inclusion of spatial correlation structures. These findings suggest that inclusion of space and time at the same time provides substantial efficiency in reducing the number of effective parameters, which also indicates the tendency of crash data to cluster spatially as well as temporally. It is noteworthy to mention that among the spatiotemporal models (Types 2 and 3), the space–time interaction tends to increase the complexity as a result of the addition of more effective parameters, which eventually corresponds to increase in DIC. Concerning the prediction capabilities, Type 1 demonstrates the best performance in most of the pertinent criteria (except MSPE). It follows that it might be difficult to predict the crashes of the large spatial unit by using a certain type of spatial correlation (distance-based in this case) applicable to all counties as a result of the enormous impacts from a large number of factors.

In summary, under the predictive evaluation criteria, both aggregated results indicate the best performance associated with the simpler modeling formulation (without spatial and/or temporal correlations). It is possible that the counties, compared with smaller spatial units, are subject to much more influential safety factors, which render it difficult to predict county-level crashes via certain types of temporal or spatial random effects. However, the results of individual models illustrate that G5T3 is superior to others in five out of seven prediction cases. Such phenomenon suggests that it is still worth utilizing specific serial or spatial random effects for better prediction, even though care should be taken for selection of the temporal treatment, as different treatments lead to a large variability in terms of prediction.

Finally, some of the previous notable trends within the results motivated the authors to conduct a correlation analysis of the ten criteria. As shown in Table 5, statistically significant correlations are observed among a set of evaluation criteria including $\bar{D}$, LPML, MAD, MSPE, R_P^2, G^2, RSS, and TRD. Such finding is illuminating given that these measures check predictive capabilities using different methods based on cross-validation, ranking difference, and deviation relative to the estimated crash counts, and so on. P_d seems to have strong reverse correlations with all these prediction-related criteria. As a combination of P_d and $\bar{D}$, DIC does not demonstrate strong correlations with all criteria for predictive capabilities, except for MSPE and TRD. Given that DIC and P_d have a statistically significant positive correlation (0.38), it appears that P_d has a heavier weight for the determination of the DIC for the county situation, as also observed in the previous discussion of evaluation results.

Conclusions and Recommendations

The later model estimates demonstrated the notable impact of inclusion of spatial correlation on improving

Table 5. The Estimate of the Correlation Coefficients among Various Evaluation Criteria

Criteria	$\bar{D}$	P_d	DIC	LPML	MAD	MSPE	R_p^2	G^2	RSS	TRD
$\bar{D}$	1.00									
P_d	−0.79	1.00								
DIC	*0.26*	0.38	1.00							
LPML	−0.93	0.85	*−0.08*	1.00						
MAD	0.93	−0.90	*−0.01*	−0.97	1.00					
MSPE	0.80	−0.50	0.42	−0.83	0.81	1.00				
R_p^2	−0.90	0.96	*0.15*	0.96	−0.98	−0.70	1.00			
G^2	0.77	−0.76	*−0.04*	−0.87	0.85	0.67	−0.85	1.00		
RSS	0.90	−0.96	*−0.15*	−0.96	0.98	0.70	−1.00	0.85	1.00	
TRD	0.75	−0.96	−0.38	−0.82	0.84	0.39	−0.92	0.73	0.92	1.00

Note: The italic values denote insignificance at 95% level.

the precision of intercept across all groups of temporal treatments. Both spatial and temporal random effects were noted to be statistically significant, reflecting the presence of spatial clustering and temporal dependency of crash counts, which accounted for the variability in models. The changing coefficients of intercept for the random intercept model further demonstrated the impact of serial changes on crash risk.

After evaluating the 27 models using various checking measures, the following conclusions are drawn:

1. In terms of prediction capabilities, the aggregation results by model types reveal that the base model without considering serial correlations outperforms others, while the aggregated results by temporal treatments indicate the top-ranked performance associated with Type 1, or, the one without considering spatial dependency. Such phenomena suggest that, because of the myriad of influential factors, it is difficult to predict county-level crash counts by using certain types of temporal and spatial correlations applicable to all counties, even though such correlations are statistically significant. However, the individual modeling results show the profound predictive benefits associated with G5T3. It follows that it is still worth addressing the serial and spatial dependency for the crash prediction of the large spatial unit, county, although caution should be exercised as there is large variability of predictive capabilities among the different temporal treatments.
2. In review of P_d, both aggregated and individual results indicate the spatiotemporal models (either Type 2 or Type 3) seem to be more efficient in reducing the effective number of parameters than Type 1, and the models accounting for temporal dependency appear to perform better than the base models in most scenarios. Interestingly, similar trends are found in the criterion of DIC, which suggest that P_d tends to have a larger impact on the DIC than does $\bar{D}$. It is noteworthy that, based on aggregation result, Group 9 (time adjacency models) are observed to have the best performance in terms of DIC compared with other groups. Therefore, it is recommended that this group should be further explored in the safety field.
3. The correlation analysis among all ten criteria confirms that P_d exerts more influence on DIC than does $\bar{D}$, although the latter demonstrates statistically significant correlations with a set of criteria which checks the predictive accuracy from different aspects. It is anticipated that such correlation analysis should provide more confidence for modeling selection for safety researchers or practitioners, especially when limited safety resources prevent a comprehensive evaluation of models.

The results observed in this paper require some caveats. First, only DVMT was included in the modeling development. The results with the inclusion of other covariates might be different. Second, an 8-year period was used as it seemed optimum to replicate the safety literature. However, the sample size of eight might be relatively small for some models, especially the time adjacency ones. Future research may be focused on different time adjacency models with a longer time period. Third, the aggregated and individual modeling results demonstrated different characteristics with respect to the prediction-related criteria. The discrepancy might be caused by the myriad of safety factors influencing the counties. Studies focusing on smaller spatial units (such as intersections, segments, traffic analysis zones) are highly recommended for verification of the results.

Acknowledgments

Support for this study was provided by 2017 UCCONNECT Research Program (Grant number: 65A0529-052.3). The authors would like to express gratitude to SCAG for providing geometric centroid distance data of counties used in this study.

Author Contributions

The authors confirm contribution to the paper as follows: study conception and design: Wen Cheng, Gurdiljot Gill; data collection: Jiao Zhou; analysis and interpretation of results: Wen Cheng, Gurdiljot Gill, Xudong Jia; draft manuscript preparation: Gurdiljot Gill. All authors reviewed the results and approved the final version of the manuscript.

References

1. Lord, D., and F. Mannering. The Statistical Analysis of Crash-Frequency Data: A Review and Assessment of Methodological Alternatives. *Transportation Research Part A: Policy and Practice*, Vol. 44, No. 5, 2010, pp. 291–305.
2. Andrey, J., and S. Yagar. A Temporal Analysis of Rain-Related Crash Risk. *Accident Analysis & Prevention*, Vol. 25, No. 4, 1993, pp. 465–472.
3. Hay, J. L., and A. N. Pettitt. Bayesian Analysis of a Time Series of Counts with Covariates: An Application to the Control of an Infectious Disease. *Biostatistics*, Vol. 2, No. 4, 2001, pp. 433–444.
4. Wang, X., and M. Abdel-Aty. Temporal and Spatial Analyses of Rear-End Crashes at Signalized Intersections. *Accident Analysis & Prevention*, Vol. 38, No. 6, 2006, pp. 1137–1150.
5. Huang, H., H. Chin, and M. Haque. Empirical Evaluation of Alternative Approaches in Identifying Crash Hot Spots. *Transportation Research Record: Journal of the Transportation Research Board*, 2009. 2103: 32–41.
6. Aguero-Valverde, J. Full Bayes Poisson Gamma, Poisson Lognormal, and Zero Inflated Random Effects Models: Comparing the Precision of Crash Frequency Estimates. *Accident Analysis & Prevention*, Vol. 50, 2013, pp. 289–297.
7. El-Basyouny, K., S. Barua, and M. Islam. Investigation of Time and Weather Effects on Crash Types Using Full Bayesian Multivariate Poisson Lognormal Models. *Accident Analysis & Prevention*, Vol. 73, 2014, pp. 91–99.
8. Song, J., M. Ghosh, S. Miaou, and B. Mallick. Bayesian Multivariate Spatial Models for Roadway Traffic Crash Mapping. *Journal of Multivariate Analysis*, Vol. 97, No. 1, 2006, pp. 246–273.
9. Guo, F., X. Wang, and M. Abdel-Aty. Modeling Signalized Intersection Safety with Corridor-Level Spatial Correlations. *Accident Analysis & Prevention*, Vol. 42, No. 1, 2010, pp. 84–92.
10. Aguero-Valverde, J., and P. Jovanis. Analysis of Road Crash Frequency with Spatial Models. *Transportation Research Record: Journal of the Transportation Research Board*, 2008. 2061: 55–63.
11. Aguero-Valverde, J., and P. Jovanis. Spatial Analysis of Fatal and Injury Crashes in Pennsylvania. *Accident Analysis & Prevention*, Vol. 38, No. 3, 2006, pp. 618–625.
12. Gill, G. S., W. Cheng, J. Zhou, and V. Park. Comparative Analysis of Cost-Weighted Site Ranking Using Alternate Distance-Based Neighboring Structures for Spatial Crash Frequency Modeling. *Journal of Transportation Safety & Security*, 2017, pp. 1–23.
13. Gill, G., W. Cheng, M. Xie, T. Vo, X. Jia, and J. Zhou. Evaluating Influence of Neighboring Structures on Spatial Crash Frequency Modeling and Site-Ranking Performance. *Transportation Research Record: Journal of the Transportation Research Board*, 2017. 2659: 117–126.
14. Cheng, W., G. Gill, R. Dasu, M. Xie, X. Jia, and J. Zhou. Comparison of Multivariate Poisson Lognormal Spatial and Temporal Crash Models to Identify Hot Spots of Intersections Based on Crash Types. *Accident Analysis & Prevention*, Vol. 99, 2017, pp. 330–341.
15. Bernardinelli, L., D. Clayton, C. Pascutto, C. Montomoli, M. Ghislandi, and M. Songini. Bayesian Analysis of Space—Time Variation in Disease Risk. *Statistics in Medicine*, Vol. 14, No. 21–22, 1995, pp. 2433–2443.
16. Dong, N., H. Huang, P. Xu, Z. Ding, and D. Wang. Evaluating Spatial-Proximity Structures in Crash Prediction Models at the Level of Traffic Analysis Zones. *Transportation Research Record: Journal of the Transportation Research Board*, 2014. 2432: 46–52.
17. Ma, X., S. Chen, and F. Chen. Multivariate Space-Time Modeling of Crash Frequencies by Injury Severity Levels. *Analytic Methods in Accident Research*, Vol. 15, 2017, pp. 29–40.
18. Abellan, J., S. Richardson, and N. Best. Use of Space–Time Models to Investigate the Stability of Patterns of Disease. *Environmental Health Perspectives*, Vol. 116, No. 8, 2008, pp. 1111–1119.
19. Abdel-Aty, M., J. Lee, C. Siddiqui, and K. Choi. Geographical Unit Based Analysis in the Context of Transportation Safety Planning. *Transportation Research Part A: Policy and Practice*, Vol. 49, 2013, pp. 62–75.
20. Cheng, W., G. S. Gill, S. Choi, J. Zhou, X. Jia, and M. Xie. Comparative Evaluation of Temporal Correlation Treatment in Crash Frequency Modelling. *Transportmetrica A: Transport Science*, 2017, pp. 1–19. https://doi.org/10.1080/23249935.2017.1418458
21. Cheng, W., G. S. Gill, J. L. Ensch, J. Kwong, and X. Jia. Multimodal Crash Frequency Modeling: Multivariate Space-Time Models with Alternate Spatiotemporal Interactions. *Accident Analysis & Prevention*, Vol. 113, 2018, pp. 159–170. https://doi.org/10.1016/j.aap.2018.01.034
22. Jung, R., M. Kukuk, and R. Liesenfeld. Time Series of Count Data: Modeling, Estimation and Diagnostics. *Computational Statistics & Data Analysis*, Vol. 51, No. 4, 2006, pp. 2350–2364.
23. Barron, D. The Analysis of Count Data: Overdispersion and Autocorrelation. *Sociological Methodology*, Vol. 22, 1992, pp. 179–220.
24. Spiegelhalter, D., A. Thomas, N. Best, and D. Lunn. *WinBUGS Version 1.4 User Manual.* MRC Biostatistics Unit, Cambridge, 2003. http://www.mrc-cam.ac.uk/bugs. Accessed June 25, 2017.
25. Lunn, D., C. Jackson, N. Best, A. Thomas, and D. Spiegelhalter. *The BUGS Book: A Practical Introduction to Bayesian Analysis.* CRC Press, Boca Raton, 2012.
26. Heydari, S., L. Fu, D. Lord, and B. Mallick. Multilevel Dirichlet Process Mixture Analysis of Railway Grade

Crossing Crash Data. *Analytic Methods in Accident Research*, Vol. 9, 2016, pp. 27–43.
27. Cheng, W., G. S. Gill, Y. Zhang, and Z. Cao. Bayesian Spatiotemporal Crash Frequency Models with Mixture Components for Space-Time Interactions. *Accident Analysis & Prevention*, Vol. 112, 2018, pp. 84–93.
28. Washington, S., M. G. Karlaftis, and F. L. Mannering. *Statistical and Econometric Methods for Transportation Data Analysis*. Chapman & Hall, Boca Raton, 2003.
29. Cheng, W., and S. Washington. New Criteria for Evaluating Methods of Identifying Hot Spots. *Transportation Research Record: Journal of the Transportation Research Board*, 2008. 2083: 76–85.
30. Earnest, A., G. Morgan, K. Mengersen, L. Ryan, R. Summerhayes, and J. Beard. Evaluating the Effect of Neighbourhood Weight Matrices on Smoothing Properties of Conditional Autoregressive (CAR) Models. *International Journal of Health Geographics*, Vol. 6, No. 1, 2007, p. 54.

The Standing Committee on Safety Data, Analysis and Evaluation (ANB20) peer-reviewed this paper (18-05071).

Article

TRR
JOURNAL OF THE TRANSPORTATION RESEARCH BOARD

Transportation Research Record
2018, Vol. 2672(38) 105–115

DOI: 10.1177/0361198118782797
journals.sagepub.com/home/trr
SAGE

Use of Bivariate Dirichlet Process Mixture Spatial Model to Estimate Active Transportation-Related Crash Counts

Wen Cheng[1], Gurdiljot Singh Gill[1], Tom Vo[2], Jiao Zhou[1], and Taha Sakrani[1]

Abstract

The current paper presents the comprehensive analysis of a bivariate Dirichlet process mixture spatial model for estimation of pedestrian and bicycle crash counts. This study focuses on active transportation at traffic analysis zone (TAZ) level by developing a semi-parametric model that accounts for the unobserved heterogeneity by combining the strengths of bivariate specification for correlation among crash modes; spatial random effects for the impact of neighboring TAZs; and Dirichlet process mixture for random intercept. Three alternate models, one Dirichlet and two parametric, are also developed for comparison based on different criteria. Bicycle and pedestrian crashes are observed to share three influential variables: the positive correlation of K12 student enrollment; the bike-lane density; and the percentage of arterial roads. The heterogeneity error term demonstrates the presence of statistically significant correlation among the bicycle and pedestrian crashes, whereas the spatial random effect term indicates the absence of a significant correlation for the area under focus. The Dirichlet models are consistently superior to non-Dirichlet ones under all evaluation criteria. Moreover, the Dirichlet models exhibit the capability to identify latent distinct subpopulations and suggest that the normal assumption of intercept associated with traditional parametric models does not hold true for the TAZ-level crash dataset of the current study.

Non-motorists are defined as road users not in or upon a motor-vehicle and consist of walking pedestrians, bicyclists, individuals in wheelchairs or motorized personal conveyances, skateboarders, and others (*1*). They are a vulnerable segment of the traveling public because of the lack of a protective structure and difference in body mass between them and motor vehicles, which renders them prone to heightened injury susceptibility in case of a collision (*2*). On the other hand, active transportation provides enormous benefits for addressing the issues of congestion, health, and environment (*3–8*). Therefore, encouraging individuals to indulge in active transportation, involving walking and bicycling, brings with it a societal obligation to protect them. In response, many research studies have investigated the factors affecting non-motorist safety on roadways (*9–14*). Although these studies are useful for identifying safety risks contributing to cyclist, pedestrian, and motor-vehicle injury occurrence, they modeled the modal crashes separately, and few attempts have been made to combine them into a multimodal approach, which allows the flexibility to determine the injury risk of different travel modes simultaneously.

A central issue to the successful implementation of the multimodal approach is the development of multivariate crash frequency models that can jointly estimate the crash risk of different modes that share some of the unobserved heterogeneity. Ignorance of such correlation among the multiple modes has been illustrated to reduce the efficiency of the models, as a result of less precise parameters (*15–17*). In comparison with many univariate models dedicated to various mode users, very few studies have used joint models to analyze the interaction between different modes of active transportation. Recently, Conway et al. performed a bivariate correlation analysis to find the locations of conflict occurrence among bicycles and pedestrians, freight, passenger cars, and cabs in an urban area (*18*). In order to simultaneously analyze the injury and traffic flow outcomes for different modes, Strauss et al. subsequently employed Bayesian multivariate Poisson models for studying safety outcomes for motor-vehicle, cyclist, and pedestrian flow at intersections (*19*).

[1]Department of Civil Engineering, California State Polytechnic University, Pomona, CA
[2]Southern California Association of Governments, Los Angeles, CA

Corresponding Author:
Address correspondence to Wen Cheng: wcheng@cpp.edu

Safety performance functions were developed and crash-contributing factors were identified for each mode.

One common limitation associated with the above two studies lies in the lack of consideration for spatial correlations, whose significance has been highlighted by many studies (*20–22*). The study by Narayanamoorthy et al. jointly analyzed pedestrian and cyclist injury severities while accounting for the spatial correlation at the census tract level using generalized ordered-response models (*23*). This study recommended the use of multivariate modeling and spatial dependency of injury counts. Similarly, Nashad et al. employed a copula-based approach for simultaneous estimation of crash counts for bicyclists and pedestrian crashes aggregated at the macro-level of traffic analysis zones (TAZs) (*24*).

Some studies in traffic safety have observed the superiority of non-parametric and semi-parametric models to address the unobserved heterogeneity (*25*, *26*). Regarding research dedicated to active transportation, the recent study by Heydari et al. (*27*) proposed the Dirichlet process mixture (*28*) to develop a flexible latent class model for joint analysis of pedestrian and cyclist injuries at the micro-level of intersections. The authors observed that the flexible approach was advantageous as it demonstrated superior predictive performance and better capability to capture the correlated crash data, which eventually provided more accurate interpretation of influential factors for improvement of safety environment.

The literature review illustrated the limited use of semi- or non-parametric models for simultaneous analysis of active transportation mode crashes. In effect, to the knowledge of the authors, research in the comprehensive analysis of flexible multivariate spatial models focusing on active transportation is non-existent in the safety literature. To fill this research gap, the authors adopted a semi-parametric formulation that accounts for the unobserved heterogeneity by combining the strengths of incorporating bivariate specification of dependency among crash modes (pedestrian and bicyclists), spatial random effects for the impact of neighboring areas, and Dirichlet process mixture for random intercepts. Four alternate models were developed for comparison based on the goodness-of-fit and predictive accuracy. The models were evaluated by employing different criteria, namely: LPML (log pseudo marginal likelihood), MSPE (mean-squared predictive error), the R_p^2 statistic, the G^2 statistic, and RSS (residual sum of squares).

Methodology

Model Specification

The full Bayesian (FB) framework was employed for estimation of 6-year bicyclist and pedestrian crashes aggregated at the TAZ level. The FB approach has been widely used for crash prediction models owing to its capability to handle the multilevel and correlated nature of crash data (*29*). Four crash frequency models were developed. The general functional form of the models is given in the following subsections, progressing from simple to sophisticated.

Model 1: Bivariate. This model assumes that the crash count of a specific modal crash j at a given location i, y_{ij}, obeys Poisson distribution, and the corresponding observation specific error term ε_{ij} follows a bivariate normal distribution:

$$y_{ij}|\lambda_{ij} \sim Poisson(\lambda_{ij}) \quad (1)$$

$$\ln(\lambda_{ij}) = X'_{ij}\beta + \varepsilon_{ij} \quad (2)$$

$$\varepsilon_{ij} \sim MVN\left(0, \sum\right) \quad (3)$$

$$\text{Where } y_{ij} = \begin{pmatrix} y_{i1} \\ y_{i2} \end{pmatrix}, \lambda_{ij} = \begin{pmatrix} \lambda_{i1} \\ \lambda_{i2} \end{pmatrix}, \varepsilon_{ij} = \begin{pmatrix} \varepsilon_{i1} \\ \varepsilon_{i2} \end{pmatrix}, \sum = \begin{pmatrix} \sigma_{11} & \sigma_{12} \\ \sigma_{12} & \sigma_{22} \end{pmatrix} \quad (4)$$

In the above equations, X' is the matrix of risk factors, β is the vector of model parameters, ε_{ij} is the independent random effect which captures the extra-Poisson heterogeneity among locations. $\sum$ is called the covariance matrix. The diagonal element σ_{jj} in the matrix represents the variance of ε_{ij}, where the off-diagonal elements represent the covariance of crash counts of different modes. The inverse of the covariance matrix represents the precision matrix and has the following distribution:

$$\sum^{-1} \sim Wishart(I, J) \quad (5)$$

Where I is the $J \times J$ identity matrix (*30*), and J is the degree of freedom, $J = 2$ here representing two crash outcomes corresponding to bicyclist and pedestrians crashes.

Model 2: Bivariate Spatial. Under Model 2, the spatial random effects were incorporated over the model represented in Equation 2. The final model takes the following form to account for spatial correlations among the TAZs:

$$\ln(\lambda_{ij}) = X'_{ij}\beta + \varepsilon_{ij} + u_{ij} \quad (6)$$

Where u_{ij} is the spatially structured random effect which follows the multivariate conditional autoregressive (MCAR) (*31*) formulation to incorporate the spatial correlation among crashes occurring at neighboring TAZs.

$$u_i|u_k, \sum i \sim N_j\left(\sum_{k\sim i} C_{ik}, u_k, \sum i\right) \quad (7)$$

Where each $\sum_i$ is a positive definite matrix representing the conditional variance matrix, and the adjacency matrix C_{ij} is of the same dimension with $\sum_i$ (*32*). The precision matrix $\sum^{-1}$ follows the Wishart distribution as shown in Equation 5.

As can be seen from the above equations, estimation of the risk in any site is conditional on risks in neighboring locations. Subscripts i and k refer to a TAZ and its neighbor, respectively, and k belongs to N_i where N_i represents the set of neighbors of TAZ i. Besides the identification of neighbors, the assigned weights also affect the risk estimation. In past studies (*33*, *34*), weight structures including various adjacency-based, distance-based models, and semi-parametric geographically weighted approaches, and so on, have been explored. The current study employs the commonly used distance-based structure to explore the spatial correlations with the following formulation:

$$w_{ij} = \frac{1}{d_{ij}} \quad (8)$$

Where w_{ij} is the weight between TAZ i and j, and d_{ij} is the distance between TAZ i and j. With this weighted structure, it is known that a larger weight was assigned to TAZs which are relatively close.

Model 3: Bivariate Dirichlet Process Mixture. The parametric model specification of the aforementioned models assumed the distribution of the parameters to be specific (normal in this study) across all concerned sites. However, the non-parametric specification removes such constraints by employing the flexible approach of the Dirichlet process that allows the incorporation of unknown random density for the parameters. The current study employs a semi-parametric approach which relaxes the restrictive distributional assumption for the intercept only, instead of all of the parameters. The removal of constraints for the intercept to follow a specific distribution represents a plausible scenario in which the TAZs are not expected to have a normal distribution. This flexible approach is expected to capture the extra variability which may escape the error terms introduced in parametric models. Equation 2 was modified to use Dirichlet process mixture over the intercept as follows (*25*):

$$ln(\lambda_{ij}) = \beta_{0rj} + X_i'\beta \quad (9)$$

$$\beta_{0rj} \approx \sum_{n=1}^{C} p_n I_{\theta_{z_i}} \sim TDP(kG_{0j}), z_i = n \text{ with probability of } p_n \quad (10)$$

$$G_0 \sim MVN\left(\mu_{G_{0j}}, \sum\right) \quad (11)$$

Where β_{0rj} is the intercept for cluster r (r ranges from 1 to C) of mode j, k is the precision parameter, and G_0 is the baseline distribution for β_{0r} which follows a bivariate normal distribution with mean μ_{G_0} and variance $\sum$, which also follows the Wishart distribution. β_{0rj} essentially represents a vector of probabilities over the space of concerned entities (203 TAZs) and follows a truncated Dirichlet process with a vector of parameters represented by kG_{0j}. The precision parameter k indicates the variability of the Dirichlet process around G_{0j}. The intercept draws random points (θ_{Z_i}) and the associated probabilities (p_n) can be obtained through the stick-breaking procedure (*26*, *28*, *35*). If one cluster is occupied, the indicator function ($I_{\theta_{z_i}}$) at θ_{Z_i} will take the value of 1, otherwise it would be 0. The number of latent clusters (r) in β_{0rj} could range from 1 to infinity, which requires immense computational effort. To reduce the computational complexity by obtaining finite dimensional approximation, a truncated Dirichlet process is utilized to fix the maximum number of possible clusters to C, where C is governed by the precision parameter k and is estimated by $5k + 2$ (*28*). As the prior distribution for precision parameter k was assumed to be $k \sim$ uniform (0.3, 9), so eventually the number of clusters were limited to be maximum of 47. The value of C used in the study can be considered in a normal range given the different C values utilized previously such as 5 (*36*), 10 (*37*), and 52 (*27*).

Model 4: Bivariate Dirichlet Process Mixture Spatial. Model 4 is distinct from Model 3 by incorporating the spatial random effects to account for the correlation among the neighboring TAZs. The model in Equation 9 takes the following form:

$$ln(\lambda_{ij}) = \beta_{0rj} + X_i'\beta + u_{ij} \quad (12)$$

Where all terms are defined as previously.

Comparison of Models Based on Cross-Validation

Many traditional cross-validation approaches are prone to overestimation as a result of double usage of data, once during model development and then again for model checking. The approach of cross-validatory predictive densities was proposed to tackle this issue (*38*), in which the full set of data was divided into two subsets (one subset for development and the other for checking). However, the splitting of two subsets posed a major problem as the selection of different subsets provides varying results. This was resolved by implementing a CV-1 (leave-one-out) technique to estimate the cross-validatory conditional predictive ordinate (CPO) (*39*). This technique removed the selection bias by employing a continuous approach of selecting all data points, except one, for model development and the left-out data point to verify the prediction accuracy of the calibrated model. Under the Markov Chain Monte Carlo (MCMC) framework, the estimate of CPO for each observation i can be calculated as:

$$CPO = \left(\frac{1}{T}\sum_{t=1}^{T}\frac{1}{f(Y_i|\beta^{(t)})}\right)^{-1} \quad (13)$$

Where Y_i is the ith observation ($i = 1, 2, 3, \ldots, n$) for all 203 TAZs and β is the vector of estimated model parameters. This harmonic mean of density (CPO) may be extended to calculate the goodness-of-fit of models by computing the product of CPOs over all observations, which is known as the pseudo marginal likelihood. For computational convenience, the LPML is calculated (*25, 40*):

$$LPML = \sum_{i=1}^{n}\log(CPO_i) \quad (14)$$

Evaluation Criteria for Predictive Accuracy

In this study, the four competing models were also evaluated based on some criteria used from previous studies: MSPE (*23*), the G^2 statistic (*41*), the R_p^2 statistic (*42*), and the chi-squared RSS (*43*). The details of each criterion are shown in the following subsections.

MSPE. As indicated by the name, such criterion is related to the average squared deviations, or, the predictive errors. Specifically, the MSPE was calculated as follows:

$$\text{MSPE} = \frac{1}{\text{n}}\sum_{\text{i}=1}^{\text{n}}(\lambda_i - y_i)^2 \quad (15)$$

Where λ_i is the Bayesian estimated crash frequency for zone i while y_i is the observed crash counts of the same zone. The smaller MSPE is preferred which indicates a better prediction performance.

RSS. MSPE is based on the deviations. A potential issue is that the larger estimated count of one zone might mask the smaller ones of multiple TAZs. To address this issue, we also calculated the chi-squared RSS to determine the deviation standardized by the estimated number of crash counts:

$$RSS = \sum_{i=1}^{n}\frac{(\lambda_i - y_i)^2}{\lambda_i} \quad (16)$$

The model with a smaller value of RSS tends to have more predictive capabilities.

The R_p^2 Statistic. The typical R-square in ordinary linear regression cannot be directly applied to the crash frequency model because of the nonlinearity of conditional mean (E[y|X]) and heteroscedasticity associated with the Poisson models. Therefore, we adopted an equivalent measure, R_p^2, which is based on standardized residuals:

$$R_p^2 = 1 - \frac{\sum_{i=1}^{n}\left[\frac{y_i-\lambda_i}{\sqrt{\lambda_i}}\right]^2}{\sum_{i=1}^{n}\left[\frac{y_i-\bar{y}}{\sqrt{\bar{y}}}\right]^2} \quad (17)$$

Where $\bar{y}$ represents the mean value of the observed counts. Similar to R-square, a smaller R_p^2 value indicates the inferior performance.

The G^2 Statistic. The sum of model deviances, G^2, is zero for a model with a perfect fit. The G^2 statistic is given as:

$$G^2 = 2\sum_{i=1}^{n} y_i LN\left(\frac{y_i}{\lambda_i}\right) \quad (18)$$

A large G^2 deviating from zero indicates that the model fits poorly as compared with the saturated model.

Data Preparation

Pedestrian and bicyclist crashes which occurred in the City of Irvine from 2007 to 2012 were analyzed for the study. Like many other research studies (*44–46*), TAZs were selected as the base units, and the crash data were aggregated at the TAZ level. Overall, there are 203 TAZs in the City. The map in Figure 1 displays the distribution of all TAZs and associated crash counts. The two transportation mode-related crashes were collected from California Statewide Integrated Traffic Records System (SWITRS). Shape file of TAZ boundary and TAZ characteristics were provided by Southern California Association of Governments (SCAG).

The variables used for model development and the associated descriptive statistics are shown in Table 1. The 6-year aggregated pedestrian and bicyclist crashes were used as the dependent variables. Daily vehicle miles traveled (DVMT) acted as a measure of exposure. The explanatory variables were the predictors commonly used in previous regional safety analyses, which include socioeconomic, transportation-related, environment-related factors, and so on. It is worth mentioning that the data from 2008 were available for explanatory variables owig to less frequent collection by the agencies, and thus it is used for model development. Also, the distance matrix containing distances among various TAZ

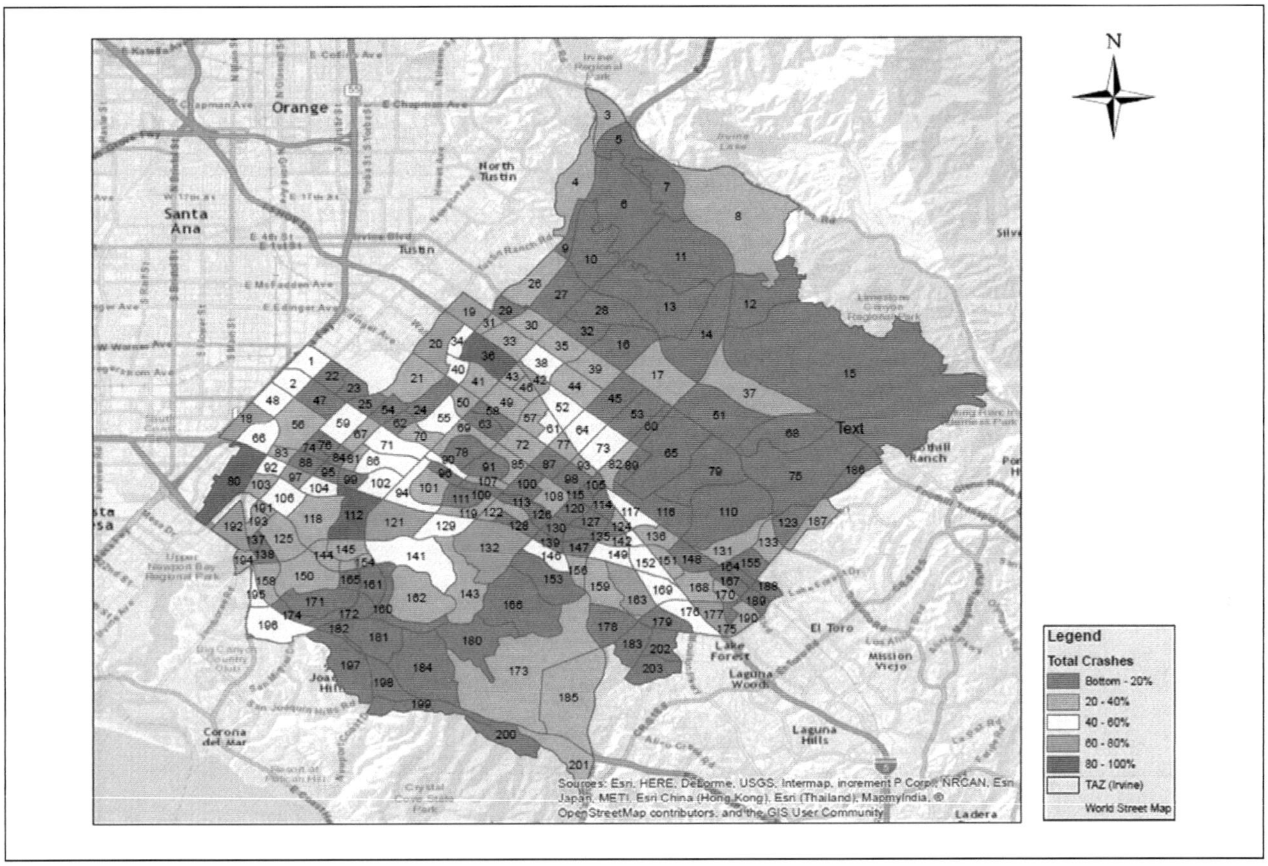

Figure 1. TAZ map with crash distributions in the City of Irvine, California.

centroids were collected from SCAG for the estimation of distance-based spatial random effect. Their descriptive statistics can be found in Table 1 as well.

Results

The crash prediction models were estimated with the freeware statistical package WinBUGS (*47*). A total of 10,000 MCMC iterations were utilized for parameter estimation after discarding the first 1,000 iterations as burn-in. The convergence was ensured by employing different approaches such as visual inspection of history plots, trace plots, and Gelman–Rubin diagram (*48*). The Pearson correlation coefficient was calculated and the covariates correlated at a significance level of 0.05 were subsequently eliminated.

Modeling Results

As shown in Table 2, the posterior inferences for influential factors for all four models demonstrate their robustness to fit the multimodal crash data at the TAZ spatial scale. All four models identify similar significant factors that affect crash frequency for a particular mode. In the case of bicycle crashes, three variables are observed to be statistically significant, namely: K12 student enrollment, percentage of arterials, and bike-lane density for the TAZ. The TAZs with higher K12 student enrollment increases the crash risk as the instances of interaction of bicyclists with other modes increases as a result of more exposure. However, the similar positive correlation for bike-lane density seems counter-intuitive, as the presence of bike lanes is expected to facilitate more usage of bicycles because of lower perceived risk of interaction with other modes. The possible rationale for this finding may be explained by the lower perceived risk which may encourage bicyclists to ride more in such areas, while conversely increasing the crash risk as a result of higher exposure of bicyclists to vehicular traffic. The negative relationship among percentage of arterial roads and bicycle crashes indicates that maybe the bicyclists tend to travel less in areas with more arterials.

For the crashes pertaining to pedestrians, college enrollment is also observed to be influential, along with other three factors shared with bicycle crashes. The increase in student population in the colleges of TAZs is

Table 1. Summary Statistics of Variables for TAZ's of the City of Irvine

Variables	Description	Mean	SD	Minimum	Maximum
Bike	Total bike-involved crashes (2007–2012)	1.82	2.45	0	12
Ped	Total pedestrian-involved crashes (2007–2012)	0.81	1.33	0	8
DVMT	Daily vehicle miles traveled	5,4262.44	56,156.84	112.57	276,079.90
Acre	TAZ Area in acre	282.90	431.75	0.69	5,062.95
Median	Median house income ($)	48,440.78	50,635.10	0	183,347
Pop_den	Population density by area	6.18	7.96	0	32.40
HH_den	Household density by area	2.34	3.15	0	13.62
Emp_den	Employment density by area	10.34	17.43	0	121.10
Ret_den	Retail job density	0.79	2.02	0	17.45
% age 5_17	% of population age 5–17	8.64%	8.78%	0	27%
% age 18_24	% of population age 18–24	5.79%	7.42%	0	40%
% age 24_64	% of population age 24–64	38.35%	36.12%	0	95%
% age 65 +	% of population age 65 or older	6.25%	10.21%	0	83%
K12	K12 student enrollment	0.39	1.00	0	5.52
College	College student enrollment	0.11	1.00	0	12.59
Int34_den	Intersection density (3- and 4-legs)	0.12	0.12	0	0.62
BKlnACC	Bike-lane access (1 = if a TAZ has bike lane)	0.92	0.28	0	1
BL_den	Bike-lane density	3.40	1.80	0	7.26
Rail	1 = at least one rail station in a TAZ	0.01	0.10	0	1
TTbus_D	Total bus stop density	0.05	0.09	0	0.53
Exbus_D	Stop density for express bus and BRT	0.002	0.007	0	0.06
HFLbus_D	High-frequency bus stop density (local bus headway ≤20 mins)	0.001	0.004	0	0.03
WalkAcc	Walk accessibility	3.87	9.46	0	74.53
% Arterial	Percent of main arterial (45–55 mph) of TAZ	10.61%	17.33%	0	80%
Distance	Distance among TAZ centroids (in miles)	4.06	2.09	0.16	11.78

Note: SD = standard deviation.

Table 2. Posterior Inference for Bicyclist and Pedestrian-Involved Crash Counts

Count type	Variables	Model 1	Model 2	Model 3	Model 4
Bicyclist	Intercept	**−10.860 (0.243)**	**−10.880 (0.246)**	**−10.780 (0.248)**	**−10.790 (0.234)**
	% age 65 +	1.532 (0.922)	1.467 (0.895)	1.413 (0.830)	1.401 (0.798)
	K12	**0.203 (0.088)**	**0.203 (0.091)**	**0.213 (0.079)**	**0.211 (0.074)**
	College	−0.013 (0.078)	−0.015 (0.077)	−0.014 (0.079)	−0.012 (0.075)
	WalkAcc	−0.007 (0.010)	−0.008 (0.010)	−0.006 (0.010)	−0.007 (0.010)
	% Arterial	**−3.517 (0.674)**	**−3.529 (0.685)**	**−3.472 (0.691)**	**−3.399 (0.655)**
	BL_den	**0.260 (0.056)**	**0.271 (0.057)**	**0.245 (0.056)**	**0.246 (0.056)**
Pedestrian	Intercept	**−12.390 (0.326)**	**−12.430 (0.357)**	**−12.360 (0.340)**	**−12.380 (0.346)**
	% age 65 +	1.205 (1.145)	1.192 (1.101)	1.097 (1.074)	1.131 (1.009)
	K12	**0.280 (0.104)**	**0.280 (0.106)**	**0.291 (0.095)**	**0.291 (0.094)**
	College	**−0.976 (0.567)**	**−0.968 (0.563)**	**−0.962 (0.562)**	**−0.957 (0.558)**
	WalkAcc	0.009 (0.010)	0.008 (0.010)	0.010 (0.010)	0.009 (0.010)
	% Arterial	**−3.826 (0.989)**	**−3.805 (0.985)**	**−3.727 (0.991)**	**−3.658 (0.996)**
	BL_den	**0.384 (0.068)**	**0.397 (0.075)**	**0.374 (0.069)**	**0.375 (0.074)**

Note: Intercept for Dirichlet Process models indicates the intercept mean from mixture points. Refer to Table 1 for detailed description of variables. Numbers in parentheses represent uncertainty estimates, or, posterior standard deviations. The statistically significant variable coefficients are shown in bold. Model 1 = bivariate; Model 2 = bivariate spatial; Model 3 = bivariate Dirichlet process mixture; Model 4 = bivariate Dirichlet process mixture spatial.

noted to be negatively linked with pedestrian crashes, though the increased pedestrian activity usually associated with the presence of college students was expected to increase crash occurrence. The probable justification may be that the known presence of students influences the vehicle drivers to be more cautious and drive sensitively, or the vehicular activity may be minimal in such areas which may help significantly reduce the possibility of interaction with pedestrians. The common significant factors (K12 student enrollment, percentage of arterials,

Table 3. Covariance Matrices for the Four Alternative Models

Models	Modes	Heterogeneity (ε_{ij})		Spatial (u_{ij})	
		Bicycle	Pedestrian	Bicycle	Pedestrian
Model 1	Bicycle	**0.896 (0.166)**	**0.854 (0.166)**		
	Pedestrian	**0.854 (0.166)**	**0.890 (0.237)**		
Model 2	Bicycle	**0.860 (0.168)**	**0.827 (0.153)**	**0.001 (2.2 × 10^{-4})**	6.7 × 10^{-5} (1.5 × 10^{-4})
	Pedestrian	**0.827 (0.153)**	**0.856 (0.213)**	6.7 × 10^{-5} (1.5 × 10^{-4})	**0.001 (2.2×10^{-4})**
Model 3	Bicycle	**0.602 (0.200)**	**0.538 (0.182)**		
	Pedestrian	**0.538 (0.182)**	**0.561 (0.226)**		
Model 4	Bicycle	**0.507 (0.231)**	**0.461 (0.234)**	**0.001 (2.1×10^{-4})**	7.4 × 10^{-5} (1.5 × 10^{-4})
	Pedestrian	**0.461 (0.234)**	**0.503 (0.270)**	7.4 × 10^{-5} (1.5 × 10^{-4})	**0.001 (2.2×10^{-4})**

Note: Numbers in parentheses represent posterior standard deviations. The statistically significant covariance values are shown in bold. Model 1 = bivariate; Model 2 = bivariate spatial; Model 3 = bivariate Dirichlet process mixture; Model 4 = bivariate Dirichlet process mixture spatial.

Table 4. Evaluation Results for Alternative Models

Model	LPML	MSPE	R_p^2	G^2	RSS
Model 1	−476.753	0.690	0.786	177.995	272.367
Model 2	−477.492	0.691	0.781	179.544	278.749
Model 3	−474.433	0.682	0.823	169.137	225.018
Model 4	−474.831	0.687	0.823	169.998	225.291

Note: Model 1 = bivariate; Model 2 = bivariate spatial; Model 3 = bivariate Dirichlet process mixture; Model 4 = bivariate Dirichlet process mixture spatial.

and bike-lane density) responsible for bicycle and pedestrian crashes support the joint estimation of such modes which are most vulnerable and impacted by similar characteristics. As shown in Table 3, the heterogeneity error term demonstrates the presence of statistically significant correlation among the bicycle and pedestrian crashes which further justifies the employment of bivariate structure for joint estimation of crashes. However, the spatial random effect term exhibits the absence of a significant correlation, as indicated by the covariance matrix. It may be possible that the explanatory variables incorporated for model development are sufficiently robust to account for the spatial characteristics that influence crash occurrence for the particular modes.

Evaluation Results

As previously stated, the four models are evaluated from different perspectives using five evaluation criteria. Unlike the traditional parametric models which usually employ deviance information criterion (DIC) for model comparison, LPML is adopted in this study as DIC is not generated by the WinBUGS owing to its sensitivity to different parameterizations (*28*, *49*). The higher value of LPML is desirable as it reflects relatively superior model fit property, and a difference of more than 5 points among two competing models help identify the model of interest (*50*). As shown in Table 4, the LPML values of all four models are close enough to not cross the threshold of 5 points for identification of the model of interest. However, the sample size also affects the numerical value of LPML. Therefore it may be worthwhile to record the model with highest LPML value and compare the observation with other criteria. As evident from the evaluation results, Model 3 demonstrates the best fit based on relatively large LPML (−474.433), closely followed by Model 4. A similar trend is observed for all other criteria, suggesting the strong correlation among the capability of a model to fit crash data and its performance at crash predictive accuracy.

Further inspection of the evaluation results reveals that the models which account for spatial correlations (Models 2 and 4) have consistently inferior performance to those with spatially unstructured heterogeneity (Models 1 and 3). This phenomenon suggests that the inclusion of spatial correlation structures raises the model complexity without notable advantage at crash prediction, which is usually expected in such cases as reduced posterior deviance compensates the increased complexity. The potential reason might be because of the insignificant spatial dependency among the two modal crashes, as shown in Table 3. Clearly, the Dirichlet models (Models 3 and 4) outperform the non-Dirichlet ones (Model 1 and 2) based on all five criteria, suggesting the use of such flexible framework.

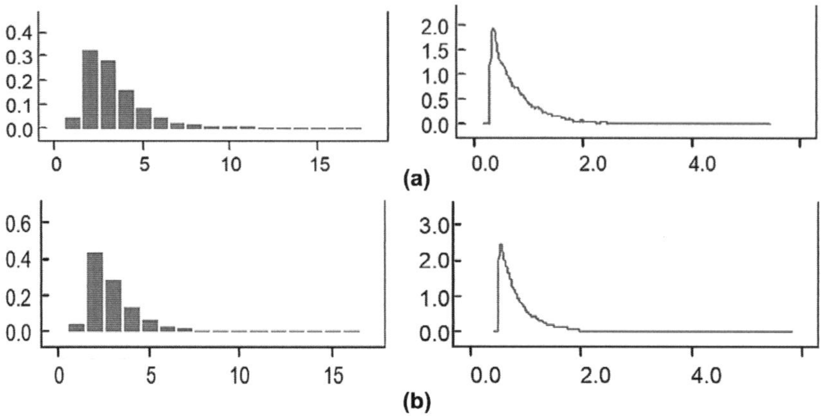

Figure 2. Kernel density plots for precision parameter and latent clusters: (*a*) kernel densities for Dirichlet Spatial (Model 4); (*b*) kernel densities for Dirichlet without Spatial (Model 3).

Apart from the above findings, the logical aspect of employing the flexible approach should also be given consideration. For a given crash dataset, the parametric approach assumes a restrictive stationary distribution of explanatory variables across all the sites under focus. As discussed in previous studies of semi-parametric models (*25*, *26*), the Dirichlet formulation allows the examination of the adequacy of standard parametric assumption. As clearly shown in Figure 2, the kernel posterior density plots of Dirichlet precision parameter *k* illustrate the closeness of the peak toward zero, which reflects that the unknown density (G) of non-parametric intercept is far from the baseline distribution (G_0). Similar plots for both Dirichlet models suggest their robustness and indicate that the normal assumption of intercept associated with traditional parametric models does not hold true for the TAZ-level crash dataset of the current study. This indicates that the 203 intercepts associated with the TAZs are not normally distributed and the standard parametric approach does not hold true for the concerned pedestrian and bicycle crashes at the planning level. This finding seems plausible, as the safety mechanisms which impact the pedestrian and bicycle crashes may vary across different TAZs as a result of diverse factors (such as driving behavior, road environment, and so forth) which may not be captured in the explanatory variables. These findings also suggest the presence of distinct subpopulations among the crash data, which was confirmed from the histogram of posterior number of latent clusters with a median of 2 clusters for most of the data. This capability of Dirichlet models to identify the latent subpopulations may prove highly beneficial for safety agencies in investigating similarities in the safety issues among different sites and allocating funding for dedicated countermeasures (*26*). This is achieved by calculating the expected probabilities of sites to fall into same clusters, which allows detection of the degree of similarity or dissimilarity among sites based on the crash risk (*25*).

The aforementioned advantages justify the use of Dirichlet process mixture with a flexible intercept, as this model specification supports more precise estimation leading to better inferences. Contrary to the parametric models which restrict the priors to a specific distribution fixed across all entities, the latent clusters capture the multimodality as a result of their unconstrained nature.

Conclusions and Recommendations

The current study contributes to the safety literature by proposing a bivariate Dirichlet process mixture spatial model and comparing its performance for crash predictions with other three competing models. The proposed semi-parametric model accounted for the unobserved heterogeneity by combining the strengths of incorporating bivariate specification to accommodate correlation among crash modes, spatial random effects for the impact of neighboring TAZs, and Dirichlet process mixture for random intercept. The present model structure allowed the flexibility to infer stochastic parameter from the crash data instead of a prespecified distribution.

All four models shared similar influential factors across both crash modes, which indicated the robustness of the models. For crashes pertaining to bicycles, K12 student enrollment, percentage of arterials, and bike-lane density for the TAZ were observed to be statistically significant at the 95% confidence interval. The positive correlation of K12 student enrollment with crash risk suggests the increased risk as a result of higher chances of physical interactions of bicyclists/pedestrians with other modes because of more exposure. However, the perceived risk appears to be the governing factor in the case of positive correlation for bike-lane density, which seems counter-intuitive. The lower perceived risk may encourage bicyclists to ride more in such areas, and therefore yield higher chances of the exposure of bicyclists to vehicular traffic. A negative correlation was observed for percentage of arterial roads and bicycle crashes, which suggests a lesser tendency of bicyclists to travel in areas with more arterials, thus reducing the exposure to possible interactions. The pedestrian crashes were observed to reduce with an increase in student population in the colleges of TAZs. This may be justified by the policies implemented in these areas for reduced vehicular traffic, which eventually reduces the possibility of interaction with pedestrians.

The heterogeneity error term demonstrated the presence of statistically significant correlation among the bicycle and pedestrian crashes, whereas the spatial random effect term exhibited the absence of a significant correlation, which might explain why models considering the spatial random effects did not yield the expected advantages compared with their non-spatial counterparts. In the comparison between Dirichlet and non-Dirichlet models, the former were consistently superior to typical bivariate ones under all criteria. These findings demonstrate the advantages associated with consideration of a flexible approach, Dirichlet process mixture in the current study, based on the goodness-of-fit and predictive accuracy of estimated crash counts. Moreover, the Dirichlet models exhibited the capability to identify the latent distinct subpopulations, and suggested the that the normal assumption of intercept associated with traditional parametric models does not hold true for the TAZ-level crash dataset of the current study. These findings justify the development of sophisticated flexible models which generate more precise estimate as a result of the unrestrictive approach, which eventually leads to better inferences.

Based on the results, this study recommends careful consideration of spatial correlations at the macro-level of TAZs, as they increased the complexity without any significant advantage at model fit or predictive accuracy. The authors also recommend exploring other spatial levels and observing whether the results of the current study hold true or if the spatial random effects prove beneficial. Similar to other studies that focus on crashes pertaining to modes of active transportation, it should be noted that both the pedestrian and bicycle crashes have been modeled by utilizing the exposure of vehicles, rather than pedestrians and bikes, owing to the unavailability of exposure data for the concerned modes. It is recommended that novel methods may be explored to account for the exposure data, such as using bike mode share, or calibrating the exposure from socioeconomic factors related to such modes (e.g., number of employees walking or cycling to work). Finally, the crash dataset utilized for model development was aggregated for a 6-year period, and future studies may incorporate temporal correlations and adopt disaggregated crash counts (*51*).

Acknowledgments

Support for this study was provided by 2017 UCCONNECT Research Program (Grant number: 65A0529-052.3).

Author Contributions

The authors confirm contribution to the paper as follows: study conception and design: Wen Cheng; data collection: Tom Vo and Taha Sakrani; analysis and interpretation of results: Gurdiljot Gill and Jiao Zhou; draft manuscript preparation: Wen Cheng and Gurdiljot Gill. All authors reviewed the results and approved the final version of the manuscript.

References

1. National Highway Traffic Safety Administration. *Traffic Safety Facts: Bicyclists and Other Cyclists*. United States Department of Transportation (DOT HS 811 624), 2012.
2. Williams, A. F. *Protecting Pedestrians and Bicyclists: Some Observations and Research Opportunities*. Insurance Institute for Highway Safety, 2013. http://www.iihs.org/frontend/iihs/documents/masterfiledocs.ashx. Accessed July 25, 2017.
3. Berrigan, D., R. Troiano, T. McNeel, C. DiSogra, and R. Ballard-Barbash. Active Transportation Increases Adherence to Activity Recommendations. *American Journal of Preventive Medicine*, Vol. 31, No. 3, 2006, pp. 210–216.
4. Frank, L., M. Greenwald, S. Winkelman, J. Chapman, and S. Kavage. Carbonless Footprints: Promoting Health and Climate Stabilization Through Active Transportation. *Preventive Medicine*, Vol. 50, 2010, pp. S99–S105.
5. Furie, G., and M. Desai. Active Transportation and Cardiovascular Disease Risk Factors in U.S. Adults. *American Journal of Preventive Medicine*, Vol. 43, No. 6, 2012, pp. 621–628.
6. Giles-Corti, B., S. Foster, T. Shilton, and R. Falconer. The Co-Benefits for Health of Investing in Active Transportation. *New South Wales Public Health Bulletin*, Vol. 21, No. 6, 2010, pp. 122–127.
7. Insall, P. Active Travel: Transport Policy and Practice for Health. *Nutrition Bulletin*, Vol. 38, No. 1, 2013, pp. 61–69.
8. Wanner, M., T. Götschi, E. Martin-Diener, S. Kahlmeier, and B. Martin. Active Transport, Physical Activity, and Body Weight in Adults. *American Journal of Preventive Medicine*, Vol. 42, No. 5, 2012, pp. 493–502.
9. Lee, C., and M. Abdel-Aty. Comprehensive Analysis of Vehicle–Pedestrian Crashes at Intersections in Florida. *Accident Analysis & Prevention*, Vol. 37, No. 4, 2005, pp. 775–786.
10. Moudon, A., L. Lin, J. Jiao, P. Hurvitz, and P. Reeves. The Risk of Pedestrian Injury and Fatality in Collisions with Motor Vehicles, a Social Ecological Study of State Routes

and City Streets in King County, Washington. *Accident Analysis & Prevention*, Vol. 43, No. 1, 2011, pp. 11–24.
11. Vivoda, J., D. Eby, R. St. Louis, and L. Kostyniuk. Cellular Phone Use While Driving at Night. *Traffic Injury Prevention*, Vol. 9, No. 1, 2008, pp. 37–41.
12. Beck, L., A. Dellinger, and M. O'Neil. Motor Vehicle Crash Injury Rates by Mode of Travel, United States: Using Exposure-Based Methods to Quantify Differences. *American Journal of Epidemiology*, Vol. 166, No. 2, 2007, pp. 212–218.
13. Wardlaw, M. Assessing the Actual Risks Faced by Cyclists. *Traffic Engineering and Control*, Vol. 43, No. 11, 2002, pp. 352–356.
14. Cai, Q., J. Lee, N. Eluru, and M. Abdel-Aty. Macro-Level Pedestrian and Bicycle Crash Analysis: Incorporating Spatial Spillover Effects in Dual State Count Models. *Accident Analysis & Prevention*, Vol. 93, 2016, pp. 14–22.
15. Bijleveld, F. The Covariance Between the Number of Accidents and the Number of Victims in Multivariate Analysis of Accident Related Outcomes. *Accident Analysis & Prevention*, Vol. 37, No. 4, 2005, pp. 591–600.
16. Park, E., and D. Lord. Multivariate Poisson-Lognormal Models for Jointly Modeling Crash Frequency by Severity. *Transportation Research Record: Journal of the Transportation Research Board*, 2007. 2019: 1–6.
17. Congdon, P. *Bayesian Statistical Modeling*. John Wiley & Sons, West Sussex, 2001.
18. Conway, A., J. Cheng, D. Peters, and N. Lownes. Characteristics of Multimodal Conflicts in Urban On-Street Bicycle Lanes. *Transportation Research Record: Journal of the Transportation Research Board*, 2013. 2387: 93–101.
19. Strauss, J., L. Miranda-Moreno, and P. Morency. Multimodal Injury Risk Analysis of Road Users at Signalized and Non-Signalized Intersections. *Accident Analysis & Prevention*, Vol. 71, 2014, pp. 201–209.
20. Guo, F., X. Wang, and M. Abdel-Aty. Modeling Signalized Intersection Safety with Corridor-Level Spatial Correlations. *Accident Analysis & Prevention*, Vol. 42, No. 1, 2010, pp. 84–92.
21. Abdel-Aty, M., and X. Wang. Crash Estimation at Signalized Intersections Along Corridors: Analyzing Spatial Effect and Identifying Significant Factors. *Transportation Research Record: Journal of the Transportation Research Board*, 2006. 1953: 98–111.
22. Gill, G., W. Cheng, M. Xie, T. Vo, X. Jia, and J. Zhou. Evaluating Influence of Neighboring Structures on Spatial Crash Frequency Modeling and Site-Ranking Performance. *Transportation Research Record: Journal of the Transportation Research Board*, 2017. 2659: 117–126.
23. Narayanamoorthy, S., R. Paleti, and C. Bhat. On Accommodating Spatial Dependence in Bicycle and Pedestrian Injury Counts by Severity Level. *Transportation Research Part B: Methodological*, Vol. 55, 2013, pp. 245–264.
24. Nashad, T., S. Yasmin, N. Eluru, J. Lee, and M. Abdel-Aty. Joint Modeling of Pedestrian and Bicycle Crashes. *Transportation Research Record: Journal of the Transportation Research Board*, 2016. 2601: 119–127.
25. Heydari, S., L. Fu, D. Lord, and B. Mallick. Multilevel Dirichlet Process Mixture Analysis of Railway Grade Crossing Crash Data. *Analytic Methods in Accident Research*, Vol. 9, 2016, pp. 27–43.
26. Shirazi, M., D. Lord, S. Dhavala, and S. Geedipally. A Semiparametric Negative Binomial Generalized Linear Model for Modeling Over-Dispersed Count Data with a Heavy Tail: Characteristics and Applications to Crash Data. *Accident Analysis & Prevention*, Vol. 91, 2016, pp. 10–18.
27. Heydari, S., L. Fu, L. Miranda-Moreno, and L. Jopseph. Using a Flexible Multivariate Latent Class Approach to Model Correlated Outcomes: A Joint Analysis of Pedestrian and Cyclist Injuries. *Analytic Methods in Accident Research*, Vol. 13, 2017, pp. 16–27.
28. Ohlssen, D., L. Sharples, and D. Spiegelhalter. Flexible Random-Effects Models Using Bayesian Semi-Parametric Models: Applications to Institutional Comparisons. *Statistics in Medicine*, Vol. 26, No. 9, 2007, pp. 2088–2112.
29. Cheng, W., G. Gill, R. Dasu, M. Xie, X. Jia, and J. Zhou. Comparison of Multivariate Poisson Lognormal Spatial and Temporal Crash Models to Identify Hot Spots of Intersections Based on Crash Types. *Accident Analysis & Prevention*, Vol. 99, 2017, pp. 330–341.
30. Congdon, P. *Bayesian Statistical Modeling*, 2nd ed. Wiley, New York, 2006.
31. Mardia, K. Multi-Dimensional Multivariate Gaussian Markov Random Fields with Application to Image Processing. *Journal of Multivariate Analysis*, Vol. 24, No. 2, 1988, pp. 265–284.
32. Jonathan, A., K. Wu, and E. Donnell. A Multivariate Spatial Crash Frequency Model for Identifying Sites with Promise Based on Crash Types. *Accident Analysis & Prevention*, Vol. 87, 2016, pp. 8–16.
33. Aguero-Valverde, J., and P. Jovanis. Bayesian Multivariate Poisson Lognormal Models for Crash Severity Modeling and Site Ranking. *Transportation Research Record: Journal of the Transportation Research Board*, 2009. 2136: 82–91.
34. Xu, P., and H. Huang. Modeling Crash Spatial Heterogeneity: Random Parameter Versus Geographically Weighting. *Accident Analysis & Prevention*, Vol. 75, 2015, pp. 16–25.
35. Ishwaran, H., and L. James. Gibbs Sampling Methods for Stick-Breaking Priors. *Journal of the American Statistical Association*, Vol. 96, No. 453, 2001, pp. 161–173.
36. Ghosh, S., and J. Norris. Bayesian Capture-Recapture Analysis and Model Selection Allowing for Heterogeneity and Behavioral Effects. *Journal of Agricultural, Biological, and Environmental Statistics*, Vol. 10, No. 1, 2005, pp. 35–49.
37. Erkanli, A., M. Sung, E. Jane Costello, and A. Angold. Bayesian Semi-Parametric ROC Analysis. *Statistics in Medicine*, Vol. 25, No. 22, 2006, pp. 3905–3928.
38. Gelfand, A. E., D. K. Dey, and H. Chang. *Model Determination Using Predictive Distributions with Implementation Via Sampling-Based Methods (No. TR-462)*. Stanford University, CA, Department of Statistics, 1992.
39. Gelfand, A. E. Model Determination Using Sampling-Based Methods. In W.R. Gilks, S. Richardson, & D. Spiegelhalter (eds.), *Markov Chain Monte Carlo in Practice*, Chapman and Hall/CRC, 1996, pp. 145–161.

40. Cheng, W., G. S. Gill, Y. Zhang, and Z. Cao. Bayesian Spatiotemporal Crash Frequency Models with Mixture Components for Space-Time Interactions. *Accident Analysis & Prevention*, Vol. 112, 2018, pp. 84–93.
41. Cheng, W., and S. Washington. Experimental Evaluation of Hotspot Identification Methods. *Accident Analysis & Prevention*, Vol. 37, No. 5, 2005, pp. 870–881.
42. Washington, S., M. G. Karlaftis, and F. L. Mannering. *Statistical and Econometric Methods for Transportation Data Analysis*. Chapman & Hall, Boca Raton, 2003.
43. Earnest, A., G. Morgan, K. Mengersen, L. Ryan, R. Summerhayes, and J. Beard. Evaluating the Effect of Neighbourhood Weight Matrices on Smoothing Properties of Conditional Autoregressive (CAR) Models. *International Journal of Health Geographics*, Vol. 6, No. 1, 2007, p. 54.
44. Ladrón, de, F. Guevara, S. Washington, and J. Oh. Forecasting Crashes at the Planning Level: Simultaneous Negative Binomial Crash Model Applied in Tucson, Arizona. *Transportation Research Record: Journal of the Transportation Research Board*, 2004. 1897: 191–199.
45. Hadayeghi, A., A. Shalaby, and B. Persaud. Safety Prediction Models: Proactive Tool for Safety Evaluation in Urban Transportation Planning Applications. *Transportation Research Record: Journal of the Transportation Research Board*, 2007. 2019: 225–236.
46. Abdel-Aty, M., C. Siddiqui, H. Huang, and X. Wang. Integrating Trip and Roadway Characteristics to Manage Safety in Traffic Analysis Zones. *Transportation Research Record: Journal of the Transportation Research Board*, 2011. 2213: 20–28.
47. Spiegelhalter, D., A. Thomas, N. Best, and D. Lunn. *WinBUGS Version 1.4 User Manual*. MRC Biostatistics Unit, Cambridge, 2003. http://www.mrc-cam.ac.uk/bugs. Accessed July 25, 2017.
48. Gelman, A., and D. Rubin. Inference from Iterative Simulation Using Multiple Sequences. *Statistical Science*, Vol. 7, No. 4, 1992, pp. 457–472.
49. Geedipally, S., D. Lord, and S. Dhavala. A Caution About Using Deviance Information Criterion While Modeling Traffic Crashes. *Safety Science*, Vol. 62, 2014, pp. 495–498.
50. Ntzoufras, I. *Bayesian Modeling in WinBugs*. Wiley Series in Computational Statistics, Hoboken, 2012.
51. Cheng, W., G. S. Gill, S. Choi, J. Zhou, X. Jia, and M. Xie. Comparative Evaluation of Temporal Correlation Treatment in Crash Frequency Modelling. *Transportmetrica A: Transport Science*, 2017, pp. 615–633. https://doi.org/10.1080/23249935.2017.1418458

The Standing Committee on Safety Data, Analysis and Evaluation (ANB20) peer-reviewed this paper (18-05685).

Article

TRR
JOURNAL OF THE TRANSPORTATION RESEARCH BOARD

Transportation Research Record
2018, Vol. 2672(38) 116–127

DOI: 10.1177/0361198118777088
journals.sagepub.com/home/trr
SAGE

Multiple Correspondence Analysis of Pedestrian Crashes in Rural Illinois

Raghunandan Baireddy[1], Huaguo Zhou[1], and Mohammad Jalayer[2]

Abstract

During the five-year period between 2010 and 2014, there were 24,178 pedestrian crashes in Illinois. Approximately 4.39% of these pedestrian crashes occurred in rural areas; and approximately 40% of the crashes resulted in a severe injury or a fatality. Thus, pedestrian safety problems exist in rural locales, and the factors contributing to these problems need to be investigated. The goal of this study is to answer the question: "Which variable categories, when acting together, contribute more to the occurrence of pedestrian crashes in rural areas?" Crashes are random events stemming from the convergence of various factors. However, traditional statistical tools can only make pairwise comparisons of dependent and independent variables. Therefore, it is necessary to apply an analytical tool that can identify complex underlying structures in crash data and spot associations among variable categories that contribute to crash occurrence. The multiple correspondence analysis (MCA) method, which is used in this study, can do just that. According to the obtained results, categories of variables such as roadway functional class, the number of lanes, lighting conditions, weather conditions, traffic control devices, driver condition, and pedestrian condition were proved to contribute to pedestrian crashes in rural Illinois.

In the year 2015, there were 5,376 fatal pedestrian crashes in the United States, accounting for 15.31% of all traffic fatalities. Approximately 26% of pedestrian fatalities in 2015 occurred in rural areas (*1*). This statistic overlooks the fact that pedestrian crashes in rural areas tend to be more severe than in urban areas (*2*). The reason for this trend is that rural roadways have a distinct set of characteristics which makes them more dangerous to pedestrian activity compared with urban roads. For instance, rural roads tend to have minimal pedestrian infrastructure and higher speed limits which create an unsafe environment for pedestrians (*3*). It is estimated that the total cost of pedestrian fatalities and injuries in the year 2000 was $20.8 billion; in today's dollars this would be equivalent to $28.63 billion. All these facts highlight the importance of the need for analyzing rural pedestrian crashes (*2*).

In this study, pedestrian crashes occurring in rural Illinois between the years 2010 and 2014 are analyzed. A preliminary analysis of crash data revealed that, during those five years, there were 24,178 pedestrian crashes in the state. But only 1,062 (4.39%) of them happened in rural areas. It also was observed that 19.68% of urban pedestrian crashes resulted in a severe injury or fatality. Contrastingly, the same statistic for rural pedestrian crashes is 40.41%. This extreme variation in severity of rural and urban pedestrian crashes calls for a detailed investigation. More specifically, the factors contributing to pedestrian crashes in rural Illinois must be identified. Doing that would aid transportation professionals in accurately targeting roadways unsafe for pedestrians and applying appropriate countermeasures.

There exists a vast body of knowledge, comprised of several past studies, concerning the identification of factors contributing to pedestrian crashes in general. However, only a few of those studies focused exclusively on pedestrian crashes along rural roadways (*4–10*). Pour-Rouholamin et al. (*4*) analyzed all types of pedestrian crashes in Illinois using a partial proportional odds model and noted that rural areas showed a 37.4% higher probability of severe injury crashes compared with urban areas. The reasons behind this statistic were speculated to be higher speed limits and lack of quick access to emergency medical services. Tulu et al. (*5*) analyzed pedestrian crashes occurring on rural two-lane, two-way roads in Ethiopia using a random parameter negative binomial model. Their results indicated that the presence of wider

[1]Department of Civil Engineering, Auburn University, Auburn, AL
[2]Center for Advanced Infrastructure and Transportation (CAIT), Rutgers, The State University of New Jersey, Piscataway, NJ

Corresponding Author:
Address correspondence to Raghunandan Baireddy: raghu.b@auburn.edu

lanes, gravel shoulders, pedestrians aged less than 19 years, and pedestrians crossing in groups contributed to the reduction of pedestrian crashes. Conversely, the presence of mountainous terrain and pedestrians crossing in pairs are likely to increase crash risks. Islam et al. (*6*) analyzed pedestrian at-fault crashes on rural and urban roadways in Alabama using a mixed logit model. Their results indicated that the presence of county roads, open country, rainy weather, mid-block location, darkness along the road, and pedestrians younger than 12 years increased the probability of minor injury crashes. They also found that pedestrians walking against traffic and wearing safety equipment decreased the likelihood of the major injury crashes. Velloso et al. (*7*) conducted an on-the-spot study of 120 pedestrian crashes occurring where rural highways cross urban areas in Brazil. The results of this empirical study indicated that negligence, lack of attention, and use of alcohol by pedestrians are significant crash contributing factors. Carter et al. (*8*) conducted an exploratory research on pedestrian and bicycle crashes on rural highways in North Carolina. This study revealed that the presence of rural two-lane roads, mid-block locations, pedestrians failing to yield, and walking along the roadway are major crash contributing factors. Zajac et al. (*9*) investigated pedestrian crashes in rural roadways in Connecticut using the ordered probit model. This study found that vehicle type, driver alcohol involvement, pedestrians aged more than 65 years, pedestrian alcohol involvement, villages, downtown fringes, and low-density residential areas would significantly increase pedestrian injury severity.

It should be noted that all the studies mentioned previously utilized statistical models, which can examine the effects of crash-related variables individually. Therefore, the results produced by such models can only tell how a given variable affects crash severity or probability. However, such statistical techniques cannot answer the question: "Which variables, when acting together, have the highest influence on crash occurrence?" It is necessary to find the answer to this issue because often a crash occurs not due to a single cause but because of convergence of multiple factors (*10*).

A statistical technique called multiple correspondence analysis (MCA) enables researchers to answer the question mentioned in the previous paragraph. The MCA method is unique because it facilitates visualization of crash data and helps in identifying groups of contributing factors that are likely to cause a crash. Application of the MCA method in transportation research is not new, but it is not as common as other frequently used statistical techniques such as logit, probit, binomial, and multinomial models. A study conducted by Das et al. (*11*) analyzed pedestrian crashes in Louisiana by using MCA. Another study (*12*) employed the MCA method to identify contributing factors related to single-vehicle fatal crashes in Louisiana. Similarly, Jalayer et al. (*13*) analyzed wrong-way driving (WWD) crashes in Alabama by using the MCA method and identified contributing factors for such crashes.

This study uses the MCA method to explore the clustering of factors contributing to rural pedestrian crashes in Illinois. The results thus obtained would be in the form of distinct groups of variable categories which, when present together, are likely to cause a crash. The benefit of identifying such category groups is that it enables transportation professionals to deploy safety treatments to locations where all the variable categories in a group are present. Per the above-mentioned literature review and to the best of authors' knowledge, no study in the past has focused on rural pedestrian crashes in Illinois. Given the renewed interest in pedestrian safety in the Illinois Department of Transportation's (IDOT) Strategic Highway Safety Plan (*14*), it is imperative that rural pedestrian crashes in their state must be studied and remedied.

Multiple Correspondence Analysis

Unlike other statistical analysis tools, MCA does not require the classification of variables as independent and dependent. Additionally, making assumptions regarding the distribution of data also is unnecessary in MCA (*15*). Conventional statistical tools are limited to comparing variables only in pairs (*13*). In contrast, MCA explores the relationships between multiple nominal variables at the same time (*16*). Such exploration is made possible by reorganizing crash data and representing them in a geographical space. An in-depth discussion of MCA is provided in Greenacre and Blasius (*17*) and Le Roux and Rouanet (*18*). The goal of the MCA method is to map all the variable categories of crash data onto an *XY* plane. MCA achieves this by implementing the following three steps:

1. Construction of indicator matrix
2. Computation of individual clouds
3. Computation of category clouds

Indicator Matrix

In an indicator matrix, each row represents an individual (i); in this study, every crash record is an individual. Similarly, columns represent variable categories (j). For instance, the "Driver Age" column in original crash data accounts for a nominal variable with three values: adult, middle, and old. In the indicator matrix, information on "Driver Age" is conveyed in three columns, one for each

Table 1. Creation of Indicator Matrix

Original crash data			Indicator matrix			
Individual	Driver age		Individual	Driver age_adult	Driver age_middle	Driver age_old
Crash 1	Old		Crash 1	0	0	1
Crash 2	Adult		Crash 2	1	0	0
Crash 3	Old		Crash 3	0	0	1
Crash 4	Middle		Crash 4	0	1	0
Crash 5	Middle		Crash 5	0	1	0

category. Consequently, an adult driver would be coded as "1 0 0," a middle-aged driver as "0 1 0," and an old driver as "0 0 1." The same process was repeated on all variables of original data to create the indicator matrix. The finished indicator matrix consists of several binary columns, with only a single column per nominal variable taking the value "1." Table 1 illustrates the creation of an indicator matrix from crash data.

If p is the number of variables and each variable p has T_p number of categories, then the total number of categories (columns) T in the indicator matrix would be $T = \sum_{p=1}^{P} T_p$.

Computing Clouds of Individuals

The cloud of individuals is based on the set of all distances between individual records for a variable, for which different categories have been selected. Clouds of individuals as well as categories can be computed based on the indicator matrix. Here, a cloud indicates a spatial representation of individuals or categories on a two-dimensional (2-D) plane. Consider a variable p with categories k and k'. Assume that n_k denotes the number of individuals that selected category k. Now, the relative frequency (f_k) is n_k/n, where n is the total number of individuals. Values of f_k for each category describe the row profile of the individual cloud.

Now, consider two individuals i and i', assume that i selected category k and that i' selected k'. The squared distance d_p^2, between records i and i' with respect to variable p can be calculated using Equation 1 (*11*, *12*) as

$$d_p^2(i,i') = \frac{1}{f_k} + \frac{1}{f_{k'}} \tag{1}$$

The total squared distance between i and i', D^2, with respect to all variables in the matrix can be calculated using Equation 2 (*11*, *12*). Values of D^2 and f_k together describe the position of a cloud of individuals.

$$D^2(i,i') = \frac{1}{P}\sum_{p \in P} d_p^2(i,i') \tag{2}$$

An origin point is required for plotting the cloud of individuals. That origin point (O) is the mean point of the cloud. If M^i is a point denoting individual record i, then the squared distance between O and M^i can be computed using Equation 3 (*11*, *12*) as

$$\left(OM^i\right)^2 = \frac{1}{P}\sum_{k \in K_i} \frac{1}{f_k} \tag{3}$$

Computing Clouds of Categories

A cloud of categories is comprised of K points (where K denotes the number of categories in the indicator matrix). Assume that, in the cloud, categories k and k' are depicted by points M^k and $M^{k'}$. These two points have weights n_k and $n_{k'}$, respectively. With n_k (or $n_{k'}$) being the number of individuals selected by category k (or k'). Consider $n_{kk'}$ denotes the number of individual records that selected both categories k and k'. Selecting two categories of the same variable will result in a null value for $n_{kk'}$ (*11*, *12*). Now, the squared distance between points M^k and $M^{k'}$ can be computed using equation 4 (*13*) as

$$\left(M^k M^{k'}\right)^2 = \frac{n_k + n_{k'} - 2n_{kk'}}{n_k n_{k'}/n} \tag{4}$$

Origin point for the cloud of categories is the same as the origin of the individual cloud (*19*). Any point M^k in the category cloud is plotted using its corresponding weight n_k, squared distance from the origin, and squared distance from other points in the cloud.

Data

For this study, information on all pedestrian crashes that occurred in Illinois between 2010 and 2014 was acquired from IDOT. The crash data maintained by IDOT comes in three separate data sets: crash data, person data, and vehicle data. Crash data includes information on spatial and temporal aspects of a crash such as time, day, season, location, severity, and so forth. Individual data set tells us about the age, gender, physical condition, visibility, and so forth, of the persons (both driver and pedestrian) involved in the crash. The vehicle data set has details on the type of vehicle involved in the crash. Vehicle data is deemed irrelevant to this study.

This study deals only with pedestrian crashes that had occurred at rural locations in the state of Illinois. The crash data matching this description were extracted from the original dataset by using the variables 'CollisionTypeCode' and 'ClassOfTrafficWayCode'. If the value of 'CollisionTypeCode' variable equals 1, it indicates a pedestrian crash. Thus, all crash records coded as 'CollisionTypeCode = 1' were filtered to extract pedestrian crashes. Rural pedestrian crashes are further extracted using the 'ClassOfTrafficWayCode' variable. According to the metadata provided by IDOT, all crash records with 'ClassOfTrafficWayCode' value of 0 to 4 have occurred on rural roadways. All pedestrian crash records coded in this way were filtered and extracted to create a spreadsheet containing exclusively rural pedestrian crashes.

Information on driver and pedestrian characteristics was available in the person dataset. The rural pedestrian crash dataset was joined with person dataset using a common variable called ICN, which is a unique identification number assigned to each crash record. The two datasets were joined based on the ICN value to create a single crash database with information on the crash, and persons involved in each crash. In total, 1,062 rural pedestrian crashes were extracted for analysis. A cross tabulation of crash data thus derived is included in Table 2. Some of the variables presented in Table 2 are overly skewed toward one of their categories. For instance, 69.77% of crashes occurred on weekdays. Mid-block crashes accounted for 82.49% of all crashes. Pedestrians who did not wear contrasting clothing features in 66.29% of crashes. In 87.19% of crashes the weather was clear or cloudy, and 81.26% crashes occurred along dry roads. Only 1.14% of motorists involved in these crashes were speeding. Uncontrolled locations saw 66.48% of crashes. Passenger cars feature prominently (66.48%) in these crashes. No road defects were present at 94.54% of crash locations, and 96.05% crashes happened along straight aligned roads. The remaining variables do not exhibit any skewing; therefore, they must be further investigated to ascertain their influence on crash occurrence.

The significance of some of the skewed variables is apparent. Previous studies found that visibility is a key factor in avoiding pedestrian crashes (*20*, *21*), so pedestrians who did not wear contrasting clothing tend to be less visible and are more prone to crashes. Similarly, it was revealed in past research that mid-block and uncontrolled location are hazardous to pedestrians (*22*, *23*). In

Table 2. Frequency Distribution of Variable Categories

Category	Frequency	Percentage
Injury severity		
Minor injury	447	42.09
No or possible injury	175	16.48
Severe injury	440	41.43
Month		
Autumn	339	31.92
Spring	231	21.75
Summer	271	25.52
Winter	221	20.81
Day of week		
Weekday	741	69.77
Weekend	321	30.23
Time of crash		
Afternoon	244	22.98
Evening	237	22.32
Morning	187	17.61
Night	394	37.10
Intersection		
No	876	82.49
Yes	186	17.51
Hit and run		
No	865	81.45
Yes	197	18.55
Number of lanes		
UNK	112	10.55
Five	36	3.39
Four	198	18.64
More than six	42	3.95
One	29	2.73
Three	30	2.82
Two	615	57.91
Work zone related		
No	1047	98.59
Yes	15	1.41
Vehicle maneuver		
Backing	35	3.30
Entering traffic or merging	13	1.22
Other or unknown	155	14.60
Slowing or stopping	59	5.56
Straight ahead	576	54.24
Turning	224	21.09
Primary cause		
Careless driving	412	38.79
DUI	71	6.69
Others	428	40.30
Speeding	72	6.78
Vision obscured	52	4.90
Weather	27	2.54
Pedestrian visibility		
Contrasting clothing	200	18.83
No contrasting clothing	704	66.29
Unknown	158	14.88
Pedestrian action		
Crossing	312	29.38
Entering or leaving a location	97	9.13
In roadway	262	24.67
Intoxicated	51	4.80
Other	161	15.16
Walking along traffic	179	16.85

(continued)

Table 2. Continued

Category	Frequency	Percentage
Pedestrian location		
In crosswalk	272	25.61
In roadway	616	58.00
Not in roadway	46	4.33
Unknown or n/a	128	12.05
Lighting condition		
Dark-lit	186	17.51
Dark-not lit	350	32.96
Dawn or dusk	37	3.48
Daylight	489	46.05
Weather condition		
Adverse	136	12.81
Clear or cloudy	926	87.19
Pedestrian		
Adult	254	23.92
Child	163	15.35
Middle-age	548	51.60
Old	97	9.13
Pedestrian gender		
Female	471	44.35
Male	580	54.61
Speeding		
Yes	12	1.14
No	1050	98.87
Driver age		
Adult	216	20.34
Middle	601	56.59
Old	245	23.07
Driver gender		
Female	338	31.82
Male	602	56.68
Unknown	122	11.49
Driver condition		
DUI	98	9.23
Illness or fatigued	9	0.85
Normal	767	72.22
Unknown	188	17.70
Road surface condition		
Dry	863	81.26
Wet	199	18.74
Functional class		
Arterial	556	52.35
Collector	157	14.78
Interstate	43	4.05
Local road	306	28.81
Road defects		
Work zone	17	1.60
No defects	1004	94.54
Other	28	2.64
Pavement defects	13	1.22
Median type		
Alley or driveway	30	2.82
Center turn lane	45	4.24
Divided (barrier)	96	9.04
Divided (no barrier)	313	29.47
Not divided	535	50.38
One-way	12	1.13
Other	31	2.92
Traffic control device		

(continued)

Table 2. Continued

Category	Frequency	Percentage
Lane use marking	144	13.56
No controls	706	66.48
Warning sign	51	4.80
Stop sign or flasher	58	5.46
Traffic signal	92	8.66
Unknown	11	1.04
Traffic control device condition		
Not functioning	14	1.32
Functioning	320	30.13
No controls	717	67.51
Other	11	1.04
Vehicle type		
Bus or van	101	9.51
Motorcycle	5	0.47
Passenger	651	61.30
SUV or pickup	203	19.11
Truck or tractor	31	2.92
Unknown	71	6.69
Road alignment		
Curve	42	3.95
Straight	1020	96.05

Note: UNK = unknown; DUI = driving under the influence; n/a = not applicable.

contrast, variables such as clear or cloudy weather, dry road surface, and passenger cars are so ubiquitous that their preponderance in crash data is expected.

In total, this study included 30 variables and 110 variables categories. MCA was performed on the crash data using R Version 3.025 statistical software and the FactoMineR package, which was proven to be suitable for multivariate exploratory analysis (*24*). The results obtained from the analysis are discussed in the next section.

Results and Discussion

The most attractive aspect of MCA is the creation of factor maps using variable categories present in data. In the factor map, each category is represented by a point on an *XY* plane. The positioning of each point reflects its association with other categories. Categories with similar distribution, that is, categories that were simultaneously selected by multiple records, would be plotted close to each other (*13*). This graphical depiction of complex patterns within crash data enables identification of an underlying association among variable categories (*18*). However, it is not possible to explain all the variance present in the data in a 2-D space. Therefore, MCA initially depicts the data points in an *n*-dimensional space. The number of dimensions in that space would be less than the difference between the number

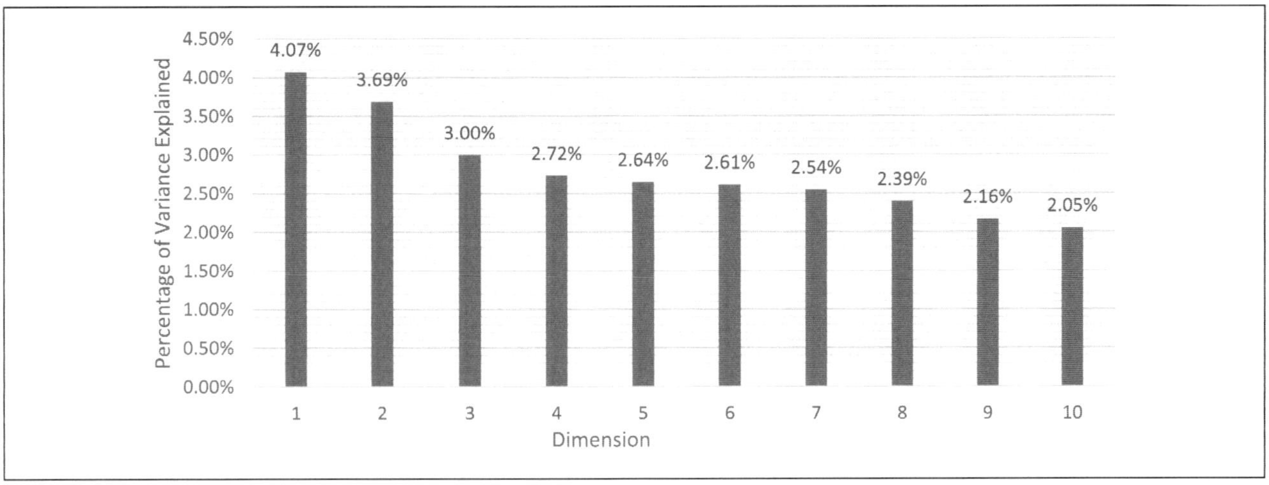

Figure 1. Percentage of variance explained by top ten dimensions.

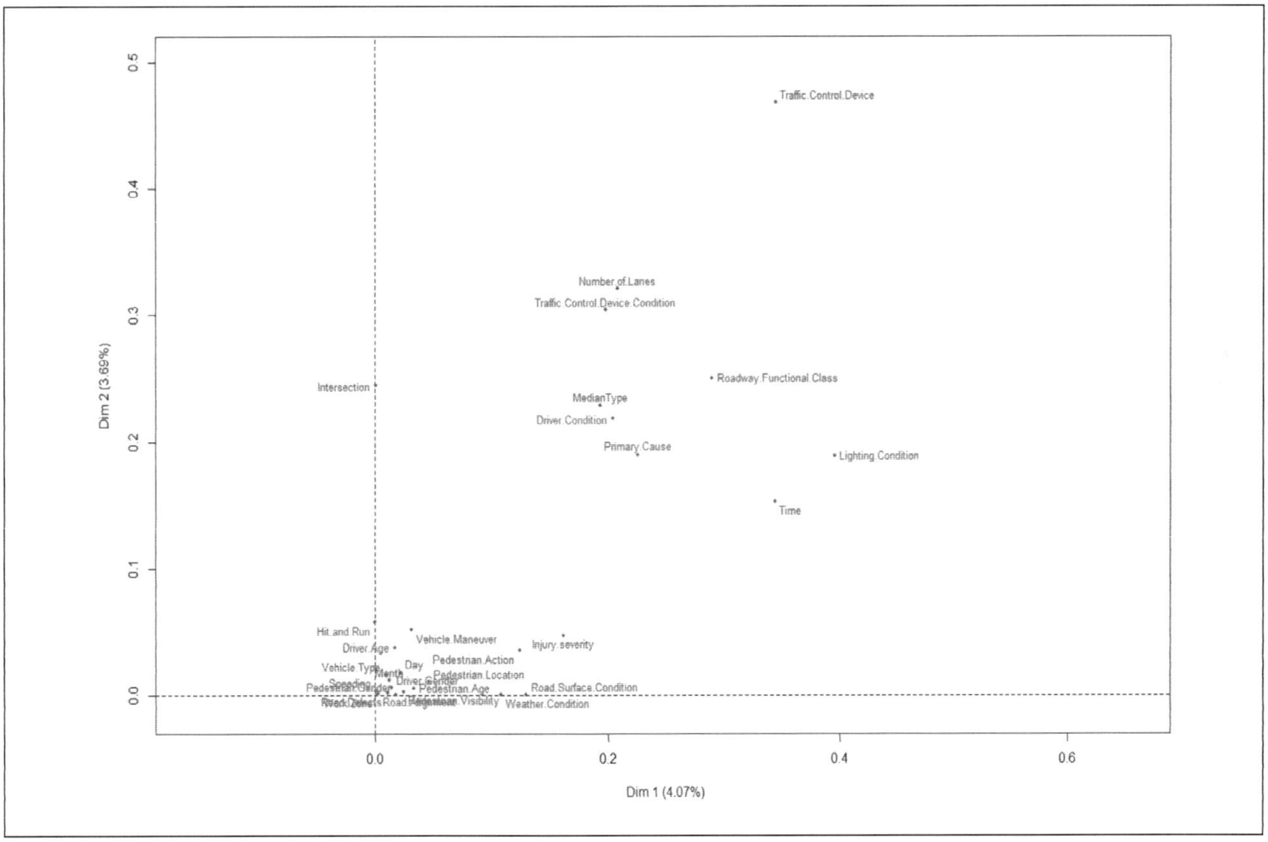

Figure 2. Factor map of study variables.

of categories and number of variables. To be specific, in this study, we have 78 dimensions. The variance explained by each dimension is measured by the eigen value of that dimension. Eigen value of a dimension can vary between 0 and 1. If the data are heterogeneous, as they are in this study, eigen values tend to be smaller (*13*). Heterogeneity of data indicates the random nature of crashes (*11*).

In this study, the eigen values of the first five dimensions are 0.11, 0.099, 0.081, 0.073, and 0.071, respectively.

To depict the factor map in a 2-D plane, the first two dimensions with the highest eigen values were selected. Together, these two dimensions can explain up to 7.57% of variance present in the data set. The eigen values of the first, second, and third dimensions are higher than the others, so a 2-D graph carries most, but not all, the information. Given this fact, the first two dimensions are usually plotted for the illustration purposes. Figure 1 describes the percentage of variance explained by the top 10 dimensions.

Based on Figure 1, the first and second principal axes on the planar surface, the MCA plot, describes 7.7 % of the total variances together. The low eigen values calculated for our database demonstrate that the variables are heterogeneous due to the random nature of the occurrence of crashes and reducing any of the variables may result in losing significant information with respect to the crash observations. Several other crash studies also found similar eigen values for the first two dimensions. For example, Das and Sun (2016) explored fatal run-off-road crashes and found out that the first two dimensions cover only 8.1% of the percentage of variance (*25*).

Figure 2 is an illustration of the associations among study variables as computed by MCA. Several variables have clustered at the origin; this means that their contributions (the eigen values) are smaller than those variables that reside away from the origin. Notable variables along the first dimension are lighting condition, time, traffic control device, and roadway functional class. Similarly, variables that contributed most to the second dimension are traffic control device, number of lanes, traffic control device condition, and intersection. The categories of variables that contributed more to the eigen values will give better insights into the underlying structures within crash data. Table 3 lists the R^2 (coefficient of determination) and p-values for each of the study variables for both dimensions.

The value of R^2 can vary between 0 and 1, where 0 indicates no relationship between the variable and MCA dimension, and 1 indicates a strong relationship (*12*). The p-value indicates the confidence level of the respective

Table 3. Significance Test Results for Study Variables in Top Two Dimensions

Variable	Dimension 1		Dimension 2	
	R^2	p-value	R^2	p-value
Crash location	0.070	<0.001	0.020	<0.001
Day of week	0.012	<0.001	0.020	<0.001
Driver age group	0.408	<0.001	0.018	<0.001
Driver condition	0.593	<0.001	0.018	<0.001
Driver gender	0.674	<0.001	0.178	<0.001
Functional class	0.010	0.014	0.021	<0.001
Hit and run	0.553	<0.001	0.275	<0.001
Injury severity	0.059	<0.001	0.005	0.028
Intersection	0.075	<0.001	0.147	<0.001
Lighting condition	0.018	0.004	0.005	0.021
Median type	0.018	0.004	0.358	<0.001
Month	Insignificant		0.219	<0.001
Number of lanes	0.015	0.012	0.022	<0.001
Pedestrian action	0.020	0.001	0.034	<0.001
Pedestrian age	Insignificant		0.240	<0.001
Pedestrian gender	0.037	<0.001	0.100	<0.001
Pedestrian location	Insignificant		0.014	0.001
Pedestrian visibility	Insignificant		0.032	<0.001
Primary cause	0.084	<0.001	0.033	<0.001
Road alignment	0.004	0.037	0.201	<0.001
Road defects	0.021	<0.001	0.006	0.011
Road surface condition	0.008	0.004	0.011	0.008
Speeding	0.096	<0.001	0.006	0.009
Traffic control device condition	0.072	<0.001	0.150	<0.001
Time	0.087	<0.001	0.154	<0.001
Traffic control device	0.101	<0.001	0.301	<0.001
Vehicle maneuver	0.178	<0.001	0.327	<0.001
Vehicle use	0.161	<0.001	0.035	<0.001
Weather condition	0.006	0.009	0.011	0.034
Work zone related	Insignificant		0.128	<0.001

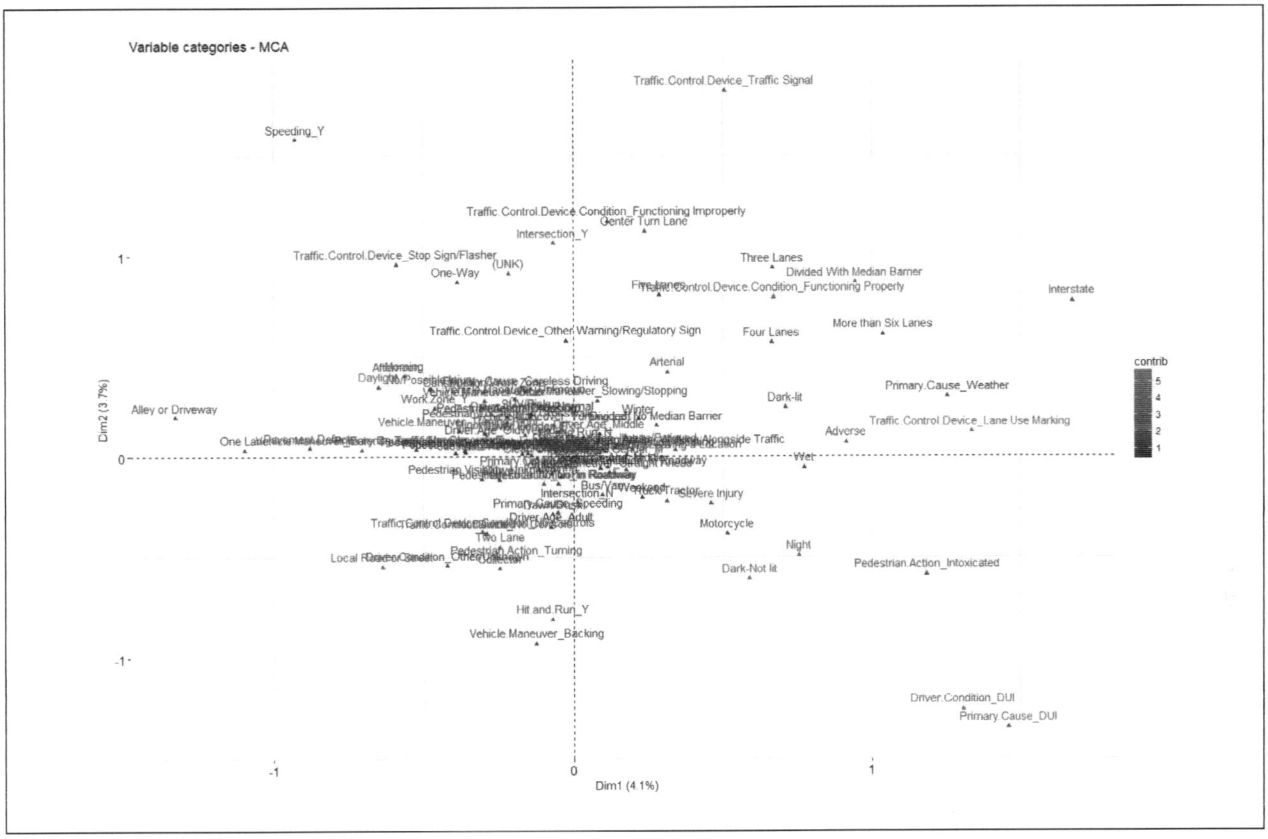

Figure 3. Factor map of all variable categories.

variable. Considering both R^2 and p-values, it can be seen from Table 3 that the variables driver gender, driver condition, driver age, vehicle maneuver, traffic control device, time, roadway functional class, median type, primary cause, and number of lanes are strongly related to one of the two dimensions. Figure 3 is the final representation of all variable categories in the form of points on a factor map. As discussed earlier, the further a category resides from the origin of the factor map, the higher is its contribution to the risk of a crash occurring. Moreover, the negatively correlated variable categories are located on opposite sides of the origin of the factor map, opposed quadrants. In Figure 3, categories with higher contribution are presented in a lighter color.

Figure 4 depicts only the top contributing categories with the remaining categories grayed out; this was done to conveniently explore the association among top contributing variable categories. The positioning of category points on the factor map enables the formation of distinct point clouds. Categories featured in each cloud are closely associated with each other; when acting together, these categories increase the risk of a pedestrian crash.

From Figure 4, several category clouds could be identified. Cloud 1 bunches the categories Four Lanes, Arterial, Divided with Median Barrier, and Traffic Control Device Condition_Functioning Properly together. This means that the majority of rural pedestrian crashes have occurred on four-lane divided arterial roads, despite traffic control devices functioning properly. On wide arterial roads, vehicles tend to travel at higher speeds, which make pedestrians more vulnerable at such locations (*26*). It is interesting to note that even the presence of an appropriate traffic control device could not prevent crashes on four-lane arterials, as indicated by these results.

Cloud 2 puts the variables Severe Injury, Adverse, Wet, Dark-Lit, and Traffic Control Device_Lane Use Marking near each other. Many of the severe injury rural pedestrian crashes have occurred during adverse weather conditions (such as cross-wind, snow, rain, hail, etc.) on dark-lit wet road surfaces with only lane use markings. As seen in previous studies, visibility is critical for pedestrian safety (*20*, *21*, *27*). Lack of traffic control devices other than lane use markings whose visibility is impeded by adverse weather conditions makes such locations hazardous to pedestrians. The problem of poor visibility during night time is compounded due to factors noted above and might lead to severe injury pedestrian crashes.

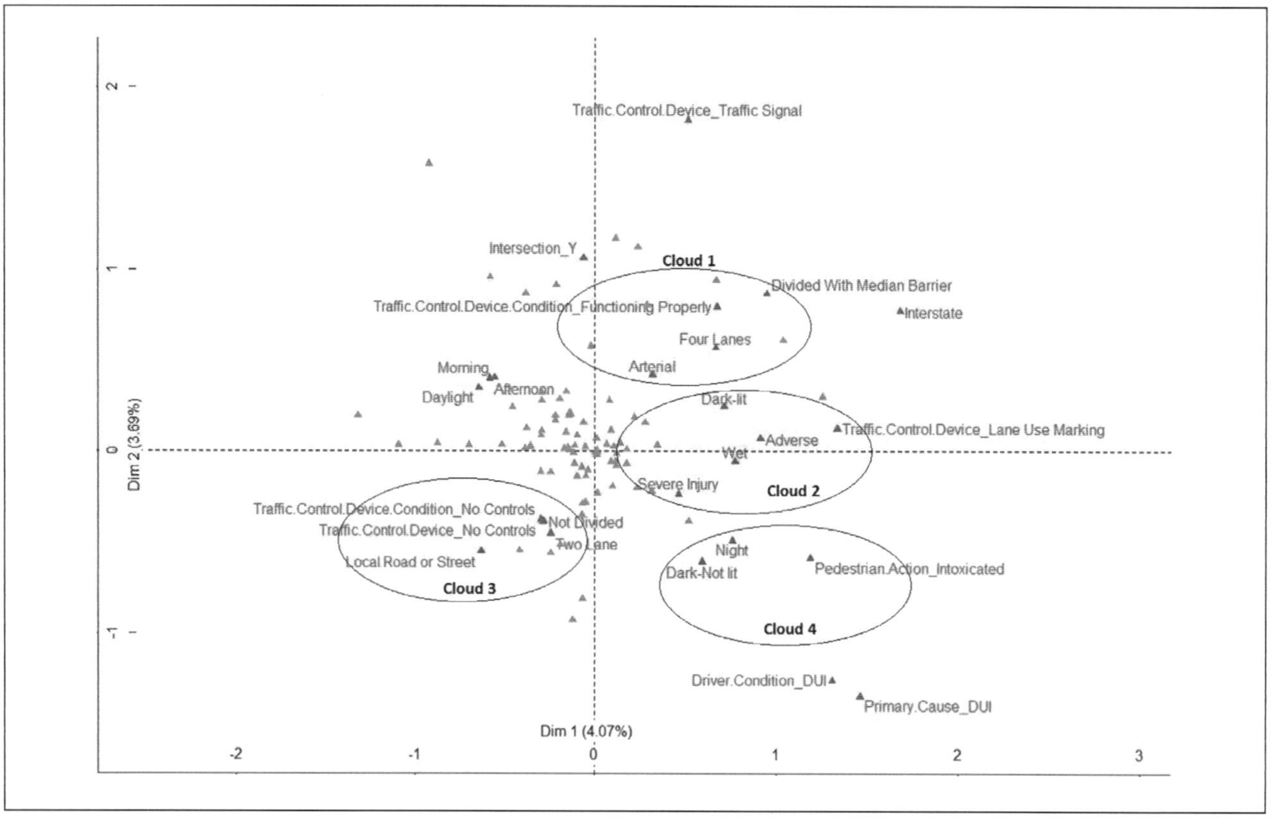

Figure 4. MCA factor map of top contributing variable categories.

Categories such as Local Road or Street, Two Lane, Not Divided, and Traffic Control Device_No Controls can be observed to have clustered together on the factor map in Cloud 3. This is a clear indication of a pedestrian safety problem at uncontrolled locations along two-lane undivided local roads in rural areas. Previous studies have separately established the hazards posed by uncontrolled locations, two-lane roads, and local streets to pedestrian safety (*6*, *28*). Traffic volumes along local roads in rural areas tend to be lower, thus encouraging higher speeds. Rural roads also might lack good pedestrian infrastructure, which forces pedestrians to walk alongside traffic, thereby exposing them to more danger (*29*). All these factors acting together contribute to undermining pedestrian safety in rural areas.

Similarly, the variables Severe Injury, Night, Dark-Not lit, and Pedestrian Action_Intoxicated are plotted close to each other in Cloud 4. The category Driver Condition_DUI and Primary Cause_DUI also can be seen lying close to this point cloud. This arrangement of variable categories underscores the dangers of intoxication among pedestrians and drivers alike, as indicated by some of the previous studies (*30*, *31*). Results of this study show that prevalence of intoxicated pedestrians or drivers during night time, along unlit roads, will contribute to the occurrence of severe pedestrian crashes in rural areas.

Finally, another point cloud on the factor map associates the variables Morning, Afternoon, and Daylight. As can be seen from Table 2, approximately 50% of crashes occurred when these conditions are prevalent. Therefore, this cloud simply indicates that the majority of crashes happened during morning and afternoon hours, in daylight conditions. Similarly, the variables Intersection, Interstate, and Traffic Control Device_Traffic Signal also appear on the factor map in isolation because only a minor portion of records have selected those variables. The fact that these categories seem rare inflates their contribution to the variance of their corresponding dimension (*11*)

Conclusions and Recommendations

In this study, MCA was employed to analyze rural pedestrian crashes that occurred in Illinois between 2010 and 2014. The aim of this study was to identify clusters of contributing factors that act together in influencing the occurrence of pedestrian crashes. MCA is the best suited statistical tool to achieve that goal because it offers a graphical representation of variable categories, thereby enabling researchers to spot associations among various categories. As opposed to traditional statistical techniques (such as logit, probit, binomial, and multinomial models), which require a hypothesis and specification of dependent and independent variables, the MCA method operates without any such constraints. The results obtained by MCA are in the form of point clouds, which reflect association among variable categories in crash data. All categories present in a cloud have a similar influence on crash occurrence. Despite other logistic regression-based studies that provide parameter estimates for each variable separately and quantify their effect individually, the method used in this study is capable of identifying the most statistically significant combinations of factors. This grouping of confounding variables can be beneficial to state departments of transportation (DOTs) and local agencies to develop effective strategies that are associated with multiple safety benefits. In other words, DOTs can use these results to identify locations in their jurisdiction, where all such categories are likely to appear together, for application of targeted countermeasures.

Results of this study indicate that the variables, roadway functional class, the number of lanes, lighting conditions, weather conditions, traffic control device type, driver condition, and pedestrian condition were among the most significant contributors to rural pedestrian crashes. Different clusters of contributing factors and potential countermeasures will be elaborated upon in this section.

First, four-lane divided arterials, which are conducive to higher travel speeds, were identified as problematic locations for pedestrian safety in rural areas. The fact that the presence of traffic control devices at such locations is unable to improve safety is noteworthy. Therefore, an effective countermeasure for these conditions should be able to address higher speeds without the aid of traffic control devices. FHWA recommends traffic calming on main roads through rural communities to address high speeds in those areas and improve pedestrian safety (*32*). According to that report, installing traffic-calming measures along rural roadways with frequent pedestrian activity will compel motorists to reduce speeds and enhance pedestrian safety. Another FHWA document suggests construction of walkable shoulders of at least 4 ft on both sides of rural roadways, where occasional pedestrian traffic is expected, to reduce their exposure to traffic (*33*).

This study also found that severe injury pedestrian crashes are prevalent in adverse weather conditions on wet road surfaces, even when lane use markings are present to guide motorists and pedestrians. Apparently, poor visibility is the underlying contributing factor in this scenario. Therefore, to address this set of contributing factors, it is imperative to identify a countermeasure that can function effectively even in poor visibility conditions. A study conducted by the Texas Transportation Institute (TTI) concluded that retro-reflective pavement markings (RRPMs) are the most efficient form of countermeasure for increasing visibility during wet-night conditions (*34*). Therefore, it is recommended that RRPMs be used at all rural locations with occasional pedestrian activity to increase visibility and improve safety.

Uncontrolled locations along rural two-lane local roads also proved to be hazardous for pedestrian safety. These locations are typically characterized by a lack of traffic controls, lower traffic volumes, and higher travel speeds. Past studies have proved that crosswalks are efficient in improving safety at these locations (*20*). However, selection of countermeasures at these type of locations must be done based on site characteristics such as speed and traffic volume. For low volumes and high speeds, an essential treatment for a crosswalk, via marking and signage, would be sufficient (*35*). For higher speeds and volumes, more sophisticated treatments such as rectangular rapid-flashing beacons (RRFB) or pedestrian hybrid beacons (PHB) can be considered (*36*).

Several past studies have proven that intoxication of either driver or pedestrian is extremely dangerous. This study arrived at the same conclusion. Along with driver or pedestrian intoxication, night time and dark unlit roads contribute to rural pedestrian crashes in Illinois. Potential countermeasures for this problem largely depend on educational or enforcement efforts rather than engineering solutions. To curb this issue, targeted mass media campaigns, stricter law enforcement initiatives, and general awareness programs, all focused on drinking and driving, must be undertaken. On the other hand, despite being proven effective in curbing driving under the influence, enforcements strategies such as random breath testing (RBT) and selective breath testing (SBT) checkpoints are still in limited use across the United States (*37*). It is necessary to bring RBT and SBT into wider usage to mitigate pedestrian or driver intoxication problems. RBT or SBT checkpoints should be set up at strategic locations during the night time to enforce anti-DUI laws.

This study has some restrictions. First, the data used in this study come from a single state in the United States. To gain deeper insight into the trends existing in pedestrian safety in rural areas, a study encompassing data from a diverse group of states is recommended. Apart from this, certain variables were not included in the analysis due to lack of information. For instance, information on land use, annual daily traffic (ADT), and posted speed limits of roadways were not available in the crash data provided by IDOT.

Acknowledgments

This study was sponsored and supported by the Illinois Department of Transportation (IDOT) and Illinois Center for Transportation (ICT).

Author Contributions

The authors confirm contribution to the paper as follows: study conception and design: RB and MJ; data collection: RB and HZ; analysis and interpretation of results: MJ and RB; draft manuscript preparation: RB. All authors reviewed the results and approved the final version of the manuscript.

References

1. National Highway Traffic Safety Administration. *Traffic Safety Facts-2015 Data*. National Highway Traffic Safety Administration, U.S. Department of Transportation, Washington, D.C., 2017.
2. Chakravarthy, B., S. Lotfipour, and F. E. Vaca. Pedestrian Injuries: Emergency Care Considerations. *The California Journal of Emergency Medicine*, Vol. 8, No. 1, 2007, pp.15–21.
3. Federal Highway Administration. *Factors Contributing to Pedestrian and Bicycle Crashes on Rural Highways*. Federal Highway Administration, U.S. Department of Transportation, Washington, D.C., 2007.
4. Pour-Rouholamin, M., and H. Zhou. Investigating the Risk Factors Associated with Pedestrian Injury Severity in Illinois. *Journal of Safety Research*, Vol. 57, 2016, pp. 9–17.
5. Tulu, G. S., S. Washington, M. M. Haque, and M. J. King. Investigation of Pedestrian Crashes on Two-way Two-lane Rural Roads in Ethiopia. *Accident Analysis and Prevention*, Vol. 78, 2015, pp. 119–126.
6. Islam, S., and S. L. Jones. Pedestrian At-fault Crashes on Rural and Urban Roadways in Alabama. *Accident Analysis and Prevention*, Vol. 72, 2014, pp. 268–276.
7. Velloso, M. S., and M. A. Jacques. On-the-spot Study of Pedestrian Crashes on Brazilian Federal District Rural Highways Crossing Urban Areas. *Transportation Research Part F*, Vol. 15, No. 5, 2012, pp. 588–599.
8. Carter, D. L., and F. M. Council. Factors Contributing to Pedestrian and Bicycle Crashes on Rural Highways. Presented at 86th Annual Meeting of the Transportation Research Board, Washington, D.C., 2007.
9. Zajac, S. S., and J. N. Ivan. Factors Influencing Injury Severity of Motor Vehicle–crossing Pedestrian Crashes in Rural Connecticut. *Accident Analysis and Prevention*, Vol. 35, No. 3, 2002, pp. 369–379.
10. Monella, A. Identifying Crash Contributory Factors at Urban Roundabouts and Using Association Rules to Explore Their Relationships to Different Crash Types. *Accident Analysis and Prevention*, Vol. 43, No. 4, 2011, pp. 1451–1463.
11. Das, S., and X. Sun. Factor Association using Multiple Correspondence Analysis in Vehicle-pedestrian Crashes. *Transportation Research Record: Journal of the Transportation Research Board*, 2015. 2519: 95–103.
12. Das, S., and X. Sun. Exploring Clusters of Contributing Factors for Single Vehicle Fatal Crashes through Multiple Correspondence Analysis. Presented at 93rd Annual Meeting of the Transportation Research Board, Washington, D.C., 2014.
13. Jalayer, M., M. Pour-Rouholamin, and H. Zhou. Multiple Correspondence Approach to Identifying Contributing Factors Regarding Wrong-Way Driving Crashes. *Traffic Injury Prevention*, Vol. 19, No. 1, 2018, pp. 35–41.
14. Illinois Department of Transportation. *Strategic Highway Safety Plan*. Illinois Department of Transportation, Illinois, 2009.
15. Nallet, N., M. Bernard, and M. Chiron. Individuals Taking a French Driving License Points Recovery Course: Their Attitudes Towards Violations. *Accident Analysis and Prevention*, Vol. 40, 2008, pp. 1836–1843.
16. Lu, S., P. Mei, J. Wang, and H. Zhang. Fatality and Influence Factors in High-casualty Fires: A Correspondence Analysis. *Safety Science*, Vol. 50, No. 4, 2012, pp. 1019–1033.
17. Greenacre, M., and J. Blasius. *Multiple Correspondence Analysis and Related Methods*. CRC Press, Boca Raton, FL, 2006.
18. Le Roux, B., and H. Rouanet. *Multiple Correspondence Analysis*. Sage Publication, Inc., Paris, 2010.
19. Costa, P. S., N. C. Santos, P. Cunha, J. Cotter, and N. Sousa. The Use of Multiple Correspondence Analysis to Explore Associations between Categories of Qualitative Variables in Healthy Ageing. *Journal of Aging Research*, Vol. 2013, 2013, pp. 1–12.
20. Baireddy, R., M. Pour-Rouholamin, H. Zhou, and Y. Qi. Factors Contributing to Injury Severity of Pedestrian Crashes at Uncontrolled Locations in Illinois. Presented at 96th Annual Meeting of the Transportation Research Board, Washington, D.C., 2017.
21. Siddiqui, N., X. Chu, and M. Guttenplan. Crossing Locations, Light Conditions, and Pedestrian Injury Severity. *Transportation Research Record: Journal of the Transportation Research Board*, 2006. 1982: 141–149.
22. Chu, X. *Pedestrian Safety at Midblock Locations*. FDOT-BD544-16. Florida Department of Transportation, Tallahassee, FL, 2006.
23. Rothman, L., A. William Howard, A. Camben, and C. Macarthur. Pedestrian Crossing Location Influences Injury Severity in Urban Areas. *Traffic Injury Prevention*, Vol. 18, No. 6, 2012, pp. 365–370.
24. Le, S., J. Josse, and F. Husson. FactoMineR: An R Package for Multivariate Analysis. *Journal of Statistical Software*, Vol. 21, No. 1, 2008, pp. 1–18.
25. Das, S., and X. Sun. Association Knowledge for Fatal Run-off-road Crashes by Multiple Correspondence Analysis. *IATSS Research*, Vol. 39, No. 2, 2016, pp. 146–155.
26. Garder, P. E. The Impact of Speed and Other Variables on Pedestrian Safety in Maine. *Accident Analysis and Prevention*, Vol. 36, No. 4, 2004, pp. 533–542.
27. Zahabi, S. A., J. Strauss, L. F. Miranda-Moreno, and K. Manaugh. Estimating Potential Effect of Speed Limits, Built Environment, and Other Factors on Pedestrian and Cyclist Injury Severity Levels in Traffic Crashes.

Transportation Research Record: Journal of the Transportation Research Board, 2011. 2247: 81–90.
28. Zeeger, C., J. Stewart, H. Huang, and P. Lagerwey. Safety Effects of Marked Versus Unmarked Crosswalks at Uncontrolled Locations: Analysis of Pedestrian Crashes in 30 Cities. *Transportation Research Record: Journal of the Transportation Research Board*, 2001. 1773: 56–68.
29. Galanis, A., G. Botzoris, and N. Eliou. Pedestrian Road Safety in Relation to Urban Road Type and Traffic Flow. *Transportation Research Procedia*, Vol. 24, 2017, pp. 220–227.
30. Kim, J.-K., G. F. Ulfarsson, V. N. Shankar, and S. Kim. Age and Pedestrian Injury Severity in Motor-vehicle Crashes: A Heteroskedastic Logit Analysis. *Accident Analysis and Prevention*, Vol. 40, No. 5, 2008, pp. 1695–1702.
31. Lee, C., and M. Adbel-Aty. Comprehensive Analysis of Vehicle–pedestrian Crashes at Intersections in Florida. *Accident Analysis and Prevention*, Vol. 37, No.4, 2005, pp. 775–786.
32. Krammes, R., and E. Sheldahl. *Traffic Calming on Main Roads Through Rural Communities*. Federal Highway Administration, U.S. Department of Transportation, Washington, D.C., 2009.
33. Furst, T. *Memorandum: Promoting the Implementation of Proven Safety Countermeasures*. Federal Highway Administration, U.S. Department of Transportation, Washington, D.C., 2012.
34. Carlson, P. J., J. D. Miles, A. M. Pike, and S. E. Park. *Evaluation of Wet-Weather and Contrast Pavement Marking Applications*. Final Report. FHWA/TX-05/0-5008-2. Texas Transportation Institute, Texas, 2007.
35. Nabros, D., and S. E. Goughnour. *Non-Motorized User Safety: A Manual for Local Rural Road Owners*. Federal Highway Administration, U.S. Department of Transportation, Washington, D.C., 2012.
36. Fitzpatrick, K., M. Brewer, and R. Avelar. Driver Yielding at Traffic Control Signals, Pedestrian Hybrid Beacons, and Rectangular Rapid-Flashing Beacons in Texas. *Transportation Research Record: Journal of the Transportation Research Board*, 2014. 2463: 46–54.
37. Fell, J. C., and V. R. Lacey. Sobriety Checkpoints: Evidence of Effectiveness is Strong, but Use is Limited. *Traffic Injury Prevention*, Vol. 5, No. 3, 2010, pp. 220–227.

The Standing Committee on Safety Data, Analysis and Evaluation (ANB20) peer-reviewed this paper (18-02996).

Article

TRR
JOURNAL OF THE TRANSPORTATION RESEARCH BOARD

Transportation Research Record
2018, Vol. 2672(38) 128–137

DOI: 10.1177/0361198118788207
journals.sagepub.com/home/trr
SAGE

Evaluating Factors Influencing the Severity of Three-Plus Multiple-Vehicle Crashes using Real-Time Traffic Data

Angela E. Kitali[1], Emmanuel Kidando[2], Paige Martz[3], Priyanka Alluri[1], Thobias Sando[4], Ren Moses[2], and Richard Lentz[5]

Abstract

Multiple-vehicle crashes involving at least two vehicles constitute over 70% of fatal and injury crashes in the U.S. Moreover, multiple-vehicle crashes involving three or more vehicles (3+) are usually more severe compared with the crashes involving only two vehicles. This study focuses on developing 3+ multiple-vehicle crash severity models for a freeway section using real-time traffic data and crash data for the years 2014–2016. The study corridor is a 111-mile section on I-4 in Orlando, Florida. Crash injury severity was classified as a binary outcome (fatal/severe injury and minor/no injury crashes). For the purpose of identifying the reliable relationship between the 3+ severe multiple-vehicle crashes and the identified explanatory variables, a binary probit model with Dirichlet random effect parameter was used. More specifically, Dirichlet random effect model was introduced to account for unobserved heterogeneity in the crash data. The probit model was implemented using a Bayesian framework and the ratios of the Monte Carlo errors were monitored to achieve parameter estimation convergence. The following variables were found significant at the 95% Bayesian credible interval: logarithm of average vehicle speed, logarithm of average equivalent 10-minute hourly volume, alcohol involvement, lighting condition, and number of vehicles involved (3, or >3) in multiple-vehicle crashes. Further analysis involved analyzing the posterior probability distributions of these significant variables. The study findings can be used to associate certain traffic conditions with severe injury crashes involving 3+ multiple vehicles, and can help develop effective crash injury reduction strategies based on real-time traffic data.

Traffic safety is a major concern for the transportation industry and the public. In 2016, over 40,000 people died in the United States as a result of motor-vehicle crashes, an increase from 32,744 in 2014. This 18% increase is the biggest two-year jump in more than 50 years (*1*). The National Safety Council estimates that the cost of deaths, injuries, and property damage attributed to crashes in 2016 totaled $432.5 billion (*2*).

A major focus of traffic safety researchers has been to identify factors contributing to severity of crashes. Researchers often analyze single- and multiple-vehicle crashes separately (*3–7*). As the name implies, crashes involving only one vehicle are termed as single-vehicle crashes. However, the definition of a multiple-vehicle crash has not been consistent across different studies. While most previous studies (*3–6, 8–10*) have defined multiple-vehicle crashes as those that involve two or more vehicles, some of the previous studies have classified crashes involving three or more vehicles as multiple-vehicle crashes (*11–13*). On the other hand, another study, conducted on I-94 in Michigan (*14*), defined multiple-vehicle crashes as those involving five or more vehicles. Moreover, some of the studies that explored multiple-vehicle collisions did not define the exact number of vehicles involved in a crash (*15–17*).

Factors contributing to two-vehicle crashes are different from the factors contributing to multiple-vehicle crashes involving three or more vehicles (*11*). Moreover,

[1]Department of Civil and Environmental Engineering, Florida International University, Miami, FL
[2]Department of Civil and Environmental Engineering, FAMU-FSU College of Engineering, Tallahassee, FL
[3]TOPS Laboratory, Department of Civil and Environmental Engineering, Madison, WI
[4]School of Engineering, University of North Florida, School of Engineering, Jacksonville, FL
[5]School of Computing, University of North Florida, Jacksonville, FL

Corresponding Author:
Address correspondence to Angela E. Kitali: akita002@fiu.edu

the injury occurrence process in a two-vehicle crash is likely to be different from that in a three-plus vehicle crash (*13*). Crashes involving two vehicles are usually rear-ending, head-on, sideswipe, or angle crashes. On the other hand, multiple-vehicle crashes involving three or more vehicles are more complex, often occur instantly following another crash, and are a result of a direct interruption of the safe traffic environment (*18*). Some of the situations that influence an occurrence of multiple-vehicle crashes involving three or more vehicles may include a driver being rear-ended and pushed forward to collide with the vehicle in front; a driver being pushed into another lane as a result of impact from a collision and possibly colliding with a vehicle in the other lane; or two vehicles collide and are stopped, then are rear-ended by following vehicles (*11*, *19*). Double-lane-changing crashes were found to result in an increase in the occurrence of crashes involving three or more vehicles (*15*).

Multiple-vehicle crashes involving three or more vehicles are usually more severe compared with the crashes involving only two vehicles. In 2014, 2.7% of all two-vehicle crashes in Florida resulted in fatal or serious injury, while the percentage increased to 5.8% for multiple-vehicle crashes involving three or more vehicles. Moreover, some of the severe multiple-vehicle collisions involved dozens of vehicles (*14*, *19*). In Florida, severe weather has been identified as one of the factors influencing the occurrence of severe multiple-vehicle collisions. For example, a multiple-vehicle pileup crash caused by reduced visibility occurred in January 2012 along I-4 in Orlando where 10 people were killed and 21 others were injured (*20–22*). Another recent multiple-vehicle crash occurred in July 2017 along the same freeway, involving a total of 12 vehicles, resulting in three injuries and more than an hour of road closure (*23*). Not only are these crashes deadly but they also lead to a prolonged roadway capacity reduction. Thus, the first step in selecting the appropriate countermeasures to address this issue is to explore factors influencing the severity of multiple-vehicle crashes involving three or more vehicles, referred to in this study as 3+ multiple-vehicle crashes.

Operating speed is one of the important factors associated with multiple-vehicle crashes (*11*, *15*, *24*). In the past, most studies have used posted speed limit as a representative variable for operating speed. Similarly, the average annual daily traffic (AADT) is the most common aggregate exposure variable used to reflect traffic conditions. Considering the fact that AADT is used to approximate vehicle-miles traveled along a roadway section, there may be a reduction in the natural variance that exists in exposure data, which may result in heavy under dispersion (*6*). In order to improve the reliability of AADT, recent studies have used traffic volume over a short period of time such as month or day, while others have used deduced hourly traffic characteristics by combining AADT and a one-day hourly traffic profile for the location analyzed (*25*). Nevertheless, even hourly measures cannot consider the short-term variation in traffic flow, especially at the time of the crash, and are not well suited for application to real-time operations (*24*).

Considering the fact that most of the freeways are equipped with continuous surveillance systems, disaggregate raw traffic data is often readily available. Therefore, in lieu of conventional posted speed limit and AADT, some recent studies (*26–34*) have used real-time speed and traffic flow as they represent the actual traffic conditions at the time of the crash. These studies have taken advantage of per-vehicle real-time data that is collected continuously. For example, in Florida, data from the microwave vehicle detectors are archived in the Regional Integrated Transportation Information System (RITIS) database. These data are often used by researchers, decision makers, planners, operations specialists, military, and homeland security officials.

In summary, the existing literature on multiple-vehicle crashes has two major limitations. First, the definition of a multiple-vehicle crash has not been consistent across different studies. While most previous studies have defined multiple-vehicle crashes as those that involve two or more vehicles (*3–6, 8–10*), some have classified crashes involving three or more vehicles as multiple-vehicle crashes (*11–13*). Multiple-vehicle crashes involving three or more vehicles have to be analyzed separately since factors contributing to these crashes are different from the factors contributing to two-vehicle crashes. The second limitation with the existing studies is related to the use of AADT and conventional posted speed limit to study the severity of multiple-vehicle crashes. Real-time speed and traffic data (instead of speed limit and AADT) provide a more representative scenario of the actual traffic conditions at the time of the crash.

This research aims to address these two limitations by evaluating factors influencing the severity of 3+ multiple-vehicle crashes using real-time data. The study findings will help enhance the performance of the measures undertaken to minimize the severity of 3+ multiple-vehicle crashes. Further, the use of real-time data will provide real-time solutions in mitigating the severity of 3+ multiple-vehicle crashes.

Study Data

The study corridor is a 111-mile section on I-4 in Orlando, Florida. Two datasets, crash and traffic-related data, are used in this research. Crash data were retrieved from Signal Four Analytics system for the years 2014–2016. Signal Four Analytics is an interactive, web-based

Table 1. Summary of Crash-Related Variables

Variable	Factor	Count	Percent (%)
Crash day	Weekend	491	26.33
	Weekday	1,374	73.67
Crash time	Off-peak hours	846	45.36
	Peak hours	1,019	54.64
Crash type	Others	300	16.09
	Rear-end	1,565	83.91
Alcohol involvement	No	1,853	99.36
	Yes	12	0.64
Weather condition	Clear	1,188	63.70
	Rain/snow/foggy/cloudy	677	36.30
Road surface condition	Dry	1,429	76.62
	Wet	436	23.38
Lighting condition	Daytime	1,414	75.82
	Nighttime	451	24.18
Number of vehicles involved	3	1,472	78.93
	>3	393	21.07
Presence of work zone	No	1,710	91.69
	Yes	155	8.31
Distraction involvement	No	1,445	77.48
	Yes	419	22.47
Severity*	Minor/no injury	984	52.76
	fatality/severe injury	881	47.24

*Dependent variable.

system designed to support the crash mapping and analysis needs for several agencies in Florida. These include law enforcement, traffic engineering, transportation planning agencies, and research institutions in the state of Florida. Only multiple-vehicle crashes involving at least three vehicles are included in the analysis. Table 1 summarizes the crash-related variables retrieved from Signal Four Analytics system. It is worth noting that the distribution of the dependent variable, crash severity, is approximately half, that is, minor/no injury = 52.76% and fatality/severe injury = 47.24%.

RITIS detectors provide high-resolution traffic data, that is, vehicle count (vehicles/30 seconds) average speed (mph), and occupancy (the percent time that the sensor is occupied) at 30-second intervals. A total of 371 RITIS detectors are located along the study area: 188 on I-4 eastbound and 183 on I-4 westbound. The average spacing between the detector stations is about 0.7 miles. For each crash, the traffic data were extracted for a period of 10 minutes before the crash occurrence time over a 1-mile radius. Traffic data just before the crash occurrence accounts for the potential inaccuracies in the reported crash time (*18*). A fixed spatial threshold of one mile captures the traffic flow characteristics in the vicinity of the crash. Note that this approach captures traffic information both upstream and downstream of the crash location. To avoid ambiguity, the process of matching crash data with RITIS detectors was performed for each direction separately, that is, eastbound and westbound directions.

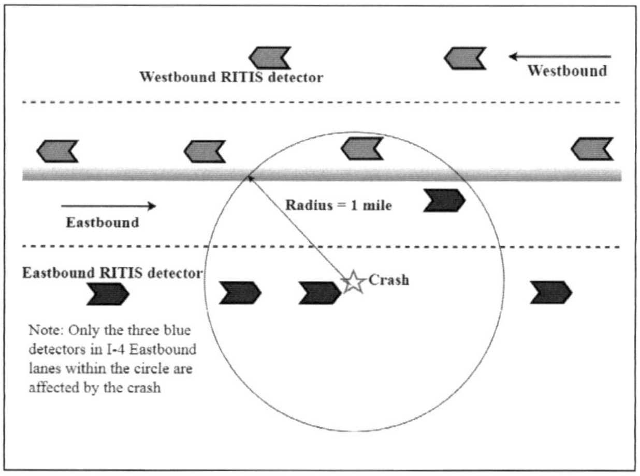

Figure 1. Detector arrangement in relation to crash location.

Figure 1 illustrates the position of the detectors on I-4 eastbound and westbound lanes. The figure also includes a crash on I-4 eastbound direction. Note that the crash affects only the detectors in the eastbound direction and only those detectors that are within one mile of the crash.

The extracted traffic data were further aggregated into 10-minute intervals and converted into the six traffic flow variables presented in Table 2. The 10-minute count data

Table 2. Summary of Real-Time Traffic-Related Variables

Variable	Description	Min	Median	Mean	Max
Average vehicle speed	Average vehicle speed during 10-min period (mph)	8.06	47.76	46.76	81.55
SD of vehicle speed	SD of vehicle speed during 10-min period (mph)	1.40	9.35	10.35	60.86
Average EHV	Average EHV during 10-min period (veh/hour)	10	206	229	1,097
SD of EHV	SD of EHV during 10-min period (veh/hour)	7	51	101	1,286
Average detector occupancy	Average detector occupancy during 10-min period (%)	0.52	11.99	13.25	43.29
SD of detector occupancy	SD of detector during 10-min period (%)	0.42	5.47	6.09	29.19

Note: All variables are continuous. SD = Standard deviation; EHV = equivalent hourly volume.

were converted into the corresponding hourly volume for easy interpretation of results. The correlation between each of the independent variables was checked. Speed and occupancy were correlated at 83%, and therefore, occupancy was dropped from the analysis. In addition, road surface condition variable was also dropped from the model since it was correlated with weather condition at 78%.

Model Development

This study developed a Bayesian probit regression model with the Dirichlet random effect parameter to evaluate factors influencing the severity of 3+ multiple-vehicle crashes on freeways. Crash data are usually asymmetrical in nature (*35, 36*), and therefore require the use of asymmetrical models such as complementary log-log model. However, in this study, the dependent variable, crash severity, is symmetrical; severe (47.24%) and non-severe (52.76%) injury classes are approximately equally distributed. Probit model, a symmetrical model, is therefore used in this study.

The crash severity of 3+ multiple-vehicle crashes is affected by not only the real-time traffic and crash-related attributes, but also several roadway characteristics (*11–13*). However, this study focuses on only the real-time traffic and crash data variables. In order to account for unobserved heterogeneity, the 3+ multiple-vehicle crash data were explored through a semi-parametric probit model. More specifically, a Dirichlet random effect parameter was incorporated into the probit model to account for the unobserved heterogeneity in the crash data. In contrast to the random effect model with a normal distribution, the Dirichlet random effect is based on the mixture distribution that can group crash variation into groups of similar features, which, in turn, improves model fitness (*37–39*).

The proposed model was implemented using Bayesian framework. Compared with the classical approach, Bayesian models do not treat the coefficients as fixed values but follow certain distributions. The motive behind Bayesian approach is that the prior distribution for each of the parameter is defined and then the data, 3+ multiple-vehicle crashes in this research, are used to update beliefs about the behavior of parameters. Furthermore, the updated probability of the parameters is used, and the posterior Bayesian credible intervals are produced. The correct interpretation is that a parameter of interest lies within the credible interval with 95% probability. In that context, the interpretation of Bayesian inference is slightly different from that in the classical approach in which the Bayesian inference derives the updated posterior probability of the parameters and constructs credibility intervals that have a natural interpretation according to their probability. More specifically, Bayesian models use Bayesian credible interval (BCI) instead of confidence intervals to identify significant variables. In this study, each parameter is considered to be statistically significant if the 95% BCI (2.5–97.5%) of the coefficient of the predictor variables does not contain zero.

The Bayesian probit regression analysis models the relationship between the dichotomous response variable (fatal/severe injury versus minor/no injury) and the real-time traffic- and crash-related variables. Specifically, when developing the model, the response variable takes two values, that is, for crash i, severe crash is $y_i = 1$, which involves fatal or severe injury; while $y_i = 0$ represents non-severe crash involving minor or no injury. The probability, $P(y_i = 1|X) = P_i$, that the involved crash is severe can be modeled using a probit regression with a covariate $\boldsymbol{X}$ as shown in the hierarchy dependencies indicated in Equations 1–7. Figure 2 shows the priors and their respective distributions, that is, gamma, Dirichlet prior (DP), uniform, and normal. It further shows the interdependency between the priors and the probit Bernoulli model:

$$y_i \sim \text{Bernoulli}(P_i) \quad (1)$$

$$P_i = \boldsymbol{\Phi}(\eta_i) \quad (2)$$

$$\eta_i = \beta_0 + \beta_j \mathbf{X} + b_i; \ \beta_0 \text{ and } \beta_j \sim \text{Normal}(0, 100) \quad (3)$$

$$b_i|G \sim G \quad (4)$$

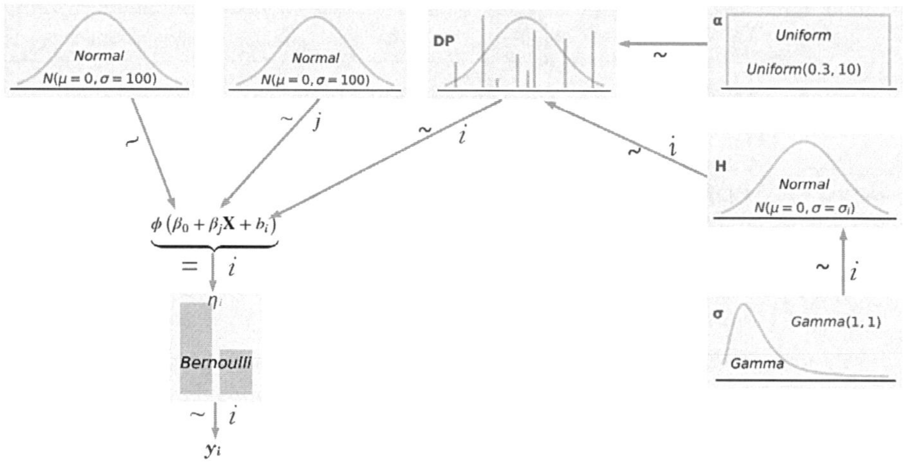

Figure 2. A graphical representation of the probit regression model with the Dirichlet random effect parameter.

$$G \sim DP(\alpha, H) \tag{5}$$

$$\alpha \sim Uniform(0.3,\ 10) \tag{6}$$

$$H \sim N\left(0, \sigma^2\right); \sigma \sim Gamma(1,\ 1) \tag{7}$$

Where,

$X$is the vector of covariates,

β_0 and β_j are the intercept and the regression coefficients,

$b_i$is the random effect of observed crash i,

α is a positive precision parameter,

H is the baseline distribution,

DP is the Dirichlet process, and

Φ is the standard normal cumulative function.

The Bayesian inference for probit regression follows the usual procedure for all Bayesian analysis. More specifically, a prior distribution for all unknown parameters has to be defined. Normally, two categories of priors are used in the Bayesian approach: informative and non-informative priors. Informative priors are based on the literature, the knowledge of the experts, or information retrieved explicitly from a previous data analysis (*40*). On the other hand, non-informative priors, also called "vague" priors, are often used in the absence of reliable prior information regarding model parameters (*35, 41*). For this study, there is no prior knowledge of the expected effect; therefore, non-informative priors are used. The most commonly used priors are normal distributions with a zero mean, expressing the prior doubt of the relationship between the predictor variable and the response variable, and large variance.

As defined in Equation 3, the prior specification for the beta parameters (β_0 and β_j) have zero mean and standard deviation of 100. With respect to the precision parameter (α), a uniform prior with lower and upper limit of 0.3 and 10, respectively, as proposed by Ohlssen et al. (*39*), is used. Lastly, the gamma parameter (σ) with shape and scale parameters equal to 1 is adopted. The specification of the priors is then followed by estimation of the likelihood function. The likelihood function for the probit regression can be expressed as shown in Equation 8:

$$\text{Likelihood}_i = \prod (X_i)^{y_i} \left(1 - \prod (X_i)\right)^{(1-y_i)} \tag{8}$$

Where $\prod (X_i)^{y_i}$ is the probability of the event for the ith crash, which has covariate vector X_i.

The priors and the likelihood function are then used to estimate the posterior distribution of the study parameters (Equation 9). The Bayes theorem is normally applied while estimating the posterior distribution of all parameters:

$$\text{Posterior} \sim \text{Prior} \times \text{Likelihood} \tag{9}$$

The above-mentioned model was implemented using PyMC3, an open source Python package (*42*). The inference regarding the parameters' posterior distributions were estimated using the Markov Chain Monte Carlo simulations. In particular, the No-U-Turn Sampler step was used to compute $\beta_0, \beta_j, H,$ and σ and the Metropolis-Hastings sampling step was used to estimate the distribution of α to accommodate for discrete nature of the parameters. Modeling involved a total of 50,000 iterations; the first 25,000 iterations as a burn-in sample, and the remaining 25,000 for parameter estimation.

The ratios of the Monte Carlo (MC) errors relative to standard deviations of the estimates, trace, density, and autocorrelation plots were monitored to achieve parameter estimation convergence. As a rule of thumb, the MC error was maintained at less than 5% of the posterior standard deviation for a parameter to converge (*41*).

Table 3. Bayesian Probit Regression Model Results

Variable	Factor	Mean	SD	MC error	BCI (%) 2.5	97.5
Constant	Continuous	−0.427	0.417	0.024	−1.229	0.375
Ln (average vehicle speed)	Continuous	0.207	0.085	0.001	0.047	0.378
Ln (average EHV)	Continuous	−0.104	0.048	0.004	−0.189	−0.005
Alcohol involvement	No					
	Yes	0.985	0.469	0.007	0.072	1.872
Lighting condition	Daytime					
	Nighttime	0.163	0.07	0.002	0.019	0.296
Number of vehicles involved	3					
	>3	0.366	0.071	0.001	0.227	0.506

Note: BCI = Bayesian credible interval; MC error = Monte Carlo error; SD = standard deviation; EHV = equivalent hourly volume.

The 95% BCI was used to determine the significance of the predictor variables, which provides probability interpretations with normality assumptions that the true parameter is inside the region with measured probability (*41*).

Results

As mentioned earlier, the Bayesian probit regression model utilizing the Dirichlet random effect parameter was used to evaluate factors influencing the severity of 3+ multiple-vehicle crashes. The posterior results of the model are presented in Table 3. Out of the 13 predictor variables used in modeling, five variables are significant at the 95% BCI. The predictor variable in the model is considered to be significant at the 95% BCI when the values of the 2.5% and 97.5% percentiles do not include zero (0), that is, they are both either negative or positive. The significant traffic-related variables include:

- logarithm of average vehicle speed
- logarithm of average EHV

The significant crash-related variables include:

- alcohol involvement
- lighting condition
- number of vehicles (3 or >3) involved in multiple-vehicle crashes

The following sections provide a detailed discussion of the above-mentioned significant predictor variables.

Average Vehicle Speed

Increase in average vehicle speed is identified as one of the significant factors (Mean 0.207, 95% BCI [0.047, 0.378]). The positive mean of this variable indicates that increase in vehicle speed increases the likelihood of persons involved in 3+ multiple-vehicle crashes to sustain fatal or severe injuries. This is expected as traffic speed has been identified as one of the variables that influence the occurrence of multiple-vehicle crashes (*24*). As a vehicle's speed increases, so does the distance traveled during the driver's reaction time (reaction distance) and the distance needed to stop (braking distance). In addition, the higher the speed, the greater the amount of kinetic energy that must be absorbed by the crash impact, increasing the crash severity.

Average EHV

The results shown in Table 3 indicate that the increase in the average vehicle count (presented as EHV) reduces the odds of persons involved in 3+ multiple-vehicle crashes sustaining fatal or severe injuries (Mean −0.104, 95% BCI [−0.189, −0.005]). Increasing traffic volume was found to have a consistently decreasing effect on crash severity (*24*). Although some studies (*43*, *44*) did not evaluate multiple-vehicle crashes, their findings are similar to the present study, implying that traffic congestion is associated with low probability of a severe injury crash.

Alcohol Involvement

Alcohol involvement in 3+ multiple-vehicle crashes was observed to increase the chance for individuals involved to sustain fatal and severe injuries (Mean 0.985, 95% BCI [0.072, 1.903]). The influence of alcohol impairs attention, judgment, vision, and coordination, resulting in significant safety concerns. This observation is consistent with numerous studies that have identified alcohol involvement to have a significant association with higher levels of crash severity (*27*, *45*).

Lighting Condition

The positive value of the lighting condition variable (Mean 0.163, 95% BCI [0.019, 0.296]) suggests that, compared with multiple-vehicle crashes occurring during daytime, crashes occurring during nighttime have a higher chance of resulting in fatal and severe injuries. Poor nighttime lighting conditions lead to reduced visibility on the roadway. Given the high operating speeds on freeways, poor lighting could result in longer perception to reaction time, limiting the drivers' collision avoidance ability and increasing the risk of severe multiple-vehicle crashes (*8*).

Number of Vehicles Involved (3 or >3)

The model results indicate that the likelihood of higher injury severity level is increased if more than three vehicles are involved in multiple-vehicle crashes. A positive coefficient value (Mean 0.366, 95% BCI [0.227, 0.506]) was observed for the categorical variable that compared crashes involving three vehicles versus crashes involving more than three vehicles. This may be attributed to the fact that crashes involving more than three vehicles will probably result in drivers or passengers trapped in a vehicle, resulting in more severe injuries. Further, as mentioned earlier in this study, a multiple-vehicle crash may involve one vehicle being hit by other vehicles several times from different directions, resulting in an increased likelihood of a severe injury crash (*13*).

Discussion on Posterior Probability Distributions

Posterior probability plots were generated to analyze the relationship between independent variables and the probability of persons involved in a 3+ multiple-vehicle crash to be severely injured. The independent variables considered in the plots are: >3 vehicles involved in a crash, nighttime condition, and alcohol involvement. Figure 3*a* and *b* provide the posterior probability plots for average vehicle speed and average EHV, respectively.

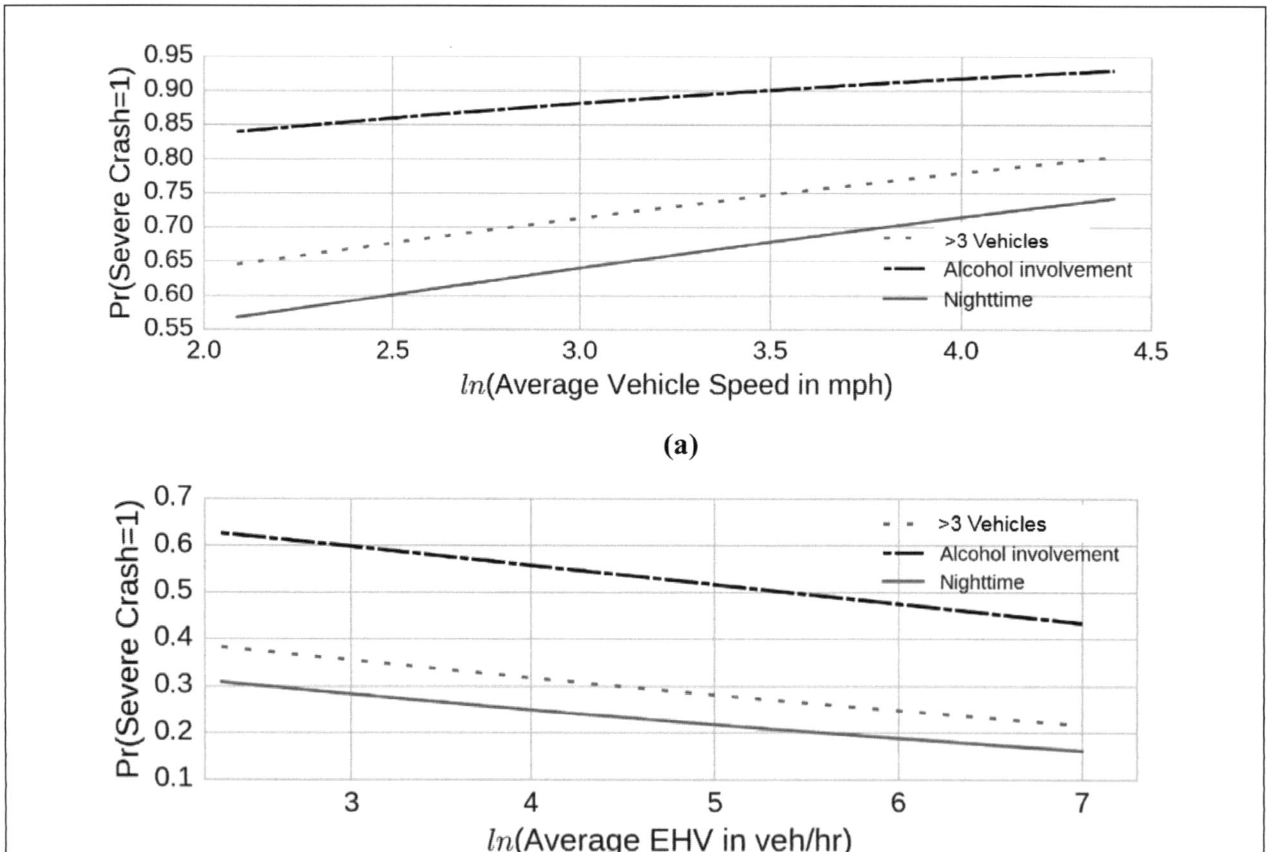

Figure 3. Posterior probability plots: (*a*) average vehicle speed posterior probability plot; (*b*) average EHV posterior probability plot.

The trends in Figure 3*a* show that, with increase in average vehicle speed, the probability of a severe injury increases for >3 vehicles, alcohol involvement, and during nighttime condition. On the other hand, the trends in Figure 3*b* show that, with increase in average EHV, the probability of a severe injury decreases for >3 vehicles, alcohol involvement, and during nighttime condition. The plots also reveal that alcohol involvement has the highest odds of resulting in severe injuries, followed by >3 vehicles involved in a crash, and nighttime condition.

As expected, the probability of alcohol-involved multiple-vehicle crashes, >3 multiple-vehicle crashes, and nighttime crashes to be severe increase with the increase in vehicle speeds and decrease with the increase in average EHV. As can be observed from Figure 3*a* and *b*, alcohol-involved multiple-vehicle crashes occurring during low speeds have over 80% probability of becoming severe, while the probability is over 60% when traffic volumes are low.

Despite being the variable with the lowest increasing probability of resulting in severe 3+ multiple-vehicle collisions, nighttime condition may be considered as one of the factors that exacerbate the severity of 3+ multiple-vehicle crashes. Multiple-vehicle crashes occurring during nighttime condition have more than 55% probability of becoming severe when speeds are low, and the probability further increases with the increase in vehicle speeds.

Conclusion

This research investigated the factors influencing the severity of multiple-vehicle crashes involving three or more vehicles using the binary probit model with the Dirichlet random effect parameter. The Dirichlet random effect parameter was incorporated in the model to account for unobserved heterogeneity existing in the data. The study area included a 111-mile corridor on I-4 in Orlando, Florida. Crash and real-time traffic-related data for the years 2014–2016 were used in the model development. Crash-related variables included crash day, crash time, crash type, alcohol involvement, weather condition, lighting condition, number of vehicles involved, presence of work zone, and distraction involvement. RITIS detectors provided high-resolution traffic data, including vehicle count, average speed, and occupancy at 30-second intervals. For each crash, real-time traffic data were extracted for a period of 10 minutes before the crash occurrence from the detectors within one mile of the crash.

The probit model was implemented using a Bayesian framework. The ratios of the MC errors relative to standard deviations of the estimates, trace, density, and autocorrelation plots were monitored to achieve parameter estimation convergence. Five predictor variables were found to be significant at the 95% BCI: logarithm of average vehicle speed, logarithm of average EHV, alcohol involvement, lighting condition, and number of vehicles involved (3 or >3). Further investigation involved analyzing the posterior probability distributions of crashes involving more than three vehicles, alcohol involvement, and nighttime condition in conjunction with the average vehicle speed and the average equivalent hourly volume. From this, alcohol involvement was found to have the highest odds of resulting in fatal and severe injuries, followed by crashes involving more than three vehicles, and nighttime conditions.

This study provides insights on the factors influencing the severity of multiple-vehicle crashes involving three or more vehicles. The study findings can be used by traffic operators and traffic management center personnel to associate certain real-time traffic conditions with severe injury crashes involving multiple vehicles. The results could help reduce the likelihood of severe multiple-vehicle crashes by developing effective countermeasures such as variable speed limits, active traffic management strategies, and so forth.

The most critical drawback of the applicability of the real-time data used in this study is its limited access. RITIS detectors are only installed on limited-access facilities. Even on the freeway system, RITIS detectors are installed on only a portion of the freeway network. Also, it is possible for a crash to occur when RITIS detectors are inactive resulting in missing real-time data. Another limitation is related to the fundamental traffic flow principles for speed, volume, and density. For example, it cannot be directly determined whether low traffic flow measurements reported by RITIS detectors represent congested or free-flow conditions. Nonetheless, this study has demonstrated that using real-time traffic data, even with its limitations, provides valuable insights on crash prediction modeling.

Author Contributions

The authors confirm contributing to the paper as follows: study conception and design: A. Kitali, P. Alluri, T. Sando; data collection: A. Kitali, R. Lentz, P. Martz; analysis and interpretation of results: A. Kitali, E. Kidando, P. Alluri, T. Sando; R. Moses; draft manuscript preparation: A. Kitali, E. Kidando, P. Martz. All authors reviewed the results and approved the final version of the manuscript.

References

1. NHTSA. *2015 Motor Vehicle Crashes: Overview*. Report DOT HS 812 318. NHTSA, Washington, D.C., 2016.
2. National Safety Council. *NSC News & Events*. National Safety Council. http://www.nsc.org/Connect/NSCNewsReleases/Lists/Posts/Post.aspx?ID=180. Accessed July 23, 2017.

3. Wu, Q., F. Chen, G. Zhang, X. C. Liu, H. Wang, and S. M. Bogus. Mixed Logit Model-Based Driver Injury Severity Investigations in Single- and Multi-Vehicle Crashes on Rural Two-Lane Highways. *Accident Analysis and Prevention*, Vol. 72, 2014, pp. 105–115. https://doi.org/10.1016/j.aap.2014.06.014.
4. Chen, F., and S. Chen. Injury Severities of Truck Drivers in Single- and Multi-Vehicle Accidents on Rural Highways. *Accident Analysis and Prevention*, Vol. 43, No. 5, 2011, pp. 1677–1688. https://doi.org/10.1016/j.aap.2011.03.026.
5. Geedipally, S. R., and D. Lord. Investigating the Effect of Modeling Single-Vehicle and Multi-Vehicle Crashes Separately on Confidence Intervals of Poisson–gamma Models. *Accident Analysis and Prevention*, Vol. 42, No. 4, 2010, pp. 1273–1282.
6. Pasupathy, R. K., J. N. Ivan., and P. J. Ossenbruggen. *Single and Multi-Vehicle Crash Prediction Models for Two-Lane Roadways*. Publication DTRS95-G-0001. New England (Region One) University Transportation Center, Massachusetts Institute of Technology, Cambridge, MA, 2000. http://www.cti.uconn.edu/pdfs/ucnr9-8_final-report.pdf.
7. Lee, J., B. Nam, and M. Abdel-Aty. Effects of Pavement Surface Conditions on Traffic Crash Severity. *Journal of Transportation Engineering*, Vol. 141, No. 10, 2015, p. 04015020.
8. Lu, G., M. Chitturi, A. Ooms, and D. Noyce. Ordinal Discrete Choice Analyses of Wisconsin Cross-Median Crash Severities. *Transportation Research Record: Journal of the Transportation Research Board*, 2010. 2148: 47–58.
9. Ouyang, Y., V. Shankar, and T. Yamamoto. Modeling the Simultaneity in Injury Causation in Multivehicle Collisions. *Transportation Research Record: Journal of the Transportation Research Board*, 2002. 1784: 143–152.
10. Abdel-Aty, M., N. Uddin, and A. Pande. Split Models for Predicting Multivehicle Crashes During High-Speed and Low-Speed Operating Conditions on Freeways. *Transportation Research Record: Journal of the Transportation Research Board*, 2005. 1908: 51–58.
11. Guarino, J., and A. Champaneri. *Factors Involved in Fatal Vehicle Crashes*. Report TR-008. RITA Bureau of Transportation Statistics Washington D.C., 2008.
12. Campbell, B. N., J. D. Smith, and W. G. Najm. *Analysis of Fatal Crashes due to Signal and Stop Sign Violations*. Report DOT HS 809 779. NHTSA, U.S. Department of Transportation, Washington, D.C., 2004.
13. Khattak, A. Injury Severity in Multivehicle Rear-End Crashes. *Transportation Research Record: Journal of the Transportation Research Board*, 2001. 1746: 59–68.
14. Michigan Department of Transportation. *Evaluation of the I-94 Corridor from the Indiana State Line to US-127 South*. Michigan Department of Transportation, 2015.
15. Chang, J. Y., and W. C. Lai. An Analysis of Pileup Accidents in Highway Systems. *Physica A: Statistical Mechanics and its Applications*, Vol. 443, 2016, pp. 423–438.
16. Nagatani, T. Chain-Reaction Crash on a Highway in High Visibility. *Physica A: Statistical Mechanics and its Applications*, Vol. 450, 2016, pp. 466–472.
17. Nagatani, T. Effect of Vehicular Size on Chain-Reaction Crash. *Physica A: Statistical Mechanics and its Applications*, Vol. 438, 2015, pp. 132–139.
18. Xu, C., P. Liu, B. Yang, and W. Wang. Real-Time Estimation of Secondary Crash Likelihood on Freeways Using High-Resolution Loop Detector Data. *Transportation Research Part C: Emerging Technologies*, Vol. 71, 2016, pp. 406–418.
19. Oliver, C. J. *Fog Related Crashes: Statewide Study of Fog Related Crashes in North Carolina*. Traffic Safety Unit, Transportation Mobility and Safety Division: North Carolina Department of Transportation, Raleigh, NC, 2013.
20. Hassan, H. M., and M. A. Abdel-Aty. Predicting Reduced Visibility Related Crashes on Freeways Using Real-Time Traffic Flow Data. *Journal of Safety Research*, Vol. 45, 2013, pp. 29–36. https://doi.org/10.1016/j.jsr.2012.12.004.
21. Wang, L., M. Abdel-Aty, Q. Shi, and J. Park. Real-Time Crash Prediction for Expressway Weaving Segments. *Transportation Research Part C: Emerging Technologies*, Vol. 61, 2015, pp. 1–10. https://doi.org/10.1016/j.trc.2015.10.008.
22. Wu, Y., M. Abdel-Aty, and J. Lee. Crash Risk Analysis during Fog Conditions Using Real-Time Traffic Data. *Accident Analysis and Prevention*, Vol. 114, 2017, pp. 4–11. https://doi.org/10.1016/j.aap.2017.05.004.
23. University of Florida. *Signal Four Analytics*. University of Florida. https://s4.geoplan.ufl.edu/. Accessed July 20, 2017.
24. Christoforou, Z., S. Cohen, and M. G. Karlaftis. Integrating Real-Time Traffic Data in Road Safety Analysis. *Procedia - Social and Behavioral Sciences*, Vol. 48, 2012, pp. 2454–2463. https://doi.org/10.1016/j.sbspro.2012.06.1216.
25. Ivan, J. N., C. Wang, and N. R. Bernardo. Explaining Two-Lane Highway Crash Rates Using Land Use and Hourly Exposure. *Accident Analysis and Prevention*, Vol. 32, No. 6, 2000, pp. 787–795. https://doi.org/10.1016/S0001-4575(99)00132-3.
26. Theofilatos, A. Incorporating Real-Time Traffic and Weather Data to Explore Road Accident Likelihood and Severity in Urban Arterials. *Journal of Safety Research*, Vol. 61, 2017, pp. 9–21. https://doi.org/10.1016/j.jsr.2017.02.003.
27. Abdelwahab, H., and M. Abdel-Aty. Development of Artificial Neural Network Models to Predict Driver Injury Severity in Traffic Accidents at Signalized Intersections. *Transportation Research Record: Journal of the Transportation Research Board*, 2001. 1746: 6–13.
28. Pande, A., A. Das, M. Abdel-Aty, and H. Hassan. Estimation of Real-Time Crash Risk. *Transportation Research Record: Journal of the Transportation Research Board*, 2011. 2237: 60–66.
29. Hourdos, J., V. Garg, P. Michalopoulos, and G. Davis. Real-Time Detection of Crash-Prone Conditions at Freeway High-Crash Locations. *Transportation Research Record: Journal of the Transportation Research Board*, 2006. 1968: 83–91.

30. Xu, C., A. P. Tarko, W. Wang, and P. Liu. Predicting Crash Likelihood and Severity on Freeways with Real-Time Loop Detector Data. *Accident Analysis and Prevention*, Vol. 57, 2013, pp. 30–39. https://doi.org/10.1016/j.aap.2013.03.035.
31. Abdel-Aty, M., A. Oloufa, Y. Peng, T. S. Shen, X. Yang, and J. Lee. *Real Time Monitoring and Prediction of Reduced Visibility Events on Florida's Highways*. Final Report Contract BDV24 962-01. Florida Department of Transportation, Tallahassee, FL, 2014.
32. Wang, L., M. Abdel-Aty, and J. Lee. Safety Analytics for Integrating Crash Frequency and Real-Time Risk Modeling for Expressways. *Accident Analysis and Prevention*, Vol. 104, 2017, pp. 58–64. https://doi.org/10.1016/J.AAP.2017.04.009.
33. Ahmed, M. M., M. Abdel-Aty, J. Lee, and R. Yu. Real-Time Assessment of Fog-Related Crashes Using Airport Weather Data: A Feasibility Analysis. *Accident Analysis and Prevention*, Vol. 72, 2014, pp. 309–317.
34. Shi, Q., M. Abdel-Aty, and J. Lee. A Bayesian Ridge Regression Analysis of Congestion's Impact on Urban Expressway Safety. *Accident Analysis and Prevention*, Vol. 88, 2016, pp. 124–137. https://doi.org/10.1016/J.AAP.2015.12.001.
35. Kitali, A. E., E. Kidando, T. Sando, R. Moses, and E. E. Ozguven. Evaluating Aging Pedestrian Crash Severity Using Bayesian Complementary Log-Log Model for Improved Prediction Accuracy. *Transportation Research Record: Journal of the Transportation Research Board*, 2017. 2659: 155–163.
36. Martin, J. L., and D. Wu. Pedestrian Fatality and Impact Speed Squared: Cloglog Modeling from French National Data. *Traffic Injury Prevention*, Vol. 19, No. 1, 2018, pp. 94–101. https://doi.org/10.1080/15389588.2017.1332408.
37. Traunmüller, R., A. Murr, and J. Gill. Modeling Latent Information in Voting Data with Dirichlet Process Priors. *Political Science*, Vol. 23, No. 1, 2015, pp. 1–20.
38. Heinzl, F., and G. Tutz. Clustering in Linear Mixed Models with Approximate Dirichlet Process Mixtures Using EM Algorithm. *Statistical Modelling*, Vol. 13, No. 1, 2013, pp. 41–67.
39. Ohlssen, D. I., L. D. Sharples, and D. J. Spiegelhalter. Flexible Random-Effects Models Using Bayesian Semi-Parametric Models: Applications to Institutional Comparisons. *Statistics in Medicine*, Vol. 26, No. 7, 2007, pp. 2088–2112.
40. Ahmed, M., M. Abdel-Aty, and R. Yu. Assessment of Interaction of Crash Occurrence, Mountainous Freeway Geometry, Real-Time Weather, and Traffic Data. *Transportation Research Record: Journal of the Transportation Research Board*, 2012. 2280: 51–59.
41. Huang, H., H. C. Chin, and M. M. Haque. Severity of Driver Injury and Vehicle Damage in Traffic Crashes at Intersections: A Bayesian Hierarchical Analysis. *Accident Analysis and Prevention*, Vol. 40, No. 1, 2008, pp. 45–54.
42. Salvatier, J., T. V. Wiecki, and C. Fonnesbeck. Probabilistic Programming in Python Using PyMC3. *PeerJ Computer Science*, Vol. 2, 2016, e55. https://doi.org/10.7717/peerj-cs.55.
43. Kidando, E., R. Moses, Y. Abdelrazig, and E. E. Ozguven. Safety Analysis Considering the Impact of Travel Time Reliability on Elderly Drivers. *Journal of Transportation Research Forum*, Vol. 56, No. 1, 2017, pp. 21–40.
44. Duncan, C., A. Khattak, and F. Council. Applying the Ordered Probit Model to Injury Severity in Truck-Passenger Car Rear-End Collisions. *Transportation Research Record: Journal of the Transportation Research Board*, 1998. 1635: 63–71.
45. Xie, Y., Y. Zhang, and F. Liang. Crash Injury Severity Analysis Using Bayesian Ordered Probit Models. *Journal of Transportation Engineering*, Vol. 35, No. 1, 2009, pp. 18–25.

The Standing Committee on Safety Data, Analysis and Evaluation (ANB20) peer-reviewed this paper (18-06389).

Article

TRR
JOURNAL OF THE TRANSPORTATION RESEARCH BOARD

Transportation Research Record
2018, Vol. 2672(38) 138–148
© National Academy of Sciences:
Transportation Research Board 2018
Article reuse guidelines:
sagepub.com/journals-permissions
DOI: 10.1177/0361198118777354
journals.sagepub.com/home/trr
SAGE

Analyzing the Injury Severity Sustained by Non-Motorists at Mid-Blocks considering Non-Motorists' Pre-Crash Behavior

Bowen Dong[1], Xiaoxiang Ma[1], and Feng Chen[1]

Abstract

Non-motorized travel is considered as one of the most beneficial transportation modes. Compared with other road users, non-motorists as a whole account for about 13% of all fatal transportation-related accidents, and from 2002 to 2009 nearly 30% of those fatalities occur at mid-blocks. In addition, there are few reported studies that investigated the impact of non-motorists' pre-crash behavior on injury severities. To examine the risk factors of non-motorist injury severity at mid-blocks, 8-year crash-related data from the General Estimates System were explored, based on the mixed logit model. The data contain various information including time characteristics, crash features, environmental conditions, roadway attributes, non-motorists' characteristics, and their pre-crash behaviors. The results show that five factors tend to have mixed effects on injury severities, including the speed limit between 30 and 55 mph, night time indicator, right-side collision, and hit-and-run action on the incapacitating injury, as well as no action of motorists on the non-incapacitating injury. Moreover, heavy and light truck, dark not lighted indicator, and age over 65 are found to increase the likelihood of fatal injury, while age below 25 decreases the likelihood of fatality. Other indicators including roadway alignment, number of lanes, and so forth also affected injury severity. After controlling for these factors, non-motorists' pre-crash behaviors such as darting or running into the road, activities in the roadway, and improper passing are found to have a significant impact on severity outcomes.

Non-motorized travel is considered as one of the most economical, convenient, and sustainable transportation modes (*1*). To reduce automobile exhaust and noise pollution, governments and transportation agencies have made significant efforts to encourage walking and cycling in most countries (*2*). Although non-motorized transportation is ecofriendly and promotes a healthy lifestyle, pedestrians and cyclists are often exposed to a higher risk of injury and fatality in traffic crashes. According to the 2013 World Health Organization road safety manual, pedestrians accounted for more than 22% of all fatalities in traffic collisions (*3*). Compared with other road users, non-motorists as a whole account for about 13% of all fatal transportation-related accidents, and from 2002 to 2009 nearly 30% of those fatalities occurred at mid-blocks (*4*, *5*).

Several studies have examined the factors that influence the frequency and severity of traffic crashes involving non-motorists (*6–12*). Many of these studies have focused on vehicle and non-motorist crashes at certain locations such as intersections (*13–15*). However, only a few of those focused on mid-blocks (*16*, *17*). The non-motorist-related crashes are more inclined to severe injury at mid-blocks as a result of faster travel speed of motor vehicles, which causes heavier casualties and more economic losses (*17*). Thus, it is of great importance to explore the contributing factors of non-motorists' crash severities at mid-blocks. Also, the pre-crash behavior of the non-motorist might have significant influences on the resulting injury severity. It is, therefore, important to investigate the impact of pre-crash behavior on injury severities. As far as the authors are aware, there are few, if any, reported studies that have examined the impact of non-motorists' pre-crash behavior on injury severities.

The objective of this study is to examine the risk factors influencing the injury severities of non-motorists at mid-block crossings. Moreover, this study attempts to explore the role of the non-motorists' pre-crash behavior

[1]College of Traffic Engineering and Key Laboratory of Road & Traffic Engineering of the Ministry of Education, Tongji University, Shanghai, China

Corresponding Author:
Address correspondence to Xiaoxiang Ma: xiaoxiang.ma@tongji.edu.cn

in their resulting injury severity outcomes. To do this, we collected eight years of non-motorists related crash data (2002–2009) from the General Estimates System (GES) which included non-motorist characteristics and behaviors, vehicle attributes, environmental factors, and crash characteristics. To capture the heterogeneity across different accidents, a mixed logit model was developed to identify significant factors and quantify their impacts on non-motorists' injury severities at mid-blocks.

The rest of this paper is organized as follows. Previous studies regarding the vehicle-nonmotorist crash severity are summarized in the next section, then a description of the data is presented, followed by a detailed explanation of the mixed logit model structures and specifications for vehicle-induced non-motorist crashes at mid-blocks. Model estimation results and their implications for reducing injury severity in crashes are then provided, and the final section summarizes the conclusions and future research recommendations.

Previous Research

This section reviews the literature on injury severity analysis of vehicle–non-motorist crashes. To be specific, it focuses on studies that investigated the contributing factors affecting the crash severity of non-motorists and those related to accidents at mid-blocks.

Zajac and Ivan studied the injury severity of pedestrian crashes on rural highways that have no traffic control devices, based on the ordered probit model (*18*). The impacts of some observable features of roadway, vehicle, and person involved were examined, and the results showed that roadway width, alcohol, vehicle type, and age over 65 are significant contributory factors. Forbes and Habib investigated the effects of the built environment contributing factors on pedestrian injury severity based on 5-year crash data in Canada, and found that personal, collision, and some environmental characteristics such as land-use type and demographic attributes may have significant impacts on the outcomes of injury severity (*19*). Numerous authors have also done much work on the injury severities of vehicle–pedestrian crashes from different perspectives (*20–23*).

Besides pedestrians, some researchers also considered another major component of non-motorists, such as the bicyclist. For instance, Kim et al. studied the injury severity of vehicle-bicycle accidents based on crash data in North Carolina from 1997 to 2002 (*24*). The results from their multinomial logit model illustrated that increase in speed, heavy trucks and pickups, alcohol, and inclement weather would increase the possibility of fatal injury. The results also indicated that the accident rate would increase with the growth of vehicle and cyclist volumes. Other researchers tend to study all types of non-motorist accidents altogether (*6, 25–27*). For example, Decker et al. used the descriptive statistical method to analyze the injury severity of non-motorist crashes including pedestrians, bicyclists, and motorcyclists, caused by reversing cars (*27*). They found that though most of these crashes were not severe, pedestrians suffered more severe injury compared with other non-motorists.

It is widely known that unobserved heterogeneity is a critical problem associated with traffic crash data. Many pedestrian-related factors, such as physical conditions and characters, cannot be observed, thus are absent from data records (*10*). These unobservable effects, if not accounted for, lead to biased estimates and erroneous inferences that follow. To solve this problem, some researchers have proposed random parameter models to study non-motorist injury severity. For example, Eluru et al. developed a mixed generalized ordered response model to explore the risk factors of pedestrian and bicyclist injury severities (*6*). Kim et al. adopted a mixed logit model to examine the impact of some potential contributing factors on the injury severities of vehicle–pedestrian crashes (*10*). The use of mixed parameter structures relaxes the fixed parameter assumption in the multinomial logit model, which offers a great capacity to address unobserved heterogeneity to a certain extent.

To obtain a deeper understanding of the mechanisms behind crashes between non-motorists and vehicles, many researchers have focused on non-motorist crashes in certain situations. Haleem et al. compared the possible determinants of pedestrian injury severity at signalized and unsignalized intersections (*14*). Diogenes and Lindau examined the risk factors of pedestrian accidents at mid-block in Brazil (*16*). The location of a bus stop, road width, and traffic volume were found to have a significant influence on the crash risk. Pour et al. developed a classification and regression tree (CART), a bagged decision tree (DT) and a boosted DT to investigate the determinants of pedestrian severities (*17*). The data set used in their study consisted of 10-year vehicle–pedestrian crashes at mid-block in Australia. The results stated that the neighborhood social characteristics were important influencing factors of pedestrian severity.

Besides the studies mentioned above, as far as we know, few studies have specialized in the injury severity of non-motorists at mid-blocks. As there are many non-motorists' traffic accidents at mid-blocks, it is crucial to investigate the mechanism of injury severity of such collisions. In addition, no study has considered the behavior of non-motorists before the crashes in injury severity studies. This study aims to fill this knowledge gap and contribute to the literature on the understanding of non-motorists' injury severity at mid-blocks by incorporating their pre-crash behaviors.

Data Description

Data Source

The data set used in this study was obtained from the GES, a data set of National Highway Traffic Safety Administration. The GES, which began operation in 1988 and has been recording crash data from police reports in about 60 areas across the USA, is used to identify dangerous highway areas and provide a basis for cost and benefit analyses (*28*, *29*). The GES now operates under the National Automotive Sampling System.

Detailed information about the crash, person, and vehicle were included in the GES. For the aim of this study, the authors extracted the data from the GES to include only vehicle–non-motorist crashes at mid-block. Eight-year crash data (from 2002 to 2009) were combined, and finally, a data set containing 4,134 accidents was obtained.

Variables Considered

Five types of variables including time characteristics, crash (including vehicle) features, environmental conditions, non-motorists' characteristics, and roadway (including traffic) attributes were incorporated as control variables in this study.

The time characteristics consist of some special temporal information which may have a significant effect on injury severity. Time of the day was processed into non-peak time, peak time and night time. Night time (8 p.m. to 7 a.m.) and peak time (7 a.m. to 10 a.m. or 5 p.m. to 8 p.m.) were included the model. In addition, day of the week (Monday to Friday) and months (January to December) are considered as control variables.

There are many types of crash and vehicle features recorded in the GES. The first point of impact (the front, the right side, and the left side of the vehicle), vehicle type, and whether hit and run have been chosen as the contributing variables. According to the original coding scheme of the GES analytical user's manual (*29*), there is a very detailed classification of vehicle type in the GES. In this study, the vehicle type variable was processed into five main categories: passenger car, light truck, bus, heavy truck, and motorcycle.

The environmental conditions controlled in this study include: (1) light condition, which was coded as daylight, dark lighted and dark not lighted; (2) road surface condition, which was coded as dry surface, wet surface, and other surface condition based on the classification of the GES. The potentially significant contributory factors of roadway and traffic attributes include the number of lanes, roadway alignment, speed limit, and so forth.

Non-motorists' characteristics, especially their behaviors right before the crash, are the main focus of this study. Age and sex were considered. It should be noted that the GES includes very detailed classification of non-motorist behaviors. Owing to limited sample sizes, the pre-crash behaviors of the non-motorists were classified into six major categories in this study: (1) Darting or running into the road; (2) improper crossing; (3) no action; (4) walking, playing, working, sitting, lying, standing in the roadway; (5) inattentive, and (6) others. The detailed explanations of these behaviors are as follows:

Darting or running into road: this behavior of non-motorists can best be described as an impulsive running or jogging across the road.

Improper crossing: the pedestrian is crossing the road, but not running or jogging, before the crash.

Walking, playing, working, sitting, lying, standing in the roadway: the non-motorists are playing a game, working because of his/her job, or standing/sitting/lying (not moving) in the roadway before the crash.

Inattentive: the non-motorists are standing, sitting, or lying inattentively before the crash.

No action: the non-motorists do not take any other actions except the behaviors record in the coding manual.

The distribution of non-motorists' pre-crash behavior by injury severity levels is shown in Table 1. Table 2 provides definitions and descriptions of the variables.

As shown in Table 1, darting or running into road accounts for a significant proportion (40.5%) of the accidents at mid-blocks. Regarding accidents involving darting or running into the roadway, non-incapacitating evident injury takes up the largest proportion. In addition, the accidents with respect to activity in road have a higher fatal ratio (30/270) compared with accidents with respect to other behavior (70/1676, 110/1256, 18/476, 1/34).

To identify the best set of contributing factors, the backward elimination method was used and 26 factors were kept in the final model.

For the injury severity of non-motorists at mid-blocks, four levels of injury severity are considered: (1) no injury or possible injury; (2) non-incapacitating evident injury; (3) incapacitating injury, and (4) fatal injury. Table 2 shows the descriptive statistics of the dependent variable and all the considered variables discussed above.

Methodology

To address unobserved heterogeneities, this paper adopted the mixed logit model to investigate the contributing factors of non-motorists' injury severity. The model structure is specified in the following section.

According to previous studies by McFadden, Washington et al. and Milton et al. (*30–32*), a propensity

Table 1. Distribution of Pre-Crash Behavior by Injury Severities

	Darting or running	Improper crossing	No action	Activity in road	Inattentive	Other	Total
NP	74 1.8%	46 1.1%	21 0.5%	11 0.3%	4 0.1%	22 0.5%	178
NI	1020 24.7%	614 14.9%	290 7.0%	144 3.5%	21 0.5%	198 4.8%	2287
I	512 12.4%	486 11.8%	147 3.6%	85 2.1%	8 0.2%	148 3.6%	1386
F	70 1.7%	110 2.7%	18 0.4%	30 0.7%	1 0.0%	54 1.3%	283
Total	1676 40.5%	1256 30.4%	476 11.5%	270 6.5%	34 0.8%	422 10.2%	4134

Note: NP = no injury and possible injury category; NI = non-incapacitating injury; I = incapacitating injury; F = fatal injury.

function expresses that non-motorist n suffers from injury severity i can be defined as

$$S_{\text{in}} = \beta_i x_{\text{in}} + \varepsilon_{\text{in}} \tag{1}$$

where x_{in} is a vector of all the explanatory variables which have effects on the injury severity, β_i is the estimable parameters of x_{in} for each category of response variables and ε_{in} is the error term which denotes the unobserved effects.

Let $P_n(i)$ be the probability of injury severity I (no injury/Possibly injury, non-incapacitating evident injury, incapacitating injury, fatal injury) of the non-motorist n, then

$$P_n(i) = P(\beta_i X_n + \varepsilon_{\text{ni}} \geq \beta_{i'} X_n + \varepsilon_{\text{ni}'}) \forall i' \in I, i' \neq i \tag{2}$$

where I is the set of all the injury severities which are mutually independent. i and i' are different injury severities. Assuming that ε_{ni} follows a type I extreme-value distribution (*30*), a multinomial logit structure which can be defined as

$$P_n(i) = \frac{e^{\beta_i X_n}}{\sum_{\forall i' \in I} e^{\beta_i{}' X_n}} \tag{3}$$

where the parameter vector β for each injury severity can be estimated by the maximum likelihood method.

To address the unobserved heterogeneity, a mixed structure is developed by making the estimable parameters β_i vary across individual accidents (*33*):

$$P_{n_{\text{jt}}}(i) = \int \frac{e^{\beta_i X_{n_{\text{jt}}}}}{\sum_{\forall i' \in I} e^{\beta_i X_{n_{\text{jt}}}}} f(\beta|\varphi) \text{d}\beta \tag{4}$$

where $f(\beta|\varphi)$ denotes the density function of β and φ denotes a parameter vector of β. This study used the NLMIXED procedure in SAS software to estimate the mixed logit model. To relax the fixed parameter assumption, a specific distribution of $f(\beta|\varphi)$ is specified on the parameter. Based on previous research (*32*, *34*), the normal distribution was found to fit the data set better than other commonly used distributions (lognormal, triangular, and uniform). Thus the mixed logit model which has the density function with normal distributions will be considered in this paper. A simulation-based maximum likelihood method using Halton draws rather than purely random draws was used to calculate the parameter vector β for higher efficiency (*35*).

For a logit model, it is noteworthy that the value of the estimated parameter does not reflect the quantitative effects of the explanatory variables. Therefore the marginal effects should be calculated. In this study, all the risk factors are processed to dummy variables. According to the previous study (*36*), a "pseudo-elasticity" can be used to quantify the effects of dummy variables. In the logit model, the formulation of pseudo-elasticity is as follows:

$$E_{x_{\text{nk}}}^{P_{\text{ni}}} = \left(e^{\beta_{\text{ik}}} \frac{\sum e^{\beta_{i'} x_n}}{\sum e^{\Delta \beta_{i'} x_n}} - 1 \right) \times 100 \tag{5}$$

where $E_{x_{\text{nk}}}^{P_{\text{ni}}}$ is the pseudo-elasticity of the kth variable on the ith category of the nth observation. $\Delta\beta_{i'} x_n$ means that the x_{nk} is set to one and $\beta_{i'} x_n$ means that the x_{nk} is set to zero. Finally the average value of the pseudo-elasticity of all the observations is calculated to be as the marginal effect of each explanatory variable.

Results

In this study, the no injury and possibly injury category is chosen as the base category. Therefore the estimable parameters of the mixed logit model denote the difference between the the corresponding injury category and the no injury and possibly injury category. Table 3 shows the model estimation results of the mixed logit model.

The McFadden R^2 is used to evaluate the performance of the mixed logit model. According to Domencich and

Table 2. Variable Definitions and Descriptions

Variable description[a]	No injury/ possibly injury		Non-incapacitating evident injury		Incapacitating injury		Fatal injury		All
Time characteristics									
Time of day									
Night time (8 p.m. to 7 a.m.)	33	2.70%	524	42.81%	489	39.95%	178	14.54%	1224
Peak time (7 a.m. to 10 a.m. or 5 p.m. to 8 p.m.)	66	4.42%	883	59.10%	477	31.93%	68	4.55%	1494
Day of week									
Sunday	20	3.99%	258	51.50%	178	35.53%	45	8.98%	501
Monday	13	2.29%	327	57.57%	192	33.80%	36	6.34%	568
Tuesday	30	5.16%	325	55.94%	183	31.50%	43	7.40%	581
Wednesday	27	4.91%	322	58.55%	177	32.18%	24	4.36%	550
Thursday	28	4.63%	338	55.87%	202	33.39%	37	6.12%	605
Friday	43	5.96%	383	53.05%	241	33.38%	55	7.62%	722
Saturday	17	2.80%	334	55.02%	213	35.09%	43	7.08%	607
Month									
January	13	4.55%	136	47.55%	114	39.86%	23	8.04%	286
February	12	4.48%	144	53.73%	78	29.10%	34	12.69%	268
March	14	4.44%	164	52.06%	118	37.46%	19	6.03%	315
April	11	3.23%	197	57.77%	111	32.55%	22	6.45%	341
May	15	4.14%	223	61.60%	100	27.62%	24	6.63%	362
June	10	2.67%	243	64.97%	98	26.20%	23	6.15%	374
July	9	2.41%	210	56.15%	135	36.10%	20	5.35%	374
August	13	3.80%	201	58.77%	111	32.46%	17	4.97%	342
September	21	5.60%	208	55.47%	119	31.73%	27	7.20%	375
October	19	4.71%	218	54.09%	138	34.24%	28	6.95%	403
November	22	5.82%	195	51.59%	134	35.45%	27	7.14%	378
December	19	6.01%	148	46.84%	130	41.14%	19	6.01%	316
Crash and vehicle characteristics									
First point of impact									
Front	104	3.92%	1400	52.77%	939	35.39%	210	7.92%	2653
Right side	37	5.89%	407	64.81%	174	27.71%	10	1.59%	628
Left side	20	5.29%	239	63.23%	109	28.84%	10	2.65%	378
Vehicle type									
Automobile	126	4.25%	1682	56.71%	997	33.61%	161	5.43%	2966
Light truck	35	3.87%	479	52.99%	297	32.85%	93	10.29%	904
Bus	0	0.00%	6	42.86%	7	50.00%	1	7.14%	14
Heavy truck	5	6.76%	23	31.08%	28	37.84%	18	24.32%	74
Motorcycle	5	10.00%	25	50.00%	16	32.00%	4	8.00%	50
Hit and run	29	4.04%	418	58.22%	228	31.75	43	5.99%	718
Environmental characteristics									
Light condition									
Daylight	121	5.15%	1479	62.99%	681	29.00%	67	2.85%	2348
Dark lighted	29	2.66%	502	46.01%	451	41.34%	109	9.99%	1091
Dark not lighted	19	3.66%	202	38.92%	196	37.76%	102	19.65%	519
Surface condition									
Dry surface	148	4.15%	1988	55.80%	1184	33.23%	243	6.82%	3563
Wet surface	25	5.40%	239	51.62%	167	36.07%	32	6.91%	463
Roadway and traffic attributes									
Number of lanes									
One lane	4	2.47%	120	74.07%	34	20.99%	4	2.47%	162
Two lanes	81	4.63%	985	56.25%	580	33.12%	105	6.00%	1751
Three lanes	10	2.51%	180	45.11%	162	40.60%	47	11.78%	399
Four lanes	13	2.78%	214	45.73%	188	40.17%	53	11.32%	468
Five or more lanes	13	3.29%	176	44.56%	162	41.01%	44	11.14%	395
Roadway alignment									
Curve	12	7.23%	72	43.37%	61	36.75%	21	12.65%	166
Straight	166	4.18%	2215	55.82%	1325	33.39%	262	6.60%	3968

(continued)

Table 2. (continued)

Variable description[a]	No injury/ possibly injury		Non-incapacitating evident injury		Incapacitating injury		Fatal injury		All
Speed limit									
Speed limit below 30	87	4.53%	1220	63.57%	561	29.23%	51	2.66%	1919
Speed limit between 30 and 55	81	3.95%	978	47.68%	782	38.13%	210	10.24%	2051
Speed limit over 60	3	3.75%	29	36.25%	26	32.50%	22	27.50%	80
Person characteristics									
Age									
Age below 25	102	4.52%	1443	63.99%	651	28.87%	59	2.62%	2255
Age between 25 and 45	36	4.06%	417	47.07%	345	38.94%	88	9.93%	886
Age between 45 and 65	29	4.50%	285	44.19%	260	40.31%	71	11.01%	645
Age over 65	11	3.16%	142	40.80%	130	37.36%	65	18.68%	348
Sex									
Female	62	4.30%	864	59.88%	445	30.84%	72	4.99%	1443
Non-motorists behavior									
Darting or running into road	74	4.42%	1020	60.86%	512	30.55%	70	4.18%	1676
Improper crossing	46	3.66%	614	48.89%	486	38.69%	110	8.76%	1256
No action	21	4.41%	290	60.92%	147	30.88%	18	3.78%	476
Activities in roadway	11	4.07%	144	53.33%	85	31.48%	30	11.11%	270
Inattentive	4	11.76%	21	61.76%	8	23.53%	1	2.94%	34

[a]All the variables are 1/0 indicators. For example, the "night time" variable is 1 for night time and 0 for day time.

Mcffadden (*37*), the McFadden R^2 value of 0.2–0.4 represents excellent model fit. Therefore, the fitted model with a McFadden R^2 of 0.202 performs relatively satisfactory. The log-likelihood at convergence, Akaike information criterion (AIC) and Bayesian information criterion (BIC) are used as the criteria to compare performance between random and fixed parameter models. As shown in Table 3, the mixed logit model has a log-likelihood of −3821.5, an AIC of 7746.9, and a BIC of 8075.0 (with 52 estimated parameters), which shows improvement compared with −3839.6, 7773.1, and 8075.0 of the corresponding multinomial logit model. Meanwhile, a likelihood ratio test was also conduct to compare the two models. The formula of likelihood ratio test (*38*) is

$$2(L_{\text{mixed}}(\beta) - L_{\text{multi}}(\beta)) \tag{6}$$

where $L_{\text{mixed}}(\beta)$ is the log-likelihood at convergence of the mixed logit model, and $L_{\text{multi}}(\beta)$ is the log-likelihood at convergence of the multinomial logit model. The likelihood ratio statistic of 36.2 with 5 degrees of freedom denotes that the log-likelihood of mixed logit model and multinomial logit model is significantly different at 99.99% confidence interval. The above results indicate that the mixed logit model is superior for modeling non-motorists' injury severity compared with the multinomial logit model. The effects of explanatory variables used in this model are discussed in the following sections.

Time Characteristics

It is shown in Table 3 that there are different contributing factors for each level of injury severity. Previous studies found that Friday has the lowest proportion of vehicle–pedestrian crashes (*7*, *20*), and the results from this study further show that the possibilities of non-incapacitating evident injury and incapacitating injury at mid-blocks tend to have a 15.43% decrease on Friday. Different indicators for months were tested, but the results show that the injury severity levels do not depend on month indicators. With regard to daily characteristics, the parameter of night time for incapacitating injury is found to have a random distribution (significant at the level of 95% with *t*-statistic 4.11) with mean 0.52 and standard deviation 1.65. This indicates that 65% of the non-motorists suffer from the higher possibility of incapacitating injury at night time, while others tend to have a lower possibility of incapacitating injury. The night time indicator is also found to be significant and positive (183%) for the fatal injury category, which means night time increases the possibility of fatal crash by 183%. Al-Ghamdi also found that the likelihood of fatal crash at night is 1.8 times those during daytime in his study investigating the injury severity of vehicle–pedestrian crashes (*20*).

Crash and Vehicle Characteristics

It is noteworthy that the data associated with the first point of impact were recorded in the GES. The point where the car first hits the non-motorists bears some valuable information on the collision outcomes of the pedestrian and bicyclist at mid-blocks. Out of the three kinds of first impact point, the right side of the vehicle for incapacitating injury is found to be better treated as a random parameter. The parameter of the right side of

Table 3. The Estimated Results of the Mixed Logit Model

Variable description	Estimated parameter	Elasticity	*t*-statistic
Constant [NI]	2.3499		15.88
Constant [I]	1.4053		5.93
Time characteristics			
Time of day			
Night time [F]	1.340	183.0%	4.38
Day of week			
Friday [NI]	−0.371	−15.4%	−2.63
Friday [I]	−0.309	−8.1%	−1.95
Crash and vehicle characteristics			
First point of impact			
Left side [I]	−0.471	−8.8%	−3.09
Left side [F]	−1.515	−71.9%	−4.2
Right side [F]	−2.0526	−80.6%	−5.68
Vehicle type			
Light truck [F]	0.801	122.8%	4.9
Heavy truck [NI]	−0.985	−80.7%	−2.84
Heavy truck [F]	1.565	571.9%	3.77
Environmental characteristics			
Light condition			
Dark not lighted [F]	1.236	244.2%	4.7
Roadway and traffic attributes			
Number of lanes			
One lane [NI]	0.501	65.0%	2.4
Three lanes [I]	0.872	18.1%	2.44
Three lanes [F]	1.146	58.7%	2.96
Four lanes [I]	0.813	10.5%	2.56
Four lanes [F]	1.247	86.3%	3.55
Five or more lanes [I]	0.550	27.7%	3.23
Five or more lanes [F]	0.751	51.9%	3.34
Roadway alignment			
Curve [NI]	−0.709	−50.8%	−3.63
Person characteristics			
Age			
Age below 25 [F]	−1.095	−66.5%	−5.98
Age over 65 [I]	0.455	−10.7%	2.76
Age over 65 [F]	1.333	188.8%	6.27
Non-motorists behavior			
Darting or running into road [NI]	0.355	66.9%	2.55
Darting or running into road [F]	−0.464	−46.3%	−2.07
Improper crossing [F]	−0.570	−47.8%	−2.73
No action [F]	−1.035	−74.4%	−3.11
Activity in roadway [NI]	0.457	57.9%	2.36
Random parameters			
Mean of no action [NI]	0.6929		3.1
Std. dev. of no action (normal distribution)	1.649		3.23
Mean of speed limit between 30 and 55 [I]	0.3749		1.6
Std. dev. of speed limit between 30 and 55 (normal distribution)	1.4218		4.9
Mean of night time [I]	0.5151		4.11
Std. dev. of night time (normal distribution)	1.3745		3.58
Mean of right-side collision [I]	−0.8252		−3.68
Std. dev. of right-side collision (normal distribution)	1.4804		3.42
Mean of hit and run [I]	−1.2997		−2.65
Std. dev. of hit and run (normal distribution)	1.5004		3.03
Fit statistics			
Log likehood at convergence	−3821.5		
Log likehood with constants only	−4787.3		
Mcfadden R^2	0.202		
AIC (smaller is better)	7746.9		
BIC (smaller is better)	8075.0		

Note: NI = non-incapacitating injury; I = incapacitating injury; F = fatal injury.

the vehicle is normally distributed with mean −0.83 and standard deviation 1.48. This means that when the right side of the vehicle hit the object first, nearly 71% of the non-motorists tend to sustain a lower likelihood of incapacitating injury, whereas the other non-motorists tend to experience a higher risk of incapacitating injury. From the results, the front-point collision is not significant for all the injury severities. In addition, the right side decreases the likelihood of incapacitating injury by 80.6%, and the left side decreases the likelihood of incapacitating injury by 8.8% and fatal injury by 71.9%.

Vehicle type, which indicates different masses, is an important contributing factor of injury severity. The results from the mixed logit model denote that when the non-motorists' collision involves a heavy truck, the possibility of non-incapacitating evident injury tends to decrease by 80.7% and the possibility of fatal injury tends to increase by 572%. This is intuitive, as collision with a heavier object would require absorbing more energy, thus leading to severe injury. In a similar vein, the light truck also increases the possibility of fatal injury, and the pseudo-elasticity is 122.8%. These findings are consistent with previous studies associated with pedestrian injury severity (*10*, *20*, *23*, *39*, *40*). Considering such a large effect of trucks on fatality rate, control measures such as speed limits for heavy vehicles should be strictly enforced at the mid-block sections.

Hit and run for fatal injury is another variable that has a random parameter. This parameter has a normal distribution, with mean −1.30 and standard deviation 1.50. Therefore, when the hit-and-run situation occurs, nearly 19% of the non-motorists tend to have a higher risk of fatal injury, and others are confronted with the opposite effect. Note that this result should not be perceived as a reflection of how the hit-and-run action affects injury severities, because hit and run is an action taken after the crash has happened. Rather, it reflects how injury severity is distributed for hit-and-run crashes in comparison with other crashes.

Environmental Characteristics

Among all the environmental characteristics of non-motorist injury severity, the light condition is found to be significant. Specifically, the dark and not lighted condition increases the possibility of fatal injury at mid-block by 244.18%. Similar results have also been found by other researchers (*6*, *10*, *14*, *24*, *41*). The dark condition without lighting can considerably compromise the sight distance of drivers, which results in longer reaction time (*13*). According to these results, to improve pedestrian safety during dark condition and night time, it is of paramount importance to install lighting facilities at mid-blocks.

Roadway and Traffic Attributes

Several roadway and traffic attributes were assessed in the mixed logit model. The results in Table 3 show that the three lanes, four lanes, and five or more lanes indicators all tend to increase the possibilities of incapacitating injury and fatal injury, but with different magnitudes. Specifically, the elasticities of three lanes, four lanes, and five or more lanes on incapacitating injury are 18.1%, 10.5%, and 27.7%, and the elasticities of three lanes, four lanes, and five or more lanes on incapacitating injury are 58.7%, 86.3%, and 51.9%. The results show that a larger number of road lanes leads to higher injury severities. A similar conclusion was also found in previous studies associated with pedestrian injury severity (*9*, *23*). This phenomenon might be because more lanes mean higher speed and longer crossing distance, both of which can lead to higher injury severity.

Only the non-incapacitating evident injury category was significantly influenced by curve road alignment. The results in Table 3 show that the likelihood of non-incapacitating evident injury tends to decrease by 50.8% on the curve road way. This is in accordance with the conclusion drawn by Kim et al. (*10*). Slower speed and greater caution shown by non-motorists at curvature might be a possible explanation for this finding.

The speed limit is one of the most important traffic characteristic related to injury severity. In this study, a speed limit between 30 and 55 mph is found to have mixed effects on non-motorist injury severities. According to previous studies on random parameters, there seems to be a problem with the estimated random parameter as the mean of the parameter is not statistically significant at the level of 95% (about 90%). However, a recent study by Behnood and Mannering (*42*) stated that if the standard deviation of a random parameter meets the condition of statistical significance, there is no need for the mean to be significant as well. According to the results, the parameter of this variable is normally distributed with mean 0.37 and standard deviation 1.42, which denotes that a speed limit between 30 and 55 mph tends to increase the likelihood of incapacitating injury of 60% non-motorists and decrease the likelihood of incapacitating injury of other 40% non-motorists. This finding demonstrates the mixed effects of the speed limit on injury severity of non-motorists at mid-block.

Non-Motorist Characteristics

Non-Motorist Attributes. The results from the mixed logit model indicate that the age of non-motorist is a significant contributing variable. Elderly pedestrians/cyclists (aged over 65) have a higher risk (188.8%) to sustain fatal injury. This result may reflect that the elderly are more physically vulnerable and are slower in reaction, as opposed to the young. This is in line with many prior studies (*5*, *12*, *18*, *23*). In contrast, the possibility of young non-motorists (aged below 25) suffering fatal injury tends to decrease by 66.5%, which confirms the finds from previous studies by Sze and Wong (*23*) and Senserrick et al. (*5*).

Pre-Crash Behaviors of Non-Motorists. This study attempts to investigate the injury severity of the vehicle–non-motorist crash at mid-block crossings, and the pre-crash behaviors of non-motorist are the factors of vital interest. According to the results, non-motorist behaviors appear to have significant and different effects on injury severity. The non-motorists have a 66.9% higher likelihood of non-incapacitating evident injury but 46.3% lower likelihood of fatal injury at mid-block when they are darting or running into the road before the crash happened. This result may seem counter-intuitive at first glance, since darting or running into normal road sections would catch the drivers off guard. However, at mid-blocks where a pedestrian crossing facility (markings, lighting, etc.) is installed, drivers become more attentive to roadside pedestrians. Therefore, this result points to the effectiveness of mid-block crossings. Walking, playing, working, sitting, lying, standing in the roadway tend to increase the likelihood of non-incapacitating evident injury by 57.9%, and this variable is not significant for other injury severities. Besides, the possibility of fatal injury decreases by 47.8% when non-motorists use an improper way to cross the road.

The variable of no action indicator associated with non-incapacitating evident injury is found to be normally distributed, with mean 0.69 and standard deviation 1.65. This indicates that before the crash happens, 66% of the non-motorists taking no action suffer from a higher risk of non-incapacitating evident injury, while other non-motorists suffer from a lower risk of non-incapacitating evident injury. Such phenomenon might be attributed to unobserved factors associated with driver characteristics, environmental characteristics, and so on. For an explanation of the internal cause, more accurate data and analysis are needed.

Conclusions and Discussions

In this paper, a mixed logit model was adopted to investigate the injury severity of vehicle–non-motorist crash at mid-block crossings using 8-year crash data from the GES. Five categories of variables were examined, including time characteristics, crash features, environmental conditions, non-motorists' characteristics, and roadway attributes. To uncover the impact of the behaviors of non-motorists, the pre-crash actions of the non-motorists at mid-blocks are incorporated into injury severity analysis for the first time, while controlling for a broad range of factors. The results show that five variables (no action before crash for non-incapacitating evident injury, speed limit between 30 and 55 mph, night time and right-side collision for incapacitating injury, hit and run for fatal injury) are found to have mixed effects across different subjects.

The findings of this study add to the literature on the injury severity outcomes of vehicle–non-motorist crashes at mid-blocks. Some of the results can help guide policy-making and design at mid-block crosswalks.

The major findings of this study are summarized as follows:

(1) Though the results show that some behavior of non-motorists, such darting or running into road and improper crossing, decrease the likelihood of fatal injury, the likelihood of non-incapacitating evident injury increases a lot as a result of these improper behaviors. In addition, no action tends to decrease the likelihood of fatal injury to a great extent. It is therefore important to reduce these improper crossing behaviors. Measures such as pedestrian warning signs could be installed at mid-block crossings to alert drivers. Law enforcement forces should target these improper behaviors of non-motorists at mid-blocks.

(2) Heavy trucks and light trucks tend to increase the likelihood fatal injury to a great extent. Three or more lanes, dark not lighted and age over 65 indicator also increase the likelihood of fatal injury. Left-side collision and age below 25 tend to decrease the likelihood of fatal injury. According to the results, heavy vehicle and large number of lanes are among the main causes of fatal crashes. Speed regulation of trucks around mid-blocks needs to be strictly enforced, and lighting installed for those without proper lighting. For mid-blocks installed on streets with three or more lanes, a pedestrian overpass or tunnel may be considered.

(3) Speed limit between 30 and 55 mph, night time and the first impact point of the right side are found to have random effects on incapacitating injury. Hit and run is found to have mixed effects on the fatal injury. Their effects on the injury severity vary across different accidents.

It should be noted that there are also some limitations to the present study. To begin with, GES data are based on police-recorded data from different states, where the coding criteria may be different. Such heterogeneity is not considered in this study because of lack of information in the GES database. In addition, though the behavior of the non-motorist is considered, the interactions between these actions and other unobserved factors are still unknown. Further study is needed to investigate the combined effects of non-motorist behaviors and other unobserved factors on injury severity.

Acknowledgments

This research was jointly sponsored by the Natural Science Foundation of China (51508409 and 71531011), the Fundamental Research Funds for the Central Universities, and the project sponsored by SRF for ROCS, SEM for supporting this research.

Author Contributions

The authors confirm contribution to the paper as follows: study conception and design: Xiaoxiang Ma; data collection: Bowen Dong, Xiaoxiang Ma; analysis and interpretation of results: Bowen Dong, Feng Chen, Xiaoxiang Ma; draft manuscript preparation: Bowen Dong, Feng Chen, and Xiaoxiang Ma. All authors reviewed the results and approved the final version of the manuscript.

References

1. Amoh-Gyimah, R., M. Saberi, and M. Sarvi. Macroscopic Modeling of Pedestrian and Bicycle Crashes: A Cross-Comparison of Estimation Methods. *Accident Analysis and Prevention*, Vol. 93, 2016, pp. 147–159. https://doi.org/10.1016/j.aap.2016.05.001.
2. Pour-Rouholamin, M., and H. Zhou. Investigating the Risk Factors Associated with Pedestrian Injury Severity in Illinois. *Journal of Safety Research*, Vol. 57, 2016, pp. 9–17. https://doi.org/10.1016/j.jsr.2016.03.004.
3. World Health Organization. *Pedestrian Safety: A Road Safety Manual for Decision-Makers and Practitioners*. World Health Organization, Geneva, 2013.
4. National Highway Traffic Safety Administration. *Traffic Safety Facts 2015*. National Highway Traffic Safety Administration, 2015.
5. Senserrick, T., S. Boufous, L. De Rome, and R. Ivers. Detailed Analysis of Pedestrian Casualty Collisions Detailed Analysis of Pedestrian Casualty Collisions in Victoria, Australia. *Traffic Injury Prevention*, Vol. 9588 (Suppl. 1), 2014, pp. S197–S205. https://doi.org/10.1080/15389588.2014.935356.
6. Eluru, N., C. R. Bhat, and D. A. Hensher. A Mixed Generalized Ordered Response Model for Examining Pedestrian and Bicyclist Injury Severity Level in Traffic Crashes. *Accident Analysis and Prevention*, Vol. 40, No. 3, 2008, pp. 1033–1054. https://doi.org/10.1016/j.aap.2007.11.010.
7. Al-Shammari, N., S. Bendak, and S. Al-Gadhi. In-Depth Analysis of Pedestrian Crashes in Riyadh In-Depth Analysis of Pedestrian Crashes in Riyadh. *Traffic Injury Prevention*, Vol. 9588, 2017, pp. 552–559. https://doi.org/10.1080/15389580903175313.
8. Sarkar, S., R. Tay, and J. D. Hunt. Logistic Regression Model of Risk of Fatality in Vehicle–Pedestrian Crashes on National Highways in Bangladesh. *Transportation Research Record: Journal of the Transportation Research Board*, 2006. 2264: 128–137.
9. Gomaa, M., N. Saunier, L. F. Miranda-Moreno, and S. V. Ukkusuri. A Clustering Regression Approach: A Comprehensive Injury Severity Analysis of Pedestrian–Vehicle Crashes in New York, US and Montreal, Canada. *Safety Science*, Vol. 54, 2013, pp. 27–37. https://doi.org/10.1016/j.ssci.2012.11.001.
10. Kim, J., G. F. Ulfarsson, V. N. Shankar, and F. L. Mannering. A Note on Modeling Pedestrian-Injury Severity in Motor-Vehicle Crashes with the Mixed Logit Model. *Accident Analysis and Prevention*, Vol. 42, No. 6, 2010, pp. 1751–1758. https://doi.org/10.1016/j.aap.2010.04.016.
11. Segni, G., S. Washington, M. Haque, and M. J. King. Investigation of Pedestrian Crashes on Two-Way Two-Lane Rural Roads in Ethiopia. *Accident Analysis and Prevention*, Vol. 78, 2015, pp. 118–126. https://doi.org/10.1016/j.aap.2015.02.011.
12. Vernez, A., L. Lin, J. Jiao, P. Hurvitz, and P. Reeves. The Risk of Pedestrian Injury and Fatality in Collisions with Motor Vehicles, a Social Ecological Study of State Routes and City Streets in King County, Washington. *Accident Analysis and Prevention*, Vol. 43, No. 1, 2011, pp. 11–24. https://doi.org/10.1016/j.aap.2009.12.008.
13. Lee, C., and M. Abdel-Aty. Comprehensive Analysis of Vehicle–Pedestrian Crashes at Intersections in Florida. *Accident Analysis and Prevention*, Vol. 37, No. 4, 2005, pp. 775–786. https://doi.org/10.1016/j.aap.2005.03.019.
14. Haleem, K., P. Alluri, and A. Gan. Analyzing Pedestrian Crash Injury Severity at Signalized and Non-Signalized Locations. *Accident Analysis and Prevention*, Vol. 81, 2015, pp. 14–23. https://doi.org/10.1016/j.aap.2015.04.025.
15. Silvano, A. P., H. N. Koutsopoulos, and X. Ma. Analysis of Vehicle-Bicycle Interactions at Unsignalized Crossings: A Probabilistic Approach and Application. *Accident Analysis and Prevention*, Vol. 97, 2016, pp. 38–48. https://doi.org/10.1016/j.aap.2016.08.016.
16. Diogenes, M. C., and L. A. Lindau. Evaluation of Pedestrian Safety at Midblock Crossings, Porto Alegre, Brazil. *Transportation Research Record: Journal of the Transportation Research Board*, 2010. 2193: 37–43.
17. Pour, A. T., S. Moridpour, R. Tay, and A. Rajabifard. Modelling Pedestrian Crash Severity at Mid-Blocks. *Transportmetrica A: Transport Science*, Vol. 13, No. 3, 2017, pp. 273–297. https://doi.org/10.1080/23249935.2016.1256355.
18. Zajac, S. S., and J. N. Ivan. Factors Influencing Injury Severity of Motor Vehicle–Crossing Pedestrian Crashes in Rural Connecticut. *Accident Analysis and Prevention*, Vol. 35, No. 3, 2003, pp. 369–379.
19. Forbes, J. J., and M. A. Habib. Pedestrian Injury Severity Levels in the Halifax Regional Municipality, Nova Scotia, Canada: Hierarchical Ordered Probit Modeling Approach. *Transportation Research Record: Journal of the Transportation Research Board*, 2015. 2519: 172–178.

20. Al-Ghamdi, A. S. Pedestrian–Vehicle Crashes and Analytical Techniques for Stratified Contingency Tables. *Accident Analysis and Prevention*, Vol. 34, No. 2, 2002, pp. 205–214.
21. Ballesteros, M. F., P. C. Dischinger, and P. Langenberg. Pedestrian Injuries and Vehicle Type in Maryland, 1995–1999. *Accident Analysis and Prevention*, Vol. 36, No. 1, 2004, pp. 73–81.
22. Roudsari, B. S., C. N. Mock, R. Kaufman, D. Grossman, B. Y. Henary, and J. Crandall. Pedestrian Crashes: Higher Injury Severity and Mortality Rate for Light Truck Vehicles Compared with Passenger Vehicles. *Injury Prevention*, Vol. 10, No. 3, 2004, pp. 154–159. https://doi.org/10.1136/ip.2003.003814.
23. Sze, N. N., and S. C. Wong. Diagnostic Analysis of the Logistic Model for Pedestrian Injury Severity in Traffic Crashes. *Accident Analysis and Prevention*, Vol. 39, No. 6, 2007, pp. 1267–1278. https://doi.org/10.1016/j.aap.2007.03.017.
24. Kim, J., S. Kim, G. F. Ulfarsson, and L. A. Porrello. Bicyclist Injury Severities in Bicycle–Motor Vehicle Accidents. *Accident Analysis and Prevention*, Vol. 39, No. 2, 2007, pp. 238–251. https://doi.org/10.1016/j.aap.2006.07.002.
25. Wedagama, D. M. P., R. N. Bird, and A. V. Metcalfe. The Influence of Urban Land-Use on Non-Motorised Transport Casualties. *Accident Analysis and Prevention*, Vol. 38, No. 6, 2006, pp. 1049–1057. https://doi.org/10.1016/j.aap.2006.01.006.
26. Vanlaar, W., M. Hing, S. Brown, H. Mcateer, and J. Crain. Fatal and Serious Injuries Related to Vulnerable Road Users in Canada. *Journal of Safety Research*, Vol. 58, 2016, pp. 67–77. https://doi.org/10.1016/j.jsr.2016.07.001.
27. Decker, S., D. Otte, D. Leslie, C. Walter, M. Omar, C. Krettek, and S. Brand. Injury Severity of Pedestrians, Bicyclists and Motorcyclists Resulting from Crashes with Reversing Cars. *Accident Analysis and Prevention*, Vol. 94, 2016, pp. 46–51. https://doi.org/10.1016/j.aap.2016.05.010.
28. Eluru, N., and C. R. Bhat. A Joint Econometric Analysis of Seat Belt Use and Crash-Related Injury Severity. *Accident Analysis and Prevention*, Vol. 39, No. 5, 2007, pp. 1037–1049. https://doi.org/10.1016/j.aap.2007.02.001.
29. National Highway Traffic Safety Administration. *National Automotive Sampling System (NASS) General Estimates System (GES) Analytical User's Manual 1988–2008*. National Highway Traffic Safety Administration, 2008.
30. McFadden, D. L. Econometric Models of Probabilistic Choice. In *A Structural Analysis of Discrete Data with Econometric Applications*, The MIT Press, Cambridge, pp. 198–272.
31. Washington, S. P., M. G. Karlaftis, and F. L. Mannering. *Statistical and Econometric Methods for Transportation Data Analysis*. Chapman & Hall/CRC, Washington, D.C., 2003.
32. Milton, J. C., V. N. Shankar, and F. L. Mannering. Highway Accident Severities and the Mixed Logit Model: An Exploratory Empirical Analysis. *Accident Analysis and Prevention*, Vol. 40, No. 1, 2008, pp. 260–266. https://doi.org/10.1016/j.aap.2007.06.006.
33. Train, K. *Discrete Choice Methods with Simulation*. Cambridge University Press, Cambridge, 2003.
34. Chen, F., and S. Chen. Injury Severities of Truck Drivers in Single- and Multi-Vehicle Accidents on Rural Highways. *Accident Analysis and Prevention*, Vol. 43, No. 5, 2011, pp. 1677–1688. https://doi.org/10.1016/j.aap.2011.03.026.
35. Bhat, C. R. Simulation Estimation of Mixed Discrete Choice Models Using Randomized and Scrambled Halton Sequences. *Transportation Research Part B*, Vol. 37, No. 9, 2003, pp. 837–855. https://doi.org/10.1016/S0191-2615(02)00090-5.
36. Shankar, V., and F. Mannering. An Exploratory Multinomial Logit Analysis of Single-Vehicle Motorcycle Accident Severity. *Journal of Safety Research*, Vol. 27, No. 3, 1996, pp. 183–194. https://doi.org/10.1016/0022-4375(96)00010-2.
37. Domencich, T. A., and D. McFadden. Urban Travel Demand: A Behavioral Analysis. *Canadian Journal of Economics*, Vol. 10, No. 4, 1975.
38. Washington, S. P., M. G. Karlaftis, and F. L. Mannering. *Statistical and Econometric Methods for Transportation Data Analysis*. CRC Press, Boca Raton, 2011.
39. Lascala, E. A., D. Gerber, and P. J. Gruenewald. Demographic and Environmental Correlates of Pedestrian Injury Collisions: A Spatial Analysis. *Accident Analysis and Prevention*, Vol. 32, No. 5, 2000, pp. 651–658.
40. Lefler, D. E., and H. C. Gabler. The Fatality and Injury Risk of Light Truck Impacts with Pedestrians in the United States. *Accident Analysis and Prevention*, Vol. 36, No. 2, 2004, pp. 295–304. https://doi.org/10.1016/S0001-4575(03)00007-1.
41. Verzosa, N., and R. Miles. Severity of Road Crashes Involving Pedestrians in Metro Manila, Philippines. *Accident Analysis and Prevention*, Vol. 94, 2016, pp. 216–226. https://doi.org/10.1016/j.aap.2016.06.006.
42. Behnood, A., and F. L. Mannering. The Effect of Passengers on Driver-Injury Severities in Single-Vehicle Crashes: A Random Parameters Heterogeneity-in-Means Approach. *Analytic Methods in Accident Research*, Vol. 14, 2017, pp. 41–54. https://doi.org/10.1016/j.amar.2017.04.001.

The Standing Committee on Safety Data, Analysis and Evaluation (ANB20) peer-reviewed this paper (18-03480).

TRR
JOURNAL OF THE TRANSPORTATION RESEARCH BOARD

Article

Transportation Research Record
2018, Vol. 2672(38) 149–159

DOI: 10.1177/0361198118798485
journals.sagepub.com/home/trr
SAGE

Performance Evaluation of Various Missing Data Treatments in Crash Severity Modeling

Fan Ye[1] and Yong Wang[2]

Abstract

Data quality, including record inaccuracy and missingness (incompletely recorded crashes and crash underreporting), has always been of concern in crash data analysis. Limited efforts have been made to handle some specific aspects of crash data quality problems, such as using weights in estimation to take care of unreported crash data and applying multiple imputation (MI) to fill in missing information of drivers' status of attention before crashes. Yet, there lacks a general investigation of the performance of different statistical methods to handle missing crash data. This paper is intended to explore and evaluate the performance of three missing data treatments, which are complete-case analysis (CC), inverse probability weighting (IPW) and MI, in crash severity modeling using the ordered probit model. CC discards those crash records with missing information on any of the variables; IPW includes weights in estimation to adjust for bias, using complete records' probability of being a complete case; and MI imputes the missing values based on the conditional distribution of the variable with missing information on the observed data. Those missing data treatments provide varying performance in model estimations. Based on analysis of both simulated and real crash data, this paper suggests that the choice of an appropriate missing data treatment should be based on sample size and data missing rate. Meanwhile, it is recommended that MI is used for incompletely recorded crash data and IPW for unreported crashes, before applying crash severity models on crash data.

With millions of drivers behind the wheel, tens of thousands of people are involved in roadway crashes each year. By retrieving causes of crashes, post-crash evaluation and crash data modeling provide traffic engineers and policymakers insights and guidance to improve roadway design and operations. In recent years, practitioners in traffic data analysis have witnessed emerging sources of traffic related data and increasing adoption of advanced statistical methods to enhance safety modeling. Such change has also brought challenges. It is widely acknowledged that applying appropriate statistical methods to good quality data is essential to attain solid statistical inferences and yield valuable information, which is supported by the "garbage in garbage out" principle (*1*).

However, data quality and selection of statistical methods have always been concerns in crash data analysis. The potential problems with data quality include record inaccuracy, incompletely recorded crashes, and crash underreporting. Inaccurate crash records include incorrectly reported crash location and time, imprecise crash severity, unreliable roadway user demographic information, incorrectly reported crash contributing factors, and so forth. (*2*). Incompletely recorded crashes refer to those partially completed records with some information missing; for instance, when a driver was killed in a crash, some of the driver's personal information or status before the crash might be unknown. Underreporting occurs mainly with crashes at lower crash severity level, which are neither officially recorded in the crash database by police nor collected in the hospital system.

The negative impact of poor quality data on crash analysis has been recognized in the literature. Benavente et al. found it challenging to identify and verify inaccurately recorded crash data for in-depth crash investigation and linkage of multiple data sources (*3*). Tefft indicated that missing data caused underestimated prevalence of crashes involving drowsy drivers (*4, 5*). In terms of data underreporting, it is stated in the Highway

[1]Civil and Environmental Engineering Department, Ohio Northern University, Ada, OH
[2]Mathematics and Statistics Department, Ohio Northern University, Ada, OH

Corresponding Author:
Address correspondence to Fan Ye: f-ye@onu.edu

Safety Manual that the situation, in which crashes with greater severity are reported more reliably, creates an issue called frequency-severity indeterminacy (*6*). This issue results in difficulty in determining whether a change in the number of reported crashes is caused by an actual change in crashes or a shift in severity proportions. Meanwhile, Imprialou and Quddus found that data underreporting caused a loss in the precision of model estimation as a result of reduced sample size and uneven weights in different crash severity levels or crash types (*2*).

Limited efforts have been made to address some specific aspects of data quality issues in crash analysis. Based on the available knowledge of underreported rates, two studies attempted to address the underreporting issue in crash data by weighting adjustments in model estimation which significantly improved model estimation performances (*7, 8*). Tefft used multiple imputation (MI) to fill in missing information on drivers' state of attention before crash (*4, 5*). MI has also been implemented in Crash Outcome Data Evaluation (CODES) and Fatality Analysis Reporting System to handle missing information on pre-crash blood alcohol content for drivers killed in crashes. However, there was no performance evaluation on the MI method in either of the studies. In addition, there is a lack of general investigation of the performance of various statistical methods, including but not limited to MI, to handle missing crash data.

As an extension of the authors' previous study on crash data underreporting (*8*), this work is intended to explore and evaluate the performance of various missing data treatments in crash severity modeling using an ordered probit (OP) model. The study focus is on some popular statistical methods that have potential to handle incompletely reported or underreported crash data. By using both simulated and real crash data, some insights are provided into the performance of these statistical methods for missing crash data at different missing levels and across varying sample sizes.

Methodology

The choice of statistical methods to handle missing data depends on the missing data mechanism. Statisticians have established the following three missing data mechanisms: missing completely at random (MCAR), missing at random (MAR), and missing not at random (MNAR) (see Ye and Lord [*8*] for more details). MAR is the most frequently used assumption for missing data. One typical example of MAR in crash reports is the status of a fatal driver right before the crash, including whether the driver was driving under the influence of alcohol, drugs, or fatigue.

Methods to Handle Missing Data

Three methods are used in this paper to handle incomplete and underreported crash data in crash severity modeling: Complete-case analysis (CC), inverse probability weighting (IPW) and MI. The performance of statistical inference based on these three methods is compared to attain the potentially best treatment of missing data in crash severity modeling. The study starts with incompletely reported data only, and then deals with a more general situation which includes both incomplete records and underreporting.

The procedures of the three missing data treatments are summarized as follows.

1. CC or list-wise deletion: Only the completely observed data are included in the analysis with the exclusion of all observations having missing values in any variable. The CC analysis is the default choice of many procedures in statistical software, such as SAS and SPSS. As ignoring incomplete crashes results in data loss, the statistical power of model estimates decreases (*9*). Meanwhile, the assumption behind CC is MCAR, which is strong but rarely stands. If data are MAR or MNAR, using CC could lead to biased estimation. For crash data, MCAR is most likely inappropriate except if there were data entry errors. For instance, Imprialou and Quddus (*2*) proved that drivers' age was not MCAR using evidence from various crash datasets in different countries.
2. IPW: Complete observations are weighted by the inverse of their probability of being a complete case in model estimation. Below are three main steps to implement IPW for missing data. See more details about IPW elsewhere (*10–14*).
   Step 1. Specify a weighting model to estimate the conditional probability (i.e., propensity score) of an individual being a complete case given a set of predictor variables;
   Step 2. Each individual is assigned a weight which is the inverse of the propensity score;
   Step 3. Run the analysis model and calculate the parameter estimates using weighted maximum likelihood.
3. MI: MI is a method for inference by imputing "complete" datasets from partially observed data, based on the assumption of MAR. MI can be carried out in three steps.
   Step 1. Impute plausible values for missing observations through proper imputation. The imputation model estimates the conditional distribution of variables with missing values and then uses random draws from the conditional distribution

to create imputed "complete" datasets. There are various imputing methods available for different data types. In this paper, MI by chained equations (MICE) (*15*) is used for all the missing data imputations.
Step 2. Each dataset created in Step 1 is analyzed separately through a substantive model. Parameter estimates are calculated based on each dataset separately.
Step 3. The model estimation results from Step 2 are combined across imputed "complete" datasets to take into account the uncertainty regarding the imputation.

Ordered Probit Model

This paper uses the OP model, which is commonly used and straightforward to implement, as an example to demonstrate the impact of various missing data treatments. The OP model uses a latent variable as shown in Equation 1 to disaggregate crash severity outcomes:

$$\text{A latent variable is}: z_i = X_i^T\beta + \varepsilon_i \qquad (1)$$

where,

$X_i = \{1, x_{i1}, \cdots, x_{ij}, \cdots, x_{im}\}$, representing each of the m explanatory variables for the ith individual crash;

$\beta = \{\beta_0, \beta_1, \cdots, \beta_j, \cdots, \beta_m\}^T$, the column vector of the coefficients for the explanatory variables;

ε_i is a random error term following standard normal distribution;

$i = 1, \cdots, n$ where n is the total number of crash events including in the model;

$j = 1, \cdots, m$ where m is the total number of explanatory variables.

The dependent variable y_i is categorical, representing crash severity that has C categories (in the paper, $C = 5$). Usually, y_i is set as 1 to C. The value of y_i is determined by Equation 2:

$$y_i = \begin{cases} 1, & if \quad -\infty < z_i \leq \gamma_1 \\ k, & if \quad \gamma_{k-1} < z_i \leq \gamma_k \\ C, & if \quad \gamma_{C-1} < z_i < \infty \end{cases} \qquad (2)$$

where,

$\gamma = \{\gamma_1, \cdots \gamma_k, \cdots, \gamma_{C-1}\}$ are the threshold values for the C crash severity categories.

The relationship between these threshold values are subject to the constraint: $-\infty < \gamma_1 \leq \cdots \leq \gamma_k \leq \cdots \leq \gamma_{C-1} < +\infty$.

Given the observed value of X_i, the probability that the crash severity of the ith individual crash belongs to each category is shown in Equation 3.

$$\begin{cases} P(y_i = 1) = \Phi(\gamma_1 - X_i^T\beta) \\ P(y_i = k) = \Phi(\gamma_k - X_i^T\beta) - \Phi(\gamma_{k-1} - X_i^T\beta) \\ P(y_i = C) = 1 - \Phi(\gamma_{C-1} - X_i^T\beta) \end{cases} \qquad (3)$$

where $\Phi(\cdot)$ stands for the cumulative distribution function of the standard normal distribution.

Simulation Design

Each individual crash record was simulated with the knowledge of true values of parameters based on the OP model structure. The bias and variability can be estimated by comparing the estimated parameters from the OP model with their true values. A similar simulation procedure may be found in other research (*8, 16*).

First of all, a categorical variable with five outcomes (i.e., y_i as shown in Equation 2) was used to replicate the five crash severity categories, that is, KABCO (K as fatal, A as incapacitating injury, B as non-incapacitating injury, C as possible injury, and O as property-damage only). In an effort to balance simulation efficiency and real-life condition representativeness, three covariates, including two dummy variables (X_1, X_2) and one continuous variable (X_3), were simulated to create the latent variable, as shown in Equation 1. As the effect of each covariate on all severity levels is restricted to be the same in OP, the coefficients of the three covariates were set to be the same for each severity level. In the paper, $\beta_1 = \beta_2 = \beta_3 = 1$ was used. Meanwhile, γ_k (threshold value, as shown in Equation 2) was set as 1, 2, 3, and 4 for the five outcomes of y_i. The probability distribution of y_i can be obtained using Equation 3. With the predetermined parameter values, the population ratios are about 2%, 5%, 12%, 18%, and 63% from Level 1 to Level 5, which are representative of the distribution of real crash severities.

For each observation i, the dummy variables X_1 and X_2 are drawn from a Bernoulli distribution with probability 0.5 and 0.6, respectively. The continuous variable X_3 is moderately correlated to X_1, generated by the linear equation, as shown in Equation 4:

$$X_{3i} = 3 + X_{1i} + \varepsilon_i \qquad (4)$$

where, ε_i is a random draw from a standard normal distribution.

Dummy variables X_1 and X_2 are designed to be MAR. Missing rates for X_1 and X_2 are set at three levels (i.e., low, medium, and high). The probabilities of missing X_1 and X_2 rely on X_3, governed by Equations 5 and 6, respectively:

$$P_i = 1/[1 + \exp(\alpha_0 + \alpha_1 X_{3i})] \qquad (5)$$

Table 1. Missing Data Design in Simulated Data

Model parameter		Designed missing level		
		Low	Medium	High
X_1	α_0	0.5	0	0
	α_1	0.5	0.3	0.1
	approximate missing rate	11%	26%	41%
X_2	α_2	1	0.5	0.2
	α_3	0.5	0.3	0.2
	approximate missing rate	7%	18%	29%
Sample size		100, 500, 1,000, 1,500, 2,000, 2,500, 5,000, 10,000		

$$P_i = 1/[1 + \exp(\alpha_2 + \alpha_3 X_{3i})] \quad (6)$$

For each observation i, a random Bernoulli draw (i.e., missing indicator) is made with probability P_i from Equation 5 for X_1 and Equation 6 for X_2. If the drawn value is 1, the underlying variable is missing, otherwise not. The parameters $(\alpha_0, \alpha_1, \alpha_2, \alpha_3)$ are set at different values to create three missing levels (low, medium and high). Continuous variable X_3 is set to be completely observed. In order to check the sample size effects on the performance of missing data techniques, eight samples sizes are used in the data generation: 100, 500, 1,000, 1,500, 2,000, 2,500, 5,000 and 10,000. The design scenarios for missing data are summarized in Table 1.

For each combination of missing data level and sample size, 100 datasets are generated based on the OP model. The three missing data treatments (i.e., CC, IPW, and MI) are applied to each dataset to produce estimates of β_1, β_2, β_3, γ_1, γ_2, γ_3, and γ_4. For each MI procedure involved in the analysis, 30 imputations are conducted. The simulation was conducted in SAS 9.4 (*17*), as well as all the other analyses in the paper.

Simulation Results

For each missing scenario and missing data treatment, the mean of the 100 parameter estimates and of the 100 estimated standard errors are calculated for each parameter, as well as the standard deviation of the parameter estimates. Meanwhile, the proportion of 95% confidence intervals containing the true value of a parameter is also provided. For illustration purpose, Table 2 summarizes the results of 100 simulated datasets with sample size of 1,000 at medium missing level.

As in Table 2, all three treatments yield similar estimates, with MI producing slightly higher bias but lower standard deviations for some parameters. As there are seven estimated parameters, it is impractical to compare the bias and variability of each parameter. The total bias square, total variance and total root mean square error (RMSE) from all seven parameters are calculated as indications of model estimation performance for each missing data treatment. Total RMSE is the sum of the RMSE of the seven parameter estimates, which includes both bias and variability of the estimates. RMSE for a parameter can be calculated by $RMSE = \sqrt{Bias^2 + Variance}$.

Effects of Imputation/Weighting Model Specification

The performance of missing data treatment methods depends on their execution, and they either start with finding the proper imputation model (for MI) (*9*) or calculating the proper weights for missing data (for IPW). A decision has to be made about variable selection in both cases to ensure reliable comparison among missing data treatments.

Figure 1 summarizes the total RMSE of four combinations (other combinations are either statistically inappropriate or infeasible for imputing the missing values or weighting the observed variables) of variables in the imputation/weighting model based on 100 simulated datasets with sample size of 10,000 at low missing level. As shown in Figure 1, the performance of MI is more sensitive to variable selection in the imputation model than IPW is to the weighting model. The smallest total RMSE comes from including all covariates (X_1, X_2, X_3) and dependent variable (y_i) in the imputation/weighting model, and largest total RMSE is caused by omitting variables related to the missingness. The results based on other sample sizes and missing levels are found to be consistent with this scenario. Based on the finding, it is suggested to include more rather than fewer variables in the imputation/weighting procedure for the OP model, even if some of the variables are not directly associated with missing variables or used in model estimation. For all the following analyses of simulated scenarios, (X_1, X_2, X_3, y_i) are included in the imputation/weighting model for MI/IPW to attain the best estimation.

Table 2. Parameter Estimates of Covariates (Medium Missing Level, Sample Size = 1,000)

Parameter	Method	True[a]	Estimate[b]	Avg. SE[c]	Std[d]	Coverage[e]	Bias Square[f]	Variance[g]	RMSE
β_1	CC	1	1.005	0.134	0.134	0.940	0.000	0.018	0.134
	IPW	1	1.006	0.134	0.134	0.960	0.000	0.018	0.134
	MI	1	0.948	0.124	0.118	0.920	0.003	0.014	0.129
β_2	CC	1	1.002	0.124	0.118	0.970	0.000	0.014	0.118
	IPW	1	1.000	0.124	0.121	0.960	0.000	0.015	0.121
	MI	1	0.947	0.105	0.101	0.930	0.003	0.010	0.114
β_3	CC	1	1.017	0.075	0.079	0.950	0.000	0.006	0.081
	IPW	1	1.017	0.076	0.082	0.940	0.000	0.007	0.083
	MI	1	0.994	0.057	0.055	0.960	0.000	0.003	0.055
γ_1	CC	1	1.013	0.266	0.258	0.970	0.000	0.067	0.258
	IPW	1	1.009	0.269	0.266	0.970	0.000	0.071	0.266
	MI	1	0.948	0.191	0.162	0.990	0.003	0.026	0.171
γ_2	CC	2	2.034	0.243	0.240	0.970	0.001	0.057	0.242
	IPW	2	2.033	0.247	0.249	0.970	0.001	0.062	0.251
	MI	2	1.942	0.182	0.156	0.970	0.003	0.024	0.167
γ_3	CC	3	3.057	0.257	0.260	0.940	0.003	0.068	0.266
	IPW	3	3.056	0.261	0.267	0.950	0.003	0.071	0.273
	MI	3	2.936	0.196	0.173	0.950	0.004	0.030	0.184
γ_4	CC	4	4.050	0.279	0.289	0.930	0.003	0.083	0.293
	IPW	4	4.047	0.284	0.295	0.940	0.002	0.087	0.299
	MI	4	3.926	0.215	0.198	0.950	0.005	0.039	0.211
Sum	CC						0.01	0.31	1.39
	IPW						0.01	0.33	1.43
	MI						0.02	0.15	1.03

[a]True value of the parameter.
[b]Mean of 100 parameter estimates.
[c]Mean of 100 standard error estimates.
[d]Standard deviation of 100 parameter estimates.
[e]Proportion of 95% confidence intervals containing the true value of the parameter.
[f]The square of the difference between a parameter estimate and its true value.
[g]The square of the standard deviation of 100 parameter estimates.

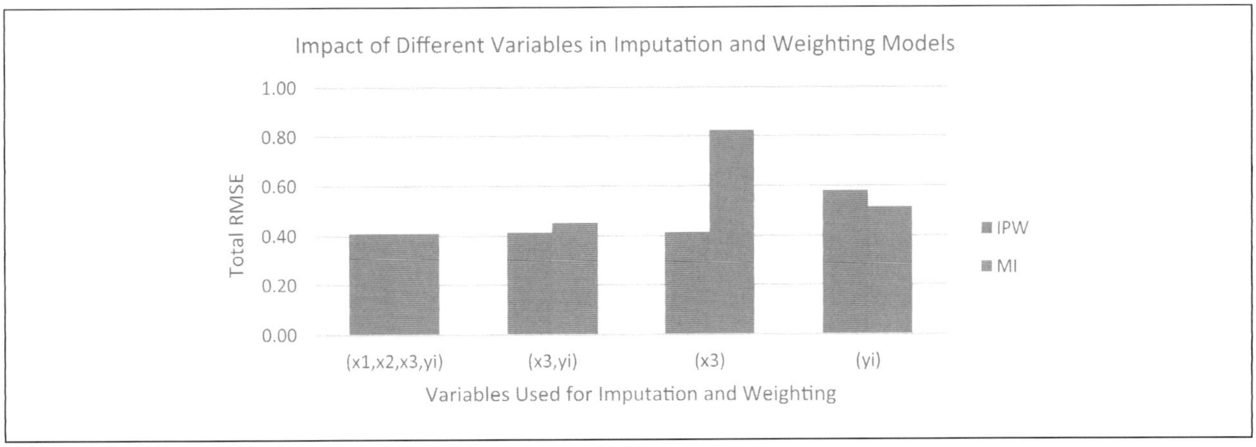

Figure 1. Model performance for imputation and weighting model specifications.
Note: Sample sizes of 100 and 500 are not shown in the above graphs because of large (out-of-scope) total RMSEs, bias and variance caused by small sample size.

Effects of Sample Size and Missing Level

By running other scenarios for simulated data with various sample sizes and missing levels for X_1 and X_2 (described in Table 1), the total RMSEs, total bias square, and total variance from each scenario can be calculated and summarized for each missing data treatment. The results are summarized in Figure 2, with a few patterns observed.

First, CC and IPW provide almost identical performance under all sample sizes and missing levels in terms of total RMSE, bias, and variance of estimation (the occasional deviations could be considered as some artifact of the simulation process). In contrast, the performance by MI seems to show a different pattern.

Second, as sample size increases, overall model performance under all three missing data treatments improves with some varying pattern. MI seems to perform better than CC and IPW when sample sizes are small, and the margin decreases as the missing rate decreases. When sample sizes are large, CC and IPW tend to perform better and the gap tends to increase when the missing level is medium to high. It is also found that a higher missing rate would worsen the overall estimation performance of all three missing data treatments, as shown in Figure 3. Figure 3 only shows the comparison under the same sample size ($n = 5000$) with varying missing levels, but the same pattern was observed under other sample sizes.

Finally, MI estimators are found to provide smaller variance than CC and IPW, which demonstrates the advantage of MI. Compared with CC and IPW, which only use complete data (either unweighted or weighted), MI estimates are calculated based on imputed values generated from a conditional distribution specified by the imputation model. Such a mechanism helps to reduce the variation of the estimates by providing a better coverage of possible values under the distribution of missing variable. It is also understandable that this advantage diminishes as the sample size increases and missing rate decreases, as can be observed in Figure 2.

The above findings indicate that MI is not always preferred to IPW or CC when it comes to dealing with incompletely reported data. A decision of choosing the appropriate missing data treatment should be based on sample size and data missing rate.

Combine Incomplete Data and Unreported Data Treatments

The above analysis only includes incompletely reported data. However, missing crash data also involves underreporting. Not addressing both issues may lead to biased model estimation.

Because of the nature of these two issues, a two-stage process is used in analysis. Stage 1 is to deal with the missingness in incompletely reported data; stage 2 is to handle the unreported data with predetermined reporting rates for each crash severity level (it is assumed that the reporting rates are 100%, 90%, 80%, 70%, and 50% for $k = 1$ to 5, which represents KABCO for crash severity levels). As concluded above, the performance of three missing treatments varies with sample size, so two sample sizes are used in the simulation: 2,000 and 10,000.

At stage 1, three treatments are applied to incompletely reported data: CC, IPW, and MI. At stage 2, two methods are used to address data underreporting: IPW and MI (imputing all covariates based on reported data). Therefore, there will be $3 \times 2 = 6$ possible combinations in total: CC+IPW, CCCC+MI, IPWCC+IPW, IPWCC +MI, MICC+IPW, and MICC+MI. For instance, CCCC+IPW uses completely reported crash observations and weights them by the inverse of their probability of being reported, based on the given reporting rates; in MICC+MI, all incomplete and unreported values are imputed.

The performances of the six combinations of missing treatments are summarized in Figure 4. As simulation results display a consistent pattern at different levels of missingness, only the result for medium missing level is shown as an illustration. As shown in Figure 4, the model estimation performance among the six treatment combinations varies with the sample size. A clear pattern is that better model performance (smaller total RMSE) is observed when using IPW instead of MI in stage 2 for unreported data. It can also be noticed that CC + MI yields the largest total RMSE regardless of sample size (and missing level, not shown in the paper). In general, MI + IPW performs the best.

This finding is consistent with the suggestion provided by other research, which imputed only isolate missing values and used weights to handle the remaining larger blocks of missing data (*13*), as unreported crash data are the extreme case of large block of missing data (i.e., missing all covariates). The MI + IPW combination takes full advantage of its individual treatment for different missing conditions and results in the best performance of handling both incomplete and underreporting issues in a dataset.

Application to Crash Data

The real data utilized in the paper include 1 year (2014) of traffic crash records from Ohio Department of Public Safety and general road inventory from Ohio Department of Transportation. In order to reduce the variability of data for analysis, data for a specific type of crash on a specific road classification are extracted for study: traffic crashes involving only single vehicle that occurred on Interstate Highways (excluding those occurring at intersections) in Ohio. Crash severity was classified into five categories: KABCO.

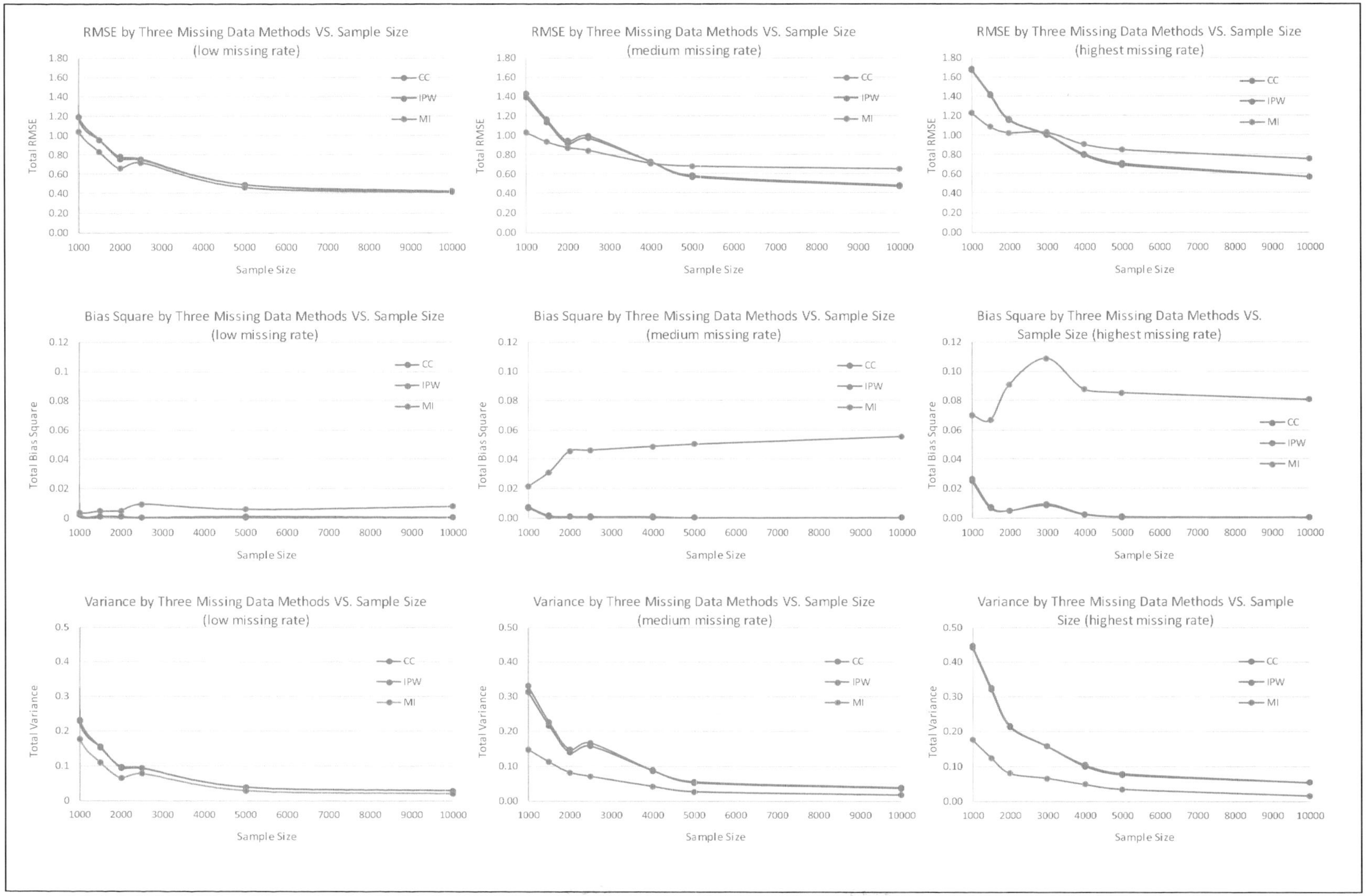

Figure 2. Model performances for three missing data treatments by various sample sizes and missing levels (top three graphs for RMSE, middle for bias square, and bottom for variance).

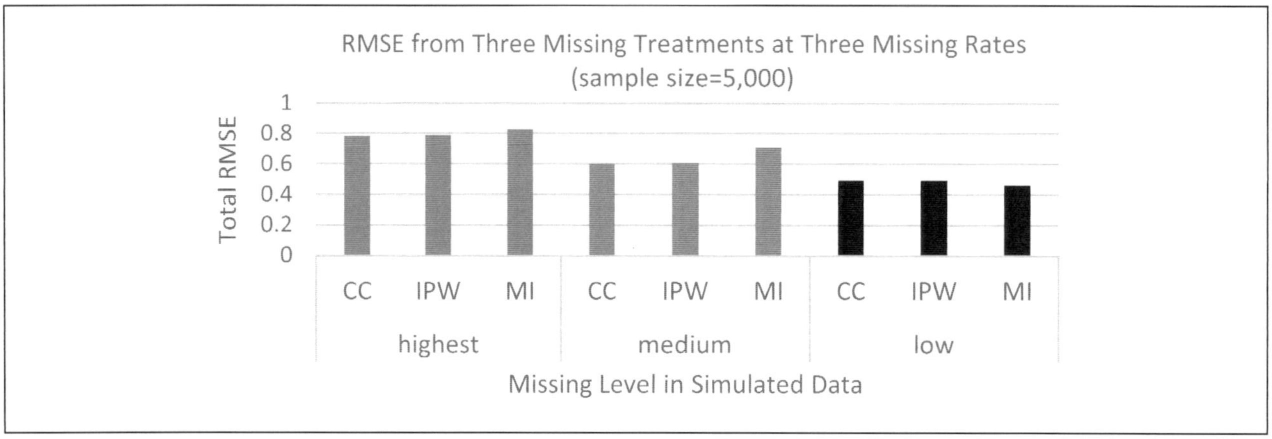

Figure 3. Total RMSE from three missing treatments at three missing levels (sample size = 5,000).

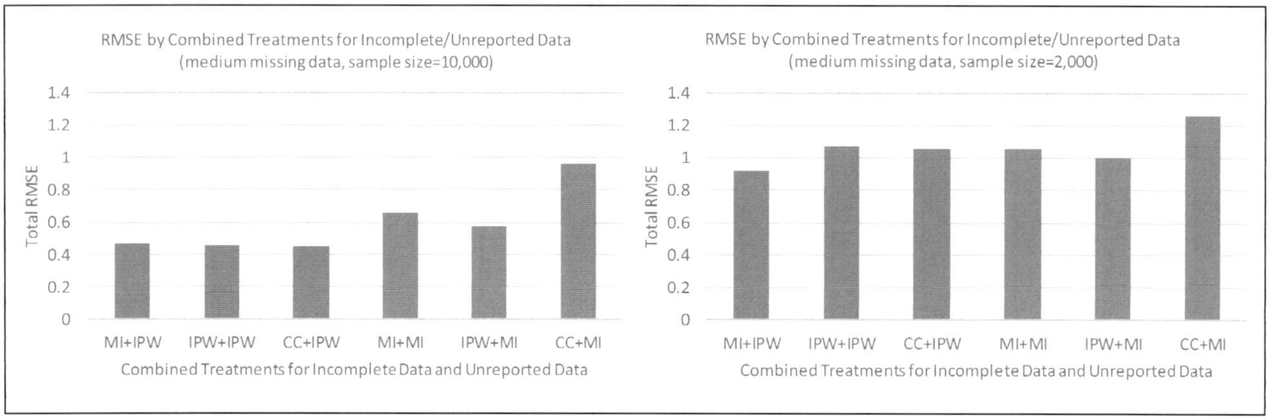

Figure 4. Model performance by treatment combinations for incomplete and unreported data at medium missing level (left is for sample size of 10,000 and right is for sample size of 2,000).

There are 10,060 records in this dataset, which contains a variety of variables including crash severity levels, contributing factors to each crash, roadway and driver's characteristics, and vehicle features, as well as weather conditions at the time of crash. The distribution of the 10,060 crashes on crash severity level is 8 (0.08%), 167 (1.66%), 826 (8.21%), 744 (7.40%), and 8,315 (82.65%) for severity K, A, B, C, and O, respectively. There are 13 independent variables of interest, as summarized in Table 3. It needs to be mentioned that some categorical variables in the original dataset were recoded into binary ones. For instance, originally there were nine categories for pavement condition when crash occurred, including dry, wet, snow, ice, sand/mud/dirt/oil/gravel, water, slush, debris, rut/holes/bumps/uneven pavement; including all nine categories would make estimation from the OP model very complicated. Therefore, the nine categories were combined into two—ideal (i.e., dry) and non-ideal—for *pavement indicator* in Table 3.

Meanwhile, not all the 13 independent variables were completely recorded for all crashes. Four (three categorical variables and one continuous variable) out of 13 variables have missing values, with the following missing number (percentage): 204 (2.0%) for *seat-belt use indicator*, 411 (4.1%) for *vehicle speed*, 711 (7.1%) for *airbag deploy indicator*, and 1,156 (11.5%) for *animal indicator*. The missing rates are comparable with those at low missing level in the simulation.

Before comparing performances among missing data treatments, MAR was tested using tests of significance in binomial logistic regression (*18*) to show the association between the missingness of a variable and the other variables in the dataset. The result of the test indicates that the probability of being missing of each of the four variables is significantly associated with some of the observed variables. Therefore, the missing mechanism for the dataset is not MCAR but rather MAR, and MI is appropriate to use.

Table 3. Summary Statistics of Independent Variables in Interest

Variable type	Description	No. missing[a]	Mean	SD
Road condition				
Lane number	Number of traffic lanes where crash occurred, between 1 and 6	–	2.76	0.83
Speed limit	Maximum speed limit is between 40 mph and 70 mph	–	65.09	5.60
Crash information				
Pavement indicator	1 = pavement in any of the non-ideal conditions: wet, snow, ice, sand, water, slush, debris, holes; 0 = otherwise	–	0.23	0.42
Dark indicator	1 = dark; 0 = daylight	–	0.50	0.50
Weather indicator	1 = any of the weather: fog, smog, smoke, rain, sleet, hail, snow, crosswinds, blowing sand; 0 = clear or cloudy	–	0.35	0.48
Vehicle speed	In mph, based on drivers statement or officers estimate	411 (4.1%)	58.97	10.64
Animal indicator	1 = animal in error; 0 = otherwise	1156 (11.5%)	0.22	0.42
Driver information				
truck indicator	1 = truck; 0 = otherwise	–	0.11	0.31
gender indicator	Drivers' gender, 1 = male; 0 = male	–	0.65	0.48
age	Drivers' age in years	–	38.32	15.37
Airbag indicator	1 = air bag deployed; 0 = otherwise	711 (7.1%)	0.18	0.39
Seat-belt indicator	1 = seat belt used; 0 = otherwise	204 (2%)	0.98	0.15
DUI indicator	1 = driving under influence (including physical and mental defect); 0 = otherwise	–	0.08	0.27

Note: SD = standard deviation; – = no missing values.
[a]Number (percentage) of missing cases for each of the variables.

Table 4. OP Model Estimation Results Using Three Missing Data Treatments

	CC			IPW			MI[a]		
Variable	Coef.	S.E.	*t*-ratio	Coef.	S.E.	*t*-ratio	Coef.	S.E.	*t*-ratio
Crash information									
Pavement indicator	0.109	0.020	5.34	0.117	0.052	2.24	0.061	0.019	3.22
Dark indicator	0.054	0.018	2.98	0.005	0.045	0.11	0.034	0.017	2.06
Animal indicator	0.433	0.030	14.68	0.446	0.081	5.53	0.415	0.029	14.45
Driver information									
Driver gender indicator	0.131	0.017	7.64	0.175	0.045	3.92	0.118	0.016	7.43
Driver age	0.002	0.001	1.99	0.004	0.003	1.46	0.002	0.001	2.11
Airbag deploy indicator	−0.415	0.019	−22.16	−0.352	0.043	−8.12	−0.434	0.018	−24.18
Seat belt use indicator	0.566	0.041	13.77	0.582	0.066	8.82	0.560	0.038	14.62
DUI indicator	−0.254	0.028	−9.24	−0.228	0.046	−4.92	−0.277	0.026	−10.77
Threshold parameters									
γ_1	−3.257	0.163	−19.93	−3.257	0.218	−14.91	−3.100	0.135	−22.96
γ_2	−1.889	0.076	−24.73	−1.932	0.145	−13.31	−1.794	0.071	−25.41
γ_3	−0.808	0.071	−11.45	−0.875	0.142	−6.16	−0.770	0.066	−11.63
γ_4	−0.378	0.071	−5.35	−0.479	0.138	−3.47	−0.350	0.066	−5.28

Note: S.E. = standard error of the estimated coefficient.
[a]The MI model estimation results are the average of estimates from 30 imputations.

As used for simulated data, three missing data treatments (CC, IPW, and MI) are applied to the crash data. After handling the missing data, the dataset is fit by the OP model with K (fatal) as the baseline. As the dataset has a large sample size and low level of missing data, the coefficients based on CC are considered to be reliable parameter estimates, and are used as the ground truth for comparisons. All 13 explanatory variables listed in Table 3 were first tested for inclusion and only eight significant variables (at significance level of 0.05, with absolute value of *t*-statistic greater than 1.94) were kept for CC case, shown in Table 4. Those eight variables were also used for OP model estimation under IPW and MI treatments.

As shown in Table 4, all the parameter estimates point to the same direction under the three missing data treatments and their values are close. A positive value of a parameter implies that an increase of variable value leads to a higher probability of less severe crashes relative to fatal crash. The coefficients of the following variables have positive signs: *pavement indicator*, *dark indicator*,

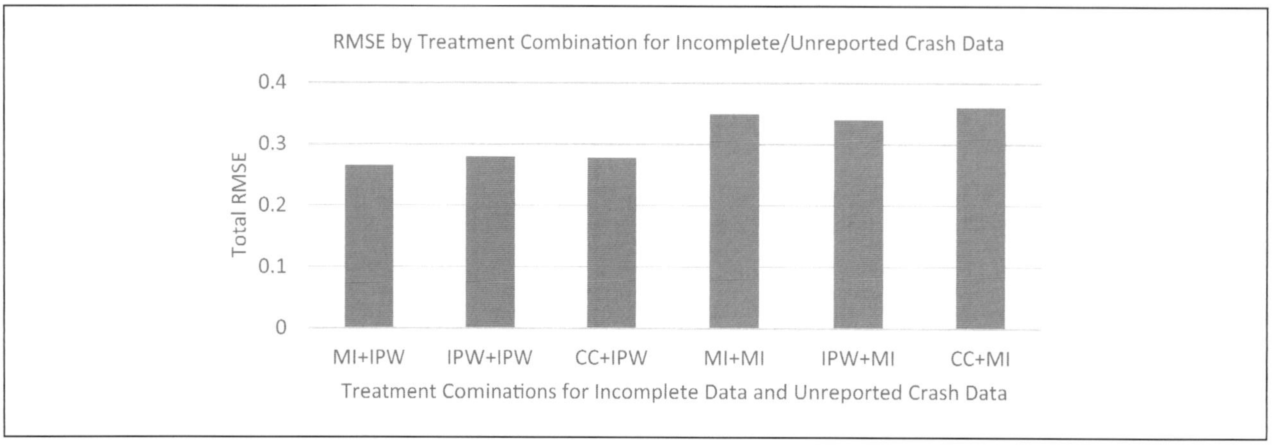

Figure 5. Model estimation performance by treatment combinations for incomplete and unreported crash data.

animal indicator, *driver gender indicator*, *driver age*, and *seat-belt use indicator*. This indicates that drivers who travel on a non-ideal pavement surface, or in a dark circumstance, or hit an animal are more likely to be involved in less severe (ABCO) crashes relative to fatal (K) crashes. Meanwhile, male drivers, or older drivers, or drivers wearing seat-belt have a higher probability to be involved in crashes that result in less severe injury. The negative coefficients for *airbag deploy indicator* and *DUI indicator* suggest that when airbags are deployed in a crash or drivers are driving under influence, the odds of drivers suffering less severe injury decreases (i.e., more likely to be involved in fatal crashes). The result overall makes sense with a few parameters open to question, but no further discussion is made here as the paper focuses on the comparison of model performances for three missing data treatments.

Meanwhile, estimates based on MI are found to have similar standard errors to those under CC, and these values are smaller than the standard errors under IPW. Using CC estimates as the baseline, the calculated total RMSEs are 0.47 and 0.23 under IPW and MI, respectively. This result indicates that MI performs slightly better than IPW for this dataset.

In addition, model performances are compared for crash data with both incomplete and underreporting issues. As the real underreporting rates are unknown for this dataset, it is assumed that the reporting rates are 100%, 90%, 80%, 70%, and 50% for KABCO. The same six treatment combinations are used to address incomplete reported data and unreported data: CC + IPW, CC + MI, IPW + IPW, IPW + MI, MI + IPW, and MI + MI, followed by fitting the data using the OP model. The result is summarized in Figure 5.

Similar to what was found in the simulation, IPW performs better than MI in terms of handling unreported data (with smaller total RMSE). Meanwhile, it also confirms that among the six treatment combinations, MI + IPW is an appealing method for handling both incomplete recording and underreporting issues in crash data.

Discussion

When applying crash severity models such as the OP model in the presence of incomplete and underreported data, the performance of different missing data treatments varies with missing levels and sample sizes. In this paper, model performance from CC, IPW, and MI were compared using simulated and real data.

The results of comparison indicate that when sample size is large and missing level is low, there is barely any difference among the three treatments. Meanwhile, CC and IPW perform similarly in almost all missing data scenarios involving incompletely recorded data. However, with a relatively small sample size and high missing rates, more attention should be paid to the selection of missing data treatments.

With few existing studies investigating MI performances with different sample sizes and missing levels, this paper fills the gap and indicates that MI is not always preferred for handling missing data in the OP model. For a dataset with relatively high missing rate and large sample size, MI leads to less reliable parameter estimates under the OP model compared with CC/IPW. This might stem from the combination of the MI algorithm and the structure of crash severity models. Crash severity models, including the OP model, are discrete choice models whose outcome is based on a discretized latent variable. As shown in Equation 1, OP models compute a continuous latent variable based on given values of covariates. The latent variable connects covariates and the dependent variable (i.e., crash severity level) which is categorical. This feature of crash severity models, combined with the stochastic nature of MI, makes the correlation between the covariates and dependent variable weaker and accordingly affects the quality of imputation, especially when missing rates are high. Therefore, it is suggested that MI be used with caution when dealing with missing crash data, especially using crash severity models like the OP model.

In situations where MI is appropriate, more attention needs to be paid to the imputation model. In order to improve the performance of MI, it is recommended to include more rather than fewer variables in the imputation model. When dealing with both incomplete and unreported crash data, this paper recommends using MI for the incomplete part and IPW for the unreported part, which takes the full advantage of MI and IPW methods. However, all the findings might be limited to the simulation design and the feature of the crash data used in the analysis, which need to be further investigated in future research. Meanwhile, the same study design can be applied to other crash severity models such as Multinomial Logit Model to generalize the findings.

Last but not the least, it needs to be emphasized that no matter which treatment is used for missing data, the goal is not to achieve correct predictions of missing values but to obtain valid inference. It is advisable to always minimize the amount of missing data up front, such as by exploring alternative data sources (naturalistic driving data, CODES, vehicles' event data recorders, or matching statewide health records with police-recorded crash reports). Meanwhile, it is important, though challenging, to have a good understanding of the missing data mechanism, causes for inaccurate/incomplete recording and underreporting, and any additional information in the data that might be used to improve the imputation or weighting model. More studies are needed to explore the quality of crash data, in order to improve the performance of model estimation with appropriate handling of missing/inaccurate data.

Author Contributions

The authors confirm contribution to the paper as follows: study conception and design: Ye; data collection: Ye; analysis and interpretation of results: Ye and Wang; draft manuscript preparation: Ye and Wang. Both authors reviewed the results and approved the final version of the manuscript.

References

1. Oliveira, P., F. Rodrigues, and P. R. Henriques. A Formal Definition of Data Quality Problems. *Proc., Tenth International Conference on Information Quality (ICIQ'05)*, MIT, Cambridge, MA, 2005.
2. Imprialou, M., and M. Quddus. Crash Data Quality for Road Safety Research: Current State and Future Directions. *Accident Analysis & Prevention*, 2017. http://www.sciencedirect.com/science/article/pii/S0001457517300842
3. Benavente, M., M. Knodler, and H. Rothenberg. Case Study Assessment of Crash Data Challenges: Linking Databases for Analysis of Injury Specifics and Crash Compatibility Issues. *Transportation Research Record: Journal of the Transportation Research Board*, 2014. 1953: 180–186.
4. Tefft, B. C. Prevalence of Motor Vehicle Crashes Involving Drowsy Drivers, United States, 1999-2008. *Accident Analysis & Prevention*, Vol. 45, No. 1, 2012, pp. 180–186.
5. Tefft, B. C. 2014. *Prevalence of motor vehicle crashes involving drowsy drivers, United States, 2009-2013*. AAA Foundation for Traffic Safety Report, Washington, D.C., 2014. https://www.aaafoundation.org/sites/default/files/AAAFoundation-DrowsyDriving-Nov2014.pdf. Accessed November 25, 2016.
6. *Highway Capacity Manual 2010*. Transportation Research Board of the National Academies, Washington, D.C., 2010.
7. Yamamoto, T., J. Hashiji, and V. Shankar. Underreporting in Traffic Accident Data, Bias in Parameters and the Structure of Injury Severity Models. *Accident Analysis and Prevention*, Vol. 40, No. 4, 2008, pp. 1320–1329.
8. Ye, F., and D. Lord. Investigation of Effects of Underreporting Crash Data on Three Commonly Used Traffic Crash Severity Models: Multinomial Logit, Ordered Probit, and Mixed Logit. *Transportation Research Record: Journal of the Transportation Research Board*, 2011. 2241: 51–58.
9. Little, R. J. A., and D. B. Rubin. *Statistical Analysis with Missing Data*, 2nd ed. Wiley, Princeton, NJ, 2002.
10. Oh, H., and F. Scheuren. Weighting Adjustment for Unit Nonresponse. Chap. 13 in vol. 2, part 4 of Incomplete Data in Sample Surveys. Academic Press, New York City, NY, 1983.
11. Kalton, G., and I. Flores-Cervantes. Weighting Methods. *Journal of Official Statistics*, Vol. 19, 2003, pp. 81–97.
12. Carpenter, J. Missing Data Example Analysis, 2006. http://www.lshtm.ac.uk/msu/missingdata/example.html. Accessed November 25, 2016.
13. Seaman, S. R., and I, R. White. Review of Inverse Probability Weighting for Dealing with Missing Data. Statistical Methods in Medical Research, Vol. 22, 2013, pp. 278–295.
14. Seaman, S. R., and I, R. White. Inverse Probability Weighting with Missing Predictors of Treatment Assignment or Missingness. *Communications in Statistics-Theory and Methods*, Vol. 43, 2014, pp. 3499–3515.
15. van Buuren, S., J. P. L. Brand, K. Groothuis-Oudshoorn, and D. B. Rubin. Fully Conditional Specification in Multivariate Imputation. *Journal of Statistical Computation and Simulation*, Vol. 76, No. 12, 2006, pp. 1049–1064.
16. Ye, F., and D. Lord. Comparing Three Commonly Used Crash Severity Models on Sample Size Requirements: Multinomial Logit, Ordered Probit and Mixed Logit Models. *Analytic Methods in Accident Research*, Vol. 1, 2014, pp. 72–85.
17. SAS Institute Inc., SAS 9.4. https://www.sas.com/en_us/software/sas9.html. Accessed August, 2017.
18. Garson, G. D. Missing Values Analysis and Data Imputation, 2015 ed. Asheboro, NC: Statistical Associates Publishers.

The Standing Committee on Safety Data, Analysis and Evaluation (ANB20) peer-reviewed this paper (18-05989).

Article

TRR
JOURNAL OF THE TRANSPORTATION RESEARCH BOARD

Transportation Research Record
2018, Vol. 2672(38) 160–170

DOI: 10.1177/0361198118783111
journals.sagepub.com/home/trr
SAGE

Analysis of Gap Acceptance Behavior for Unprotected Right and Left Turning Maneuvers at Signalized Intersections using Data Mining Methods: A Driving Simulation Approach

Somayeh Mafi[1], Yassir Abdelrazig[1], and Ryan Doczy[1]

Abstract

Gap acceptance predictability has become a vital area of interest for traffic safety and operations due to its complexity and significance in understanding a population's driving behavior. Recent studies have implemented statistical modeling techniques, such as binary logit model (BLM), to predict drivers' gap acceptance behaviors. However, these models have inherent presumptions and pre-set correlations that, if contravened, can produce erroneous results. The use of non-parametric data mining techniques, such as decision trees, avoids these deficiencies, thus resulting in improvements to the predictive capability of the models. In this study, the feasibility of C4.5 decision trees, instance-based (IB), and random forest (RF) models for predicting drivers' gap decisions was examined by comparing their results with BLM. To accomplish this objective, 66 study participants drove through ten driving simulation scenarios requiring the navigation of unprotected right and left-turning maneuvers at four-legged, signalized intersections. The data collected from these tests will provide means to directly compare and rank the data mining and statistical models, while also allowing for the identification of variables that are significantly influencing gap acceptance. Results produced from the models indicated that data mining models were superior to BLM at accurately predicting a participant's gap decisions. RF models outperformed the C4.5 and IB models in predicting gap acceptance behaviors for both the left and right turning scenarios. Because of its superior performance, the authors recommend the implementation of the RF model for predicting gap decisions at unprotected signalized intersections.

When performing unprotected left or right turning maneuvers at an intersection, it is vital for drivers to see and identify adequate gaps among vehicular and pedestrian traffic. This behavior, known as gap acceptance behavior, is believed to have a close relationship with traffic safety and operation (*1*, *2*). Research demonstrated that erroneous gap acceptance may cause about 30 percent of left-turn crashes (*3*).

Several studies have investigated the parameters that affect drivers' gap acceptance behavior. Hamed et al. (*4*) reported the following significant parameters in predicting left-turn gap acceptance behaviors: age and gender of the driver, vehicle age, type of the trip (e.g. shopping, work), time that the driver waited at the front of the queue, flow of opposing-through traffic, existence of a median, time of day, type of maneuver, number of lanes, and traffic speed on the main stream. Some research has shown age-related declines in older drivers' cognitive and perceptual abilities, with this decline being a strong predictor of crash rate (*5–7*). In contrast, Kazazi et al. (*8*) found no significant difference in the collision rates of older and younger age groups. Research relating to the significance of high-visibility crosswalks on drivers' gap acceptance behavior has been mixed in its conclusions (*9*, *10*). In a study examining the gap acceptance behavior of drivers making left-turning maneuvers in the presence of pedestrians, Alhajyaseen et al. (*11*) found drivers to be less conservative when accepting gaps between individual pedestrians versus multiple pedestrians crossing shoulder-to-shoulder.

Literature shows that a variety of estimation methods have been used to investigate drivers' gap acceptance behavior. Statistical models, such as binary probit (*4*) and binary logit model (BLM) (*12–15*), have previously been used to analyze drivers' gap decision behavior. Most statistical models have predefined functional form that might confine their

[1]Department of Civil and Environmental Engineering, FAMU-FSU College of Engineering, Tallahassee, FL

Corresponding Author:
Address correspondence to Somayeh Mafi: sm14ay@my.fsu.edu

prediction accuracy, and if contravened, can produce erroneous results. Statistical models can be negatively influenced by outliers and missing values in the dataset. Contrarily, data mining models are non-parametric tools that do not require any pre-established correlations among the dependent variable and predictors, while being able to manage outliers and missing values. Data mining techniques have been used in different engineering problems (*16–18*). Recently, researchers have employed these techniques for predicting drivers' gap decisions, which are comparable or sometimes superior to statistical models. Artificial neural networks (ANN), generally known as a biologically inspired method, are one such method that has been used for gap acceptance analysis (*13*). However, ANN models have the drawbacks of greater interpretation complexity and the potential for overfitting the training data. Pawar et al. (*14*) implemented support vector machines (SVMs) to analyze drivers' gap acceptance behavior at uncontrolled intersections and pedestrian midblock crossings. SVM performance was determined to be comparable with that of the BLM in all cases and superior in a few. Nagalla et al. (*19*) used SVM, random forest (RF), and decision trees (DT) to predict left-turn gap acceptance behaviors at four-legged un-signalized intersections. The study concluded that SVM and DT models had highly comparable outputs; however, the RF model outperformed both the SVM and DT models.

Several studies illustrated the validity of a driving simulator for measuring driving performance (*20–22*). De Winter et al. (*23*) presented several advantages of using driving simulation technology for research including, "controllability, reproducibility, and standardization, ease of data collection, possibility of encountering dangerous driving conditions without being physically at risk, novel opportunity for feedback and instruction." Despite their advantages, the use of driving simulators carries the risk of developing simulator adaption syndrome (SAS), which is the potential for individuals to develop motion-sickness-like symptoms (*24*).

The primary focus of this paper is evaluating the application of C4.5, RF and IB models for classifying and predicting drivers' gap acceptance decisions while performing unprotected left and right turning maneuvers at four-legged signalized intersections. Thorough literature review showed that this is the first application of the IB model to predict driver gap acceptance decisions. This paper identified potentially significant factors in drivers' gap acceptance behavior based on previous literature, while also considering the limitations of the study's driving simulation technology. Therefore, drivers' characteristics (age, gender, etc.), gap size, number of lanes, crosswalk type, the presence and quantity of pedestrians, light conditions, and weather conditions were selected as potentially significant parameters. Using the data extracted from the simulator, the performances of C4.5, IB, and RF models were evaluated and compared with the BLM models for each turning maneuver.

Background

Data Mining Models Characteristics

C4.5. Decision tree (DT) is a popular machine learning method exhibiting substantial processing power. It presents rules of classification, which are simple and interpretable by humans. DTs are capable of processing outliers or datasets with missing values, and they prevent overfitting by utilizing pruning (*25*). The C4.5 algorithm, used for generating a type of DT, was first developed by Quinlan (*26*) utilizing a tree-shaped graph comprised of a root node, internal nodes (representing test conditions), branches, and leaf nodes (representing the final classes). Generally, the tree starts from the root nodes, moves downward through branches and internal nodes, and ends at the leaves. Classification in a DT is defined as the mapping from a set of features (inputs) to a particular class (output). At each node (except the leaves), the division into classes is carried out based on the features that best divide its set of samples into subsets. The attribute with the greatest normalized information gain is selected to make the decision. This procedure is applied to each split subset of the data recursively until that subset's size is small enough, or all the instances in the current subset belong to the same class.

Random Forest (RF). Despite many advantages of DT algorithm, it has some deficiencies. DT sorts all numerical features whenever it decides to split a node. For overcoming both this issue and, in some cases, overfitting problems, Breiman (*27*) introduced the concept of RF. RF is comprised of multiple trees generated from randomly selected samples of training data. Each tree is grown to the largest possible extent (without pruning). Prediction accuracy of the RF is calculated by aggregating the results of DTs, which is the majority vote for classification or averaging for regression. Every tree in a RF is built by the random sampling of a training data subset with replacement (the notion of Bagging) and randomly sampling a variable subset (the notion of Randomization), which results in improved prediction power and greater model efficiency. In an RF, m features are chosen at each node from a total of M input features ($m<<M$). The best split of these m features is utilized for splitting the node. The value of m remains constant during the forest generation. RF generally retains the advantages of the DT while producing superior results, when compared with individual tree classifier such as C4.5.

Instance-Based (IB). In the nearest-neighbor IB learning method, all attributes have equal contributions to the decision-making process. An advantage of IB learning over most other machine learning schemes is that new instances can be added to the training set at any time. In IB models, the training instances are stored literatim and a distance function, usually Euclidean distance, is utilized to specify the member

Table 1. Scenario Description

Scenario	Turning maneuver	Number of lanes	Presence of pedestrians	Crosswalk type	Environmental condition
1	Left	4	Yes	Ladder	No
2	Left	2	Yes	Ladder	No
3	Left	4	No	Ladder	No
4	Left	4	Yes	Standard	No
5	Right	2	Yes, multiple	Standard	No
6	Right	2	Yes	Standard	No
7	Right	4	Yes	Ladder	No
8	Right	4	Yes	Standard	No
9	Left	4	No	Ladder	Night
10	Left	4	No	Ladder	Fog

of the training set that is closest to an unknown test instance. When the closest training instance has been specified, its class is predicted for the test instance. IB learning is simple and efficient, however, it is computationally intensive compared with the other data mining methods. To make a single prediction for a testing instance, it calculates the distance from every member of the training set and selects the one with the smallest distance. This model was notoriously slow until it was improved by application of kD-trees in the early 1990s (*28*). It considers the training set as a tree-shaped structure, such as a kD-tree. This is a binary tree that splits the input space using a hyperplane and then splits each part again, recursively. All divisions are done parallel to one of the axes, in the two-dimensional case. The structure is named a kD-tree since it stores a set of instances in k-dimensional space, which k is the number of attributes. In an attempt to find the nearest neighbor of the specified target instance, the model flows from the tree's root down to the zone containing the target instance.

Methodology

Development of the Scenarios

In this study, the simulation software HyperDrive Authoring Suite was used to create ten highly-controlled scenarios capable of being driven by participants using the experiment's driving simulator. Scenarios were constructed using items known as tiles, which are a combination of roadway sections, buildings, vegetation, etc. Each scenario included an intersection tile, several roadway tiles leading to the intersection, preprogrammed simulation vehicles, and (for most scenarios) pre-programmed pedestrian entities. Both the pedestrian and vehicular entities were programmed to cross in front of the participant's direction of travel to simulate real-life scenarios that would require the participants to make gap acceptance decisions. These entities were set to start their maneuvers at different times, so that a series of gaps between entities were developed. The participant's decision to accept or reject these gaps provided a metric that enabled the comparison of drivers' behavior between different scenarios.

The scenarios were developed so that discrepancies between certain scenarios could be attributed to the change in one isolated variable (e.g. number of lanes) while all other variables are held constant (Table 1). Data for driver specific variables (i.e. age, gender, and driving experience) were collected for each participant prior to conducting experiments.

Left-turn scenarios required the participants to drive through the intersection and perform a permissive left-turn when there was no vehicle attempting to pass through the intersection and when there was no pedestrian located in the crosswalk (Figure 1). In these scenarios, once the participant approaches the intersection, the light changes to solid green. Opposing through-traveling vehicles, driving at 30 mph, will then begin to cross the roadway. Table 2 shows the order of gaps between consecutive vehicles that the participants encountered for different scenarios of this study. The last gap between vehicles was set to 550 ft (12.50 second) because it was assumed that all participants would accept a gap of this size (*29*). Scenarios 9 and 10 were the same as scenario 3 in all aspects excluding the environmental conditions. Scenario 9 was designed for night conditions, while scenario 10 included heavy fog with a quarter-mile of visibility.

In right-turn scenarios, the participants were instructed to perform a right-turn maneuver at the intersection while navigating through pedestrian traffic (Figure 1). In these scenarios, the participants will drive to an intersection where they will stop behind a vehicle already waiting to perform a right turn. Once they are close to the intersection, the intersection's light changes to solid green. This will prompt pedestrians to start crossing the roadway at 3 mph, and for the vehicle stopped ahead of the participant's vehicle to perform a right-turn maneuver. The order of gaps between the crossing pedestrians is shown in Table 2. For scenario 5, pedestrians cross the roadway in groups of two to four moving roughly shoulder-to-shoulder.

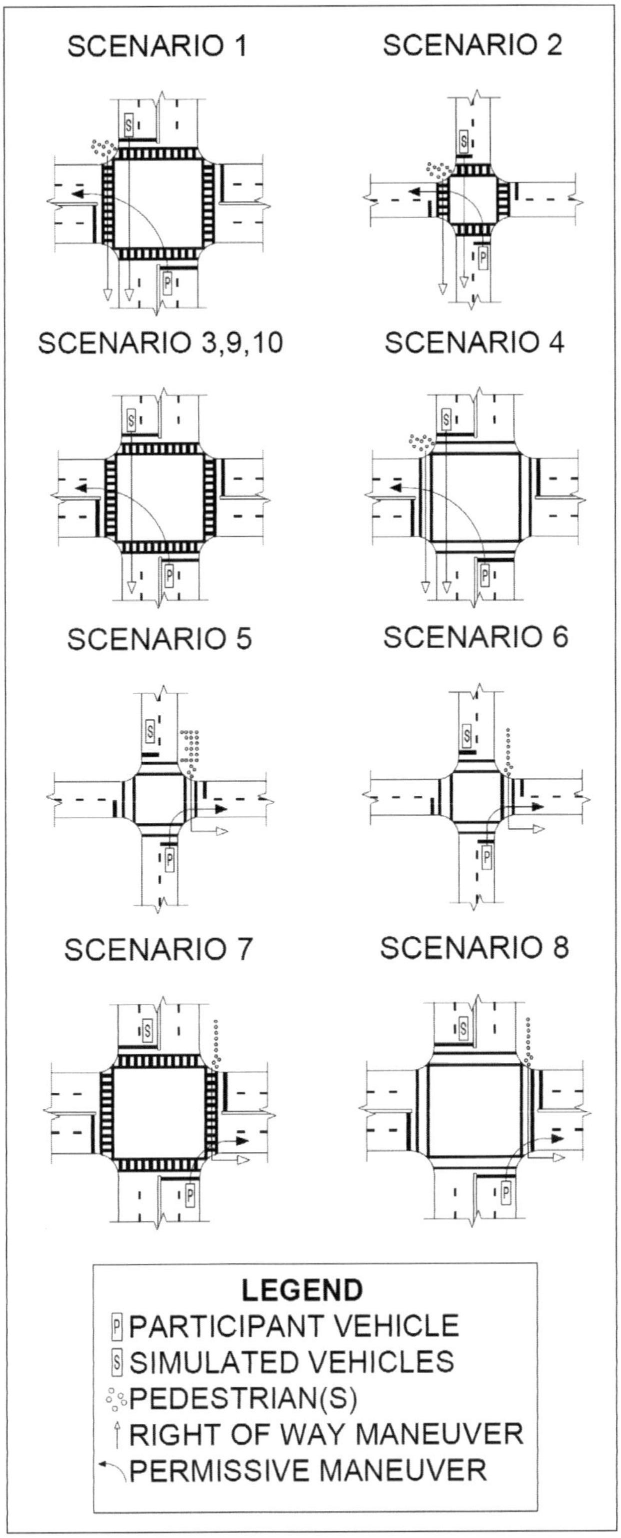

Figure 1. Left-turn (top two rows) and right-turn (bottom two rows) driving scenarios.

Conducting Experiments

The experiment was conducted using a driving simulator, DriveSafety CDS 250, which consists of a partial cab (based on a Ford Focus) and three monitors that provide a 110-degree field of vision. A total of 66 drivers participated in the study, of which 32 were +65 years old. Upon each participant's arrival, they were asked to fill out and sign a consent form reviewed and approved by the university Institutional Review Board (IRB). Each participant was informed that they could quit the experiment at any time if they began to feel ill. Due to SAS, three of the older participants were not able to complete all scenarios. Also, two younger individuals were unable to complete enough of the scenarios to be able to utilize their data in this study. Therefore, the data of 64 individuals were utilized in this study; 32 older drivers with the mean age of 71 ± 5 and the mean driving experience of 50 ± 10 years, and 32 younger drivers with the mean age of 32 ±12 and the mean driving experience of 13 ± 12 years. Since participants had previously participated in driving simulation experiments, the number of individuals who dropped out due to the effects of SAS were relatively minimal when compared with similar studies (*24*).

One of the issues related to the driving simulation studies is the participants' struggle to become familiarized with how the simulator responds to the inputs (e.g. the brake force applied when pressing on the brake pedal). These problems could lead to participants performing significantly better in scenarios towards the end of the experiment. To mitigate this problem, a training simulation was developed to allow the participants to get familiar with the simulator and mitigate bias in the results. The training scenario for this paper took roughly seven minutes and required the participants to deal with the following actions: accelerating to a set speed, decelerating to a set speed, braking to a stop, performing left and right turning maneuvers, following an automated vehicle, and changing lanes. Upon completion of the pre-experiment activity, the participants were given a minimum of a five-minute break to minimize the potential of developing symptoms related to SAS. Next, participants drove ten scenarios (each scenario taking roughly 1–1.5 minutes) broken up with mandatory two-minute breaks after the third and sixth scenarios. After going through all the scenarios, the participants completed a survey answering questions about their exact age, driving experience, driving history, etc.

Training and Testing of the Models

To evaluate the performance of classifiers, stratified holdout procedure was used. In this method, the random sampling was carried out in a manner that guaranteed each class (i.e. acceptance and rejection) was appropriately represented in both training and test sets. Another advantage of the stratified holdout method is that it allows for swapping the roles of the testing and training data to reduce the effect of uneven representation in training and testing data sets. In n-fold cross-validation, the whole data set is divided into an approximately equal fixed number of folds (n). Each partition is held out in turn (test data) and the classifier trained on the remaining data; then its error rate is computed on the test set. Therefore, the classification is performed a total of n times on different training sets which have many commonalities. The overall error is calculated by taking the average of n error estimates. Based on the theoretical evidence, 10 is approximately the most appropriate number of folds to get the best estimate of error. Previous tests have also indicated that the use of stratification improves model results (*30*).

Results and Discussions

Table 2 depicts the sequential order of gaps the participants encountered for any given scenario. Each gap required the participant to accept or reject it based on if they felt the gap was large enough to safely permit the appropriate turning maneuver. Once a gap was accepted, the sizes of the accepted and rejected gaps were recorded and the scenario was ended. The nature of the experiment leads to a preponderance of rejected gaps in the data set, which will cause the final models to be overly biased towards rejecting gaps. To counteract this problem, each scenario only included the largest rejected gap and the accepted gap.

The following variables were considered in the development of the driving scenarios: drivers' characteristics (age, gender, and driving experience), the presence of pedestrians in or nearby the crosswalk, number of lanes, different crosswalk configurations (ladder or standard) and contextual conditions (heavy fog, and night conditions). Also, the influence of multiple pedestrians crossing shoulder-to-shoulder, as opposed to individually, was investigated for right-turn scenarios. The driving experience was excluded from the list of explanatory variables due to the concerns of collinearity with the age variable.

Table 2. Gap Sizes for All Scenarios

	Gap size in ft (second)		
Gap number	Scenarios 1,2, and 4	Scenarios 3,9, and 10	Scenarios 5,6,7, and 8
1	200 (4.55)	150 (3.41)	15 (3.42)
2	150 (3.41)	150 (3.41)	20 (4.55)
3	125 (2.84)	200 (4.55)	25 (5.61)
4	150 (3.41)	250 (5.68)	30 (6.75)
5	150 (3.41)	300 (6.82)	35 (7.88)
6	125 (2.84)	350 (7.95)	40 (9.02)
7	200 (4.55)	400 (9.09)	45 (10.16)
8	200 (4.55)	450 (10.23)	60 (13.57)
9	125 (2.84)	500 (11.36)	n/a
10	300 (6.82)	550 (12.50)	n/a
11	200 (4.55)	n/a	n/a
12	250 (5.68)	n/a	n/a
13	300 (6.82)	n/a	n/a
14	350 (7.95)	n/a	n/a
15	400 (9.09)	n/a	n/a
16	450 (10.23)	n/a	n/a
17	500 (11.36)	n/a	n/a
18	550 (12.50)	n/a	n/a

Note: n/a= not applicable.

Table 3. Confusion Matrix

	Predicted decision	
Observed decision	Rejection	Acceptance
Rejection	Correct negatives (z)	False alarms (f)
Acceptance	Misses (m)	Hits (h)

Skill Scores for Gap Acceptance Prediction

To evaluate the performance of gap acceptance models, skill scores need to be calculated based on the confusion matrix (Table 3). Bias is a ratio comparing the predicted number of accepted gaps to the total number of observed accepted gaps ($f+h/m+h$). It represents the potential for the model to underestimate or overestimate the accepted gap outcomes. The probability of detection (POD) or recall is the ratio of the number of correctly classified accepted gaps to the total number of observed accepted gaps ($h/m+h$), which shows the fraction of observed accepted gaps that are correctly predicted. The probability of non-detection (POND) is the ratio of the number of correctly classified rejected gaps to the total number of observed rejected gaps ($z/z+f$), which indicates the fraction of observed rejected gaps that are correctly predicted (*19*). Precision is the ratio comparing the number of correctly classified accepted gaps to the total number of predicted accepted gaps ($h/f+h$). F-measure is a weighted harmonic mean of the precision and recall ($2\,precision \times recall/[precision+recall]$). An F-measure close to 1 is optimal (*14*).

Binary Logit Models (BLMs)

Two distinct BLMs were developed for the left and right turning maneuvers. The results of the models, which have been reduced to only include significant parameters at 95% significance level, are shown in Table 4.

In Table 4, AGE is the age of participant, PED describes the presence of pedestrians at or nearby the crosswalk (=0 absence; and =1 presence), LNECNT describes the number of opposing through lanes (=0 one; and =1 two), MLTPED explains the presence of multiple pedestrians crossing the street shoulder-to-shoulder (0: sets of individual pedestrians; 1: sets of multiple pedestrians), GAP in left-turn model is the

Table 4. BLMs for Gap Decision in Left and Right Turning Maneuvers

Maneuver	Variable	Coefficient	SE	p-value
Left-turn	Intercept	−0.7205	0.2982	0.0157
	AGE	−0.5165	0.1599	0.0012
	PED	−0.5294	0.1780	0.0029
	LNECNT	−0.6010	0.2362	0.0109
	GAP	0.2520	0.0464	0.0000
	Logit(Decision) ~ 1 + AGE + PED + LNECNT + GAP			
	p-value = 3.52e-10			
Right-turn	Intercept	−2.0045	0.32645	8.23E-10
	AGE	0.41616	0.19963	0.037097
	LNECNT	−0.88331	0.27465	0.001299
	MLTPED	−0.73382	0.28952	0.011256
	GAP	0.40894	0.053579	2.31E-14
	Logit(Decision) ~ 1 + AGE + LNECNT + MLTPED + GAP			
	p-value = 4.03e-16			

gap size (in seconds) between consecutive opposing through vehicles, however in right-turn model it is the gap size (in seconds) between consecutive pedestrians crossing in front of the participant.

The results illustrated that the participant's age was a significant parameter for gap acceptance behavior. Older participants were more cautious in left-turn gap acceptance; whereas, they were less conservative in accepting a gap in right-turn maneuvers compared with the younger cohort. Both cohorts were more conservative in performing left-turning maneuvers at four-lane intersections than at two-lane intersections, most likely because of the increased distance required to perform the left-turn maneuver at a four-lane intersection. In the left-turn models, the presence of pedestrians in the crosswalk led participants to be more conservative in their gap selections. The results also showed that the participants were more conservative in performing right-turning maneuvers at four-lane intersections than at two-lane intersections. This behavior could be attributed to the increased size of the four-lane crosswalks, thus allowing for a greater number of pedestrians to be present on the crosswalks at any given time. In right-turn scenarios, the presence of multiple pedestrians crossing the street shoulder-to-shoulder reduced the probability that a participant would accept gap between pedestrians. This is consistent with previous research related to this subject (*11*). The results of the BLM models showed that there was no significant difference in predicting gap acceptance behavior when accounting for participants' gender, crosswalk types, and different environmental conditions.

Table 5 depicts the left and right turning gap decision confusion matrices for the BLM, C4.5, IB, and RF models. The numbers in bold represent the total percentage of gaps that were correctly predicted to be either accepted or rejected. The left-turn BLM model correctly predicted the acceptance

Table 5. Confusion Matrices of BLMs, C4.5, IB, and RF Models for Gap Decision in Left and Right Turning Maneuvers (in Percent)

Maneuver	Observed data	Predicted decision: Rejection	Predicted decision: Acceptance	Total observed
BLMs				
Left-turn	Rejection	24.27	22.82	47.09
	Acceptance	19.91	32.99	52.91
	Total predicted	44.19	55.81	100.00
	Prediction accuracy = 57%			
Right-turn	Rejection	33.54	15.63	49.17
	Acceptance	19.58	31.25	50.83
	Total predicted	53.13	46.88	100.00
	Prediction accuracy = 65%			
C4.5 Models				
Left-turn	Rejection	35.61	11.48	47.09
	Acceptance	6.69	46.22	52.91
	Total predicted	42.30	57.70	100.00
	Prediction accuracy = 82%			
Right-turn	Rejection	40.00	9.17	49.17
	Acceptance	6.04	44.79	50.83
	Total predicted	46.04	53.96	100.00
	Prediction accuracy = 85%			
IB Models				
Left-turn	Rejection	40.70	6.40	47.09
	Acceptance	3.34	49.56	52.91
	Total predicted	44.04	55.96	100.00
	Prediction accuracy = 90%			
Right-turn	Rejection	46.25	2.92	49.17
	Acceptance	3.96	46.88	50.83
	Total predicted	50.21	49.79	100.00
	Prediction accuracy = 93%			
RF Models				
Left-turn	Rejection	40.70	6.40	47.09
	Acceptance	1.89	51.02	52.91
	Total predicted	42.59	57.41	100.00
	Prediction accuracy = 92%			
Right-turn	Rejection	46.25	2.92	49.17
	Acceptance	3.33	47.50	50.83
	Total predicted	49.58	50.42	100.00
	Prediction accuracy = 94%			

and rejection of a gap for 24.27% and 32.99% of the total data, respectively. Likewise, the BLM right-turn model correctly predicted the acceptance and rejection of a gap for 33.54% and 31.25% of the total data, respectively. The overall prediction accuracy of left and right turn models are 57% and 65% respectively, which will be seen as relatively poor when compared with the data mining models.

Data Mining Models

Data mining models were implemented in this paper for their ability to find patterns that would be difficult to detect via regression models. The confusion matrices of C4.5, IB, and

(a)

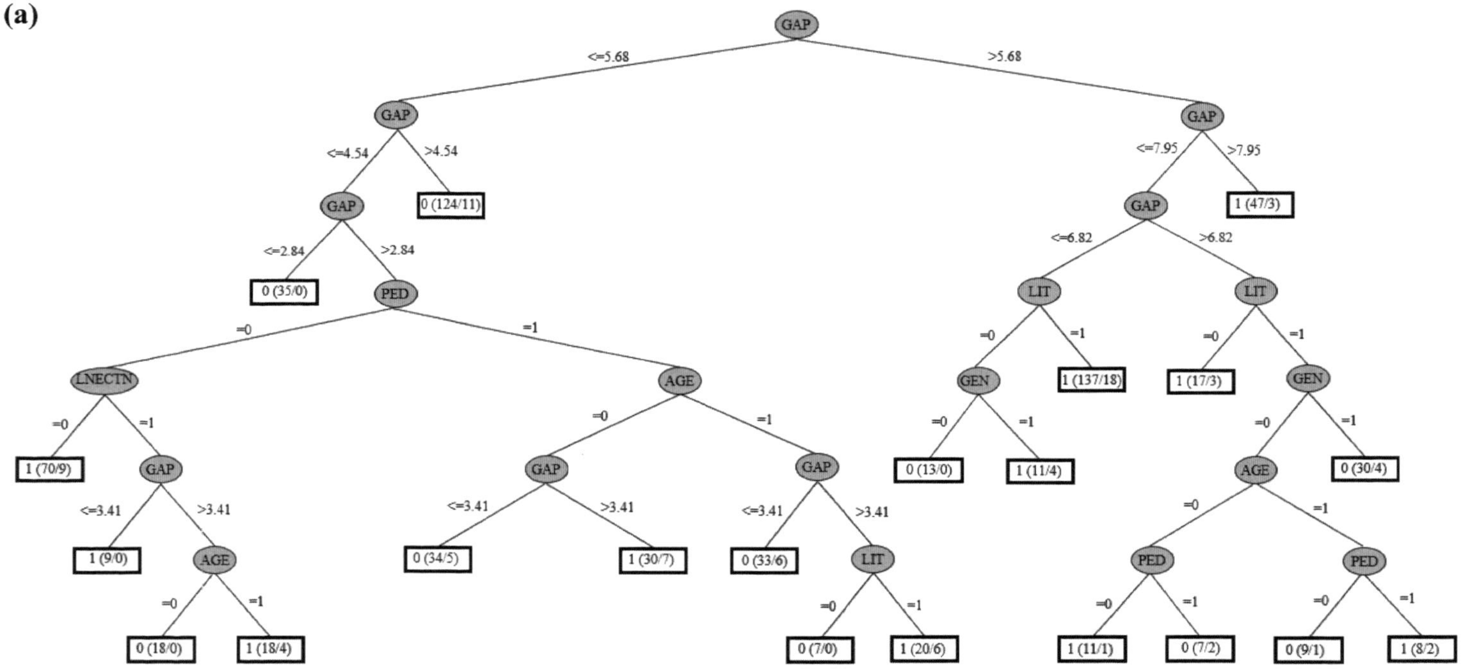

Variable	Symbol	Definition
Driver's age	AGE	0: aged under 65 (younger); 1:aged 65 and above (old)
Driver's gender	GEN	0: female; 1:male
Lane count	LNECNT	0: one opposite through lane; 1:two opposite through lanes
Gap size	GAP	The gap size between consecutive opposite through vehicles (second)
Light condition	LIT	0: night; 1:day
Pedestrian presence	PED	0: in absence of pedestrians; 1:in presence of pedestrians

(continued)

(b)

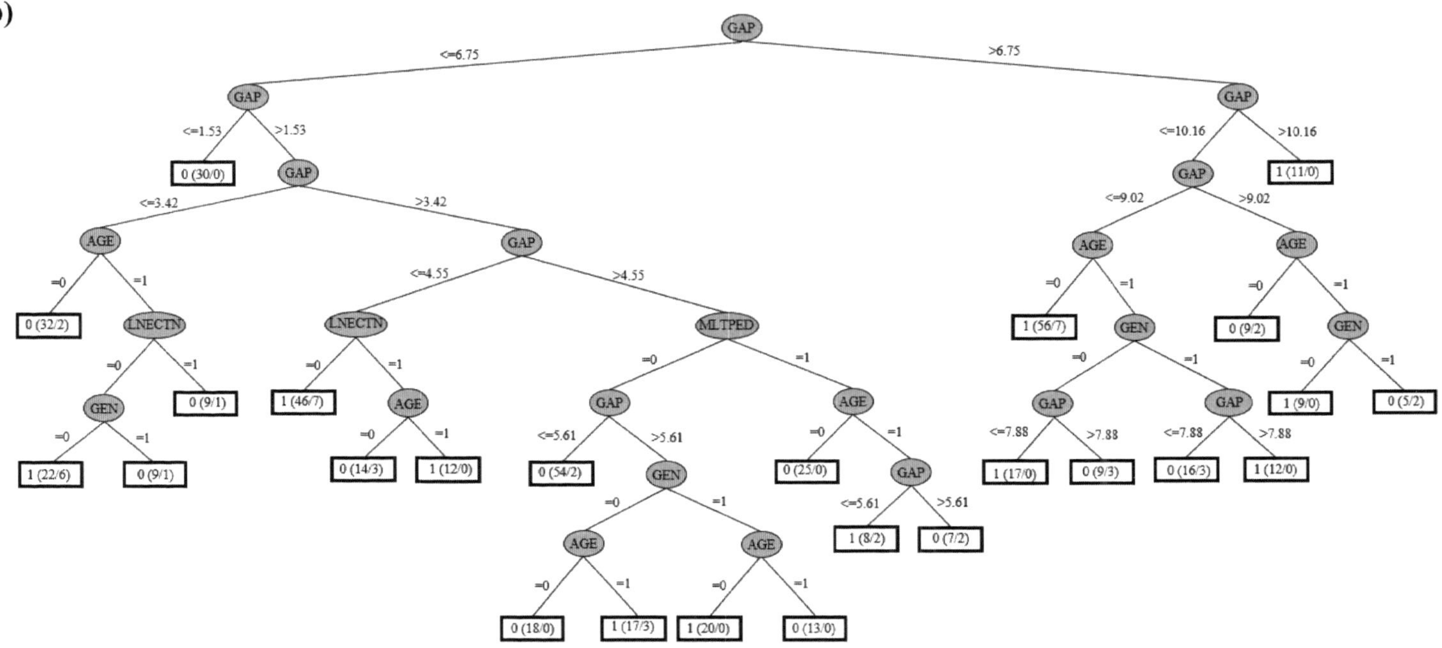

Variable	Symbol	Definition
Driver's age	AGE	0: aged under 65 (younger); 1:aged 65 and above (old)
Driver's gender	GEN	0: female; 1:male
Lane count	LNECNT	0: one opposite through lane; 1:two opposite through lanes
Gap size	GAP	The gap size between consecutive pedestrians crossing in front of the participant (second)
Multiple pedestrians	MLTPED	0: individual pedestrian crossing; 1: multiple pedestrian crossing shoulder-to-shoulder

Figure 2. The structure of C4.5 gap acceptance models for (*a*) left, and (*b*) right turning maneuvers.

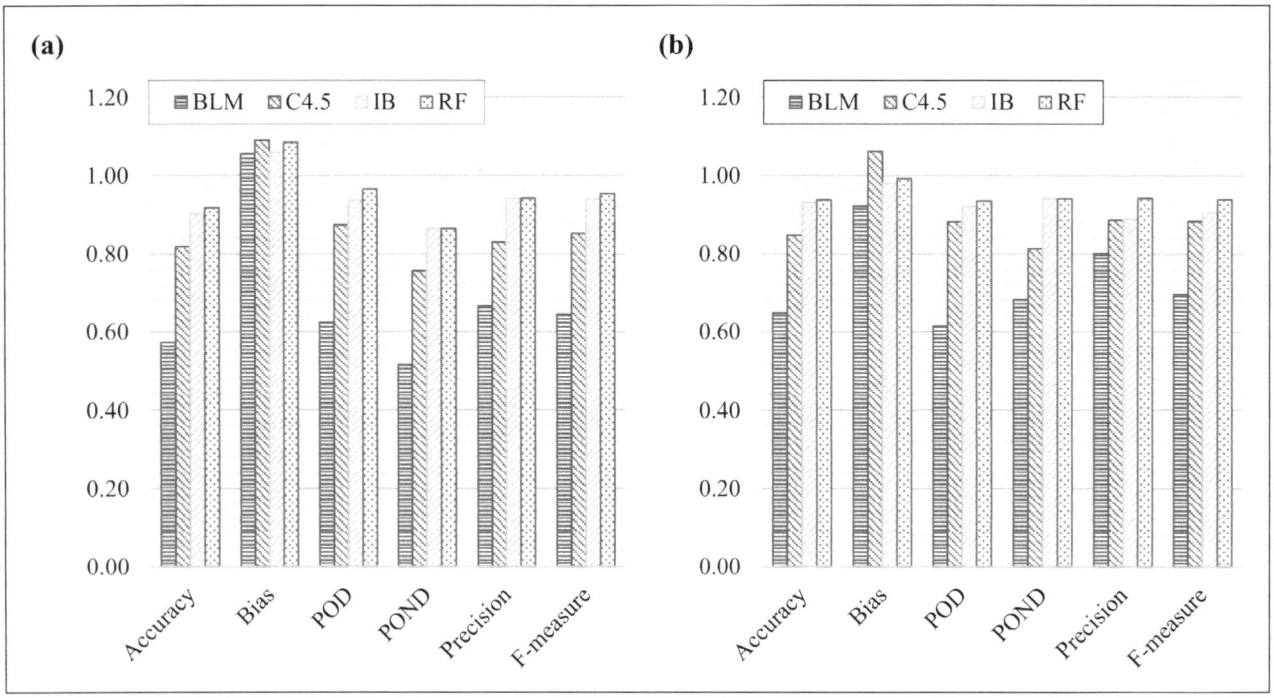

Figure 3. Performance of four models in predicting driver gap decision in (*a*) left-turning and (*b*) right-turning maneuvers.

RF classifiers for left and right turning maneuvers are shown in Table 5. The confusion matrices reveal that data mining models are superior to BLM models in terms of prediction accuracy, which is consistent with previous research (*13*, *14*). Among data mining models, RF models greatly outperformed the C4.5 models, which is consistent with Nagalla et al. (*19*), and slightly outperformed the IB models.

The structure of the C4.5 gap acceptance models for left and right turning maneuvers are shown in Figure 2. These models are a series of "if-then" questions at each node (e.g. GAP <= 1.53 seconds) that allow for pursuing a path down the tree to a leaf, located at the end of the tree, where the driver's gap decision is predicted. Each leaf contains two numbers in parentheses depicting the total number of instances reaching the leaf and the number of misclassified instances, respectively. The variables that the model finds most significant for predicting gap acceptance are found in the questions among the highest nodes of the DT. The DTs found the size of the gap to be the most significant variable in predicting gap acceptance behavior, while the crosswalk type (CRSWLK) and the weather condition (WTHR) variables were not found to be significant and were excluded entirely from the tree.

Figure 3 illustrates the comparison of BLM, C4.5, IB and RF models utilizing skill scores of bias, POD, POND, precision, and F-measure. The data mining models were found to be superior to the BLM models in each of these metrics and in their overall ability to predict gap decision behaviors. The RF models were found to outperform all of the other models in the aforementioned metrics, excepting bias. The values of POD and POND confirmed a similar trend, thus demonstrating that RFs are more accurate at predicting accepted and rejected gaps when compared with C4.5 and IB models. The F-measure values for the RF models are closer to the optimal value of 1 than the C4.5 and IB models. However, for the left-turn models, the bias score for the IB model was closer to 1 than for the RF model. This indicates that the RF model tends to over-estimate the number of predicted accepted gaps to a greater extent when compared with the IB model. It is noteworthy that the discrepancy in the RF and IB model accuracy scores was not significant.

Summary and Conclusions

This paper utilized the C4.5, IB, and RF data mining methods and the BLM statistical modeling technique to classify and predict drivers' gap decisions in unprotected left and right turning maneuvers at four-legged signalized intersections. After conducting a literature review on factors that could potentially influence gap acceptance decisions, the following variables were considered in the development of the driving scenarios: drivers' characteristics (age, gender, and driving experience), presence of pedestrians in or nearby the crosswalk, number of lanes, different crosswalk configurations (ladder or standard), and contextual conditions (heavy fog, and night conditions). Also, the influence of multiple pedestrians crossing shoulder-to-shoulder, as opposed to individually, was investigated for right-turn scenarios. Ten driving simulation scenarios were developed to test for the significance these factors might play on gap acceptance behavior. Data from 64 participants were collected to develop the models. The skill scores of accuracy, bias, POD, POND, precision, and F-measure were utilized to evaluate the performance of the models. Results indicated that the performance of data mining models was superior to BLMs in predicting drivers' gap decision for both left and right turning maneuvers. The best prediction performance was achieved by the RF models, followed by the IB models, and lastly the C4.5 models. Moreover, RF models lack problems related to over-fitting and can manage numerous determinants without requiring any parameter selection. Given its performance, the RF model is recommended to be an alternative model for predicting drivers' gap acceptance behaviors. Although the C4.5 models exhibited lower prediction accuracies than their data mining counterparts, these models had superior predictive capabilities when compared with the BLM models while remaining easily interpretable. The DTs, produced by the C4.5 method, provided visual outputs that reveal the relative importance of variables in predicting the drivers' gap acceptance decisions.

The paper mainly encountered limitations relating to the model-building capabilities of the driving simulation authoring software. The software only had a limited number of intersections available and didn't allow for the modification of certain parameters (e.g. adding or modifying the size of a median, adding glare). The simulator's rain conditions were unable to reproduce the same problems that a driver would normally expect to encounter (e.g. rain on the windshield, ponding of water). Future research focusing on the addition of these parameters will potentially add to the output produced from this paper while also identifying how and when these parameters will affect gap acceptance behavior.

Acknowledgments

This project was supported by the United States Department of Transportation (U.S.DOT) grant DTRT13-G-UTC42 and administered by the Center for Accessibility and Safety for an Aging Population (ASAP) at the Florida State University (FSU), Florida A&M University (FAMU), and University of North Florida (UNF). We also thank the Florida Department of Transportation (FDOT) for providing the crash data.

Author Contributions

The authors confirm contribution to the paper as follows: study conception and design: Somayeh Mafi, Yassir AbdelRazig.data collection: Somayeh Mafi, Ryan Doczy. analysis and interpretation of results: Somayeh Mafi, Yassir AbdelRazig, Ryan Doczy. draft manuscript preparation: Somayeh Mafi, Yassir AbdelRazig, Ryan Doczy. All authors reviewed the results and approved the final version of the manuscript.

References

1. Alexander, J., P. Barham, and I. Black. Factors Influencing the Probability of an Incident at a Junction: Results from an Interactive Driving Simulator. *Accident Analysis & Prevention*, Vol. 34, No. 6, 2002, pp. 779–792.
2. Spek, A., P. A. Wieringa, and W. Janssen. Intersection Approach Speed and Accident Probability. *Transportation Research Part F: Traffic Psychology and Behaviour*, Vol. 9, No. 2, 2006, pp. 155–171.
3. Chovan, J. D., L. Tijerina, J. H. Everson, J. A. Pierowicz, and D. L. Hendricks. *Examination of Intersection, Left Turn across Path Crashes and Potential IVHS Countermeasures*. Report DOT-VNTSC-NHTSA-94-4. Volpe National Transportation Systems Center, Cambridge, MA, 1994, p. 50.
4. Hamed, M., S. Easa, and R. Batayneh. Disaggregate Gap-Acceptance Model for Unsignalized T-Intersections. *Journal of Transportation Engineering*, Vol. 123, No. 1, 1997, pp. 36–42.
5. Ball, K., and C. Owsley. Identifying Correlates of Accident Involvement for the Older Driver. *Human Factors*, Vol. 33, No. 5, 1991, pp. 583–595.
6. Ball, K., C. Owsley, M. E. Sloane, D. L. Roenker, and J. R. Bruni. Visual Attention Problems as a Predictor of Vehicle Crashes in Older Drivers. *Investigative Ophthalmology & Visual Science*, Vol. 34, No. 11, 1993, pp. 3110–3123.
7. Owsley, C., G. McGwin, Jr., , and K. Ball. Vision Impairment, Eye Disease, and Injurious Motor Vehicle Crashes in the Elderly. *Ophthalmic Epidemiology*, Vol. 5, No. 2, 1998, pp. 101–113.
8. Kazazi, J., S. Winkler, and M. Vollrath. The Influence of Attention Allocation and Age on Intersection Accidents. *Transportation Research Part F: Traffic Psychology and Behaviour*, Vol. 43, 2016, pp. 1–14.
9. Koepsell, T., L. McCloskey, M. Wolf, A. V. Moudon, D. Buchner, J. Kraus, and M. Patterson. Crosswalk Markings and the Risk of Pedestrian–Motor Vehicle Collisions in Older Pedestrians. *Jama*, Vol. 288, No. 17, 2002, pp. 2136–2143.
10. Zegeer, C. V., J. R. Stewart, H. H. Huang, P. A. Lagerwey, J. Feaganes, and B. Campbell. *Safety Effects of Marked Versus Unmarked Crosswalks at Uncontrolled Locations: Final Report and Recommended Guidelines*. FHWA-HRT-04-100. McLean, Va., 2005, p. 112.
11. Alhajyaseen, W. K., M. Asano, and H. Nakamura. Left-turn Gap Acceptance Models Considering Pedestrian Movement Characteristics. *Accident Analysis & Prevention*, Vol. 50, 2013, pp. 175–185.
12. Abou-Henaidy, M., S. Teply, and J. D. Hunt. Gap Acceptance Investigations in Canada. *Proceedings of the 2nd International Symposium on Highway Capacity*, Vol. 1, 1994, pp. 1–19.
13. Pant, P. D., and P. Balakrishnan. Neural Network for Gap Acceptance at Stop-Controlled Intersections. *Journal of Transportation Engineering*, Vol. 120, No. 3, 1994, pp. 432–446.
14.Pawar, D. S., G. R. Patil, A. Chandrasekharan, and S. Upadhyaya. Classification of Gaps at Uncontrolled Intersections and Midblock Crossings Using Support Vector Machines. *Transportation Research Record: Journal of the Transportation Research Board*, 2015. 2515: 26–33.
15. Yan, X., and E. Radwan. Effect of Restricted Sight Distances on Driver Behaviors During Unprotected Left-Turn Phase at Signalized Intersections. *Transportation Research Part F: Traffic Psychology and Behaviour*, Vol. 10, No. 4, 2007, pp. 330–344.
16Fatehnia, M., and G. Amirinia. A Review of Genetic Programming and Artificial Neural Network Applications in

Pile Foundations. *International Journal of Geo-Engineering*, Vol. 9, No. 2, 2018, pp. 1–20.
17. Mafi, S., and G. Amirinia. Forecasting Hurricane Wave Height in Gulf of Mexico Using Soft Computing Methods. *Ocean Engineering*, Vol. 146, 2017, pp. 352–362.
18. Mafi, S., A. Yeganeh-Bakhtiary, and M. H. Kazeminezhad. Prediction Formula for Longshore Sediment Transport Rate with M5'Algorithm. *Journal of Coastal Research*, Vol. 2, No. 65, 2013, pp. 2149–2154.
19. Nagalla, R., P. Pothuganti, and D. S. Pawar. Analyzing Gap Acceptance Behavior at Unsignalized Intersections Using Support Vector Machines, Decision Tree and Random Forests. *Procedia Computer Science*, Vol. 109, 2017, pp. 474–481.
20. Mayhew, D. R., H. M. Simpson, K. M. Wood, L. Lonero, K. M. Clinton, and A. G. Johnson. On-Road and Simulated Driving: Concurrent and Discriminant Validation. *Journal of Safety Research*, Vol. 42, No. 4, 2011, pp. 267–275.
21. Shechtman, O., S. Classen, K. Awadzi, and W. Mann. Comparison of Driving Errors between On-The-Road and Simulated Driving Assessment: A Validation Study. *Traffic Injury Prevention*, Vol. 10, No. 4, 2009, pp. 379–385.
22. Underwood, G., D. Crundall, and P. Chapman. Driving Simulator Validation with Hazard Perception. *Transportation Research Part F: Traffic Psychology and Behaviour*, Vol. 14, No. 6, 2011, pp. 435–446.
23. De Winter, J., P. Van Leuween, and P. Happee. Advantages and Disadvantages of Driving Simulators: A Discussion. *Proceedings of Measuring Behavior*, Citeseer, Utrecht, Netherlands, 2012, pp. 47–50.
24. Matas, N. A., T. Nettelbeck, and N. R. Burns. Dropout during a Driving Simulator Study: A Survival Analysis. *Journal of Safety Research*, Vol. 55, 2015, pp. 159–169.
25. Bhargava, N., G. Sharma, R. Bhargava, and M. Mathuria. Decision Tree Analysis on J48 Algorithm for Data Mining. *Proceedings of International Journal of Advanced Research in Computer Science and Software Engineering*, Vol. 3, No. 6, 2013.
26. Quinlan, J. R. *C4. 5: Programs for Machine Learning*. Morgan Kaufmann, 1993.
27. Breiman, L. Random Forests. *Machine Learning*, Vol. 45, No. 1, 2001, pp. 5–32.
28. Aha, D. W., D. Kibler, and M. K. Albert. Instance-Based Learning Algorithms. *Machine Learning*, Vol. 6, No. 1, 1991, pp. 37–66.
29. Gattis, J., and S. T. Low. Gap Acceptance at Atypical Stop-Controlled Intersections. *Journal of Transportation Engineering*, Vol. 125, No. 3, 1999, pp. 201–207.
30. Witten, I. H., E. Frank, M. A. Hall, and C. J. Pal. *Data Mining: Practical Machine Learning Tools and Techniques*. Morgan Kaufmann, 2016.

The Standing Committee on Safety Data, Analysis and Evaluation (ANB20) peer-reviewed this paper (18-03400).

The opinions, results, and findings expressed in this paper are those of the authors and do not necessarily represent the views of the United States Department of Transportation, the Florida Department of Transportation, the center for Accessibility and Safety for an Aging Population, the Florida State University, the Florida A&M University, or the University of North Florida.

TRR
JOURNAL OF THE TRANSPORTATION RESEARCH BOARD

Article

Transportation Research Record
2018, Vol. 2672(38) 171–183

DOI: 10.1177/0361198118794292
journals.sagepub.com/home/trr
SAGE

Machine Learning Methods to Analyze Injury Severity of Drivers from Different Age and Gender Groups

Somayeh Mafi[1], Yassir AbdelRazig[1], and Ryan Doczy[1]

Abstract

Access to non-biased and accurate models capable of predicting driver injury severity of collision events is vital for determining what safety measures should be implemented at intersections. Inadequate models can underestimate the potential for collision events to result in driver fatalities or injuries, which can lead to improperly assessing the safety criteria of an intersection. This study investigates how injury severity differs between drivers of various ages and gender groups using cost-sensitive data-mining models. Previous research efforts have used machine learning methods for predicting injury severity; however, these studies did not consider the consequences (cost) of incorrect predictions. This paper addresses this shortfall by considering the monetary cost of incorrect injury severity predictions when developing C4.5, instance-based (IB), and random forest (RF) machine-learning models. One model of each method was developed for four distinct cohorts of drivers (i.e., younger males, younger females, older males, and older females). Each model considered a selection of driver, vehicular, road/traffic, environmental, and crash parameters for determining if they significantly influenced driver injury severity. A five-year period of two-vehicle crash data collected at signalized intersections in the metropolitan area of Miami, Florida was used in the models. Results indicated that cost-sensitive learning classifiers were superior to regular classifiers at accurately predicting injuries and fatalities of crashes. Among cost-sensitive models, RF outperformed C4.5 and IB models in predicting driver injury severity for four groups of drivers. The models displayed substantial differences in injury severity determinants across the age/gender cohorts.

The ability to predict the injury severity of collision events and identify significant risk parameters has received much attention in an effort to reduce the number of fatalities and injuries resulting from vehicular collisions at intersections. The random nature associated with injury severity, resulting from vehicular collisions, leads to the use of statistical models (e.g., logit, ordered probit models) designed to predict injury severity and identify parameters influencing the severity of collision events. These statistical models require a predefined functional form and can also be negatively influenced by outliers and missing values in the dataset. By contrast, data-mining models are non-parametric tools capable of managing outliers and missing values, while also being able to select the most influential predictors to classify the target variable.

Few studies have implemented data mining models for predicting the injury severity of collision events, and none have considered the cost of making incorrect predictions. Moreover, differences in injury severity determinants across age/gender groups are still not fully understood. Given the limitations of existing research resulting from previous data analysis methods and the limited number of risk parameters, a knowledge gap exists in accurately predicting crash severity. This paper presents cost-sensitive C4.5, random forest (RF), and instance-based (IB) models with the following objectives: (a) to predict driver injury severity at signalized intersections; and (b) to identify the risk factors affecting injury severity across age/gender cohorts. This paper contributes to the field of intersection safety through its implementation of the IB method and cost-sensitive data mining models to predict injury severity, the use of which allows for improved accuracy and lower bias than was previously possible.

[1]Department of Civil & Environmental Engineering, FAMU-FSU College of Engineering, Tallahassee, FL

Corresponding Author:
Address correspondence to Somayeh Mafi: sm14ay@my.fsu.edu

Literature Review

The immense economic and societal expense related to improper traffic safety measures (e.g., medical costs, excessive delays for road users, and property damage) have highlighted the importance of the field of traffic safety. Two important aspects of traffic safety, predicting injury severity and identifying related influential parameters, have been the focus of previous research. Various approaches have been developed to investigate injury severity, empirically and methodologically (see [*1–3*] for an exhaustive review of these studies).

From an empirical viewpoint, several studies have investigated the risk parameters that might affect injury severity. Several studies identified driver characteristics (e.g., age, alcohol consumption, seatbelt usage) as significant parameters in predicting injury severity (*4–14*). The influence of roadway characteristics (e.g., area type, speed limit, average annual daily traffic [AADT]) on injury severity has been explored previously (*4, 7, 10, 11, 13, 15–17*). Previous studies discovered that environmental features (e.g., weather condition, roadway surface condition, light condition, crash time) influenced the severity of crashes (*6, 8, 9, 12, 15, 16, 18–20*). Additional research found vehicle type to be an influential parameter in predicting injury severity (*5, 6, 12, 14, 18, 19, 21, 22*), while others found that crash characteristics (e.g., contributing factors, driver/vehicle maneuver, driver ejection from vehicle) influence the injury severity (*12, 16, 21, 22*).

From a methodological viewpoint, various statistical approaches have been employed to predict the injury severity of crashes. Some studies developed binary discrete outcome models such as binary logit and probit (*21, 23–25*). Binary discrete outcome models have evolved to take several outcomes to consider several injury severity classes as dependent variables. Some multiple outcome models, such as multinomial logit models (*26–31*) and nested logit models (*18, 24, 32–37*), don't take the ordering of injury outcome into account, while others lack this limitation, such as the ordered probit and logit model (*5, 6, 9, 16, 19, 24, 31, 38–42*). Kockelman and Kweon (*5*) applied ordered probit models using single- and two-vehicle crash data to investigate the risk parameters influencing the probabilities of a crash resulting in a severe injury or fatality. Data mining methods have recently been applied in different engineering fields (*43–45*) including traffic safety. Some studies utilized Artificial Neural Networks (ANN) to estimate injury severity (*10, 11, 22, 46–48*). Abdel-Aty and Abdelwahab (*11*) utilized multilayer perceptron and fuzzy adaptive resonance theory neural networks to predict driver injury severity, finding that ANNs exhibited superior prediction accuracies compared with traditional statistical models. Other studies implemented tree-based models to analyze injury severities (*21, 46, 47, 49–53*). Chang and Yeh (*54*) developed a classification and regression tree (CART) model to predict injury severity based on driver, vehicle, highway, environmental, and crash characteristics. They found CART to be a good alternative for predicting injury severity when compared with traditional statistical models, due to its ability to search for the best input variables and threshold values.

Methodology

Figure 1 outlines this study's methodology. Using previous literature, this paper identified multiple subgroups of parameters that are potentially influential in predicting driver injury severity during a collision event. These subgroups include driver, vehicle, roadway/traffic, environmental, and crash characteristics. A five-year period of two-vehicle crash data at signalized intersections in the metropolitan area of Miami, Florida was used.

Several studies illustrated that injury severity varies considerably across different age/gender groups (*12, 26, 55, 56*). This paper looks to evaluate these differences by developing separate models based on gender/age subsets of the collision data. This study separated subjects into two age-related sub-groups, younger and older (+ 65), based on previous research (*56, 57*). The four cohorts considered in this study are older female (O-F), older male (O-M), younger female (Y-F), and younger male (Y-M). C4.5, IB, and RF were implemented for each cohort of drivers, after improving their prediction performance via cost sensitivity analysis. For developing these models, the open source software package Weka (*58*) was used. The most accurate models and the most significant variables associated with each driver group will be presented for predicting injury severity.

Models Characteristics

C4.5. Decision Tree (DT) is a popular machine learning method with substantial processing power. It presents rules of classification, which are simple and interpretable by humans. DTs are capable of processing outliers or datasets with missing values, and they prevent overfitting by utilizing pruning (*59*). The C4.5 algorithm, used for generating a type of DT, was first developed by Quinlan (*60*) utilizing a tree-shaped graph comprised of a root node, internal nodes that represent test conditions, branches, and leaf nodes that represent the final classes. Generally, the tree starts from the root nodes, moves downward through branches and internal nodes, and ends at the leaves. Classification in a DT is defined as mapping from a set of features (inputs) to a particular class (output). At each node (except the leaves), the division into classes is carried out based on the features that

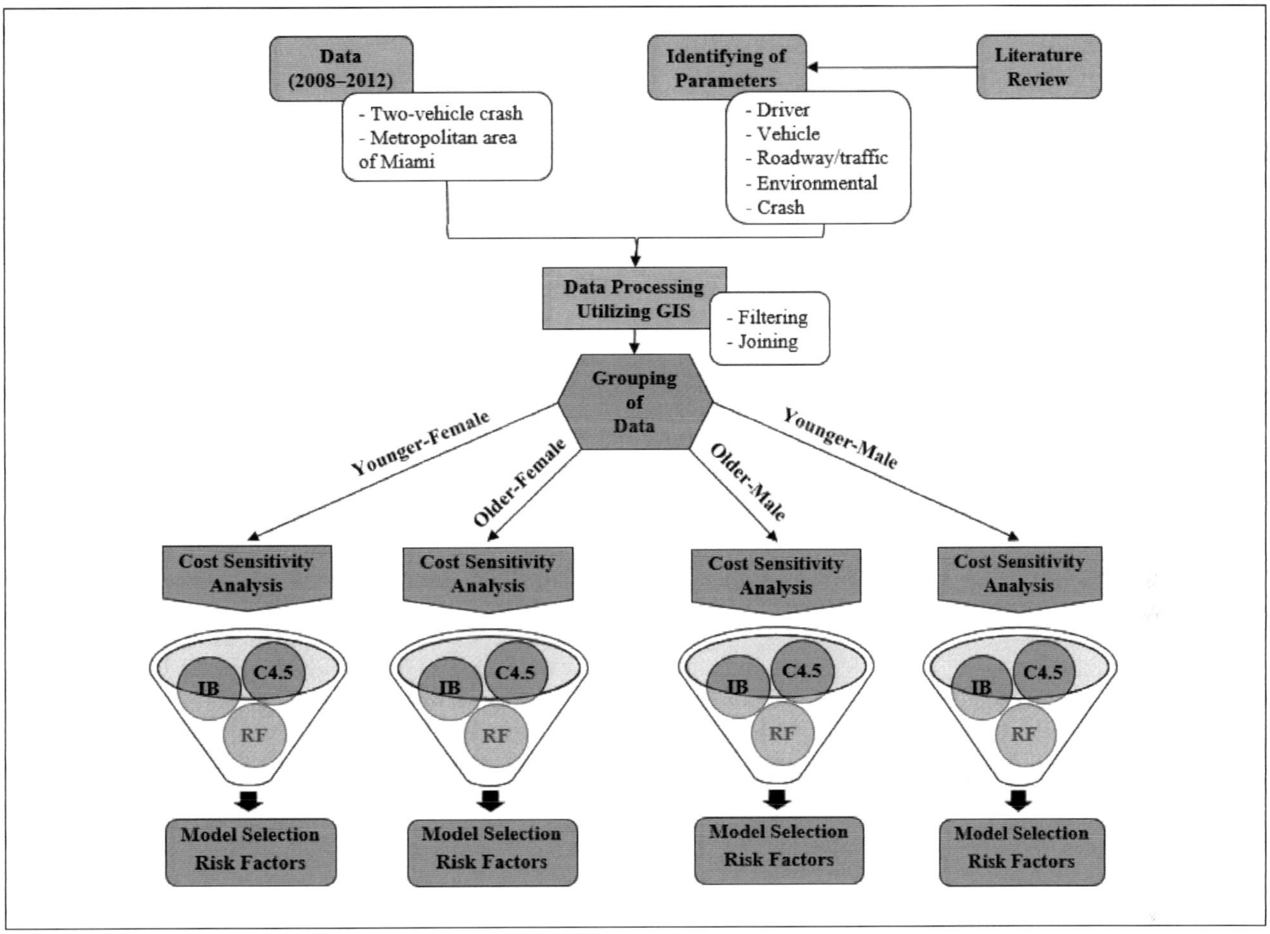

Figure 1. Methodology of the study.

best divide that node's set of samples into subsets. The attribute with the greatest normalized information gain is selected to make the decision. This procedure is applied to each split subset of the data recursively until that subset's size is small enough, or all the instances in the current subset belong to the same class.

Random Forest (RF). Despite many advantages attributed to the DT algorithm, it has some deficiencies. DT sorts all numerical features whenever it decides to split a node. This procedure becomes taxing in terms of efficiency, running time, or memory size, especially with large datasets. For overcoming both this issue and, in some cases, over-fitting problems, Breiman (*61*) introduced the concept of RF. RF is comprised of multiple trees generated from randomly selected samples of training data. Each tree is grown to the largest possible extent (without pruning). Prediction accuracy of the RF is calculated by aggregating the results of DTs, which is the majority vote for classification or averaging for regression. Every tree in a RF is built by the random sampling of a training data subset with replacement (the notion of Bagging) and randomly sampling a variable subset (the notion of randomization), which results in improved prediction power and greater model efficiency. In a RF, m features are chosen at each node from a total of M input features ($m < < M$). The best split on these m features is utilized for splitting the node. The value of m remains constant during the forest generation. RF generally retains the advantages of the DT while producing superior results, when compared with individual tree classifiers such as C4.5.

Instance-Based (IB). In the nearest-neighbor instance-based (IB) learning method, all attributes have equal contributions to the decision-making process. An advantage of IB learning over most other machine-learning schemes is that new instances can be added to the training set at any time. In IB models, the training instances are stored *literatim* and a distance function, usually Euclidean distance, is used to specify the member of the training set that is closest to an unknown test instance. When the closest training instance has been specified, its class is predicted for the test instance. IB learning is simple and efficient; however, it is computationally intensive compared with the other data-mining methods. To make a single prediction for a testing instance, it calculates the distance from every member of the training set and selects the one with the smallest distance. This model was notoriously slow until it was improved by application of kD-trees in the early 1990s (*62*). It considers the training set as a tree-shaped structure, such as a kD-tree. This is a binary tree that splits the input space using a hyperplane and then splits each part again, recursively. All divisions are done parallel to one of the axes, in the two-dimensional case. The structure is named a kD-tree since it stores a set of instances in k-dimensional space, in which k is the number of attributes. In an attempt to find the nearest neighbor of the specified target instance, the model flows from the tree's root down to the zone containing the target instance.

Making the Models Cost-Sensitive

Machine learning methods implicitly consider equal error costs for all incorrect predictions; however, this assumption doesn't always hold true. In an example provided by Witten et al. (*63*), the lost-business cost for a bank lending to a defaulter is far greater when compared with the cost of refusing a loan to a non-defaulter. In multiclass problems, costs can be presented in the form of a square matrix whose size is the number of classes and the off-diagonal elements represent the cost of incorrect classification. By considering different costs for incorrect predictions, the success rate of each model is replaced by its average cost. A general way to make any classifier cost sensitive is to generate training data with a different proportion of instances in each class. For example, in a two-class problem with outcomes of "yes" and "no," where the cost of errors for the "no" instances is tenfold those of the "yes" instances, the number of "no" instances was increased artificially by a factor of 10 and the resulting dataset was used for training. The resulting classifier is biased toward avoiding errors on the "no" instances since such errors are effectively penalized tenfold (*63*).

Training and Testing of the Models

To evaluate the performance of classifiers, a stratified holdout procedure was used. In this method, the random sampling was carried out in a manner that guaranteed each class (i.e., "no-injury," "minor" injury, and "severe" injury) was appropriately represented in both training and test sets. The stratified holdout method also allows for swapping the roles of the testing and training data to reduce the effect of uneven representation in training and testing data sets. In n-fold cross-validation, the whole data set is divided into an approximately equal fixed number of folds (n). Each partition is held out in turn (test data) and the classifier trained on the remaining data; then its error rate is computed on the test set. Therefore, the classification is performed a total of n times on different training sets which have many commonalities. The overall error is calculated by taking the average of n error estimates. Based on theoretical evidence, 10 is approximately the most appropriate number of folds to get the best estimate of error. Previous tests have also indicated that the use of stratification improves the results of the models (*63*).

Study Data

In this study, the Miami metropolitan area in Florida was selected for analysis due to its population size and concentration. The collision data was provided by the Florida Department of Transportation (FDOT) databases (*64*). The data was obtained in shape-file format, which included coded information on reported crashes during the period from 2008 to 2012. ESRI's ArcGIS software was used to filter the crash data to only include two-vehicle crashes occurring at signalized intersections on state roadways. The target variable was the injury severity (1: "no-injury"; 2: "minor" injury; 3: "severe" injury) of each driver involved in a crash. The independent and dependent variables have either numerical values or categorical codes, both of which are explained in Table 1. Incomplete and questionable data records were screened out, resulting in a database of 32,730 records. The driver injury distribution was 28,309 (86.5%) "no-injury", 4,205 (12.8%) "minor" injury, and 216 (0.7%) "severe" injury.

Results and Discussions

Since most traffic crashes resulted in "no-injury," classifiers were biased toward predicting each collision event to result in "no-injury." To remove this bias, the cost of various injury level was considered based on the KABCO injury classification scale produced by the Federal Highway Administration (FHWA) (*65*) (Table 2). Fatalities and incapacitating injuries, terms used by the FHWA, are represented in this study by the term "severe" injury level. In the same vein, the term "minor" injury level is used in this study to be representative of non-incapacitating and possible injuries. For the combined terms (e.g., fatality and incapacitating as "severe"), the higher of the two KABCO costs controlled the

Table 1. The Definition and Descriptive Statistics of Variables

Variable	Type	Description	Mean	SD	Min	Max
Explanatory variables						
Driver characteristics						
Sobriety condition	Qualitative	1: drinking; 2: not drinking	1.980	0.139	1	2
Use of seat belt	Qualitative	1: yes; 2: no	1.019	0.137	1	2
Air bag deployment	Qualitative	1: yes; 2: no	1.990	0.101	1	2
Physical defects	Qualitative	1: yes; 2: no	1.987	0.111	1	2
Roadway/traffic characteristics						
Speed limit	Continuous	Posted speed limit	39.951	6.088	15	70
Work zone involvement	Qualitative	1: yes; 2: no	1.960	0.196	1	2
Area type	Qualitative	1: rural; 2: urban	1.808	0.394	1	2
AADT	Continuous	Average annual daily traffic	42,275	19,107	2,035	194,500
Median width	Continuous	The width of the median in feet	18.190	10.271	0	70
Road surface width	Continuous	Total road width of all through lanes in a single direction in feet	31.280	7.512	10	72
Visibility	Qualitative	1: not obstructed; 2: obstructed by stopped/parked car, load on vehicle, fixed objects or etc.	1.028	0.164	1	2
Environmental features						
Hour	Qualitative	1: peak; 2: else	1.779	0.415	1	2
Day of week	Qualitative	1: Monday; 2: Tuesday; 3: Wednesday; 4: Thursday; 5: Friday; 6: Saturday; 7: Sunday	3.942	1.964	1	7
Month	Qualitative	Month of accident (1–12)	6.494	3.454	1	12
Weather	Qualitative	1: clear; 2: else	1.252	0.434	1	2
Light	Qualitative	1: daylight; 2: dusk/dawn; 3: dark (street light); 4: dark (no street light)	1.644	0.926	1	4
Roadway surface	Qualitative	1: dry; 2: else	1.144	0.351	1	2
Vehicle characteristics						
Vehicle type	Qualitative	1: automobile; 2: van; 3: light/medium truck	1.376	0.740	1	3
Crash characteristics						
Ejection of driver from vehicle	Qualitative	1: yes; 2: no	1.996	0.067	1	2
Lane departure event	Qualitative	1: yes; 2: no	1.815	0.388	1	2
Driver/vehicle maneuver	Qualitative	Vehicle maneuver before crash occurrence (1: straight ahead; 2: slowing/stopped/stalled; 3: making left turn; 4: backing; 5: making right turn; 6: changing lanes; 7: making u-turn)	2.028	1.417	1	7
Contributing circumstance	Qualitative	1: no improper driving/action; 2: careless driving; 3: failed to yield right-of-way; 4: improper backing; 5: improper lane change; 6: improper turn; 7: followed too closely; 8: disregarded traffic control signal	3.844	2.365	1	8
Target variable						
Driver injury severity level	Qualitative	1: no-injury; 2: minor injury; 3: severe injury	1.142	0.368	1	3

Note: SD = standard deviation; Min = minimum; Max = maximum.

representative costs (i.e., fatality cost was assigned for "severe" injury level), to be more conservative.

The driver injury severity cost matrix (Table 3) was formed based on the information from Table 2. The cost-sensitive classifier avoids predicting a "no-injury" instead of a "severe" or a "minor" injury since such errors are effectively penalized 541.74 ($4,008,900/$7,400) and 10.68 ($79,000/$7,400)-fold, respectively. This classifier is biased toward avoiding errors on predicting "minor" injuries instead of "severe" injuries since such error is penalized 50.75 ($4,008,900/$79,000)-fold.

In the implemented models, the dimensions of the training and testing data are reduced by removing non-significant attributes before applying classifiers (using

Table 2. Classification of Crash Costs Based on Injury Severity Level by FHWA and by This Study

Injury severity level	Crash costs
FHWA	
Fatality (K)	$4,008,900
Incapacitating injury (A)	$216,000
Non-incapacitating injury (B)	$79,000
Possible injury (C)	$44,900
PDO (O)	$7,400
This study	
Severe	$4,008,900
Minor	$79,000
No-injury	$7,400

Table 3. Driver Injury Severity Cost Matrix

	Predicted		
Observed	No-injury	Minor	Severe
No-injury	0	1	1
Minor	10.68	0	1
Severe	541.74	50.75	0

the AttributeSelectedClassifier filter in Weka). Area type (e.g., urban, rural), median width, and crash contributing factors were identified as key determinants of driver injury severity in all four cohorts of drivers. Seatbelt usage and physical disability are important determinants associated only with the older driver cohort; whereas, airbag deployment is only significant for the younger driver cohort. Additionally, the driver/vehicle maneuver variable plays an influential role for both female cohorts.

Table 4 shows the confusion matrices of C4.5, IB, and RF classifiers for four driver groups with (CSLC) and without (C) cost-sensitive learning. The bolded numbers represent the total percentages of "no-injury", "minor" and "severe" injuries correctly predicted. On examining the confusion matrix of the RF model for the O-F driver group, the implementation of cost-sensitive classifiers reduced prediction accuracy from 87.15% to 84.39% but increased the correctly predicted percentage of "minor" and "severe" injuries from 17.60% to 20.66% and from 0.15% to 0.61%, respectively.

The classifiers without cost-sensitive learning averaged accuracies 8.6% higher than their cost-sensitive counterparts. Despite this drawback, the cost-sensitive classifiers are superior in their ability to accurately predict "minor" and "severe" injury events. Table 5 provides the percentages that the severity of an injury event was correctly predicted (averaged across all cohorts). Comparing the cost-sensitive and regular classifiers across all models reveal that the cost-sensitive classifiers averaged 23.5% and 42.4% better accuracies when predicting collision events resulting in "minor" or "severe" injuries, respectively, and 14% worse accuracies when predicting "no-injury" events. The prevalence of "no-injury" events leads to the reduced overall accuracies produced by the cost-sensitive classifiers, however, these classifiers have made strides toward being able to accurately predict injury severities of collision events.

When averaged across the four cohorts, the cost-sensitive RF models maintained an accuracy of 83.4%. This outperformed both the IB and C4.5 models, which averaged accuracies of 82.8% and 75.7%, respectively. Despite the lower accuracies associated with the C4.5 models, this method doesn't suffer from the same limitations inherent to RF and IB models. The C4.5 method provides a visual output that reveals the relative importance of variables in predicting the outcome of an event. The structure of the C4.5 model for the Y-F cohort is shown in Figure 2. This model can be used to predict the driver injury severity of a collision event, located at the end of the tree, by moving down the tree and answering the "if-then" questions at each node (e.g., if MED $<=$ 18 Ft). The variables most significant for predicting injury severity are found in the questions among the highest nodes of the decision tree. For example, the Y-F model decision tree found the contributing circumstance factor (CAUSE) to be the most significant variable in predicting driver injury severity. Parameters entirely excluded from a tree played little role in affecting the driver injury severity.

The Geographic Information System (GIS) tool has been used in different scientific fields to visualize the results (*66–68*). Figure 3 is a GIS-based map that displays the results of RF injury severity models for four driver groups in a sample region. The left images show the observed injury severities and the right ones show the predicted injury severities. The location of study area and sample region is shown in Figure 3*a*.

Conclusions

This paper introduces a novel methodology for producing models capable of predicting the driver injury severity of collision events located at intersections. Taking the C4.5, IB, and RF data-mining methods and integrating them with a cost matrix (developed from the FHWA's KABCO injury classification scale), produced models capable of accounting for the greater importance placed on accurately predicting collision events resulting in "minor" or "severe" injuries. Four different cohorts of drivers, divided by gender and age, were modeled separately to identify significant parameters in predicting the driver injury severity of collision events. The data used to train and test the models consisted of collision data

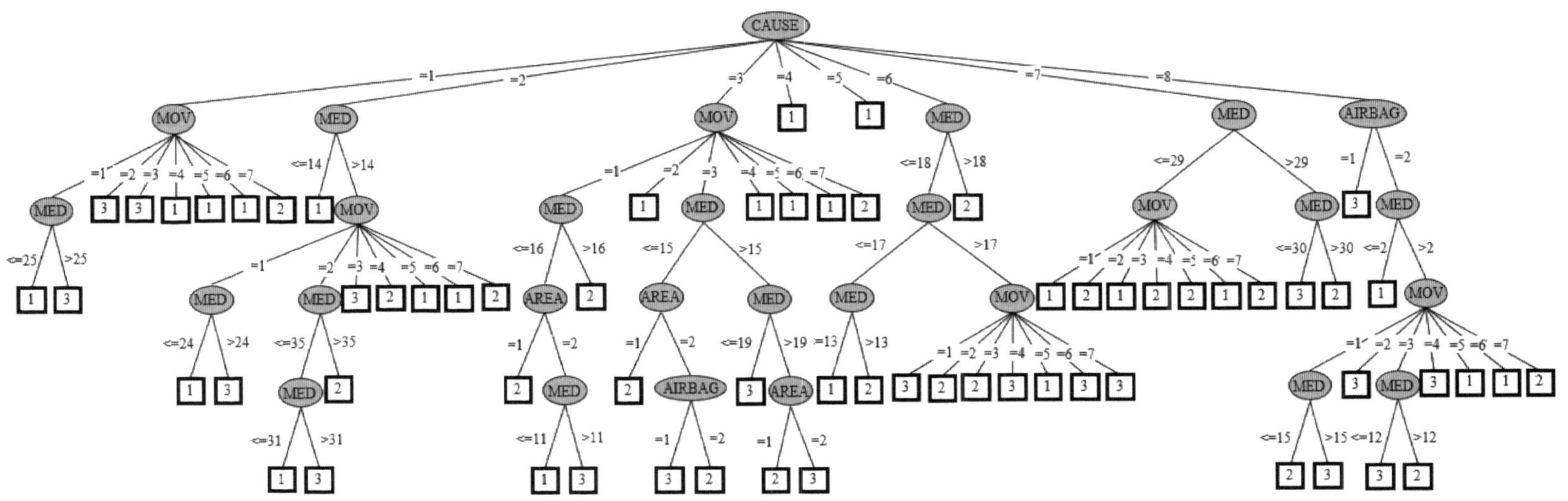

Variable	Symbol	
Air bag deployment	AIRBAG	1: yes; 2: no
Area type	AREA	1: rural; 2: urban
Median width	MED	The width of the median in feet
Contributing factor	CAUSE	1: no improper driving/action; 2: careless driving; 3: failed to yield right-of-way; 4: improper backing; 5: improper lane change; 6: improper turn; 7: followed too closely; 8: disregarded traffic control signal
Driver/vehicle maneuver	MOV	Vehicle maneuver before crash occurrence (1: straight ahead; 2: slowing/stopped/stalled; 3: making left turn; 4: backing; 5: making right turn; 6: changing lanes; 7: making u-turn)
Driver injury severity level.	-	1: no-injury; 2: minor injury; 3: severe injury

Figure 2. The structure of the C4.5 model for driver group of Y-F.

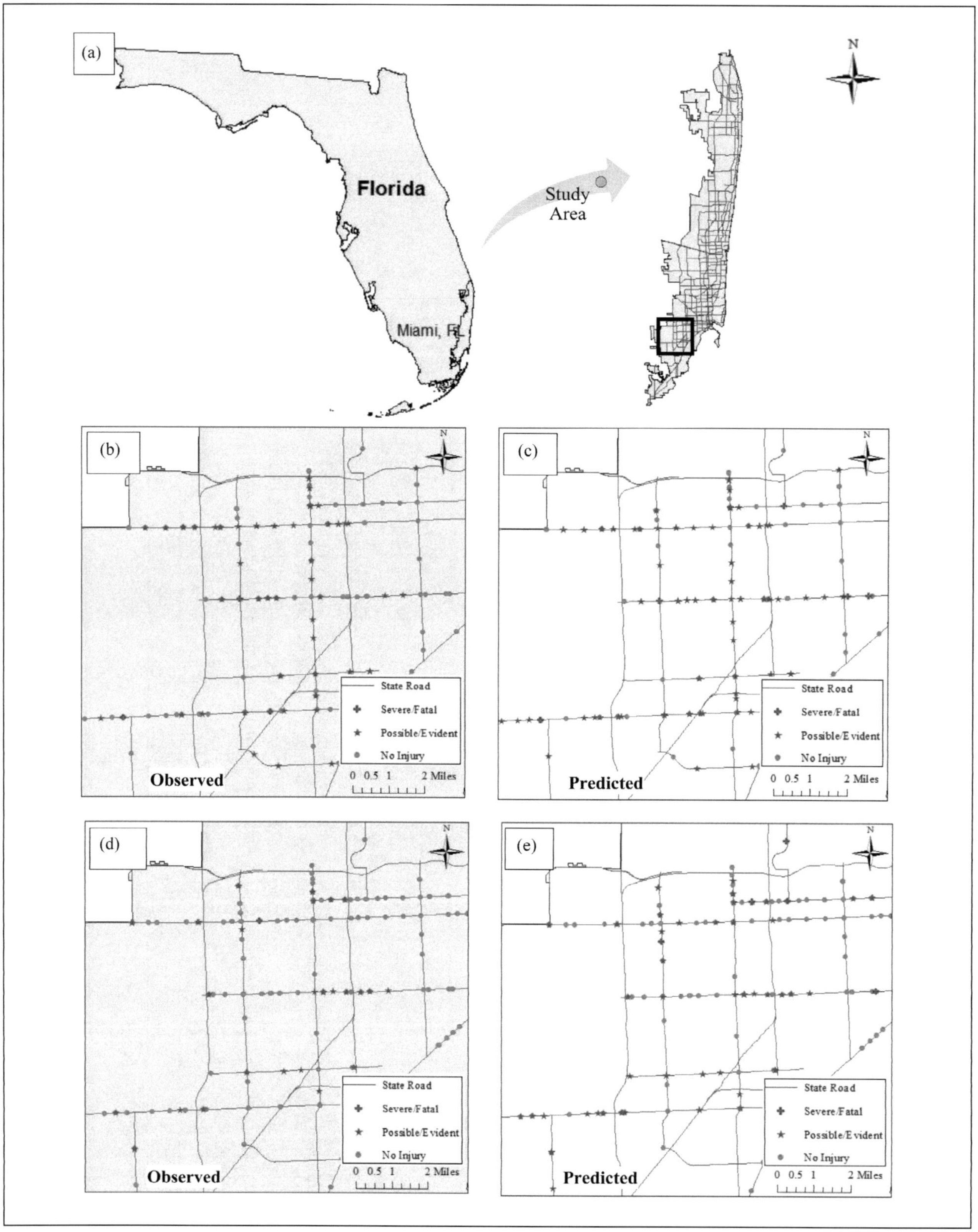

Figure 3. (continued)

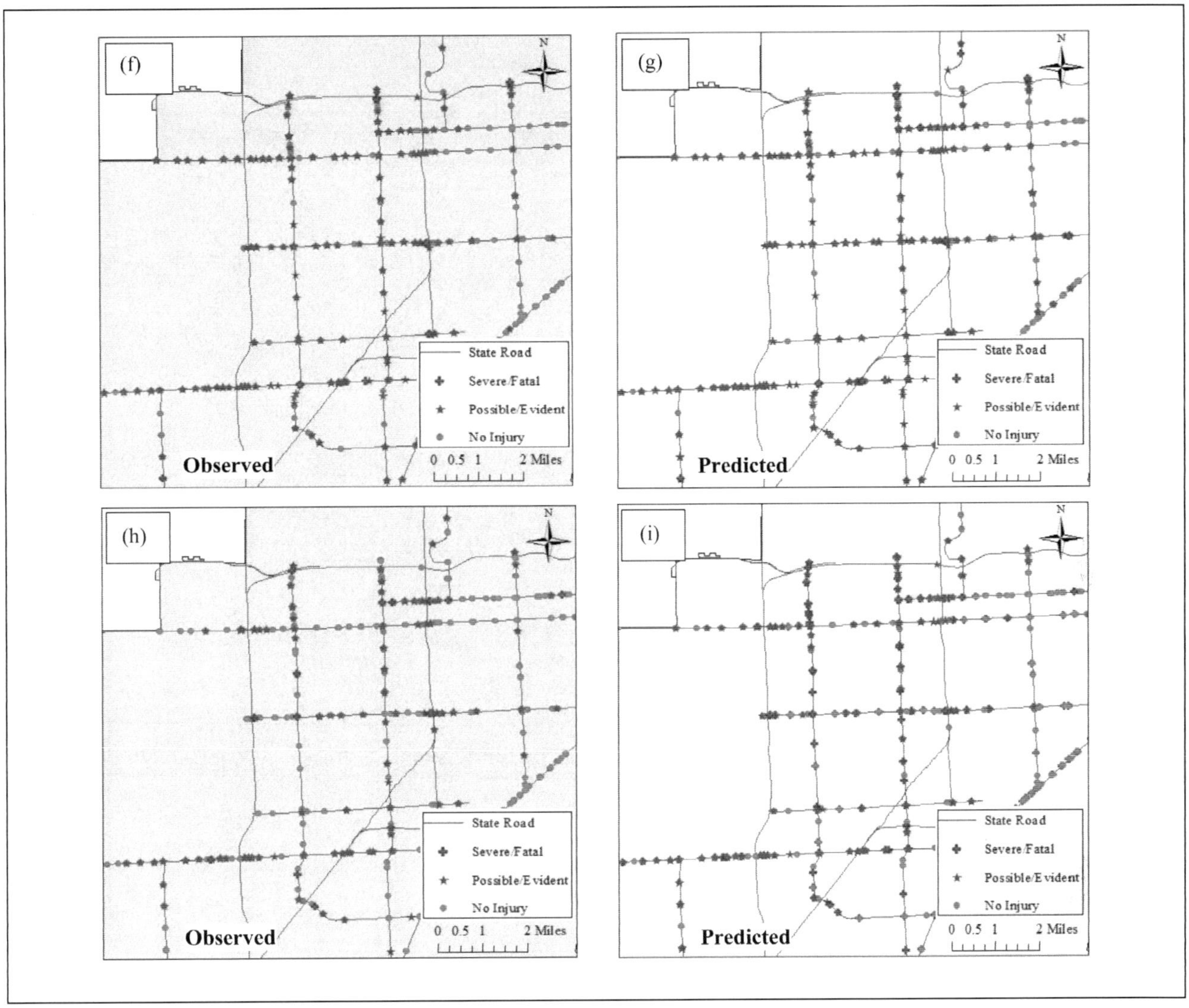

Figure 3. (*a*) Study area and sample region, results of driver injury severities using cost-sensitive RF models in the sample region; (*b*) observed OF; (*c*) predicted OF; (*d*) observed OM; (e) predicted OM; (*f*) observed YF; (*g*) predicted YF; (*h*) observed YM; (*i*) predicted YM.

from the metropolitan area of Miami, Florida, taken over a five-year period.

The cost-sensitive classifiers have overall average accuracies that were 8.6% lower than their regular classifier counterparts, however, they also produced 23.5% and 42.4% higher accuracies when correctly predicting events that would cause "minor" and "severe" driver injury severities. Given that the cost-sensitive classifiers exhibited lower biases compared with regular classifiers and superior accuracies when predicting if a collision event will result in an injury, the authors recommend the use of cost-sensitive classifiers when producing models tasked with injury severity prediction. Of the three data mining methods assessed by this paper, the accuracy of Random Forest (RF) models was superior to the Instance-Based (IB) and C4.5 models across all injury severity levels and cohorts. The authors advocate for the use of RF models when predicting the driver injury severity of collision events. Results illustrated the substantial differences in injury severity determinants across age/gender group combinations. Area type, median width, and crash contributing factors were identified as the key determinants of driver injury severity in all four driver cohorts. Seatbelt usage and physical disability are important determinants associated with the older cohorts, while airbag deployment is significant with the younger cohorts. The driver/

Table 4. Confusion Matrices of C4.5, IB, and RF Classifiers for Four Groups of Driver, with (CSLC) and without (C) Cost-Sensitive Learning (in Percent)

		Predicted decision						
		No-injury		Minor		Severe		
Driver group	Observed data	CSLC	C	CSLC	C	CSLC	C	Total observed
C4.5 models								
O-F	No-injury	55.16	69.17	15.30	5.74	4.44	0.00	74.90
	Minor	4.36	9.41	17.98	14.54	1.61	0.00	23.95
	Severe	0.54	1.07	0.46	0.08	0.15	0.00	1.15
	Total predicted	60.06	79.65	33.74	20.36	6.20	0.00	100.00
	Prediction accuracy = 73.30 (CSLC); 83.71 (C)							
O-M	No-injury	74.93	86.52	12.17	3.67	3.20	0.10	90.30
	Minor	3.51	9.12	5.45	0.00	0.16	0.00	9.12
	Severe	0.31	0.58	0.16	0.00	0.10	0.00	0.58
	Total predicted	78.76	96.22	17.78	3.67	3.46	0.10	100.00
	Prediction accuracy = 80.49 (CSLC); 86.52 (C)							
Y-F	No-injury	58.33	75.50	14.55	3.68	6.33	0.02	79.20
	Minor	3.26	7.65	14.80	12.40	2.01	0.02	20.08
	Severe	0.18	0.49	0.12	0.19	0.41	0.03	0.71
	Total predicted	61.77	83.64	29.48	16.27	8.75	0.07	100.00
	Prediction accuracy = 73.55 (CSLC); 87.93 (C)							
Y-M	No-injury	71.67	90.64	9.29	1.85	11.64	0.11	92.60
	Minor	2.07	4.96	3.57	1.85	1.17	0.01	6.81
	Severe	0.14	0.50	0.11	0.08	0.34	0.01	0.59
	Total predicted	73.88	96.1	12.98	3.78	13.14	0.13	100.00
	Prediction accuracy = 75.58 (CSLC); 92.50 (C)							
IB models								
O-F	No-injury	62.97	69.32	9.03	5.36	2.91	0.23	74.90
	Minor	3.14	6.50	20.20	17.29	0.61	0.15	23.95
	Severe	0.23	0.69	0.38	0.31	0.54	0.15	1.15
	Total predicted	66.34	76.51	29.61	22.96	4.06	0.53	100.00
	Prediction accuracy = 83.70 (CSLC); 86.76 (C)							
O-M	No-injury	78.76	87.31	8.60	2.99	2.94	0.00	90.30
	Minor	3.15	6.40	5.82	2.62	0.16	0.10	9.12
	Severe	0.16	0.52	0.16	0.05	0.26	0.00	0.58
	Total predicted	82.07	94.23	14.58	5.66	3.36	0.10	100.00
	Prediction accuracy = 84.85 (CSLC); 89.93 (C)							
Y-F	No-injury	63.19	75.85	10.11	3.29	5.91	0.06	79.20
	Minor	2.63	6.27	15.94	13.76	1.52	0.05	20.08
	Severe	0.20	0.44	0.09	0.17	0.42	0.11	0.71
	Total predicted	66.02	82.56	26.14	17.22	7.84	0.22	100.00
	Prediction accuracy = 79.55 (CSLC); 89.72 (C)							
Y-M	No-injury	78.94	91.06	5.89	1.51	7.77	0.03	92.60
	Minor	1.90	4.44	3.95	2.32	0.97	0.04	6.81
	Severe	0.14	0.46	0.08	0.06	0.37	0.07	0.59
	Total predicted	80.98	95.96	9.91	3.89	9.11	0.14	100.00
	Prediction accuracy = 83.26 (CSLC); 93.45 (C)							
RF models								
O-F	No-injury	63.12	69.40	8.57	5.43	3.21	0.08	74.90
	Minor	2.68	6.35	20.66	17.60	0.61	0.00	23.95
	Severe	0.23	0.84	0.31	0.15	0.61	0.15	1.15
	Total predicted	66.03	76.59	29.53	23.18	4.44	0.23	100.00
	Prediction accuracy = 84.39 (CSLC); 87.15 (C)							
O-M	No-injury	78.87	87.41	8.55	2.88	2.88	0.00	90.30
	Minor	2.83	6.24	6.19	2.78	0.10	0.10	9.12
	Severe	0.10	0.47	0.16	0.10	0.31	0.00	0.58
	Total predicted	81.80	94.12	14.89	5.76	3.30	0.10	100.00
	Prediction accuracy = 85.37 (CSLC); 90.19 (C)							

(continued)

Table 4. (continued)

		Predicted decision						
		No-injury		Minor		Severe		
Driver group	Observed data	CSLC	C	CSLC	C	CSLC	C	Total observed
Y-F	No-injury	63.25	75.87	10.06	3.26	5.89	0.08	79.20
	Minor	2.26	6.13	16.28	13.90	1.54	0.05	20.08
	Severe	0.16	0.43	0.12	0.18	0.43	0.11	0.71
	Total predicted	65.67	82.43	26.47	17.34	7.86	0.24	100.00
	Prediction accuracy = 79.96 (CSLC); 89.88 (C)							
Y-M	No-injury	78.98	91.09	5.95	1.48	7.67	0.03	92.60
	Minor	1.69	4.33	4.29	2.44	0.82	0.04	6.81
	Severe	0.12	0.46	0.06	0.07	0.41	0.07	0.59
	Total predicted	80.79	95.88	10.31	3.99	8.90	0.14	100.00
	Prediction accuracy = 83.68 (CSLC); 93.60 (C)							

Table 5. Average Percentage of Correct Predictions across All Driver Cohorts

	C4.5		IB		RF	
Injury severity	CSLC	C	CSLC	C	CSLC	C
No-injury	76.9%	95.3%	84.1%	95.8%	84.2%	95.9%
Minor	65.2%	37.4%	71.4%	50.9%	74.6%	52.3%
Severe	36.4%	1.5%	53.4%	10.1%	59.1%	10.1%

vehicle maneuver plays a significant role in determining the injury severity for both female cohorts.

Author Contributions

The authors confirm contribution to the paper as follows: study conception and design: Somayeh Mafi, Yassir AbdelRazig; data collection: Somayeh Mafi; analysis and interpretation of results: Somayeh Mafi, Yassir AbdelRazig, Ryan Doczy; draft manuscript preparation: Somayeh Mafi, Yassir AbdelRazig, Ryan Doczy. All authors reviewed the results and approved the final version of the manuscript.

References

1. Christoforou, Z., S. Cohen, and M. G. Karlaftis. Vehicle Occupant Injury Severity on Highways: An Empirical Investigation. *Accident Analysis & Prevention*, Vol. 42, No. 6, 2010, pp. 1606–1620.
2. Mannering, F. L., and C. R. Bhat. Analytic Methods in Accident Research: Methodological Frontier and Future Directions. *Analytic Methods in Accident Research*, Vol. 1, 2014, pp. 1–22.
3. Mannering, F. L., V. Shankar, and C. R. Bhat. Unobserved Heterogeneity and the Statistical Analysis of Highway Accident Data. *Analytic Methods in Accident Research*, Vol. 11, 2016, pp. 1–16.
4. Abdel-Aty, M. A., C. L. Chen, and J. R. Schott. An Assessment of the Effect of Driver Age on Traffic Accident Involvement Using Log-Linear Models. *Accident Analysis & Prevention*, Vol. 30, No. 6, 1998, pp. 851–861.
5. Kockelman, K. M., and Y.-J. Kweon. Driver Injury Severity: An Application of Ordered Probit Models. *Accident Analysis & Prevention*, Vol. 34, No. 3, 2002, pp. 313–321.
6. Abdel-Aty, M. Analysis of Driver Injury Severity Levels at Multiple Locations Using Ordered Probit Models. *Journal of Safety Research*, Vol. 34, No. 5, 2003, pp. 597–603.
7. de Lapparent, M. Willingness to Use Safety Belt and Levels of Injury in Car Accidents. *Accident Analysis & Prevention*, Vol. 40, No. 3, 2008, pp. 1023–1032.
8. Huang, H., H. C. Chin, and M. M. Haque. Severity of Driver Injury and Vehicle Damage in Traffic Crashes at Intersections: A Bayesian Hierarchical Analysis. *Accident Analysis & Prevention*, Vol. 40, No. 1, 2008, pp. 45–54.
9. Xie, Y., Y. Zhang, and F. Liang. Crash Injury Severity Analysis using Bayesian Ordered Probit Models. *Journal of Transportation Engineering*, Vol. 135, No. 1, 2009, pp. 18–25.
10. Chimba, D., and T. Sando. The Prediction of Highway Traffic Accident Injury Severity with Neuromorphic Techniques. *Advances in Transportation Studies*, Vol. 2009, No. 19, 2009, pp. 17–26.
11. Abdel-Aty, M. A., and H. T. Abdelwahab. Predicting Injury Severity Levels in Traffic Crashes: a Modeling Comparison. *Journal of Transportation Engineering*, Vol. 130, No. 2, 2004, pp. 204–210.
12. Behnood, A., A. M. Roshandeh, and F. L. Mannering. Latent Class Analysis of the Effects of Age, Gender, and Alcohol Consumption on Driver-Injury Severities.

Analytic Methods in Accident Research, Vol. 3, 2014, pp. 56–91.
13. Boufous, S., C. Finch, A. Hayen, and A. Williamson. The Impact of Environmental, Vehicle and Driver Characteristics on Injury Severity in Older Drivers Hospitalized as a Result of a Traffic Crash. *Journal of Safety Research*, Vol. 39, No. 1, 2008, pp. 65–72.
14. Chong, M., A. Abraham, and M. Paprzycki. Traffic Accident Analysis Using Machine Learning Paradigms. *Informatica*, Vol. 29, No. 1, 2005, pp. 89–98.
15. Gray, R. C., M. A. Quddus, and A. Evans. Injury Severity Analysis of Accidents Involving Young Male Drivers in Great Britain. *Journal of Safety Research*, Vol. 39, No. 5, 2008, pp. 483–495.
16. Wang, X., and M. Abdel-Aty. Analysis of Left-turn Crash Injury Severity by Conflicting Pattern Using Partial Proportional Odds Models. *Accident Analysis & Prevention*, Vol. 40, No. 5, 2008, pp. 1674–1682.
17. Zajac, S. S., and J. N. Ivan. Factors Influencing Injury Severity of Motor Vehicle–Crossing Pedestrian Crashes in Rural Connecticut. *Accident Analysis & Prevention*, Vol. 35, No. 3, 2003, pp. 369–379.
18. Abdel-Aty, M., and H. Abdelwahab. Modeling Rear-End Collisions Including the Role of Driver's Visibility and Light Truck Vehicles Using a Nested Logit Structure. *Accident Analysis & Prevention*, Vol. 36, No. 3, 2004, pp. 447–456.
19. Yamamoto, T., and V. N. Shankar. Bivariate Ordered-Response Probit Model of Driver's and Passenger's Injury Severities in Collisions with Fixed Objects. *Accident Analysis & Prevention*, Vol. 36, No. 5, 2004, pp. 869–876.
20. Chimba, D., and T. Sando. Neuromorphic Prediction of Highway Injury Severity. *Advances in Transportation Studies*, Vol. 19, No. 1, 2009, pp. 17–26.
21. Chang, L.-Y., and H.-W. Wang. Analysis of Traffic Injury Severity: An Application of Non-parametric Classification Tree Techniques. *Accident Analysis & Prevention*, Vol. 38, No. 5, 2006, pp. 1019–1027.
22. Delen, D., R. Sharda, and M. Bessonov. Identifying Significant Predictors of Injury Severity in Traffic Accidents Using a Series of Artificial Neural Networks. *Accident Analysis & Prevention*, Vol. 38, No. 3, 2006, pp. 434–444.
23. Bedard, M., G. H. Guyatt, M. J. Stones, and J. P. Hirdes. The Independent Contribution of Driver, Crash, and Vehicle Characteristics to Driver Fatalities. *Accident Analysis & Prevention*, Vol. 34, No. 6, 2002, pp. 717–727.
24. Haleem, K., and M. Abdel-Aty. Examining Traffic Crash Injury Severity at Unsignalized Intersections. *Journal of Safety Research*, Vol. 41, No. 4, 2010, pp. 347–357.
25. Krull, K., A. Khattak, and F. Council. Injury effects of rollovers and events sequence in single-vehicle crashes. *Transportation Research Record: Journal of the Transportation Research Board*, 2000. 1717: 46–54.
26. Islam, S., and F. Mannering. Driver Aging and its Effect on Male and Female Single-Vehicle Accident Injuries: Some Additional Evidence. *Journal of Safety Research*, Vol. 37, No. 3, 2006, pp. 267–276.
27. Malyshkina, N., and F. Mannering. Effect of increases in speed limits on severities of injuries in accidents. *Transportation Research Record: Journal of the Transportation Research Board*, 2008. 2083: 122–127.
28. Malyshkina, N. V., and F. L. Mannering. Empirical Assessment of the Impact of Highway Design Exceptions on the Frequency and Severity of Vehicle Accidents. *Accident Analysis & Prevention*, Vol. 42, No. 1, 2010, pp. 131–139.
29. Shankar, V., and F. Mannering. An Exploratory Multinomial Logit Analysis of Single-vehicle Motorcycle Accident Severity. *Journal of Safety Research*, Vol. 27, No. 3, 1996, pp. 183–194.
30. Ulfarsson, G. F., and F. L. Mannering. Differences in Male and Female Injury Severities in Sport-utility Vehicle, Minivan, Pickup and Passenger Car Accidents. *Accident Analysis & Prevention*, Vol. 36, No. 2, 2004, pp. 135–147.
31. Ye, F., and D. Lord. Comparing Three Commonly Used Crash Severity Models on Sample Size Requirements: Multinomial Logit, Ordered Probit and Mixed Logit Models. *Analytic Methods in Accident Research*, Vol. 1, 2014, pp. 72–85.
32. Shankar, V., F. Mannering, and W. Barfield. Statistical Analysis of Accident Severity on Rural Freeways. *Accident Analysis & Prevention*, Vol. 28, No. 3, 1996, pp. 391–401.
33. Chang, L.-Y., and F. Mannering. Predicting Vehicle Occupancies from Accident Data: An Accident Severity Approach. *Transportation Research Record: Journal of the Transportation Research Board*, 1998. 1635: 93–104.
34. Chang, L.-Y., and F. Mannering. Analysis of Injury Severity and Vehicle Occupancy in Truck-and Non-Truck-Involved Accidents. *Accident Analysis & Prevention*, Vol. 31, No. 5, 1999, pp. 579–592.
35. Hu, W., and E. T. Donnell. Median Barrier Crash Severity: Some New Insights. *Accident Analysis & Prevention*, Vol. 42, No. 6, 2010, pp. 1697–1704.
36. Patil, S., S. R. Geedipally, and D. Lord. Analysis of Crash Severities Using Nested Logit Model—Accounting for the Underreporting of Crashes. *Accident Analysis & Prevention*, Vol. 45, 2012, pp. 646–653.
37. Yasmin, S., and N. Eluru. Evaluating Alternate Discrete Outcome Frameworks for Modeling Crash Injury Severity. *Accident Analysis & Prevention*, Vol. 59, 2013, pp. 506–521.
38. Khattak, A. Injury Severity in Multivehicle Rear-End Crashes. *Transportation Research Record: Journal of the Transportation Research Board*, 2001. 1746: 59–68.
39. Khattak, A. J., M. D. Pawlovich, R. R. Souleyrette, and S. L. Hallmark. Factors Related to More Severe Older Driver Traffic Crash Injuries. *Journal of Transportation Engineering*, Vol. 128, No. 3, 2002, pp. 243–249.
40. Lee, C., and M. Abdel-Aty. Comprehensive Analysis of Vehicle–Pedestrian Crashes at Intersections in Florida. *Accident Analysis & Prevention*, Vol. 37, No. 4, 2005, pp. 775–786.
41. Gårder, P. Segment Characteristics and Severity of Head-on Crashes on Two-Lane Rural Highways in Maine. *Accident Analysis & Prevention*, Vol. 38, No. 4, 2006, pp. 652–661.
42. Yasmin, S., N. Eluru, and S. V. Ukkusuri. Alternative Ordered Response Frameworks for Examining Pedestrian Injury Severity in New York City. *Journal of Transportation Safety & Security*, Vol. 6, No. 4, 2014, pp. 275–300.

43. Fatehnia, M., and G. Amirinia. A Review of Genetic Programming and Artificial Neural Network Applications in Pile foundations. *International Journal of Geo-Engineering*, Vol. 9, No. 2, 2018, pp. 1–20.
44. Mafi, S., and G. Amirinia. Forecasting Hurricane Wave Height in Gulf of Mexico using Soft Computing Methods. *Ocean Engineering*, Vol. 146, 2017, pp. 352–362.
45. Mafi, S., A. Yeganeh-Bakhtiary, and M. H. Kazeminezhad. Prediction Formula for Longshore Sediment Transport Rate with M5'Algorithm. *Journal of Coastal Research*, Vol. 2, No. 65, 2013, pp. 2149–2154.
46. Sohn, S. Y., and H. Shin. Pattern Recognition for Road Traffic Accident Severity in Korea. *Ergonomics*, Vol. 44, No. 1, 2001, pp. 107–117.
47. Sohn, S. Y., and S. H. Lee. Data Fusion, Ensemble and Clustering to Improve the Classification Accuracy for the Severity of Road Traffic Accidents in Korea. *Safety Science*, Vol. 41, No. 1, 2003, pp. 1–14.
48. Rungrattanaubol, J., A. Na-udom, and A. Harfield. An Exploratory Neural Network Model for Predicting Disability Severity from Road Traffic Accidents in Thailand. *Proc., 3rd International Conference on Knowledge and Smart Technologies*, No. 2011, 2011, p. 23.
49. Kuhnert, P. M., K.-A. Do, and R. McClure. Combining Non-parametric Models with Logistic Regression: An Application to Motor Vehicle Injury Data. *Computational Statistics & Data Analysis*, Vol. 34, No. 3, 2000, pp. 371–386.
50. Kashani, A. T., and A. S. Mohaymany. Analysis of the Traffic Injury Severity on Two-lane, Two-way Rural Roads Based on Classification Tree Models. *Safety Science*, Vol. 49, No. 10, 2011, pp. 1314–1320.
51. Chang, L.-Y., and J.-T. Chien. Analysis of Driver Injury Severity in Truck-involved Accidents Using a Non-parametric Classification Tree Model. *Safety Science*, Vol. 51, No. 1, 2013, pp. 17–22.
52. Abellán, J., G. López, and J. De OñA. Analysis of Traffic Accident Severity Using Decision Rules via Decision Trees. *Expert Systems with Applications*, Vol. 40, No. 15, 2013, pp. 6047–6054.
53. Harb, R., X. Yan, E. Radwan, and X. Su. Exploring Precrash Maneuvers Using Classification Trees and Random Forests. *Accident Analysis & Prevention*, Vol. 41, No. 1, 2009, pp. 98–107.
54. Chang, H.-L., and T.-H. Yeh. Risk Factors to Driver Fatalities in Single-Vehicle Crashes: Comparisons Between Non-Motorcycle Drivers and Motorcyclists. *Journal of Transportation Engineering*, Vol. 132, No. 3, 2006, pp. 227–236.
55. Richardson, J., K. Kim, L. Li, and L. Nitz. Patterns of Motor Vehicle Crash Involvement by Driver Age and Sex in Hawaii. *Journal of Safety Research*, Vol. 27, No. 2, 1996, pp. 117–125.
56. Awadzi, K. D., S. Classen, A. Hall, R. P. Duncan, and C. W. Garvan. Predictors of Injury Among Younger and Older Adults in Fatal Motor Vehicle Crashes. *Accident Analysis & Prevention*, Vol. 40, No. 6, 2008, pp. 1804–1810.
57. U. S. National Highway Traffic Safety Administration. *Safe Mobility for a Maturing Society: Challenges and Opportunities*. National Highway Traffic Safety Administration, Washington, D.C., 2003.
58. Bouckaert, R. R., E. Frank, M. Hall, R. Kirkby, P. Reutemann, A. Seewald, and D. Scuse. WEKA Manual for Version 3-7-8, 2013.
59. Bhargava, N., G. Sharma, R. Bhargava, and M. Mathuria. Decision Tree Analysis on J48 Algorithm for Data Mining. *Proc., International Journal of Advanced Research in Computer Science and Software Engineering*, Vol. 3, No. 6, 2013.
60. Quinlan, J. *C4. 5: Programs for Machine Learning*. Morgan Kaufmann Publishers, San Francisco, CA, 1993.
61. Breiman, L. Random Forests. *Machine Learning*, Vol. 45, No. 1, 2001, pp. 5–32.
62. Aha, D. W., D. Kibler, and M. K. Albert. Instance-based Learning Algorithms. *Machine Learning*, Vol. 6, No. 1, 1991, pp. 37–66.
63. Witten, I. H., E. Frank, M. A. Hall, and C. J. Pal. *Data Mining: Practical Machine Learning Tools and Techniques*. Morgan Kaufmann, 2016.
64. Florida Department of Transportation. *Traffic Data*. Florida Department of Transportation, Tallahassee, FL.
65. Herbel, S., L. Laing, and C. McGovern. *Highway Safety Improvement Program Manual*. US Department of Transportation, Federal Highway Administration, Office of Safety, Washington, D.C., 2010.
66. Amirinia, G., B. Kamranzad, and S. Mafi. Wind and Wave Energy Potential in Southern Caspian Sea Using Uncertainty Analysis. *Energy*, Vol. 120, 2017, pp. 332–345.
67. Wang, X. Integrating GIS, Simulation Models, and Visualization in Traffic Impact Analysis. *Computers, Environment and Urban Systems*, Vol. 29, No. 4, 2005, pp. 471–496.
68. Amirinia, G., S. Mafi, and S. Mazaheri. Offshore Wind Resource Assessment of Persian Gulf Using Uncertainty Analysis and GIS. *Renewable Energy*, Vol. 113, 2017, pp. 915–929.

The Standing Committee on Safety Data, Analysis and Evaluation (ANB20) peer-reviewed this paper (18-03427).

Article

Multivariate Poisson Lognormal Modeling of Weather-Related Crashes on Freeways

Transportation Research Record
2018, Vol. 2672(38) 184–198
© National Academy of Sciences: Transportation Research Board 2018
Article reuse guidelines: sagepub.com/journals-permissions
DOI: 10.1177/0361198118776523
journals.sagepub.com/home/trr
SAGE

Kai Wang[1], Shanshan Zhao[1], and Eric Jackson[1]

Abstract

Adverse weather conditions are one of the primary causes of motor vehicle crashes. To identify the factors contributing to crashes during adverse weather conditions and recommend cost-effective countermeasures, it is necessary to develop reliable crash prediction models to estimate weather-related crash frequencies. To account for the variations in crash count among different adverse weather conditions, crash types, and crash severities for both rain- and snow-related crashes, crash data on freeways was collected from the State of Connecticut, and crash prediction models were developed to estimate crash counts by crash type and severity for each weather condition. To account for the potential correlations among crash type and severity counts due to the common unobserved factors, integrated nested Laplace approximation (INLA) multivariate Poisson lognormal (MVPLN) models were developed to estimate weather-related crashes counts by crash type and severity simultaneously (four MVPLN models were estimated in total). To verify the model prediction ability, univariate Poisson lognormal (UPLN) models were estimated and compared with the MVPLN models. The results show that the effects of factors contributing to crashes, including median width, horizontal curve, lane width, and shoulder width, vary not only among different adverse weather conditions, but also among different crash types and severities. The crash types and severities are shown to be highly correlated and the model comparison verifies that the MVPLN models significantly improve the model prediction accuracy compared with the UPLN models. Therefore, the MVPLN model is recommended to provide more unbiased parameter estimates when estimating weather-related crashes by crash type and severity.

In the United States, motor vehicle crashes have been responsible for significant economic and societal losses over the past few decades. According to the National Highway Traffic Safety Administration (*1*), 35,092 people were killed in motor vehicle crashes in 2015, and more than 10% of fatal crashes occurred during hazardous weather conditions. The Road Safety Strategy (*2*) developed in Canada shows that hazardous weather conditions, such as heavy rain and snow, are one of the dominant contributors to fatal crashes, as they reduce driver visibility and cause loss of vehicle control. Given the substantial economic losses caused by motor vehicle crashes during hazardous weather conditions, it is imperative to develop crash prediction models to estimate weather-related crash counts, identify contributing factors, and help both national and state agencies to develop cost-effective countermeasures and implement roadway safety campaigns to mitigate crashes during hazardous weather conditions.

In recent years, a variety of research has been generated on estimating weather-related crashes and identifying factors contributing to crashes. To account for the disparities between different weather conditions, crash counts have been estimated for rain-related crashes (*3–16*), snow-related crashes (*5–21*), smoke/fog-related crashes (*4–7*, *22–25*) and wind-related crashes (*5*, *11*, *12*, *19*, *23–28*). Among these, in order to account for variations among crash types, some studies have estimated weather-related crashes by crash type (*7*, *11*, *19*). Similarly, in order to accommodate the crash variations by severity, crash prediction models have been developed by severity level (*4*, *7*, *9*, *10*, *12*, *13*, *18*, *20*, *29*). Meteorological conditions (e.g., temperature, precipitation, and wind speed), traffic information (e.g., annual average daily traffic (AADT) and vehicle miles traveled (VMT) and roadway characteristics (e.g., number of lanes, lane width, shoulder width, and median width) are the most common factors used in crash prediction models, and the Poisson and negative binomial regression models are two approaches commonly used to predict

[1]Connecticut Transportation Safety Research Center, Connecticut Transportation Institute, University of Connecticut, Storrs, CT

Corresponding Author:
Address correspondence to Kai Wang: kai.wang@uconn.edu

total crash counts under hazardous weather conditions, or crash counts by crash type and crash severity (*30, 31*). Koetse and Rietveld (*32*) and Theofilatos and Yannis (*33*) reviewed and summarized numerous studies associated with weather-related roadway safety and offered additional insights about crash patterns during different hazardous weather conditions, and Lord and Mannering (*30*) provided comprehensive guidance on model selection and evaluation in estimating crash frequencies.

However, all of the aforementioned studies assume crash types and severities under the same weather condition to be independent. When considering crash types and severities in relation to weather conditions simultaneously, estimating crash type or severity counts separately and independently might be questionable since weather-related crash counts might be correlated with crash types and severities due to the presence of common unobserved factors associated with different crash type or severity counts. Neglecting to account for their potential correlations might lead to biased model estimation and result in making misleading policy decisions regarding roadway safety improvements (*31*). Therefore, it is necessary to explore more robust models to estimate crash counts by crash type and severity jointly under different adverse weather conditions.

A large number of methodologies have been utilized to estimate crash counts by crash type and severity jointly under normal weather conditions. These methods include, but are not limited to, simultaneous equation model (*34, 35*), two-stage mixed multivariate model (*36*), multinomial generalized Poisson with spatiotemporal correlation model (*37–39*), Bayesian joint probability model (*40, 41*), and artificial neural network (*42*). With the development of flexible statistical software, WinBUGS (*43*), the Bayesian multivariate regression models using the Markov chain Monte Carlo (MCMC) method have been increasingly applied by researchers in recent years to simultaneously estimate crash counts by crash type or severity by accounting for the potential correlations between crash type or severity count (*31*). Among these multivariate models, Ma and Kockelman (*44*) applied a Bayesian multivariate Poisson (MVP) model to estimate crashes by severity and found the crash counts to be significantly correlated among varied severity levels. To address the overdispersion issue included in crash data, the multivariate Poisson lognormal (MVPLN) model (*45–51*), multivariate negative binomial (MVNB) model (*52*) and multivariate Tobit (MVT) model (*52*) have been implemented over the MVP model. Furthermore, in order to account for the excess zero, unobserved heterogeneity, and temporal-spatial issues in crash data, as well as to provide more unbiased and consistent parameter estimates, numerous studies have been conducted to estimate crashes by crash type or severity using the multivariate random-parameter zero-inflated negative binomial model (*53*), multivariate Poisson lognormal spatial model (*54*), multivariate Poisson lognormal conditional-autoregressive model (*55*), and multivariate random-parameter spatial Poisson lognormal model (*56*).

The Bayesian approach mentioned above using the MCMC simulation method, however, is computationally challenging and time consuming, especially for large data sets or data sets with a dependent variable containing many categories (*31*). Alternatively, the integrated nested Laplace approximation (INLA) Bayesian method (*57*) has been increasingly used in multivariate models to estimate crashes among different crash types or severities. Serhiyenko et al. (*58*) used the INLA multivariate Poisson lognormal (MVPLN) model to estimate simultaneously the crash type counts in Connecticut, and the study verified that the INLA approach can dramatically reduce the model running time compared with the MCMC simulation approach. Wang et al. (*59*) used the INLA MVPLN model to estimate crash type and severity counts on rural two-lane highways. The study verified that the crash counts are significantly correlated between different crash types and crash severities, and the effects of roadway geometric factors on crash frequencies can be inconsistent between crash type and severity levels.

However, to date, no study has been conducted to investigate the use of MVPLN in simultaneously estimating crash counts by crash type and severity in different adverse weather conditions. Bridging this gap might be helpful to transportation agencies for implementing effective countermeasures to mitigate weather-related crashes, as the effects of contributing factors on crash frequencies might be different among crash types and severities (*59*), and among different adverse weather conditions (*8, 9*). Overall, to verify the different effects of crash contributors on different crash types and severities in different adverse weather conditions, and to identify the potential correlations among crash type and severity counts, this paper presents the MVPLN using the INLA Bayesian framework to simultaneously estimate crash counts by crash type and severity for two types of common adverse weather conditions on Connecticut freeways, i.e., rain and snow. Furthermore, the univariate Poisson lognormal (UPLN) models were also estimated to compare with the MVPLN models.

The remaining parts of this paper are organized as follows: the next section presents the INLA MVPLN model formula and the model estimation method, the third section describes the data used in model estimation, and the fourth section presents the model outputs. Comparisons of models and contributing factors are presented in the fifth section, and conclusions are discussed in the final section.

Methodologies

Framework for INLA MVPLN Model

Assume $y_i = (Y_{1i}, Y_{2i} \ldots, Y_{Ji})'$ for $i = 1, 2, \ldots, n$ be a J-dimensional vector (which represents J crash types or crash severities under one specific weather condition) of crash counts across all n freeway segments. With the assumption that crash type counts or severity counts under each weather condition are correlated, they can be simultaneously estimated by the MVPLN model described as follows (*58*, *59*):

$$Y_{ji}|\lambda_{ji} \sim \text{Poisson}(\lambda_{ji}) \quad (1)$$

where λ_{ji} ($j = 1, 2, \ldots, J$ and $i = 1, 2, \ldots, n$) represents the mean of the Poisson distribution, which is assumed to be randomly distributed and can be estimated as:

$$\ln(\lambda_{ji}) = \pi + \boldsymbol{x}'_{ji}\boldsymbol{\beta}_j + \varepsilon_{ji} \quad (2)$$

where π is an offset which equals to [log(365*7*segment length)] (seven-year crash data were used in this study and the details are shown in next section). $\boldsymbol{x}_{ji}$ is a vector of exploratory variables, and $\boldsymbol{\beta}_j$ is a vector of coefficients to be estimated. ε_{ji} is a random disturbance. Assume $\varepsilon_i = (\varepsilon_{1i}, \varepsilon_{2i} \ldots, \varepsilon_{Ji})'$ is a vector of random effects at freeway segment i, which follows a J-dimensional normal distribution,

$$\varepsilon_i \sim \text{Normal}(0, \boldsymbol{\Sigma}) \quad (3)$$

where 0 is a J-dimensional zero vector, and $\boldsymbol{\Sigma}$ is a $J*J$ variance-covariance matrix. Assume $\boldsymbol{\Sigma} = (\boldsymbol{\sigma}_{rs})_{1 \le r < s \le J}$, in the MVPLN model, the expectation, variance, and covariance for the crash counts among different crash types or severities (noted as r and s) for a specific weather condition can be written as (*60*, *61*):

$$E[Y_{ji}] = \exp(\pi + \boldsymbol{x}'_{ji}\boldsymbol{\beta}_j)\exp\left(\frac{\sigma_{jj}}{2}\right) \quad (4)$$

$$\text{Var}[Y_{ji}] = \exp(\pi + \boldsymbol{x}'_{ji}\boldsymbol{\beta}_j)\exp\left(\frac{\sigma_{jj}}{2}\right) + \exp(2(\pi + x'_{ji}\beta_j)(\exp^2(\sigma_{jj}) - \exp(\sigma_{jj})) \quad (5)$$

$$\text{Cov}[Y_{ri}, Y_{si}] = \exp(\pi + \boldsymbol{x}'_{ji}\boldsymbol{\beta}_j) \exp\left(\frac{\sigma_{rr}}{2}\right)\exp\left(\frac{\sigma_{ss}}{2}\right)(\exp(\sigma_{rs}) - 1) \quad (6)$$

The potential correlations among crash type or severity counts can be accommodated by the MVPLN model, through the covariance part which is represented by Equation 6. σ_{rs} is the off-diagonal element of the variance-covariance matrix $\boldsymbol{\Sigma}$, which shows whether the correlation between the rth and sth components of the response vector (i.e., the crash counts for the rth and sth crash type or severity levels) is positive or negative, and can be written as,

$$\sigma_{rs} = \rho_{rs}\sqrt{\sigma_{ss} * \sigma_{rr}} \quad (7)$$

where σ_{ss} and σ_{rr} are diagonal elements of the variance-covariance matrix $\boldsymbol{\Sigma}$, and ρ_{rs} represents an estimated correlation coefficient between -1 and 1. The marginal distribution of the crash counts y_i in the MVPLN model can be written as:

$$g(\boldsymbol{y}_i|\boldsymbol{x}'_{ji}\boldsymbol{\beta}_j, \boldsymbol{\Sigma}) = \int \ldots \int f_{\textbf{Normal}, J}(\varepsilon_i|0, \boldsymbol{\Sigma}) \prod_{j=1}^{J} f_{\textbf{Poisson}}(y_{ij}|\varepsilon_{ji}, \boldsymbol{x}'_{ji}\boldsymbol{\beta}_j)d\varepsilon_i. \quad (8)$$

where $f_{\text{Normal},J}$ is a J-dimensional normal probability distribution function, and f_{Poisson} is a Poisson probability distribution function. Since the above marginal distribution has no closed algebraic solution to the J-dimensional integral (*58*), the approximate Bayesian framework was used to estimate the coefficients in the MVPLN model. First, assume the $\boldsymbol{\beta}_j$'s in Equation 2 follows a normal prior distribution, and the $\boldsymbol{\Sigma}^{-1}$ in Equation 3 follows a Wishart prior distribution. Then the posterior distribution of the coefficients was estimated using the Bayesian inference in the MVPLN model (*58*). In order to decrease the computational time, the INLA approach proposed by Rue et al. (*57*) was applied in this study to carry out the Bayesian inference in the MVPLN model. Detailed discussions of the INLA approach and estimation process are presented in the previous paper published by the lead author (*59*) and several relevant papers (*57–59*).

Framework for UPLN Model

In the UPLN model, the crash counts by crash type or severity under one specific weather condition are independent at each freeway segment. Similar to the MVPLN model, assume $y_i = (Y_{1i}, Y_{2i} \ldots, Y_{Ji})'$ for $i = 1, 2, \ldots, n$ be a J-dimensional vector (which represents J crash types or crash severities under one specific weather condition) of crash counts across all n freeway segments. The crash counts can still be written as Equations 1 and 2. However, in Equation 2, instead of assuming ε_{ji} s are correlated across different crash type or severity counts, we now assume that ε_{ji} is independent for $j = 1, \ldots, J$, and can be written as:

$$\varepsilon_{ji} \sim \text{Normal}(0, \xi_j^2) \quad (9)$$

Other terms are the same as the MVPLN model. Then the mean and variance of the different crash counts by crash type or severity under a specific weather condition can be derived as (*58*):

$$E[Y_{ji}] = \exp(\pi + x'_{ji}\beta_j)\exp\left(\frac{\xi_j^2}{2}\right) \quad (10)$$

$$\text{Var}[Y_{ji}] = \exp(\pi + x'_{ji}\beta_j)\exp\left(\frac{\xi_j^2}{2}\right) + \exp(2(\pi + x'_{ji}\beta_j)(\exp^2(\xi_j^2) - \exp(\xi_j^2)) \quad (11)$$

Similar to the MVPLN model, the marginal distribution of the crash counts y_i in the UPLN model can be derived as:

$$g(y_i|x'_{ji}\beta_j, \Sigma) = \int \ldots \int f_{\text{Normal}}(\varepsilon_i|0, \Sigma) \prod_{j=1}^{J} f_{\text{Poisson}}(y_{ij}|\varepsilon_{ji}, x'_{ji}\beta_j)d\varepsilon_i \quad (12)$$

All parameters are the same as Equation 8, and the coefficients were also estimated using the INLA Bayesian approach discussed earlier.

Data Collection and Analysis

Freeway segment data and roadway geometric information was collected from the Connecticut Department of Transportation (*62*). After removing freeway sections for which information was missing, and all sections related to speed-change lanes, 3,915 freeway segments were used in this study. Four continuous variables, i.e., AADT, inside shoulder width, outside shoulder width, and median width, and four categorical variables, i.e., number of lanes, area type, roadway alignment, and average lane width, were used as predictors in crash prediction models. The climate information (e.g., historical precipitation amounts, average temperature, and wind speed) was initially collected from the National Center for Environmental Information (*63*) and considered as a predictor as well. However, there are only 10 weather stations in Connecticut, and the historical data for several of them were missing; therefore, the climate information was ignored in model estimation.

Seven-year (2010–2016) crash data for freeways was collected from the Connecticut Crash Data Repository (*64*) and assigned to all freeway segments. In order to obtain sufficient crash samples by crash type and severity for each adverse weather condition, only rain- and snow-related crashes were used in this study because rain and snow are the two most common adverse weather conditions in the State of Connecticut. Crash type counts were categorized into five categories based on the Model Minimum Uniform Crash Criteria (MMUCC) (*65*): (a) head-on crashes; (b) rear-end crashes; (c) sideswipe crashes; (d) other multi-vehicle crashes, and (e) single-vehicle crashes. Crash severity counts were categorized into three levels: (a) fatal crashes; (b) injury crashes, and (c) property damage only (PDO) crashes. The explanations and descriptive characteristics for both crash data and roadway geometric data are summarized in Table 1.

Model Outputs

Tables 2 to 5 show the MVPLN and UPLN model estimation results for both rain- and snow-related crashes by crash type and severity, and the UPLN models for total rain- and snow-related crashes on freeways. The upper part of each table shows the parameter estimates and the lower part of each table shows the correlation coefficients among crash type or severity counts estimated by the MVPLN model.

Table 2 shows the model estimation results for rain-related crashes by crash type and in total. In terms of the correlation coefficients estimated by the MVPLN model, the rain-related crashes are highly correlated among different crash types, with the highest correlation coefficient being 0.828 between sideswipe crashes and single-vehicle crashes. With regard to the parameter estimates, the coefficients of variables are very similar between the MVPLN model and the UPLN model. However, the standard errors estimated by the MVPLN model are slightly lower than the UPLN model. Some variables that are significant in one of the models may not be significant in the other model, such as five- and six-lane freeways, seven- and eight-lane freeways, outside shoulder width, and horizontal curve. This verifies the finding in previous studies conducted by Serhiyenko et al. (*58*) and Wang et al. (*59*) that accounting for the potential correlations among dependent variables in the MVPLN model can lead to insignificant variables in the UPLN model becoming significant and vice versa. In terms of the effects of variables on crash type frequencies, AADT is shown to have a positive association with all crash type counts. Rural area is associated with decreased crash counts for all crash types. When the coefficients are statistically significant, freeways with more than four lanes are associated with increased crash counts for all crash types. Lane width is only associated with increased sideswipe crashes, and both inside and outside shoulder width are associated with decreased crash counts for all crash types. Comparing the models that estimate rain-related crashes by crash type with the model that estimates rain-related crashes in total, one important finding is that median width is associated with decreased rain-related crashes when crashes are estimated in total. However, median width is associated with increased single-vehicle crashes in rainy weather conditions when crashes are estimated by crash type. This might be because rain conditions reduce the driving visibility and the driver's lane control ability, which results

Table 1. Descriptive Characteristics of Roadway and Crash Data

Continuous variables	Min.	Max.	Mean	Std dev.
Rain-related crashes				
Head-on crashes	0.0	1.0	0.1	0.1
Rear-end crashes	0.0	69.0	1.7	3.9
Sideswipe crashes	0.0	18.0	0.7	1.5
Multi-other crashes	0.0	12.0	0.2	0.6
Single-vehicle crashes	0.0	88.0	2.1	3.9
PDO crashes	0.0	88.0	3.7	5.9
Injury crashes	0.0	35.0	1.1	2.1
Fatal crashes	0.0	2.0	0.1	0.1
Snow-related crashes				
Head-on crashes	0.0	1.0	0.1	0.1
Rear-end crashes	0.0	8.0	0.3	0.8
Sideswipe crashes	0.0	8.0	0.3	0.7
Multi-other crashes	0.0	5.0	0.1	0.4
Single-vehicle crashes	0.0	37.0	1.2	1.9
PDO crashes	0.0	45.0	1.6	2.4
Injury crashes	0.0	9.0	0.3	0.8
Fatal crashes	0.0	1.0	0.1	0.1
Freeway characteristics				
AADT	4,600	169,900	66,800	37,423
Inside shoulder width (ft)	0.0	30.0	5.0	4.0
Outside shoulder width (ft)	1.0	26.0	9.8	3.0
Median width (ft)	0.0	550.0	49.6	55.4
Categorical variables	**Freq.**	**% of Total**		
Area type				
Rural	238	6.1%		
Urban	3,677	93.9%		
Number of lanes				
≤4	2,444	62.4%		
5–6	1,208	30.9%		
7–8	249	6.4%		
9–10	14	0.3%		
Roadway alignment				
Horizontal curve	2,441	62.4%		
Tangent	1,474	37.6%		
Average lane width (ft)				
≤12	3,847	98.3%		
>12	68	1.7%		

in median-related run-off-road crashes with the presence of a wider median (*66*). The presence of horizontal curve is associated with decreased rain-related crash counts in total, but is associated with increased single-vehicle crashes. This might be due to the fact that rain conditions reduce the pavement friction, and vehicles are more likely to run off the road at horizontal curves (*67*).

Table 3 shows the model estimation results for rain-related crashes by crash severity and in total. Similar to the crash model estimation by crash type, the correlation coefficients estimated by the MVPLN model are all highly significant among crash severities, which indicates the rain-related crashes are dependent among different crash severity levels. The estimated coefficients are similar between the MVPLN model and UPLN model, but the standard errors in the MVPLN model are slightly lower than the UPLN model. AADT is associated with increased crash counts for all crash severity levels. Rural area is associated with decreased crash severity counts. Freeways with higher numbers of lanes are associated with increased crash severities, but are mostly insignificant in both MVPLN and UPLN models. Lane width is only significant for sideswipe crashes and is associated with increased sideswipe crashes. Both shoulder width and median width are associated with decreased crash counts for all crash severities. The presence of horizontal curve is shown to have a negative association with total rain-related crashes, but it is positively associated with injury-only crashes. While horizontal curve increases single-vehicle crashes on rainy days, drivers may tend to drive at a slower speed which might result in fewer serious crash outcomes.

Table 2. Estimated Rain-related Crashes by Crash Type

Coefficient estimates

Variables	Head-on MVPLN	Head-on UPLN	Rear-end MVPLN	Rear-end UPLN	Sideswipe MVPLN	Sideswipe UPLN	Multi-other MVPLN	Multi-other UPLN	Single-vehicle MVPLN	Single-vehicle UPLN	Total rain-related crashes UPLN
Intercept	**−16.667**	**−16.088**	**−25.287**	**−25.848**	**−24.257**	**−25.026**	**−19.366**	**−19.258**	**−11.078**	**−11.199**	**−14.933**
	1.910	2.171	1.019	1.064	1.071	1.197	1.749	1.824	0.752	0.783	0.563
Log (AADT)	**0.784**	**0.674**	**1.860**	**1.901**	**1.633**	**1.684**	**1.093**	**1.066**	**0.550**	**0.552**	**1.001**
	0.172	0.194	0.092	0.096	0.096	0.107	0.157	0.163	0.068	0.071	0.051
Rural area	**−1.135**	**−1.063**	**−0.830**	**−0.761**	0.026	0.136	−0.188	−0.094	**−0.267**	**−0.237**	**−0.251**
	0.488	0.491	0.240	0.252	0.177	0.191	0.319	0.305	0.110	0.114	0.088
Lane counts											
(≤4)						Base level					
Lane counts (5–6)	**0.579**	**0.642**	−0.065	−0.056	**0.228**	**0.249**	**0.326**	0.333*	**0.190**	**0.180**	**0.235**
	0.198	0.226	0.098	0.099	0.097	0.106	0.160	0.170	0.085	0.088	0.061
Lane counts (7–8)	0.420	0.557*	−0.216	−0.205	0.213	**0.290**	0.238	0.344	0.179	0.176	**0.204**
	0.284	0.323	0.147	0.148	0.134	0.146	0.231	0.235	0.132	0.137	0.097
Lane counts (9–10)	1.118	1.423	−0.198	−0.110	0.234	0.308	−6.234	−6.310	**0.975**	0.832*	0.380
	0.753	0.858	0.476	0.472	0.407	0.436	11.777	11.830	0.430	0.454	0.326
Lane width	0.330	0.436	0.547*	0.589*	**0.860**	**1.065**	−0.393	−0.287	0.112	0.050	**0.457**
	0.594	0.648	0.318	0.328	0.299	0.313	0.756	0.749	0.251	0.263	0.185
Inside shoulder width	−0.008	−0.005	**−0.021**	**−0.026**	**−0.034**	**−0.037**	**−0.032**	**−0.035**	**−0.046**	**−0.044**	**−0.029**
	0.017	0.020	0.009	0.009	0.008	0.009	0.014	0.015	0.008	0.009	0.006
Outside shoulder width	−0.042*	−0.030	**−0.078**	**−0.070**	−0.014	−0.007	**−0.057**	**−0.048**	**−0.024**	−0.020	**−0.050**
	0.024	0.031	0.013	0.013	0.013	0.014	0.021	0.022	0.011	0.015	0.008
Median width	**−0.003**	**−0.004**	**−0.002**	**−0.003**	**−0.002**	**−0.003**	**−0.004**	**−0.005**	**0.002**	**0.002**	**−0.002**
	0.001	0.002	0.001	0.001	0.001	0.001	0.001	0.002	0.001	0.001	0.001
Horizontal curve	**−0.346**	**−0.435**	**−0.172**	**−0.179**	−0.052	−0.055	0.063	0.074	**0.045**	0.021	−0.062
	0.140	0.160	0.071	0.072	0.069	0.076	0.121	0.125	0.021	0.025	0.045
Overdispersion	**2.597**	**3.226**	**1.653**	**1.566**	**0.906**	**0.887**	**1.416**	**1.067**	**1.239**	**1.473**	**1.236**
	0.038	0.036	0.023	0.034	0.075	0.098	0.062	0.414	0.026	0.031	0.024

Correlation coefficients from MVPLN model

	Head-on	Rear-end	Sideswipe	Multi-other	Single-vehicle
Head-on	1.000				
Rear-end	**0.753 ($p < 0.001$)**	1.000			
Sideswipe	**0.749 ($p < 0.001$)**	**0.811 ($p < 0.001$)**	1.000		
Multi-other	**0.305 ($p < 0.001$)**	**0.637 ($p < 0.001$)**	**0.723 ($p < 0.001$)**	1.000	
Single-vehicle	**0.611 ($p < 0.001$)**	**0.626 ($p < 0.001$)**	**0.828 ($p < 0.001$)**	**0.813 ($p < 0.001$)**	1.000

Note: The first row is the coefficient estimate; the second row is the standard error; bold coefficients are statistically significant at the 5% significance level; asterisked coefficients are statistically significant at the 10% significance level; italicized variables have different coefficient signs between the model that estimates crashes by crash type and the model that estimates crashes in total.

Table 3. Estimated Rain-related Crashes by Crash Severity

Coefficient estimates	PDO		Injury		Fatal		Total rain-related crashes
Variables	MVPLN	UPLN	MVPLN	UPLN	MVPLN	UPLN	UPLN
Intercept	−15.370	−15.341	−18.074	−19.427	−10.873	−10.328*	−14.933
	0.685	0.704	0.880	0.966	5.520	5.772	0.563
Log (AADT)	1.018	1.016	1.152	1.257	0.014	0.058	1.001
	0.062	0.064	0.079	0.087	0.499	0.518	0.051
Rural area	−0.657	−0.642	−0.704	−0.684	−0.769	−0.730	−0.706
	0.130	0.133	0.169	0.183	1.088	1.082	0.105
Lane counts							
(≤4)				Base level			
Lane counts (5–6)	0.188	0.185	0.162*	0.135	0.930	0.900	0.235
	0.074	0.077	0.090	0.095	0.610	0.614	0.061
Lane counts (7–8)	0.124	0.122	0.143	0.117	0.689	0.798	0.204
	0.115	0.118	0.134	0.139	0.920	0.920	0.097
Lane counts (9–10)	0.561	0.578	0.112	−0.074	2.632	2.660	0.380
	0.385	0.395	0.439	0.470	1.205	1.224	0.326
Lane width	0.452	0.433*	0.268	0.416	−4.659	−4.302	0.457
	0.219	0.226	0.287	0.299	13.463	13.879	0.185
Inside Shoulder width	−0.031	−0.031	−0.029	−0.031	−0.026	−0.019	−0.029
	0.007	0.008	0.008	0.009	0.057	0.057	0.006
Outside Shoulder width	−0.045	−0.048	−0.049	−0.035	−0.028	−0.022	−0.050
	0.009	0.010	0.011	0.012	0.086	0.089	0.008
Median width	−0.002	−0.002	−0.002	−0.003	−0.002	−0.003	−0.002
	0.001	0.001	0.001	0.001	0.004	0.005	0.001
Horizontal curve	−0.078	−0.079	0.023	0.010	0.306	0.188	−0.062
	0.054	0.056	0.006	0.009	0.510	0.507	0.045
Overdispersion	1.212	1.267	1.176	1.180	0.981	0.985*	1.236
	0.003	0.033	0.037	0.050	0.402	0.543	0.024

Correlation coefficients from MVPLN model

	PDO	Injury	Fatal
PDO	1.000		
Injury	0.949 ($p < 0.001$)	1.000	
Fatal	0.883 ($p < 0.001$)	0.889 ($p < 0.001$)	1.000

Tables 4 and 5, respectively, show the model estimation results for crash counts by crash type and severity on snowy days. The estimated correlation coefficients verified that snow-related crash counts are significantly correlated among all crash types, but the crashes are only correlated between PDO and injury severity levels when crashes are simultaneously estimated by severity. AADT is fairly significant for crash counts by crash type and severity, and it is positively associated with all crash types and severity counts. Rural area is statistically significant across both the MVPLN and UPLN models for head-on, rear-end, and single-vehicle crashes, but it is only significant in MVPLN model for injury crashes. Freeways with more lanes are fairly associated with increased head-on, rear-end, and sideswipe crashes, and are positively correlated with both PDO and injury crashes. Lane width is positively related to the rear-end crashes, but it is negatively related to the single-vehicle crashes and both PDO and injury crashes only in the MVPLN model. This verifies the finding in a previous study conducted by Wang et al. (*59*) that wider lane width might increase the driving speed and decrease the distance headway to the leading vehicle, which results in more rear-end crashes rather than run-off-road crashes. Inside shoulder width is only significant for single-vehicle crashes and PDO crashes. Outside shoulder width is associated with decreased rear-end, other multi-vehicle crashes, and injury crashes, but it is associated with increased single-vehicle crashes and fatal crashes. This finding presents a phenomenon that the presence of a wider outside shoulder on freeways might increase the tendency to drive close to the shoulder and result in more

Table 4. Estimated Snow-related Crashes by Crash Type

Coefficient estimates											
	Head-on		Rear-end		Sideswipe		Multi-other		Single-vehicle		Total snow-related crashes
Variables	MVPLN	UPLN	MVPLN	UPLN	MVPLN	UPLN	MVPLN	UPLN	MVPLN	UPLN	UPLN
Intercept	−12.865	−12.618	−15.519	−15.742	−15.878	−16.352	−16.951	−16.264	−6.605	−6.574	−8.782
	2.349	2.394	1.263	1.325	1.258	1.330	2.116	2.093	0.660	0.689	0.514
Log (AADT)	0.359*	0.331	0.805	0.814	0.764	0.791	0.821	0.764	0.170	0.160	0.321
	0.210	0.214	0.114	0.120	0.113	0.119	0.190	0.188	0.060	0.063	0.047
Rural area	−1.135	−1.063	−0.830	−0.761	0.026	0.136	−0.188	−0.094	−0.267	−0.237	−0.251
	0.488	0.491	0.240	0.252	0.177	0.191	0.319	0.305	0.110	0.114	0.088
Lane counts											
(≤4)						Base level					
Lane counts (5–6)	0.566	0.602	0.268	0.267	0.591	0.625	0.159	0.200	0.135*	0.126	0.249
	0.236	0.234	0.125	0.132	0.118	0.124	0.194	0.185	0.075	0.079	0.058
Lane counts (7–8)	0.678	0.837	0.305*	0.354*	0.529	0.622	−0.083	0.104	−0.126	−0.123	0.175*
	0.330	0.323	0.181	0.200	0.168	0.176	0.288	0.272	0.126	0.132	0.092
Lane counts (9–10)	1.090	1.083	0.769	0.886*	1.109	0.126	−0.536	−0.417	0.715	0.759	0.648
	0.739	0.730	0.462	0.487	0.386	0.425	0.793	0.738	0.340	0.359	0.278
Lane width	−0.382	−0.397	0.890	0.993	−0.108	−0.083	−0.431	−0.405	−1.099	−1.113	−0.432
	1.039	1.048	0.361	0.372	0.528	0.539	1.024	1.016	0.307	0.316	0.195
Inside shoulder width	−0.008	−0.006	−0.008	−0.005	−0.014	−0.013	0.017	0.022	−0.014	−0.014*	−0.007
	0.020	0.021	0.011	0.012	0.009	0.011	0.015	0.015	0.007	0.008	0.005
Outside shoulder width	−0.001	−0.001	−0.056	−0.062	0.011	0.011	−0.055	−0.058	0.020	0.022	−0.007
	0.033	0.034	0.016	0.017	0.016	0.017	0.025	0.025	0.009	0.010	0.008
Median width	−0.002	−0.001	0.002	0.001	−0.001	−0.001	−0.002	−0.002	0.002	0.001	0.002
	0.001	0.002	0.001	0.001	0.001	0.001	0.001	0.002	0.001	0.001	0.001
Horizontal curve	−0.216	−0.294*	−0.235	−0.240	−0.097	−0.129	0.039	−0.042	−0.138	−0.149	−0.172
	0.169	0.170	0.088	0.094	0.084	0.090	0.141	0.136	0.055	0.058	0.042
Overdispersion	0.951	1.189	0.960	1.202	0.461	0.640	0.540	0.703	0.598	0.731	0.717
	0.091	0.092	0.094	0.086	0.227	0.239	0.199	0.210	0.119	0.094	0.064

Correlation coefficients from MVPLN model

	Head-on	Rear-end	Sideswipe	Multi-other	Single-vehicle
Head-on	1.000				
Rear-end	0.850 ($p < 0.001$)	1.000			
Sideswipe	0.664 ($p < 0.001$)	0.802 ($p < 0.001$)	1.000		
Multi-other	0.781 ($p < 0.001$)	0.811 ($p < 0.001$)	0.882 ($p < 0.001$)	1.000	
Single-vehicle	0.750 ($p < 0.001$)	0.740 ($p < 0.001$)	0.791 ($p < 0.001$)	0.894 ($p < 0.001$)	1.000

Table 5. Estimated Snow-related Crashes by Crash Severity

Coefficient estimates

Variables	PDO MVPLN	PDO UPLN	Injury MVPLN	Injury UPLN	Fatal MVPLN	Fatal UPLN	Total snow-related crashes UPLN
Intercept	−9.712	−9.558	−13.249	−12.827	−25.000	−32.233	−8.782
	0.562	0.669	0.977	1.168	10.278	14.895	0.514
Log (AADT)	0.382	0.374	0.569	0.530	0.826	1.036	0.321
	0.051	0.061	0.088	0.105	0.715	0.819	0.047
Rural area	−0.123	−0.181	−0.271*	−0.318	−5.867	−5.997	−0.251
	0.093	0.112	0.155	0.193	12.170	12.053	0.088
Lane counts							
(≤4)				Base level			
Lane counts (5–6)	0.205	0.171	0.342	0.337	0.868	0.867	0.249
	0.062	0.073	0.099	0.117	0.908	0.997	0.058
Lane counts (7–8)	0.152	0.190	0.346	0.353	−4.331	−3.849	0.175*
	0.098	0.116	0.145	0.177	13.802	14.614	0.092
Lane counts (9–10)	0.611	0.759	0.819	0.899	−2.662	−2.444	0.648
	0.291	0.315	0.360	0.398	16.739	17.397	0.278
Lane width	−0.739	−0.318	−1.235	−0.733	−4.670	−4.720	−0.432
	0.238	0.253	0.600	0.607	13.850	13.746	0.195
Inside shoulder width	−0.009*	−0.016	0.010	0.010	0.007	−0.001	−0.007
	0.005	0.007	0.008	0.010	0.089	0.109	0.005
Outside shoulder width	−0.010	−0.009	−0.028	−0.034	0.284	0.320	−0.007
	0.008	0.010	0.013	0.016	0.126	0.134	0.008
Median width	0.002	0.001	0.002	0.001	−0.004	−0.006	0.002
	0.001	0.001	0.001	0.001	0.008	0.009	0.001
Horizontal curve	−0.190	−0.182	−0.169	−0.228	5.421	6.352	−0.172
	0.045	0.053	0.071	0.085	7.089	11.850	0.042
Overdispersion	0.741	0.716	0.755	0.686	0.260	0.182	0.717
	0.071	0.087	0.125	0.203	0.032	0.041	0.064

Correlation coefficients from MVPLN model

	PDO	Injury	Fatal
PDO	1.000		
Injury	0.905 ($p < 0.001$)	1.000	
Fatal	0.336 ($p = 0.736$)	0.409 ($p = 0.683$)	1.000

run-off-road and fatal crashes under snow conditions. Median width is only significant in the MVPLN model for head-on, rear-end, other multi-vehicle and single-vehicle crashes when snow-related crashes are estimated by crash type, and is only significant in the MVPLN model for PDO and injury crashes when considering crashes by crash severity. Furthermore, median width is shown to have a positive relationship with rear-end and single-vehicle crashes, but it has a negative relationship with head-on and other multi-vehicle crashes. The presence of horizontal curve is associated with decreased crash counts for all crash types, and both PDO and injury crashes, but it is only fairly significant in both MVPLN and UPLN models for rear-end and single-vehicle crashes when estimating snow-related crashes by crash type, and PDO and injury crashes when estimating snow-related crashes by crash severity.

Model Comparison

Table 6 summarizes the effects of variables on rain- and snow-related crashes by crash type and severity identified in the last section, respectively, as well as the model prediction performance of the MVPLN and UPLN models.

In order to evaluate and compare the model prediction performance between the MVPLN and UPLN models in predicting each of the weather-related crash counts by crash type and severity, 70% of the data was randomly selected to estimate the model coefficients, and the remaining 30% was used to compare the model prediction

Table 6. Model Estimation Comparison

	Crash type										Crash severity						Total crash	
	Head-on		Rear-end		Sideswipe		Multi-other		Single-vehicle		PDO		Injury		Fatal			
Variable	Rain	Snow	Rain	Snow	Rain	Snow	Rain	Snow	Rain	Snow	Rain	Snow	Rain	Snow	Rain	Snow	Rain	Snow
AADT	+	+	+	+	+	+	+	+	+	+	+	+	+	+			+	+
Rural area	–	–	–	–	–				–	–	–		–	–			–	–
Lane counts (5–6)	+	+		+	+	+	+		+	+	+	+	+	+			+	+
Lane counts (7–8)	+	+		+	+	+								+			+	+
Lane counts (9–10)				+		+			+	+		+		+	+			+
Lane width			+	+	+					–	+	–		–			+	–
Inside shoulder width			–		–		–		–	–	–	–	–				–	
Outside shoulder width	–		–	–			–	–	–	+	–		–	–		+	–	
Median width	–	–	–	+	–		–	–	+	+	–	+	–	+			–	+
Horizontal curve	–	–	–	–	–				+	–		–	+	–				–

Rain-related crashes by crash type	Head-on		Rear-end		Sideswipe		Multi-other		Single-vehicle	
	MVPLN	UPLN	MVPLN	UPLN	MVPLN	UPLN	MVPLN	UPLN	MVPLN	UPLN
APMAE	0.10	0.14	1.38	1.66	0.35	0.67	0.17	0.24	1.45	2.01
AMSPE	0.14	0.17	1.59	1.69	0.65	0.64	0.22	0.20	1.75	1.93

Snow-related crashes by crash type	Head-on		Rear-end		Sideswipe		Multi-other		Single-vehicle	
	MVPLN	UPLN	MVPLN	UPLN	MVPLN	UPLN	MVPLN	UPLN	MVPLN	UPLN
APMAE	0.10	0.11	0.26	0.34	0.28	0.32	0.13	0.16	0.52	1.14
AMSPE	0.10	0.10	0.34	0.38	0.32	0.35	0.12	0.15	0.99	1.14

Rain-related crashes by crash severity	PDO		Injury		Fatal	
	MVPLN	UPLN	MVPLN	UPLN	MVPLN	UPLN
APMAE	0.53	0.73	0.41	0.56	0.02	0.02
AMSPE	2.94	2.98	1.03	1.08	0.02	0.02

Snow-related crashes by crash severity	PDO		Injury		Fatal	
	MVPLN	UPLN	MVPLN	UPLN	MVPLN	UPLN
APMAE	0.57	1.61	0.30	0.53	0.01	0.01
AMSPE	1.18	1.38	0.36	0.49	0.01	0.01

Note: + = positive relationship; – = negative relationship; blank cells represent variables that are not significant in either the MVPLN model or the UPLN model; underlined italicized variables have inconsistent coefficient signs among models for crashes by weather condition, crash type, or crash severity.

accuracy based on two statistical criteria, i.e., the predicted mean absolute error (PMAE) and the mean squared predictor error (MSPE). This process was duplicated 30 times to calculate the average predicted mean absolute deviation (APMAE) and average mean squared prediction error (AMSPE) (68), which can be measured as:

$$APMAE_j = \frac{1}{30}\sum_{l=1}^{L}\left(\frac{1}{n}\sum_{i=1}^{n}\left|\hat{Y}_{ji,l} - Y_{ji,l}\right|\right) \quad (13)$$

$$AMSPE_j = \frac{1}{30}\sum_{l=1}^{L}\left[\frac{1}{n}\sum_{i=1}^{n}\left(\hat{Y}_{ji,l} - Y_{ji,l}\right)^2\right] \quad (14)$$

where j ($j = 1, 2, \ldots\ldots, J$) represents the jth crash type or crash severity for rain- or snow-related crashes, i ($i = 1, 2, \ldots\ldots, n$) represents the ith freeway segment, l ($l = 1, 2, \ldots\ldots, L$) represents the duplication time, which is equal to 30 in this study, $\hat{Y}_{ji,l}$ represents the predicted rain- or snow-related crash counts for jth crash type or crash severity at ith freeway segment, and $Y_{ji,l}$ represents the observed rain- or snow-related crash counts for jth crash type or crash severity at ith freeway segment. Lower APMAE and AMSPE values indicate better model prediction accuracy.

The lower part of Table 6 shows the model prediction comparison between the MVPLN and UPLN model in estimating both rain- and snow-related crashes by crash type and severity. In terms of the models that estimate

Table 7. Summary of Contributing Factors

Variables with consistent effects	
AADT	AADT is associated with increased crash counts for total rain- and snow-related crashes, and is associated with increased crash counts for all crash types and severities under each weather condition, except for fatal crashes in which the AADT is not significant.
Rural area	Rural area is associated with decreased total rain- and snow-related crashes, and is associated with decreased head-on, rear-end single-vehicle, and injury crashes for both rain and snow conditions. It is only associated with decreased sideswipe and PDO crashes for rain-related crashes.
Freeways (5–6 lanes)	Freeways with five to six lanes are associated with increased total rain- and snow-related crashes and are associated with increased head-on, sideswipe, single-vehicle, PDO and injury crashes for both rain and snow conditions; but they are only associated with increased rear-end crashes for snow conditions and other multi-vehicle crashes for rain conditions.
Freeways (7–8 lanes)	Freeways with seven to eight lanes are shown to be positively correlated with total rain- and snow-related crashes, as well as head-on and sideswipe crashes for both weather conditions. They are only associated with increased rear-end and injury crashes in snow conditions.
Freeways (9–10 lanes)	Freeways with nine to ten lanes are only positively related to total snow-related crashes. They were found to be positively correlated with both rain- and snow-related single-vehicle crashes, but they are only associated with increased snow-related rear-end, sideswipe, PDO and injury crashes, and only associated with increased rain-related fatal crashes.
Inside shoulder width	Inside shoulder width is only associated with decreased total rain-related crashes and is associated with decreased single-vehicle and PDO crashes for both rain- and snow-related crashes. It is found to be associated with decreased rear-end, sideswipe, other multi-vehicle, and injury crashes only for crashes occurring on rainy days.
Variables with inconsistent effects	
Lane width	Lane width is associated with increased rain-related crashes, but is associated with decreased snow-related crashes when crashes are estimated in total. When crash counts are estimated by crash type and severity in terms of snow-related crashes, lane width is associated with increased rear-end crashes, but is associated with decreased single-vehicle crashes.
Outside shoulder width	Outside shoulder width is associated with decreased snow-related rear-end crashes, but is associated with increased snow-related single-vehicle crashes. Moreover, outside shoulder width is associated with decreased snow-related injury crashes, but is associated with increased snow-related fatal crashes.
Median width	Median width is negatively correlated with total rain-related crashes, but is positively correlated with total snow-related crashes. In terms of rain-related crashes, median width is associated with decreased head-on, rear-end, sideswipe, and other multi-vehicle crashes, but it is associated with increased single-vehicle crashes. When considering the snow-related crashes, median width is associated with decreased head-on, other multi-vehicle crashes, but it is associated with increased rear-end and single-vehicle crashes.
Horizontal curve	Horizontal curve is verified to be negatively associated with rain-related head-on, rear-end, and sideswipe crashes, but it is found to be positively associated with rain-related single-vehicle crashes.

crash counts by crash type, both APMAE and AMSPE show that the MVPLN model slightly outperforms the UPLN model in predicting crash type counts for rain- and snow-related crashes. For the models that estimate crash counts by severity, the MVPLN model is shown to significantly improve the prediction accuracy, especially for the prediction of PDO and injury crashes.

Discussion and Conclusions

This paper has attempted to estimate weather-related crash counts by crash type and severity on freeways in the State of Connecticut, and identify the critical factors contributing to specific crash types or severities under different adverse weather conditions. Rain- and snow-related crashes were considered, the crash type was categorized as head-on, rear-end, sideswipe, other multi-vehicle or single-vehicle, and the crash severity was categorized as PDO, injury, or fatal. The roadway geometric factors of AADT, area type, number of through lanes, lane width, inside and outside shoulder width, median width, and presence of horizontal curve were used as predictors. Both the MVPLN and UPLN models were developed, respectively, to estimate weather-related crash counts by crash type and severity, and the UPLN models were developed to estimate weather-related crashes in total. To finalize, the prediction performance of MVPLN models was compared to the UPLN models based on the APMAE and AMSPE statistics.

In terms of the effects of contributing factors on crash counts, most variables are consistent between rain- and snow-related crashes, and among different crash types and severities for each weather condition, some variables were found to vary. The effects of variables on weather-related crashes are summarized in Table 7. These different impacts of variables on crash types and severities cannot be identified when weather-related crashes are estimated in total, which verifies that estimating crashes by crash type and severity might be more helpful in identifying critical contributing factors on weather-related crashes, and more reliable in selecting relevant countermeasures to mitigate the targeted crash types or severities. For instance, horizontal curve is found to be highly correlated with single-vehicle crashes under rainy weather. Implementing warning signs at horizontal curves might be a potential strategy in reducing single-vehicle crashes. Shoulder width is found to be associated with snow-related fatal crashes. Measures such as implementing shoulder rumble strips and driver education and training about proper vehicle control are recommended to reduce severe crashes in snowy conditions. The estimated correlation coefficients in Tables 1 to 5 show that weather-related crashes are highly correlated among different crash types and severities, and the standard errors of estimated coefficients in the MVPLN models are lower than the UPLN models, which indicates that the MVPLN model can estimate more unbiased parameters than the UPLN model by accounting for the potential correlations among crash types and severities. Furthermore, Table 6 presents the prediction performance comparison between the MVPLN models and UPLN models. It is demonstrated that in terms of both rain- and snow-related crashes, the MVPLN models significantly improve the model accuracy compared with the UPLN models in predicting crashes by crash type and severity. In summary, one can conclude that the MVPLN model can lead to more reliable parameter estimates and more accurate prediction accuracy in simultaneously estimating weather-related crashes by crash type and severity.

It is expected that this study can shed light on the selection of rigorous approaches to estimate crash counts by crash type and severity in adverse weather conditions, identify the impacts of roadway geometric characteristics on weather-related crash types and severities, and offer additional insight about implementing cost-effective countermeasures to reduce crashes during adverse weather conditions. One limitation of this study is that the climate information (e.g., historical precipitation amounts, average temperature, and wind speed) was not considered in model estimations due to limited weather stations in Connecticut. Future work could focus on collecting reliable climate data and estimating new MVPLN models to identify the effects of climate change on roadway safety.

Acknowledgments

The authors thank the Connecticut Department of Transportation (CTDOT) and the Connecticut Transportation Crash Data Repository (CTCDR) for providing the roadway geometric and crash data.

Author Contributions

The authors confirm contribution to the paper as follows: study conception and design: Kai Wang; data collection: Kai Wang, Shanshan Zhao; analysis and interpretation of results: Kai Wang, Shanshan Zhao; draft manuscript preparation: Kai Wang, Shanshan Zhao, Eric Jackson. All authors reviewed the results and approved the final version of the manuscript.

References

1. NHTSA. *National Highway Traffic Safety Administration.* http://www.nhtsa.gov/NCSA
2. Transport Canada, 2011. *Road Safety in Canada.* http://www.tc.gc.ca/eng/roadsafety/tp-tp15145-1201.htm
3. Brodsky, H., and A. Hakkert. Risk of a Road Accident in Rainy Weather. *Accident Analysis and Prevention*, Vol. 20, 1988, pp. 161–176.

4. Edward, J. The Relationship Between Road Accident Severity and Recorded Weather. *Journal of Safety Research*, Vol. 29, No. 4, 1998, pp. 249–262.
5. Naik, B., L. Tung, S. Zhao, and A. Khattak. Weather Impacts on Single-Vehicle Truck Crash Injury Severity. *Journal of Safety Research*, Vol. 58, 2016, pp. 57–65.
6. Khan, G., X. Qin, and D. Noyce. Spatial Analysis of Weather Crash Patterns. *Journal of Transportation Engineering*, Vol. 134, No. 5, 2008, pp. 191–202.
7. Khattak, A., P. Kantor, and F. Council. Role of Adverse Weather in Key Crash Types on Limited-Access Roadways. Implications for Advanced Weather Systems. *Transportation Research Record: Journal of the Transportation Research Board*, 1998. 1621: 10–19.
8. Amin, M., A. Zareie, and L. Amador-Jimenez. Climate Change Modeling and the Weather-related Road Accidents in Canada. *Transportation Research Part D*, Vol. 32, 2014, pp. 171–183.
9. Eisengerg, D. The Mixed Effects of Precipitation on Traffic Crashes. *Accident Analysis and Prevention*, Vol. 36, 2004, pp. 637–647.
10. Abdel-Aty, M., A. Ekram, H. Huang, and K. Choi. A Study on Crashes Related to Visibility Obstruction Due to Fog and Smoke. *Accident Analysis and Prevention*, Vol. 43, No. 5, 2011, pp. 1730–1737.
11. Chen, S., C. Cai, and B. Wolshon. From Normal Operation to Evacuation: Single-Vehicle Safety Under Adverse Weather, Topographic, and Operational Conditions. Technical Papers. *Natural Hazards Review*, Vol. 10, No. 2, 2009, pp. 68–76.
12. Qiu, L, and W. Nixon. Effects of Adverse Weather on Traffic Crashes. *Transportation Research Record: Journal of the Transportation Research Board*, 2008. 2055: 139–146.
13. Andrey, J., B. Mills, M. Leahy, and J. Suggett. Weather as a Chronic Hazard for Road Transportation in Canadian Cities. *Natural Hazards*, Vol. 28, No. 2–3, 2003, pp. 319–343.
14. Brijs, T., D. Karlis, and G. Wets. Studying the Effect of Weather Conditions on Daily Crash Counts Using a Discrete Time-Series Model. *Accident Analysis & Prevention*, Vol. 40, No. 3, 2008, pp. 1180–1190.
15. Ahmed, M., M. Abdel-Aty, and R. Yu. Assessment of Interaction of Crash Occurrence, Mountainous Freeway Geometry, Real-Time Weather, and Traffic Data. *Transportation Research Record: Journal of the Transportation Research Board*, 2012. 2280: 51–59.
16. El-Basyouny, K., S. Barua, and M. T. Islam. Investigation of Time and Weather Effects on Crash Types Using Full Bayesian Multivariate Poisson Lognormal Models. *Accident Analysis & Prevention*, Vol. 73, No. 26, 2014, pp. 91–99.
17. Qin, X., D. Noyce, C. Lee, and J. Kinar. Snowstorm Event-based Crash Analysis. *Transportation Research Record: Journal of the Transportation Research Board*, 2006. 1948: 135–141.
18. Khattak, A., and K. Knapp. Interstate Highway Crash Injuries During Winter Snow and Non-Snow Events. *Transportation Research Record: Journal of the Transportation Research Board*, 2001. 1746: 30–36.
19. Young, R., and J. Liesman. Estimating the Relationship Between Measured Wind Speed and Overturning Truck Crashes Using a Binary Logit Model. *Accident Analysis and Prevention*, Vol. 39, No. 3, 2007, pp. 574–580.
20. Eisenberg, D., and K. Warner. Effects of Snowfalls on Motor Vehicle Collisions, Injuries and Fatalities. *American Journal of Public Health*, Vol. 95, 2005, pp. 120–124.
21. Andersson, A., and L. Chapman. The Impact of Climate Change on Winter Road Maintenance and Traffic Accidents in West Midlands, UK. *Accident Analysis and Prevention*, Vol. 43, No. 1, 2011, pp. 284–289.
22. Oh, S., K. Chung, and C. Chan. Analysis of Wet Weather Related Collision Concentration Locations: Empirical Assessment of Continuous Risk Profile. Presented at 88th Annual Meeting of the Transportation Research Board, Washington, D.C., 2009.
23. Wu, Y., M. Abdel-Aty, and J. Lee. Crash Risk Analysis During Fog Conditions Using Real-Time Traffic Data. *Accident Analysis & Prevention*, Vol. 114, 2017, pp. 4–11.
24. Ahmed, M. M., M. Abdel-Aty, J. Lee, and R. Yu. Real-Time Assessment of Fog-Related Crashes Using Airport Weather Data: A Feasibility Analysis. *Accident Analysis & Prevention*, Vol. 72, 2014, pp. 309–317.
25. Wu, Y., M. Abdel-Aty, Q. Cai, J. Lee, and J. Park. Rear-End Crash Risk Algorithm under Fog Condition Based on Kinematics Analysis. Presented at 96th Annual Meeting of the Transportation Research Board, Washington, D.C., 2017.
26. Mulinazzi, T., and S. Schrock. *Mitigating Wind Induced Truck Crashes*. Report # MATC-KU: 263, Mid-America Transportation Center, 2010.
27. Hermans, E., T. Brijs, T. Stiers, and C. Offermans. The Impact of Weather Conditions on Road Safety Investigated on an Hourly Basis. Presented at 85th Annual Meeting of the Transportation Research Board, Washington, D.C., 2006.
28. Baker, C.J., and S. Reynolds. Wind-Induced Accidents of Road Vehicles. *Accident Analysis & Prevention*, Vol. 24, No. 6, 1992, pp. 559–575.
29. Wanvik, P. Effects of Road Lighting: An Analysis Based on Dutch Accident Statistics 1987–2006. *Accident Analysis and Prevention*, Vol. 41, No. 1, 2009, pp. 123–128.
30. Lord, D., and F. Mannering. The Statistical Analysis of Crash-Frequency Data: A Review and Assessment of Methodological Alternatives. *Transportation Research Part A*, Vol. 44, No. 5. 2010, pp. 291–305.
31. Mannering, F., and C. Bhat. Analytic Methods in Accident Research: Methodological Frontier and Future Directions. *Analytic Methods in Accident Research*, Vol. 1. 2014. pp. 1–22.
32. Koetse, M., and P. Rietveld. The Impact of Climate Change and Weather on Transport: An Overview of Empirical Findings. *Transportation Research Part D*, Vol. 14, No. 3. 2009, pp. 205–221.
33. Theofilatos, A., and G. Yannis. A Review of the Effect of Traffic and Weather Characteristics on Road Safety. *Accident Analysis & Prevention*, Vol. 72, 2014, pp. 244–256.
34. Ye, X., R. Pendyala, S. Washington, K. Konduri, and J. Oh. A Simultaneous Equations Model of Crash Frequency by Collision Type for Rural Intersections. *Safety Science*, Vol. 47, No. 3, 2009, pp. 443–452.

35. Ye, X., R. Pendyala, V. Shankar, and K. Konduri. A Simultaneous Equations Model of Crash Frequency by Severity Level for Freeway Sections. *Accident Analysis and Prevention*. Vol. 57, 2013, pp. 140–149.
36. Wang, C., M. A. Quddus, and S. G. Ison. Predicting Accident Frequency at Their Severity Levels and Its Application in Site Ranking Using a Two-Stage Mixed Multivariate Model. *Accident Analysis and Prevention*, Vol. 43, 2011, pp. 1979–1990.
37. Chiou, Y., and C. Fu. Modeling Crash Frequency and Severity Using Multinomial Generalized Poisson Model with Error Components. *Accident Analysis and Prevention*. Vol. 50, 2013, pp. 73–82.
38. Chiou, Y., C. Fu, and H. Chih-Wei. Incorporating Spatial Dependence in Simultaneously Modeling Crash Frequency and Severity. *Analytic Methods in Accident Research*. Vol. 2, 2014, pp. 1–11.
39. Chiou, Y., and C. Fu. Modeling Crash Frequency and Severity with Spatiotemporal Dependence. *Analytic Methods in Accident Research*. Vol. 5–6, 2015, pp. 43–58.
40. Cai, Q., M. Abdel-Aty, and J. Lee. Macro-level Vulnerable Road Users Crash Analysis: A Bayesian Joint Modeling Approach of Frequency and Proportion. *Accident Analysis & Prevention*, Vol. 107, 2017, pp. 11–19.
41. Pei, X., S. Wong, and N. Sze. A Joint-probability Approach to Crash Prediction Models. *Accident Analysis and Prevention*, Vol. 43, No. 3, 2011, pp. 1160–1166.
42. Zeng, Q., H. Huang, X. Pei, and S. Wong. Modeling Nonlinear Relationship Between Crash Frequency by Severity and Contributing Factors by Neural Networks. *Analytic Methods in Accident Research*. Vol. 10, 2016, pp. 12–25.
43. WinBUGS. MRC Biostatistics Unit. University of Cambridge. https://appdb.winehq.org/objectManager.php?sClass=version&iId=17849&iTestingId=44665
44. Ma, J., and K. Kockelman. Bayesian Multivariate Poisson-Lognormal Regression for Models of Injury Count by Severity. *Transportation Research Record: Journal of the Transportation Research Board*, 2006. 1950: 24–34.
45. Ma, J., K. Kockelman, and P. Damien. A Multivariate Poisson-Lognormal Regression Model for Prediction of Crash Counts by Severity, Using Bayesian Methods. *Accident Analysis and Prevention*. Vol. 40, No. 3. 2008, pp. 964–975.
46. Park, E. S., and D. Lord. Multivariate Poisson-Lognormal Models for Jointly Modeling Crash Frequency by Severity. *Transportation Research Record: Journal of the Transportation Research Board*, 2007. 2019: 1–6.
47. Lee, J., M. Abdel-Aty, and X. Jiang. Multivariate Crash Modeling for Motor Vehicle and Non-motorized Models at the Macroscopic Level. *Accident Analysis and Prevention*, Vol. 78, 2015, pp. 146–154.
48. Chib, S., and R. Winkelmann. Markov Chain Monte Carlo Analysis of Correlated Count Data. *Journal of Business and Economic Statistics*, Vol. 19, No. 4, 2001, pp. 428–435.
49. Aguero-Valverde, J., and P. P. Jovanis. Bayesian Multivariate Poisson Log-normal Models for Crash Severity Modeling and Site Ranking. Presented at 88th Annual Meeting of the Transportation Research Board, Washington, D.C., 2009.
50. El-Basyouny, K., and T. Sayed. Collision Prediction Models Using Multivariate Poisson-Lognormal Regression. *Accident Analysis and Prevention*, Vol. 41, No. 4, 2009, pp. 820–828.
51. Zhao, M., C. Liu, W. Li, and A. Sharma. Multivariate Poisson-Lognormal Model for Analysis of Crashes on Urban Signalized Intersections Approach. *Journal of Transportation Safety & Security*, Vol. 10, No. 3, 2018, pp. 251–265.
52. Anastasopoulos, P., V. Shankar, J. Haddock, and F. Mannering. A Multivariate Tobit Analysis of Highway Accident-Injury-Severity Rates. *Accident Analysis and Prevention*, Vol. 45, 2012, pp. 110–119.
53. Dong, C., D. Clarke, X. Yan, A. Khattak, and B. Huang. Multivariate Random-Parameter Zero-Inflated Negative Binomial Regression Model: An Application to Estimate Crash Frequencies at Intersections. *Accident Analysis and Prevention*, Vol. 70, 2014, pp. 320–329.
54. Barua, S., K. El-Basyouny, and M. Islam. A Full Bayesian Multivariate Count Data Model of Collision Severity with Spatial Correlation. *Analytic Methods in Accident Research*, Vol. 3–4. 2014, pp. 28–43.
55. Wang, Y., and K. Kockelman. A Poisson-Lognormal Conditional-Autoregressive Model for Multivariate Spatial Analysis of Pedestrian Crash Counts Across Neighborhoods. *Accident Analysis and Prevention*, Vol. 60, 2013, pp. 71–81.
56. Barua, S., K. El-Basyouny, and M. Islam. Multivariate Random Parameters Collision Count Data Models with Spatial Heterogeneity. *Analytic Methods in Accident Research*, Vol. 9, 2016, pp. 1–15.
57. Rue, H., and S. Martino. Approximate Bayesian Inference for Hierarchical Gaussian Markov Random Field Models. *Journal of Statistical Planning and Inference*, Vol. 137, No. 10., 2007, pp. 3177–3192.
58. Serhiyenko, V., S. Mamun, J. Ivan, and N. Ravishanker. Fast Bayesian Inference for Modeling Multivariate Crash Counts. *Analytic Methods in Accident Research*, Vol. 10, 2016, pp. 44–53.
59. Wang, K., J. Ivan, N. Ravishanker, and E. Jackson. Multivariate Poisson Lognormal Modeling of Crashes by Type and Severity on Rural Two Lane Highways. *Accident Analysis and Prevention*, Vol. 99, 2017, pp. 6–19.
60. Aitchison, J., and C.H. Ho. The Multivariate Poisson Lognormal Distribution. *Biometrika*. Vol. 76, No. 4, 1989, pp. 643–653.
61. Chib, S., and R. Winkelmann. Markov Chain Monte Carlo Analysis of Correlated Count Data. *Journal of Business and Economic Statistics*, Vol. 19, No. 4, 2001, pp. 428–435.
62. *Connecticut Department of Transportation*. http://www.ct.gov/dot/site/default.asp
63. *National Center for Environmental Information*. National Oceanic and Atmospheric Administration. Asheville, NC. https://www.ncdc.noaa.gov/cdo-web/datatools/lcd
64. *Connecticut Crash Data Repository*. University of Connecticut. http://www.ctcrash.uconn.edu/
65. *Model Minimum Uniform Crash Criteria (MMUCC)*. Governors Highway Safety Association, National Highway Traffic Safety Administration. https://www.transport

ation.gov/government/traffic-records/model-minimum-uniform-crash-criteria-mmucc-0

66. Roy, U., and S. Dissanayake. Comparison of Factors Associated with Run-off-road and Non-run-off-road Crashes in Kansas. *Journal of Transportation Research Forum*, Vol. 50, No. 2, 2011, pp. 69–86.
67. Liu, C., and T. Ye. Run-off-road Crashes: An On-scene Perspective. *National Highway Traffic Safety Administration*. Final Report of National Center for Statistics and Analysis. Report No. *DOT HS 811 50*. 2011. US Department of Transportation.
68. Wang, K., J. Ivan, A. Burnicki, and S. Mamun. Predicting Local Road Crashes Using Socioeconomic and Land Cover Data. *Journal of Transportation Safety & Security*. Vol. 9, No. 3, 2017, pp. 301–318.

The Standing Committee on Safety Data, Analysis and Evaluation (ANB20) peer-reviewed this paper (18-01509).

Article

TRR
JOURNAL OF THE TRANSPORTATION RESEARCH BOARD

Transportation Research Record
2018, Vol. 2672(38) 199–209

DOI: 10.1177/0361198118773576
journals.sagepub.com/home/trr
SAGE

Influence of Type of Traffic Control on Injury Severity in Bicycle–Motor Vehicle Crashes at Intersections

Rabbani Rash-ha Wahi[1], Narelle Haworth[1], Ashim Kumar Debnath[2], and Mark King[1]

Abstract

Many studies have identified factors that contribute to bicycle–motor vehicle (BMV) crashes, but little is known about determinants of cyclist injury severity under different traffic control measures at intersections. Preliminary analyses of 5,388 police-reported BMV crashes from 2002 to 2014 from Queensland, Australia revealed that cyclist injury severity differed according to whether the intersection had a Stop/Give-way sign, traffic signals or no traffic control. Therefore, separate mixed logit models of cyclist injury severity (fatal/hospitalized, medically treated, and minor injury) were estimated. Despite similar distributions of injury severity across the three types of traffic control, more factors were identified as influencing cyclist injury severity at Stop/Give-way controlled intersections than at signalized intersections or intersections with no traffic control. Increased injury severity for riders aged 40–49 and 60 + and those not wearing helmets were the only consistent findings across all traffic control types, although the effect of not wearing helmets was smaller at uncontrolled intersections. Cyclists who were judged to be at fault were more severely injured at Stop/Give-way and signalized intersections. Speed zone influenced injury severity only at Stop/Give-way signs and appears to reflect differences in intersection design, rather than speed limits per se. While most BMV crashes occurred on dry road surfaces, wet road surfaces were associated with an increased cyclist injury severity at Stop/Give-way intersections. The results of this study will assist transport and enforcement agencies in developing appropriate mitigation strategies to improve the safety of cyclists at intersections.

Intersections vary in their complexity, and more complex intersections are associated with increased severity of injury for cyclists involved in crashes with motor vehicles (*1*). Traffic control measures at intersections are designed to maximize traffic flow and minimize crash risk. This study was conducted in Australia where bicycles are required to obey the same rules in relation to traffic controls as motor vehicles. At "Stop" signs, all road users must come to a complete stop and yield to all oncoming traffic. "Give-way" signs correspond to "Yield" signs in other countries and road users must slow down or stop, if required, and yield to all oncoming traffic. Traffic signals are of the conventional three-phase design, with only a small number of intersections allowing traffic to turn left on red (traffic travels on the left in Australia). Intersections on minor and residential roads may have "Give-way" signs or no traffic control.

Although many studies have looked at the factors contributing to cyclist injury severity and provide abundant guidance, few studies have specifically focused on the influence of type of traffic control. For example, studying the cyclist characteristics shows that males were more likely to be severely injured than female cyclists (*2*, *3*) and older cyclists were more likely to sustain severe injury in BMV crashes than their younger cyclist counterparts (*4*, *5*). In relation to the environmental factors, adverse weather conditions such as wet surfaces and fog were associated with an increase in the probability of cyclist injury and its severity (*6*). Land use variables, such as proportions of industrial and commercial land use, were critical factors associated with cyclist injury severity (*7*) but the land use mixture and population density were not significantly associated with cyclist injury severity (*8*). Urban roads also increased cyclist injury severity relative to rural roads (*9*). Other infrastructure

[1]Queensland University of Technology (QUT), Centre for Accident Research and Road Safety- Queensland (CARRS-Q), QLD, Australia
[2]Victoria University (VU), VIC, Australia

Corresponding Author:
Address correspondence to Rabbani Rash-ha Wahi:
rabbani.wahi@hdr.qut.edu.au

characteristics such as bicycle paths (*10*), intersections (*11*), and roundabouts (*12*) were identified as significant factors associated with more severe cyclist injuries. Apart from infrastructure characteristics, increasing posted speed limits was identified as a further significant factor influencing cyclist injury severity (*13*).

Some studies have attempted to identify the factors that contribute to cyclist injury severity at intersections. For example, Eluru, Bhat and Hensher (*14*) found higher cyclist injury severities at unsignalized intersections than at signalized intersections. Another recent study by Wang, Lu and Lu (*15*) focused on unsignalized intersections, and found that stop controlled intersections were associated with higher injury severities for cyclists compared with uncontrolled intersections. This suggests the need to separately examine cyclist injury severity for each type of intersection control.

There are several reasons why type of traffic control at intersections would influence injury severity. First, cyclists may have difficulties in clearing signalized intersections before motorized cross traffic receives a green indication, and whenever the traffic signals go green, motor-vehicles are more likely to proceed through intersections at high speed with less visual search than they would at unsignalized intersections (*16*). Second, cyclists at uncontrolled intersections can misjudge the distance and speed of a vehicles on the through approach (*15*). Finally, drivers are more likely to be at fault in failing to stop or reduce motor vehicle speed before entering intersections controlled by a stop or give-way sign (*17*).

Therefore, the main objective of this study is to identify the significant factors that affect cyclist injury severity resulting from BMV crashes under various traffic control measures at intersections to better inform measures to improve the safety of cyclists.

Method

Setting

This research was conducted in Australia's northeastern state of Queensland where the climate varies from subtropical to tropical, allowing cyclists to ride throughout the year. The most recent national cycling participation survey reported that about 24% of the Queensland population rode a bicycle in the previous month and among them 40% used a bicycle for transport (*18*). Most urban roads in Queensland have a speed limit of 60 km/h, while the default speed limit in built-up areas is 50 km/h. Vehicles drive on the left side of the road. In Queensland, it is legal for cyclists of any age to ride on the sidewalk (footpath) with pedestrians (they are required to keep left and give way to pedestrians) unless it is signed "no bikes", and wearing helmets for all ages is mandatory. Off-road bicycle facilities are not consistently available across Queensland, requiring cyclists to ride on the road in many areas.

Data Description

A dataset of police-reported multi-vehicle bicycle crashes (n = 5,772) at intersections from January 1, 2002 to December 31, 2014 in Queensland was supplied by the State Department of Transport and Main Roads. It contains data on crash-involved bicyclists and drivers, crash characteristics, roadway geometry, and environmental conditions associated with each crash. Crashes occurring on public roadways leading to injury or property damage greater than $2,500 or a vehicle being towed away are required to be reported to Police in Queensland. Crashes occurring on private property (e.g., car parks) and road reserves (e.g., off-road paths) are not included. BMV crashes are under-reported in police data compared with hospital data (*19*), but the police data is the only source of information about the crash itself (e.g., crash location, type of traffic control, road surface condition). BMV crashes are more likely to be reported than single bicycle crashes given their high impact force, higher likelihood of causing injuries, and associated compensation issues (*20, 21*).

Police officers code injury severity into the following categories: minor injury (first-aid only or extent of injury unknown), medically treated injury (injury crash requiring medical treatment from a general practitioner or hospital but no overnight stay), hospitalized injury (injury crash requiring hospitalization), and fatality (person died from injuries within thirty days of a crash) (*22*).

The unit judged by the police officer to be most at fault is labelled as "Unit 1". Alcohol involvement by a rider was coded as a contributing circumstance, but there was no distinction between "alcohol not involved" and "not reported."

Bicycle-bicycle crashes (n = 62) were excluded as the severity of collisions between non-motorized vehicles was expected to be different from that of collisions with motorized vehicles. Bicycle-pedestrian crashes were not included in the dataset of BMV crashes. BMV crashes which occurred in very remote or unknown locations (n = 49) were also excluded owing to high levels of under-reporting (*23*). Bicycle–motor vehicle crashes involving more than two units (n = 197) were excluded from the dataset to keep the analysis focused on two-unit crashes. Where a bicycle pillion passenger was injured, their data was excluded. Cyclist casualties were excluded (n = 76) in cases where an injury had occurred other than under the three different traffic control types. The resulting dataset contains 5,388 two-vehicle crashes involving a bicycle and a motorized vehicle (car, bus, truck, motorcycle, utility or panel van) resulting in 1,086 minor injuries, 2,182 medical treatment injuries, 2,081 hospitalized injuries, and 39 bicyclist fatalities.

Statistical Methodology

This section presents the modelling methodology adopted in this study. The severity levels of BMV crashes were grouped into three discrete categories: minor, medically treated, and hospital/fatal (given the low frequency of fatalities the single "hospital/fatal" category was used). Discrete outcome models which are either ordinal or nominal, such as the ordered logit/probit and multinomial logit models (MNL), have been successfully applied to analyze crash injury severity (e.g., *14*, *24*). However, researchers have discussed the limitations and strengths of the ordered and multinomial models (for more details see *25*). For example, traditional ordered models restrict the effect of variables either to increasing the probability of the highest severity category and decreasing the lowest, or vice versa. On the other hand, multinomial logit and probit models neglect the data's ordinality, require estimation of more parameters, and the effect of unobserved heterogeneity in the parameters is not considered. To overcome such limitations, the mixed logit model (MXL) is used in this study to identify contributing factors to each severity category. The major advantage of the mixed logit approach is that it relaxes the independence from irrelevant alternatives (IIA) property and can also accommodate unobserved heterogeneity by allowing parameters for the exogenous variables to vary across observations (*26*). Following the approach of Pai, Hwang, and Saleh (*27*), and Islam, Jones, and Dye (*28*), the injury severity function determining injury severity outcome is defined as

$$R_{in} = \beta_i X_{in} + \varepsilon_{in} \quad (1)$$

where R_{in} is a function of severity category i (e.g., minor injury, medically treated injury) for crash n, X_{in} is a vector of explanatory factors (e.g., roadway characteristics, environmental characteristics, rider characteristics) which affect the injury outcome for crash n,β_i is the vector of estimable coefficients for injury severity outcome i, and ε_{in} is the error term. If ε_{in} is assumed to be independently distributed and following an extreme value distribution, then the standard MNL can be expressed as

$$P_n(i) = \frac{\exp[\beta_i X_{in}]}{\sum_{\forall I} \exp[\beta_i X_{in}]} \quad (2)$$

where $P_n(i)$ is the probability that cyclist n suffers injury outcome i, and i is the full set of all injury severity categories (e.g., minor injury, medical treated injury, and fatality). The MXL is a generalization of the MNL structure. It allows the parameter vector β_i to be either fixed or randomly distributed for each element of the parameters with fixed means, allowing for heterogeneity within the observed crash data set (*29*)

$$P_n(i|\varphi) = \int \frac{\exp[\beta_i X_{in}]}{\sum_{\forall I} \exp[\beta_i X_{in}]} f(\beta|\varphi) d\beta \quad (3)$$

where $P_n(i|\varphi)$ is the outcome probability for condition $f(\beta|\varphi)$, which represents the density function of β and φ denotes a vector of parameters describing the density function (mean and variance). The values of β are used to capture the individual cyclist-specific variation of the effect of X_{in} on injury severity across each observation, with the density function $f(\beta|\varphi)$.

Elasticity analysis was conducted in this study to assess the effects of individual parameters on rider injury outcome probabilities. As shown in Table 1, all explanatory variables are indicator variables (those with the value of 0 or 1), where pseudo-elasticity can be calculated as

$$E_{X_{in}}^{P_n(i)} = \frac{P_n(i|X_{in} = 1) - P_n(i|X_{in} = 0)}{P_n(i|X_{in} = 0)} \quad (4)$$

The pseudo-elasticity of a variable specific to an injury severity outcome indicates the percentage effect on the probability when an indicator variable is switched from a value of zero to one (*30*), where X_{in} is the nth independent variable associated with probability $P_n(i)$ of n individual crash experiencing i injury severity outcome. The average pseudo-elasticity (APE) is widely utilized for its computational cost-efficiency (*1*, *31*).

Mixed logit models were estimated for intersections controlled by three different control measures – Model 1: Stop/Give-way sign, Model 2: No traffic control, and Model 3: Operating traffic signals. In all models, a backward elimination fitting method was performed to drop the non-significant variables in an iterative process so that the Akaike Information Criterion (AIC) was minimized. When the estimated random parameters were non-significant, then the parameters become fixed parameters if their standard deviations are not statistically significant ($p < 0.05$). As demonstrated in previous research (*32*, *33*), a simulation-based maximum likelihood with 200 Halton draws was used for accurate parameter estimates. For the random parameters' density functional forms, the normal distribution gave the best fit results among the normal, uniform, and lognormal distributions for injury severity data, which is in line with past studies (*27*, *31*). In this study, the statistical software tool, NLOGIT 6, was used for model parameter estimation.

Empirical Results

Descriptive Statistics

A total of 58 explanatory variables were examined in the mixed logit model. The details of these variables, including frequencies for each type of traffic control, are

Table 1. Rider Injury Frequency and Percentage under Different Traffic Control Measures at Intersections

Variable name	Stop/Give-way sign- frequency (%)	No traffic control- frequency (%)	Operating traffic signals- frequency (%)
Injury severity			
Fatal & Hospitalized	908 (36.93%)	815 (41.35%)	397 (41.44%)
Medical treatment	1051 (42.74%)	767 (38.91%)	364 (38.00%)
Minor	500 (20.33%)	389 (19.74%)	197 (20.56%)
Gender of rider			
Male	1,966 (79.95%)	1,606 (81.48%)	771 (80.48%)
Female	490 (19.93%)	364 (18.64%)	184 (19.21%)
Unknown	3 (0.12%)	1 (0.05%)	4 (0.42%)
Age of rider			
0–15	274 (11.14%)	464 (23.54%)	192 (20.04%)
16–20	214 (8.70%)	200 (10.15%)	137 (14.30%)
21–24	195 (7.93%)	166 (8.42%)	98 (10.23%)
25–39	749 (30.46%)	523 (26.53%)	308 (32.15%)
40–49	475 (19.32%)	314 (15.93%)	106 (11.06%)
50–59	309 (12.57%)	168 (8.52%)	59 (6.16%)
60 and over	231 (9.39%)	118 (5.99%)	31 (3.24%)
Unknown	12 (0.49%)	18 (0.91%)	17 (1.77%)
Helmet use			
Worn	2060 (83.73%)	1,514 (76.81%)	718 (74.95%)
Not worn	158 (6.38%)	218 (11.06%)	113 (11.80%)
Unknown	241 (9.80%)	239 (12.13%)	127 (13.26%)
Alcohol involvement			
Yes	28 (1.14%)	43 (2.18%)	19 (1.98%)
No/Unknown	2,431 (98.86%)	1,928 (97.82%)	939 (98.02%)
Day			
Weekday	1,989 (80.89%)	1,615 (81.94%)	785 (81.94%)
Weekend	470 (19.11%)	356 (18.06%)	173 (18.06%)
Time period			
00.00 am–5.59 am	147 (5.98%)	94 (4.77%)	40 (4.18%)
6.00 am–11.59 am	1,208 (49.13%)	794 (40.28%)	357 (37.27%)
12.00 pm–5.59 pm	845 (34.36%)	838 (42.52%)	424 (44.15%)
6.00 pm–11.59 pm	259 (10.53%)	245 (12.43%)	137 (14.30%)
Road type			
Cross-intersection	463 (18.83%)	253 (12.84%)	604 (63.05%)
T junction	923 (37.54%)	1,578 (80.01%)	330 (34.45%)
Roundabout	1,041 (42.33%)	104 (5.23%)	0 (0.00%)
Other intersections	32 (1.30%)	36 (1.73%)	24 (2.40%)
Road condition			
Dry	2,256 (91.74%)	1,836 (93.15%)	888 (92.69%)
Wet	202 (8.21%)	134 (6.80%)	69 (7.20%)
Unknown	1 (0.05%)	1 (0.05%)	1 (0.10%)
Horizontal alignment			
Straight	2,311 (93.98%)	1,871 (94.93%)	925 (96.56%)
Curved obscured	16 (0.65%)	16 (0.81%)	3 (0.31%)
Curved open	132 (5.37%)	84 (4.26%)	30 (3.13%)
Vertical alignment			
Crest	65 (2.64%)	58 (2.94%)	18 (1.88%)
Dip	67 (2.72%)	73 (3.70%)	16 (1.67%)
Grade	321 (13.05%)	383 (19.43%)	127 (13.26%)
Level	2,006 (81.58%)	1,457 (73.92%)	797 (83.19%)
Speed zone (km/h)			
0≤50	707 (28.75%)	658 (33.38%)	143 (14.93%)
60	1,604 (65.23%)	1,214 (61.59%)	708 (73.80%)
70	71 (2.89%)	54 (2.74%)	80 (8.35%)
80–90	69 (2.81%)	32 (1.62%)	22 (2.40%)
100–110	8 (0.33%)	13 (0.66%)	5 (0.52%)
Crash type			
Intersection from adjacent approaches	956 (38.88%)	542 (27.50%)	238 (24.84%)

(continued)

Table 1. Continued

Variable name	Stop/Give-way sign- frequency (%)	No traffic control- frequency (%)	Operating traffic signals- frequency (%)
Opposing vehicle turning	189 (7.69%)	228 (11.57%)	123 (12.84%)
Lane changes	167 (6.79%)	128 (6.49%)	70 (7.31%)
Parallel lane turning	170 (6.91%)	150 (7.61%)	85 (8.87%)
Vehicle leaving driveway	447 (18.18%)	411 (20.85%)	210 (21.92%)
Others	530 (21.55%)	512 (25.98%)	232 (24.22%)
Light condition			
Daylight	1,916 (77.92%)	1,558 (79.08%)	759 (79.23%)
Darkness-lighted	252 (10.25%)	199 (10.10%)	134 (13.99%)
Darkness-unlighted	47 (1.91%)	51 (2.59%)	3 (0.31%)
Dusk	243 (9.88%)	160 (8.12%)	60 (6.26%)
Unknown	1 (0.04%)	3 (0.15%)	2 (0.21%)
At-fault status			
Cyclist at fault	671 (27.31%)	814 (41.32%)	471 (49.16%)
Cyclist not at fault	1,788 (72.69%)	1,157 (58.68%)	487 (50.84%)
Remoteness classification			
Major city	1,511 (61.45%)	1,342 (68.09%)	635 (66.28%)
Inner regional	403 (16.39%)	286 (14.51%)	133 (13.88%)
Outer regional	524 (21.31%)	327 (16.59%)	183 (19.10%)
Remote	21 (0.85%)	16 (0.81%)	7 (0.73%)

presented in Table 1. The final merged dataset includes 5,388 crashes, with 2,459 (45.6%) at Stop/Give-way sign controlled intersections; 1,971 (36.5%) at intersections with no traffic control; and 958 (17.7%) at signalized intersections. About 80% of the riders were male and almost one-third were aged from 25 to 39 years (29.5%). Just over the half of the crashes (52.5%) occurred at T-intersections. About two-thirds (65.4%) of the crashes occurred in areas with a speed limit of 60 km/h, and almost all (94.4%) occurred while the cyclist was riding straight ahead. The most common crash type involved adjacent approaches (32.1%). Further examination of the 'other' crashes revealed they were composed of less than 50 of each of the following crash types: hit parked vehicle, head-on, off-carriageway on curve, off-carriageway on straight, rear-end, U-turn, and other. Daytime (78.5%) crashes resulted in more severe injuries compared with night-time. A total of 64.7% of all crashes occurred in major cities, followed by 19.1% in outer regional areas. Drivers were most at fault in 63.6% of the crashes, while cyclists were at fault in 36.3% of the crashes. Few (1.6%) of the cyclists were intoxicated.

Model Estimation Results

Table 2 shows the estimation results of the Stop/Give-way sign, No traffic control, and Operating traffic signals mixed logit models, including parameter coefficients and *P*-values, that is, levels of significance of a parameter different from zero. The three categories of injury severity, *minor injury, medical treatment injury*, and *fatality/hospitalized injury*, are labelled as MI, M, and F/H, respectively. The log-likelihood values of the Stop/Give-way sign, No traffic control, and Operating traffic signals mixed logit models are −2,488.11, −2,039.50, and −951.29, respectively. The corresponding McFadden pseudo- ρ^2 values are 0.07, 0.05, and 0.09; indicating a reasonable level of statistical fit for the injury severity models. All of the estimated parameters included in these three models are statistically significant at a 0.05 significance level or better.

Once the mixed logit models were developed, likelihood ratio tests (*30*) were performed to justify the necessity to estimate different traffic control measures in this study. The $\chi2$ statistic ($\chi2$ = 237.84), with degrees of freedom equal to the summation of the number of estimated parameters in all traffic control measures models minus the number of estimated parameters in the overall model, provides the confidence level at which we can reject the null hypothesis. The null hypothesis states that full model parameter estimates are no better than the separate model estimate. The Chi square statistics for the likelihood ratio test with 58 degrees of freedom resulted in a value greater than the 99.99% confidence limit ($\chi2$ = 97.03), suggesting that the estimation of separate models is statistically warranted.

Model Discussions

The significant variables for each severity outcome are discussed in the following sections, along with the elasticities of each explanatory variable with respect to the injury severity outcomes presented in Table 3.

Table 2. Mixed Logit Injury Severity Models for BMV Crashes under Various Traffic Control Measures at Intersections

Variable name	Stop/Give-way sign coefficients (*P*-value)	No traffic control coefficients (*P*-value)	Operating traffic signals coefficients (*P*-value)
Dependent variable[a]			
Constant [fatal/hospitalized]	−0.255 (0.360)	−0.073 (0.368)	−0.049 (0.868)
Constant [minor]	−0.367 (0.542)	−0.559 (0.000)	0.002 (0.987)
Age of rider			
16–20 [M]	–	0.362 (0.018)	–
40–49 [F/H]	0.334 (0.007)	0.384 (0.003)	0.764 (0.002)
50–59 [F/H]	–	0.489 (0.003)	–
60 and over [F/H]	0.765 (0.000)	–	1.195 (0.000)
60 and over [MI]	–	−0.966 (0.006)	–
Helmet use			
Worn [MI]	−1.287(0.000)	–	−0.932 (0.000)
Worn [M]	−0.414 (0.033)	–	–
Not worn [F/H]	0.989 (0.000)	0.685 (0.000)	0.844 (0.000)
Alcohol involvement			
Yes [F/H]	0.961 (0.038)	–	–
Road condition			
Wet [F/H]	0.483 (0.019)	–	–
Vertical alignment			
Crest [F/H]	–	0.587 (0.031)	–
Dip (F/H)	–	–	1.328 (0.043)
Speed zone (km/h)			
0–50 [M]	0.575 (0.009)	–	–
60 [F/H]	−0.564 (0.005)	–	–
Crash type			
Intersection from adjacent approaches [F/H]	0.354 (0.009)	–	–
Vehicle leaving driveway [F/H]	–	–	0.355 (0.032)
Others [M]	–	0.246 (0.036)	–
Light condition			
Daylight [F/H]	−0.409 (0.001)	–	–
At-fault status			
At fault [F/H]	0. 582 (0.000)	–	1.042 (0.000)
Model statistics			
Log-likelihood	−2488.11	−2039.50	−951.29
Restricted log-likelihood	−2690.50	−2163.16	−1050.27
McFadden, ρ^2	0.07	0.05	0.09
Number of observations	2459	1971	958

Note: –, not found significant.
[a]Letters in parentheses indicate variable coefficients are significant specific to: [MI] Minor injury, [M] Medical treatment injury, and [F/H] Fatality/Hospitalized injury. For the constants, the medical treatment outcome has, without loss of generality, its coefficient normalized to zero.

Rider Characteristics. There were several significant results related to the age group of the rider, but the elasticities were generally less than 5%, indicating relatively small effects. Young riders (aged 16–20) were associated with an increased probability of medical treatment injury (1.95%) at no traffic control intersections, but not at Stop/Give-way and signalized intersections. This result is consistent with earlier findings in Taiwan (*34*). Riders aged 40–49 were associated with an increased probability of fatal/hospitalized injury at Stop/Give-way sign, un-controlled, and signalized intersections by 3.55%, 3.26%, and 3.85%, respectively. The marginal effects further show that cyclists aged 50–59 years had a 2.11% increased probability of fatal/hospitalized injury at no traffic control intersections. These results indicate that riders aged 40–59 are especially vulnerable. This may reflect the greater propensity for riders of this age group in Queensland to ride racing bikes (*35*), with faster riding speeds resulting in less time to respond by both rider and driver, and higher impact speeds, both increasing the severity of injury. Older cyclists (over 60) were 3.39% more likely to experience fatal/hospitalized injury severity at Stop/Give-way signs; and 1.50% more likely to experience fatal/hospitalized injury severity at signalized intersections. This could be because some older cyclists have delayed perceptions, slower reaction times, and physical frailty which increase the risk and severity of injury (*36*). Older cyclists (over 60) were 5.30% less likely to experience minor injury at uncontrolled intersections. Further analysis identified that drivers were more often at

Table 3. Comparisons of Variable Elasticities between Different Traffic Control Measures at Intersections

Explanatory variables	Stop/Give-way sign	No traffic control	Operating traffic signals
Age of rider			
16–20 [M]	–	1.95%	–
40–49 [F/H]	3.55%	3.26%	3.85%
50–59 [F/H]	–	2.11%	–
60 and over [F/H]	3.39%	–	1.50%
60 and over [MI]	–	–5.30%	–
Helmet use			
Worn [MI]	–86.39%	–	–54.71%
Worn [M]	–17.65%	–	
Not worn [F/H]	2.89%	3.62%	4.32%
Alcohol involvement			
Yes [F/H]	0.36%	–	–
Road condition			
Wet [F/H]	1.61%	–	–
Vertical alignment			
Crest [F/H]	–	0.78%	–
Dip [F/H]	–	–	0.67%
Speed zone (km/h)			
0–50 [M]	8.31%	–	–
60 [F/H]	–22.2%	–	–
Crash type			
Intersection from adjacent approaches [F/H]	8.58%	–	–
Vehicle leaving driveway [F/H]	–	–	4.32%
Others [M]	–	2.07%	–
Light condition			
Daylight [F/H]	–18.4%	–	–
At-fault status			
At fault [F/H]	7.89%	–	24.15%
At fault [MI]	3.15%	–	–

Note: –, not found significant; Variables are defined for outcomes: [MI] Minor injury, [M] Medical treatment injury, [F/H] Fatality/Hospitalized injury.

fault (70%) in such locations in crashes involving a vehicle that was leaving a driveway.

Helmet use is mandatory in Queensland and observed wearing rates are in excess of 98% (*37*). The results show that not wearing a helmet increased the likelihood of fatal/hospitalized injury by 2.89%, 3.62%, and 4.32% at Stop/Give-way sign, uncontrolled, and signalized intersections, respectively. Conversely, wearing a helmet reduced minor injury by 86.39% and 54.71% at Stop/Give-way sign and signalized intersections, respectively. It is well established that helmets protect against head, brain, neck, and facial injuries for all ages of cyclists (*38*) and the results are consistent with those of numerous past studies (*31*, *39*, *40*).

Based on police officers' judgement of at-fault status, the most-at-fault party involved in a crash was considered as the at-fault party whereas the other parties were considered as 'not-at-fault.'Table 1 shows that motor vehicle drivers are more likely to be at fault in BMV crashes at Stop/Give-way intersections than other intersections. When cyclists were at fault, the probability of fatal/hospitalized injury increased by 7.89% and 24.15% at Stop/Give-way and signalized intersections, respectively. In addition, when cyclists were at fault in BMV crashes at Stop/Give-way intersections the probability of minor injury increased by 3.15%. Schramm, Rakotonirainy and Haworth (*41*) noted that when the cyclist was found to be at fault in BMV crashes, the most frequent contributing factors were inattention, inexperience, and disobeying traffic signals. This suggests it is important to decrease illegal behavior among riders and drivers, as well as improving levels of attention and inexperience.

As shown in Table 3, cycling under the influence of alcohol was found to increase the probability of fatal/hospitalized injury by 0.36% at Stop/Give-way intersections. When cyclists have been drinking, they may be more likely to ride without lights at night or behave unpredictably which can increase the probability of injury (*42*). The low elasticity found for the presence of alcohol may have resulted from few people being tested (*43*) and the police-reported crash data not distinguishing between 'alcohol absent' and 'alcohol status unknown.' It is recommended that more data be collected on alcohol, which can provide greater insights into the prevalence and consequences of these behaviors.

Roadway Characteristics. Roadway conditions also had an effect on cyclist injury outcomes. The elasticity estimates suggest that at Stop/Give-way intersections, wet surfaces increased the risk of fatal and hospitalized injury (1.61%). Given that two-thirds of the BMV crashes occurred where the speed limit was 60 km/h, the combination of wet roads and relatively high speeds may be particularly perilous. On such wet surfaces, on roads with a relatively high posted speed limit (compared with the default 50 km/h urban speed limit), it would be difficult for the driver and cyclist to stop or slow down shortly before entering or approaching intersections given the high probability of skidding. Typically, skidding is more likely to happen on a wet surface because of the decreased friction between tire and road surface (*44*).

The models developed in this study show that crest alignment increases (0.78%) the probability of fatal and hospitalized injury at uncontrolled intersections, possibly owing to insufficient sight distance to avoid potential conflicts between vehicles turning or crossing. If sufficient sight distance is unable to be provided by the road designer, then more restrictive control should be implemented. Dips also increased (0.67%) the probability of fatal and hospitalized injury at signalized intersections. Most BMV crashes occurred in city areas which are generally hilly (*45*), resulting in poor visibility because drivers and cyclists cannot be sure whether or not there is an oncoming vehicle hidden beyond the rise.

Environmental Characteristics. The results of the models show that daylight was associated with reduced likelihood of fatal injury and severe injury at Stop/Give-way intersections. Generally the flow of traffic is higher during daylight (*18*), so travel speeds are lower and also visibility is better.

Crash Characteristics. BMV crashes at Stop/Give-way signs in 0–50 km/h speed limit zones were associated with an 8.31% increased probability of medically treated injury. Further analysis showed that most of the BMV crashes (58.22%) in zones with a speed limit of 50 km/h or lower occurred at T-junctions. This finding reinforces the need to provide an unobstructed view to drivers and cyclists approaching a give way condition or leaving from a stopped position at a T-junction to avoid potential BMV crashes (*46*). In 60 km/h speed zones, the likelihood of fatal/hospitalization injuries was 22.1% lower at Stop/Give-way intersections. Additional analysis identified about half of the crashes in 60 km/h zones (53%) occurred at roundabouts, a particular type of Stop/Give-way intersection. Roundabouts provide drivers more time to react and stop in a shorter distance (*47*), which is likely to reduce the severity of injury. It has been recommended that speed limits in areas of high cycling activity should be set at 30 km/h, because in lower-speed streets motor vehicles can stop more quickly, and severity of injury would be lower (*48*).

The modelling results also revealed that vehicles leaving driveways at signalized intersections increased the probability of hospitalized/fatal injury by 4.32%. Cyclists were more likely to be at fault (65.5%) in crashes that involved leaving driveways and may have failed to notice motorists on the road. Restricting turns or improving visibility at certain signalized intersections may be useful to avoid potential conflicting points among motorists and cyclists. In this study, the other most common crash type was collision at an intersection between vehicles from adjacent approaches. This variable is found to increase the probability of hospitalized/fatal injury by 8.58% at intersections controlled by Stop/Give-way signs. In this cyclist–motorist crash pattern, approximately 63% of cyclists with hospitalized/fatal injury had exposure to speeds between 50 km/h and 60 km/h. Results are consistent with past research (*45*), which shows that cyclists have difficulties in judging gap sizes and speed before deciding whether to initiate a roadway entry or a turning manoeuver.

Limitations

Despite having a systematic process of crash data collection and processing, the police-reported crash data suffer from issues related to under-reporting (*49*, *50*), which may lead to inaccurate model estimates. This study does not include single bicycle crashes, cyclist–cyclist crashes, and cyclist–pedestrian crashes as they are under-reported and it would be interesting to see if traffic control type is associated with the severity of these crashes at intersections.

Statistical models provide a hint about the prevalence of crashes with the different combinations of independent variables. Analyzing crash data with exposure data may provide more comprehensive predictions of injury risk. However, this may not be a problem because the dependent variable in this study is the injury severity.

Police crash data provides only limited detail on the explanatory variables associated with injury severity such as characteristics of crashes, roadway features, and riders. For example, details of road design (e.g., turning radius, presence of shared lanes or shoulder), road rules (e.g., warning or advisory signs), and rider-behavior related information (e.g., visual search skills, fluorescent clothing) might influence the crash characteristics, but are not recorded in policed-reported crash data.

A final limitation of the study is that it is restricted to crashes involving only two units (a bicycle and a motor vehicle). Given that the number of crashes involving more than two-units (n = 197) is low, this does not have much of an effect on the results, however it is recommended that future research should consider bicycle crashes involving more than two units to more comprehensively examine the risk of cyclist injury severity in BMV crashes.

Summary and Conclusions

More than half of the BMV crashes analyzed occurred at signalized intersections or intersections without traffic control but few factors were identified as significant contributors to cyclist injury severity at these locations. Despite similar distributions of injury severity across the three types of traffic control, more factors were identified as influencing cyclist injury severity at Stop/Give-way controlled intersections than at signalized intersections or intersections with no traffic control. Increased injury severity for riders aged 40–49 and 60 + and those not wearing helmets were the only consistent findings across all traffic control types, although the effect of not wearing helmets was smaller at uncontrolled intersections. Cyclists who were judged to be at fault were more severely injured at Stop/Give-way and signalized intersections. Speed zone influenced injury severity only at Stop/Give-way signs and appears to reflect differences in intersection design, rather than speed limits per se. Injuries tended to be more serious on wet surfaces, therefore providing skid resistant pavements on approaches to intersections controlled by Stop/Give-way signs would likely to be an effective countermeasure. Given that cyclist alcohol use also significantly influenced severity of injury, enforcement and education about the potential effects of alcohol use could influence safer behavior in cyclists. Regarding geometric variables, dips and crests were found to increase the probability of higher cyclist injury severity.

The differences in the models of injury severity for BMV crashes under various traffic controls at intersections suggest that not all intersections are equal when it comes to cyclist safety and that the type of traffic control is an important factor influencing cyclist safety. Further study needs to incorporate intersection design, land-use characteristics, and driver and cyclist behavior to provide a more detailed picture of BMV crashes at intersections.

Acknowledgments

The authors would like to acknowledge the Queensland Department of Transport and Main Road for the crash data provided for this study. The corresponding author sincerely thanks Dr Alexander Hainen of the University of Alabama for his helpful suggestions on assessing model performance as well as the anonymous reviewer for the insightful comments.

Author Contributions

The authors confirm contribution to the paper as follows: study conception and design: RRW, NH, AKD; analysis and interpretation of results: RRW, NH, AKD, MK; draft manuscript preparation: RRW; draft manuscript reviewing and editing: RRW, NH, AKD, and MK. All authors reviewed the results and approved the final version of the manuscript.

References

1. Kim, J.-K., S. Kim, G. F. Ulfarsson, and L. A. Porrello. Bicyclist Injury Severities in Bicycle–Motor Vehicle Accidents. *Accident Analysis & Prevention*, Vol. 39, No. 2, 2007, pp. 238–251.
2. Kim, K., and L. Li. Modeling Fault Among Bicyclists and Drivers Involved in Collisions in Hawaii, 1986-1991. *Transportation Research Record: Journal of the Transportation Research Board*, 1996. 1538: 75–80.
3. Stipancic, J., S. Zangenehpour, L. Miranda-Moreno, N. Saunier, and M.-A. Granie. Investigating the Gender Differences on Bicycle-Vehicle Conflicts at Urban Intersections Using an Ordered Logit Methodology. *Accident Analysis & Prevention*, Vol. 97, 2016, pp. 19–27.
4. Bíl, M., M. Bílová, and I. Müller. Critical Factors in Fatal Collisions of Adult Cyclists with Automobiles. *Accident Analysis & Prevention*, Vol. 42, No. 6, 2010, pp. 1632–1636.
5. Islam, S., and A. B. Hossain. Comparative Analysis of Injury Severity Resulting from Pedestrian–Motor Vehicle and Bicycle–Motor Vehicle Crashes on Roadways in Alabama. *Transportation Research Record: Journal of the Transportation Research Board*, 2015. 2514: 79–87.
6. Klop, J. R., and A. J. Khattak. Factors Influencing Bicycle Crash Severity on Two-Lane, Undivided Roadways in North Carolina. *Transportation Research Record: Journal of the Transportation Research Board*, 1999. 1674: 78–85.
7. Narayanamoorthy, S., R. Paleti, and C. R. Bhat. On Accommodating Spatial Dependence in Bicycle and Pedestrian Injury Counts by Severity Level. *Transportation Research Part B: Methodological*, Vol. 55, 2013, pp. 245–264.
8. Zahabi, S., J. Strauss, K. Manaugh, and L. Miranda-Moreno. Estimating Potential Effect of Speed Limits, Built Environment, and Other Factors on Severity of Pedestrian and Cyclist Injuries in Crashes. *Transportation Research Record: Journal of the Transportation Research Board*, 2011. 2247: 81–90.
9. Kaplan, S., and C. G. Prato. A Spatial Analysis of Land Use and Network Effects on Frequency and Severity of Cyclist–Motorist Crashes in the Copenhagen Region. *Traffic Injury Prevention*, Vol. 16, No. 7, 2015, pp. 724–731.
10. Washington, S., N. Haworth, and A. Schramm. Relationships between self-reported bicycling injuries and perceived risk of cyclists in Queensland, Australia. *Transportation Research Record: Journal of the Transportation Research Board*, 2012. 2314: 57–65.

11. Carter, D., W. Hunter, C. Zegeer, J. Stewart, and H. Huang. Bicyclist Intersection Safety Index. *Transportation Research Record: Journal of the Transportation Research Board*, 2007. 2031: 18–24.
12. Jensen, S. Safety Effects of Converting Intersections to Roundabouts. *Transportation Research Record: Journal of the Transportation Research Board*, 2013. 2389: 22–29.
13. Chen, P., and Q. Shen. Built Environment Effects on Cyclist Injury Severity in Automobile-Involved Bicycle Crashes. *Accident Analysis & Prevention*, Vol. 86, 2016, pp. 239–246.
14. Eluru, N., C. R. Bhat, and D. A. Hensher. A Mixed Generalized Ordered Response Model for Examining Pedestrian and Bicyclist Injury Severity Level in Traffic Crashes. *Accident Analysis & Prevention*, Vol. 40, No. 3, 2008, pp. 1033–1054.
15. Wang, C., L. Lu, and J. Lu. Statistical Analysis of Bicyclists' Injury Severity at Unsignalized Intersections. *Traffic Injury Prevention*, Vol. 16, No. 5, 2015, pp. 507–512.
16. Figliozzi, M., N. Wheeler, and C. Monsere. Methodology for Estimating Bicyclist Acceleration and Speed Distributions at Intersections. *Transportation Research Record: Journal of the Transportation Research Board*, 2013. 2387: 66–75.
17. Hunter, W. W., J. C. Stutts, W. E. Pein, and C. L. Cox. *Pedestrian and Bicycle Crash Types of the Early 1990's*. Publication Number FHWA-RD-95-163. Federal Highway Administration, Washington, D.C., 1996.
18. Austroads. *National Cycling Participation Survey 2017*. National Results Report No. AP-C91-17. Austroads, Australia, 2017.
19. Watson, A., B. C. Watson, and K. Vallmuur. How Accurate is the Identification of Serious Traffic Injuries by Police? The Concordance between Police and Hospital Reported Traffic Injuries. *Proc., 2013 Australasian Road Safety Research, Policing & Education Conference*, Australasian College of Road Safety (ACRS), Brisbane Convention and Exhibition Center, Brisbane, Australia, 2013.
20. Vandenbulcke, G., I. Thomas, B. de Geus, B. Degraeuwe, R. Torfs, R. Meeusen, and L. I. Panis. Mapping Bicycle Use and the Risk of Accidents for Commuters Who Cycle to Work in Belgium. *Transport Policy*, Vol. 16, No. 2, 2009, pp. 77–87.
21. Kaplan, S., K. Vavatsoulas, and C. G. Prato. Aggravating and Mitigating Factors Associated with Cyclist Injury Severity in Denmark. *Journal of Safety Research*, Vol. 50, 2014, pp. 75–82.
22. TMR. *Data Analysis Unit Road Crash Glossary: Data Analysis Unit, Department of Transport and Main Roads*. https://www.webcrash.transport.qld.gov.au/webcrash2/external/daupage/docs/glossary.pdf.
23. Watson, A., B. Watson, and K. Vallmuur. Estimating Under-Reporting of Road Crash Injuries to Police Using Multiple Linked Data Collections. *Accident Analysis & Prevention*, Vol. 83, 2015, pp. 18–25.
24. Milton, J. C., V. N. Shankar, and F. L. Mannering. Highway Accident Severities and the Mixed Logit Model: An Exploratory Empirical Analysis. *Accident Analysis & Prevention*, Vol. 40, No. 1, 2008, pp. 260–266.
25. Savolainen, P. T., F. L. Mannering, D. Lord, and M. A. Quddus. The Statistical Analysis of Highway Crash-Injury Severities: A Review and Assessment of Methodological Alternatives. *Accident Analysis & Prevention*, Vol. 43, No. 5, 2011, pp. 1666–1676.
26. Hensher, D. A., and W. H. Greene. The Mixed Logit Model: The State of Practice. *Transportation*, Vol. 30, No. 2, 2003, pp. 133–176.
27. Pai, C.-W., K. P. Hwang, and W. Saleh. A Mixed Logit Analysis of Motorists' Right-of-Way Violation in Motorcycle Accidents at Priority T-Junctions. *Accident Analysis & Prevention*, Vol. 41, No. 3, 2009, pp. 565–573.
28. Islam, S., S. L. Jones, and D. Dye. Comprehensive Analysis of Single-and Multi-Vehicle Large Truck At-Fault Crashes on Rural and Urban Roadways in Alabama. *Accident Analysis & Prevention*, Vol. 67, 2014, pp. 148–158.
29. Train, K. E. *Discrete Choice Methods with Simulation*. Cambridge University Press, New York, 2009.
30. Washington, S. P., M. G. Karlaftis, and F. L. Mannering. *Statistical and Econometric Methods for Transportation Data Analysis*. Chapman and Hall/CRC, Boca Raton, Fla., 2010.
31. Moore, D. N., W. H. Schneider IV, P. T. Savolainen, and M. Farzaneh. Mixed Logit Analysis of Bicyclist Injury Severity Resulting from Motor Vehicle Crashes at Intersection and Non-Intersection Locations. *Accident Analysis & Prevention*, Vol. 43, No. 3, 2011, pp. 621–630.
32. Bhat, C. R. Simulation Estimation of Mixed Discrete Choice Models Using Randomized and Scrambled Halton Sequences. *Transportation Research Part B: Methodological*, Vol. 37, No. 9, 2003, pp. 837–855.
33. Gkritza, K., and F. L. Mannering. Mixed Logit Analysis of Safety-Belt Use in Single-and Multi-Occupant Vehicles. *Accident Analysis & Prevention*, Vol. 40, No. 2, 2008, pp. 443–451.
34. Doong, J.-L., and C.-H. Lai. Risk Factors for Child and Adolescent Occupants, Bicyclists, and Pedestrians in Motorized Vehicle Collisions. *Traffic Injury Prevention*, Vol. 13, No. 3, 2012, pp. 249–257.
35. Schramm, A. J., N. L. Haworth, K. Heesch, A. Watson, and A. K. Debnath. *Evaluation of the Queensland Minimum Passing Distance Road Rule*. Research Report. Transport and Main Roads, Brisbane, Queensland, 2016. https://www.tmr.qld.gov.au/Safety/Road-safety/Road-safety-research-reports
36. Maring, W., and I. Van Schagen. Age Dependence of Attitudes and Knowledge in Cyclists. *Accident Analysis & Prevention*, Vol. 22, No. 2, 1990, pp. 127–136.
37. Debnath, A. K., N. Haworth, A. Schramm, and A. Williamson. Observational Study of Compliance with Queensland Bicycle Helmet Laws. *Accident Analysis & Prevention*, Vol. 97, 2016, pp. 146–152.
38. Attewell, R. G., K. Glase, and M. McFadden. Bicycle Helmet Efficacy: A Meta-Analysis. *Accident Analysis & Prevention*, Vol. 33, No. 3, 2001, pp. 345–352.
39. Haworth, N. L., A. J. Schramm, M. J. King, and D. A. Steinhardt. *Bicycle Helmet Research: CARRS-Q*

Monograph 5. Queensland University of Technology, Australia, 2010.
40. Boufous, S., L. D. Rome, T. Senserrick, and R. Ivers. Risk Factors for Severe Injury in Cyclists Involved in Traffic Crashes in Victoria, Australia. *Accident Analysis & Prevention*, Vol. 49, 2012, pp. 404–409.
41. Schramm, A. J., A. Rakotonirainy, and N. L. Haworth. The Role of Traffic Violations in Police-Reported Bicycle Crashes in Queensland. *Journal of the Australasian College of Road Safety*, Vol. 21, No. 3, 2010, pp. 61–67.
42. Crocker, P., O. Zad, T. Milling, and K. A. Lawson. Alcohol, Bicycling, and Head and Brain Injury: A Study of Impaired Cyclists' Riding Patterns R1. *The American Journal of Emergency Medicine*, Vol. 28, No. 1, 2010, pp. 68–72.
43. Juhra, C., B. Wieskötter, K. Chu, L. Trost, U. Weiss, M. Messerschmidt, A. Malczyk, M. Heckwolf, and M. Raschke. Bicycle Accidents–Do We Only See the Tip of the Iceberg?: A Prospective Multi-Centre Study in a Large German City Combining Medical and Police Data. *Injury*, Vol. 43, No. 12, 2012, pp. 2026–2034.
44. Mayora, J. M. P., and R. J. Piña. An Assessment of the Skid Resistance Effect on Traffic Safety Under Wet-Pavement Conditions. *Accident Analysis & Prevention*, Vol. 41, No. 4, 2009, pp. 881–886.
45. Haworth, N., and A. K. Debnath. How Similar are Two-Unit Bicycle and Motorcycle Crashes? *Accident Analysis & Prevention*, Vol. 58, 2013, pp. 15–25.
46. Vandenbulcke, G., I. Thomas, and L. I. Panis. Predicting Cycling Accident Risk In Brussels: A Spatial Case–Control Approach. *Accident Analysis & Prevention*, Vol. 62, 2014, pp. 341–357.
47. Räsänen, M., and H. Summala. Attention and Expectation Problems in Bicycle–Car Collisions: An In-Depth Study. *Accident Analysis & Prevention*, Vol. 30, No. 5, 1998, pp. 657–666.
48. Steriu, M. *Raising the Bar: Review of Cycling Safety Policies in the European Union*. European Transport Safety Council, Brussels, 2012.
49. Yamamoto, T., J. Hashiji, and V. N. Shankar. Underreporting in Traffic Accident Data, Bias in Parameters and the Structure of Injury Severity Models. *Accident Analysis & Prevention*, Vol. 40, No. 4, 2008, pp. 1320–1329.
50. Elvik, R., and A. Mysen. Incomplete Accident Reporting: Meta-Analysis of Studies Made in 13 Countries. *Transportation Research Record: Journal of the Transportation Research Board*, 1999. 1665: 133–140.

The Standing Committee on Safety Data, Analysis and Evaluation (ANB20) peer-reviewed this paper (18-02266).

The contents and opinions of this paper reflect the personal views of the authors, who are responsible for the accuracy of the data presented here.

Article

TRR
JOURNAL OF THE TRANSPORTATION RESEARCH BOARD

Transportation Research Record
2018, Vol. 2672(38) 210–222

DOI: 10.1177/0361198118776759
journals.sagepub.com/home/trr
SAGE

Exploring the Effect of Different Neighboring Structures on Spatial Hierarchical Joint Crash Frequency Models

Saif A. Alarifi[1], Mohamed A. Abdel-Aty[1], Jaeyoung Lee[1], and Xuesong Wang[2]

Abstract
Corridor safety analysis is a primary interest of many road safety studies. Corridors typically contain intersections and roadway segments. Having both components while analyzing corridors in addition to corridor-level variables in a hierarchical joint model would provide a comprehensive understanding of the existing corridor safety problems. There will probably be spatial correlation among road entities along a corridor, especially if the distance between the road entities is not large. Therefore, it is crucial to consider spatial effects in the model. However, this data structure is relatively new, and the best spatial weight matrix for this hierarchical spatial joint model has yet to be investigated. Therefore, this study estimates a hierarchical Poisson-lognormal (HPLN) joint model with spatial effects and explores the effect of different neighboring structures. A total of thirteen HPLN joint models are estimated: one HPLN joint model with corridor random effect and twelve HPLN joint models with spatial effects. Four types of conceptualization of spatial relationships were considered: (a) adjacency-based, (b) adjacency-route, (c) distance-order, and (d) distance-based spatial weight features. The results show the importance of incorporating spatial effects in the model. It was found that having a joint model is important since one of the best models is the adjacency-based first-order model, where the feeding road entities in addition to the directly adjacent road entity of the same type as the road entity of interest are considered. The results confirm the importance of spatial autocorrelation between road entities along the same corridor.

Safety analysis of corridors is a major concern in transportation research. Many corridor studies have accounted for either intersections (*1–4*) or roadway segments (*5–7*) to provide a better understanding of the factors contributing to crashes on the targeted road entity. Having both of these components of corridors in addition to corridor-level variables would, however, provide a much better understanding of existing corridor safety problems and the risk factors associated with intersections, roadway segments, and corridors in general. Establishing and understanding this relationship between corridor component predictors and risk variables is an important step to improve the safety and efficiency of travel along corridors.

There is a high probability of the existence of spatial correlation among the road entities along corridors, especially when the distance between them is not large, since they have similar roadway and driver characteristics (*2*). Therefore, it is crucial to consider the spatial effects in the model since the spatial effects can be surrogates for unknown and relevant covariates (*6*, *8*). However, the best spatial weight matrix to be considered for the hierarchical joint model, which contains intersection, roadway segment, and corridor variables, needs to be investigated. To do so, exploring different neighboring structures between the road entities in the suggested model is required.

This study aims to gain a better understanding of the associated risk factors of intersections and roadway segments along corridors by estimating a hierarchical spatial

[1]Department of Civil, Environmental, and Construction Engineering, University of Central Florida, Orlando, FL
[2]Key Laboratory of Road and Traffic Engineering and the Engineering Research Center for Road Traffic Safety and Environment, College of Transportation Engineering, Tongji University, Shanghai, China

Corresponding Author:
Address correspondence to Saif A. Alarifi: Saif.a@knights.ucf.edu

joint model that contains intersection, roadway segment, and corridor variables. It also aims to explore the effect of different neighboring structures on the hierarchical joint model and to identify the best fit of the joint model to the data.

Previous Studies

Accounting for spatial correlation has become a key element in the development of traffic crash models with close proximity of road entities, and ignoring this issue may lead to a biased estimation of model parameters (*2*, *9*). Crashes at adjacent locations are likely to be spatially correlated for several reasons. First, intersections and roadway segments along the same corridors tend to have similar through-traffic, and they may have similar land use, roadway characteristics, and driver characteristics (*2*). In addition, adjacent signals along certain corridors, when the distances between them are 0.5 miles or less, are likely to be coordinated in most situations (*10*). Therefore, spatial correlation exists between adjacent road entities, and many studies have proved that considering the spatial effect has helped to improve the model performance (*6*, *11–13*).

Scholars have investigated the safety of intersections or roadway segments along corridors (*1–7*, *14*). Abdel-Aty and Wang (*1*) proposed generalized estimating equations (GEEs) to deal with the spatially correlated adjacent intersections along certain corridors. However, the GEEs set the same correlation matrix for different intersection groups, so the discrepancies in correlations among different group of intersections cannot be reflected (*3*). El-Basyouny and Sayed (*5*) analyzed a three-year dataset (1994–1996) for 392 roadway segments, which were clustered into 58 corridors, and they applied the traditional Poisson-lognormal model with two extended Poisson-lognormal models. It was found that the Poisson-lognormal model with random corridor parameters provides the best fit. Guo et al. (*2*) analyzed a six-year dataset for 170 four-legged signalized intersections along 25 corridors in the state of Florida. They estimated several models: Poisson, NB, NB mixed effect, NB conditional autoregressive (CAR) prior, and Poisson CAR models, and found that the Poisson CAR prior model provides the best fit for the data. Xie et al. (*4*) studied a year of crash data for 195 signalized intersections along 22 corridors in Shanghai, China, and they developed several models: NB, random effect NB, random parameter NB, Bayesian hierarchical NB (HNB), and hierarchical CAR (HCAR). The results indicated that the HCAR model provides the best fit for the data. Wang et al. (*7*) analyzed 176 roadway segments along 18 urban arterials in Shanghai, China. Four different models have been estimated: PLN, PLN with maximum likelihood priors, hierarchical PLN, and hierarchical PLN with maximum likelihood priors. It was found that the hierarchical PLN model with maximum likelihood priors provided the best fit for the data. Many of the previous studies (*3*, *4*, *7*) have tried corridor-level variables in the model. All of those studies have analyzed signalized intersections or roadway segments along corridors, and none of them have considered both components in a spatial joint model.

Incorporating spatial correlations to the model is important; yet different weight matrices are possible. Some studies have investigated the effect of different neighboring structures in traffic crash models. Aguero-Valverde and Jovanis (*6*) investigated the effect of different roadway segment neighboring structures in road crash models. A full Bayes hierarchical CAR model was estimated to account for the spatial autocorrelation. Many weight matrices were compared, and the authors found that the spatial correlation becomes more significant within distances of one mile or less. Also, roadway segments that belong to the same corridor have stronger spatial correlation. Regarding intersections, a commonly used inverse distance function was adopted in many studies (*2*, *15*). Castro et al. (*16*) investigated the effect of spatial correlation in traffic crash models. They studied a sample drawn from files between 2003 and 2009 to estimate a latent variable model. The authors used distance-based neighbors weight matrix. Dong et al. (*17*) assessed the spatial proximity effects in predicting traffic crashes at the level of traffic analysis zones. They compared different weight matrices: first-order adjacency-based, common boundary-length-based, geometric centroid-distance-based, and crash-weighted centroid-distance-based. Lastly, Zeng and Huang (*18*) estimated a spatial joint model of traffic crashes on a road network. With respect to the spatial proximity matrix for the joint model, the authors only accounted for the directly connected road entities in the weight matrix without examining different neighboring structures. These findings are important for future research to decide which spatial weight features should be explored at the desired micro or macro level.

To sum up, while previous corridor studies have investigated many issues on a single type of road entity, this study aims to estimate a Bayesian hierarchical Poisson-lognormal joint model with spatial effect that contains main corridor components—segments and intersections—in addition to corridor-level variables. In addition, it aims to explore the effect of different neighboring structures on the joint model.

Data Collection and Preparation

A total of 255 signalized intersections and 220 roadway segments along 20 principal and minor arterials in

Orange and Broward counties in the state of Florida have been investigated to test the effect of spatial correlation. The roadway segments and corridors have an average length of 0.372 and 4.976 miles, respectively. The number of roadway segments for corridors varies from 4 to 25 with an average of 11 segments, and the number of signalized intersections on each corridor varies from 5 to 28 with an average of 12.75 intersections.

The crash data were obtained from the Crash Analysis Reporting System (CARS), which is a data resource maintained by the Florida Department of Transportation (FDOT). A three-year period (2010–2012) of crash data was collected for the selected road entities. The safety of intersections and roadway segments was measured by the total number of crashes. It should be noted that crashes that occurred at intersections or within 250 feet from the intersection sign were considered as intersection-related crashes. Geographic information system (GIS) helped to distinguish between intersection and roadway segment crashes.

Data was collected on roadway characteristics, traffic volume, and crash information for the selected roadway segments and intersections as well as corridors. The Roadway Characteristics Inventory, which is one of FDOT's resources, was used mainly to retrieve some geometric characteristics and traffic volume information. Google Earth was also used to collect detailed geometric characteristics for the roadway segments and intersections. The information collected on intersections includes: major and minor traffic volumes, number of lanes, intersection configuration, posted speeds of major and minor roads, number of through, left, and right lanes of major and minor approaches, number of access points within the influence area of intersection (access points within 250 feet from the center of an intersection), distance to the nearest intersection, and median type and width of major and minor roads. The information collected on roadway segments includes: traffic volume, truck traffic volume, number of lanes, posted speed, median type and width, length, driveway density, and median opening density. Table 1 shows the descriptive statistics of roadway segment-, intersection-, and corridor-related variables. Regarding corridor variables, many variables have been considered, including: roadway classification, intersection density, driveway density, average posted speed, median opening density, county, truck percentage, and average segment length.

Although a large number of variables were collected, only some of them have been considered in the final model since some of the variables are highly correlated and thus excluded from the final model. If the absolute value of correlation coefficient is 0.6 or above, then the correlation between the two independent variables is considered high (*19*). Also, some of the variables were included in a categorical form, as shown in Table 1, and other variables have been tried in different forms, especially the correlated ones, such as difference of absolute speed limit between major and minor roads. The correlation between this variable and traffic volume is low; however, major posted speed is highly correlated with major annual average daily traffic (AADT), and minor posted speed is highly correlated with minor AADT. As a result, these two variables have been combined into one meaningful variable.

Research Methodology

Hierarchical Poisson-lognormal (HPLN) Joint Model

Poisson-lognormal (PLN) models incorporate REs to a typical Poisson model to address the over-dispersion for unobserved or unmeasured heterogeneity (*5, 7*). Many recent highway safety studies have employed PLN models using the Bayesian approach (*5, 7, 12, 14*). This study proposes a hierarchical model to deal with the multilevel structure of the data, where intersections and roadway segments are at the bottom level and corridors are at the top level. A proposed Bayesian hierarchical joint model, which includes roadway segment, intersection, and corridor-related variables, was implemented before testing the spatial effects, and it explains variation between corridors. In this study, a HPLN joint model with corridor random effect (RE) is implemented, and it is specified as follows:

$$y_{ij}|\lambda_{ij} \sim Poisson(\lambda_{ij}) \quad (1)$$

$$\log(\lambda_{ij}) = (\beta_{Int}X_{Int,ij}) \times \nu_{ij} + (\beta_{Seg}X_{Seg,ij} + \ln(l_{ij})) \times (1 - \nu_{ij}) + \beta_{cor}X_{cor,j} + \varepsilon_{ij} + u_j \quad (2)$$

$$\varepsilon_{ij} \sim N(0, \tau_\varepsilon) \quad (3)$$

where y_{ij} is the crash frequencies at the targeted road entity (either an intersection or a roadway segment) $i = 1, 2, \ldots, n$ from corridor $j = 1, 2, \ldots, m$, and λ_{ij} is the Poisson parameter, which is equal to the expected number of crashes for the road entity i in corridor j. $(\beta_{Int}X_{Int,ij})$, $(\beta_{Seg}X_{Seg,ij})$, and $(\beta_{cor}X_{cor,j})$ represent the intersection, roadway segment, and corridor significant explanatory variables, respectively. $X_{Int,ij}$, $X_{Seg,ij}$, and $X_{cor,j}$ denote a vector of intersection, roadway segment, and corridor explanatory variables. β_{Int}, β_{Seg}, and β_{cor} denote a vector of estimable parameters for intersections, roadway segments, and corridors respectively. ν_{ij} is a dummy variable used to differentiate between intersection and roadway segment data; it has a value of 1 at intersection data points and a value of 0 at roadway segment data points. l_{ij} is the length of roadway segment i in corridor j, and $\ln(l_{ij})$ is used as an offset for roadway segments (*20*). $\exp(\varepsilon_{ij})$ is a multiplicative RE, and it follows a lognormal distribution. The τ_ε is the precision,

Table 1. Summary Statistics of Intersections, Roadway Segments, and Corridors

Variable name		Mean	S.D.	Min.	Max.
Intersection variables (N = 255)	Major AADT (veh/day)	36,996	15,465	2,100	76,500
	Minor AADT (veh/day)	12,354	10,181	1,600	54,000
	Number of legs (4 legs = 1, 3 legs = 0)	0.902	0.298	0	1
	Major posted speed (mph)	41.274	5.598	30	50
	Minor posted speed (mph)	33.196	7.160	20	50
	Absolute speed limit difference between major and minor roads (mph)	8.549	7.116	0	25
	Intersection size (1 if number of lases from all approaches>=19, 0 otherwise)	0.204	0.404	0	1
	Presence of exclusive right-turn lane on major roads (1 = Yes, 0 = No)	0.553	0.498	0	1
	Presence of exclusive right-turn lane on minor roads (1 = Yes, 0 = No)	0.623	0.485	0	1
	Number of through lanes at major roads	5.016	1.270	2	7
	Number of through lanes at minor roads	2.698	1.539	0	6
	Presence of exclusive left-turn lane on major roads (1 = Yes, 0 = No)	0.886	0.318	0	1
	Presence of exclusive left-turn lane on minor roads (1 = Yes, 0 = No)	0.831	0.375	0	1
	Presence of median at major roads (1 = Yes, 0 = No)	0.882	0.322	0	1
	Presence of median at minor roads (1 = Yes, 0 = No)	0.616	0.487	0	1
	Distance to nearest intersection (mile)	0.351	0.283	0.046	3.452
	Total number of access points within the influence area of intersection	2.843	2.514	0	12
	Total number of crashes	37.894	39.395	0	260
Roadway segment variables (N = 220)	AADT (veh/day)	36,974	14,450	2,500	67,000
	Truck AADT (veh/day)	1498.1	846	110	4544
	Posted speed (mph)	41.932	5.081	30	50
	Median width (ft)	19.150	10.180	0	44.00
	Presence of bike lane (1 = Yes, 0 = No)	0.414	0.494	0	1
	Driveway density (No. of driveways/mile)	24.583	37.067	0	310.59
	Median opening density (No. of median openings/mile)	4.872	4.257	0	29.663
	Segment length (mile)	0.372	0.325	0.003	3.357
	Number of crashes	18.145	18.620	0	117
Corridor variables (N = 20)	Signalized intersection density (No. of intersections/mile)	3.362	2.496	1.304	10.780
	Average posted speed (mph)	40.700	5.882	30	50
	Median opening density (No. of median openings/mile)	4.223	2.270	0	9.83
	Driveway density (No. of driveways/mile)	11.593	3.670	5.21	17.89
	Average length of segments per corridor (mile)	0.366	0.187	0.031	0.805
	Roadway classification (Principal arterial = 1, minor arterial = 0)	0.678	0.468	0	1
	Length (mile)	4.976	2.497	1.205	9.239
	Number of intersections	12.75	4.940	4	28
	Number of roadway segments	11	4.877	5	25
	Number of crashes	679.8	422.23	59	1516

which is the inverse of the variance, and it controls the extra-Poisson variation (*21*) and has a hyper-prior gamma (0.001, 0.001). Lastly, u_j represents the RE across corridors, and it has a mean of 0 and a variance of σ_u^2 ($u_j \sim N(0, \sigma_u^2)$). σ_u^2 represents the additional variation across corridors. The variance parameter for u_j is specified as a gamma prior distribution (0.001, 0.001). A non-informative Normal (0, 100,000) prior was given for all parameter coefficients.

HPLN Joint Model with Spatial Effects

This model incorporates the spatial effect and will be used to test the effect of different neighboring structures.

Many spatial models are estimated under the Bayesian approach due to its flexibility in estimating complicated modeling structures (*22*). This model is specified as follows:

$$\log(\lambda_{ij}) = (\beta_{Int} X_{Int,ij}) \times \nu_{ij} + (\beta_{Seg} X_{Seg,ij} + \ln(l_{ij})) \times (1 - \nu_{ij}) + \beta_{cor} X_{cor,j} + \varepsilon_{ij} + \varphi_{ij} \quad (4)$$

$$\alpha = \frac{sd(\varphi)}{sd(\varepsilon) + sd(\varphi)} \quad (5)$$

φ_{ij} is a spatially correlated RE term, and a Gaussian conditional autoregressive (CAR) prior was assigned to φ_{ij} to incorporate the spatial correlation among road entities. α is the proportion of variation in the REs due to spatial autocorrelation, and sd is the empirical marginal standard deviation function. The conditional distribution of CAR prior has the following form (*23*):

$$\varphi_{ij} | \varphi_{-ij} \sim N\left(\sum_{i'j' \sim ij} \frac{w_{i'j',ij}}{w_{i+}} \varphi_{i'j'}, \frac{1}{\tau_c w_{i+}} \right) \quad (6)$$

where φ_{-ij} is the neighbor of φ_{ij}, τ_c is the precision parameter which controls the Poisson extra-variation due to spatial correlation, $i'j' \sim ij$ indicates that road entity $i'j'$ is a neighbor of ij, $w_{i'j',ij}$ represents the weight of the $i'j'$ neighbor of the ij road entity, and w_{i+} indicates the sum of the neighbor's weights of road entity ij.

As shown in Equation 6, $w_{i'j',ij}$ denotes the spatial weight between road entity $i'j'$ and road entity ij. The spatial relationship structures are formally expressed in a symmetric $n \times n$ weight matrix $\boldsymbol{W}$ in the CAR model, and $\boldsymbol{W}$ contains zeros on the diagonal and the assigned weight values in the off diagonal. In the present study, different weight specifications have been tested to examine their effects on the model fit of the data. Four types of conceptualization of spatial relationships are explored in this study:

(a) Adjacency-based spatial weight feature: Four different neighboring structures are proposed under the adjacency-based structure. The first neighboring structure is the adjacency-based connected entities only (AB-CO), where only the feeding road entities of the other type to the road entity of interest are considered. The second neighboring structure is the adjacency-based first-order (AB-FO), which has been employed in many traffic safety studies (*6*, *18*, *20*). In this study, since we are modeling roadway segments and intersections simultaneously, the definition of the AB-FO neighbors should be clarified. In terms of roadway segments, all roadway segments that are directly adjacent to the segment of interest are considered first-order in addition to the feeding intersections to the segment of interest. Regarding intersections, all intersections that are directly adjacent to the intersection of interest are considered first-order in addition to the feeding segments to the intersection of interest. The third proposed structure is the adjacency-based first-order without connected entities (AB-FOWC), where only the directly adjacent road entities of the same type as the road entity of interest are considered. The last neighboring structure under this category is the adjacency-based second-order (AB-SO). Here, with respect to roadway segments, the segments that connect directly to the first-order segments are additionally added as second-order neighbors. In terms of intersections, the intersections that are neighbors to the first-order intersections are additionally added as second-order neighbors. The assigned weight is the inverse of the order, as proposed by Aguero-Valverde and Jovanis (*24*), and many previous studies have considered this approach to assigning the weight (*6*, *25*). Therefore, the assigned weight is 1 if the road entity is connected to the road entity of interest or considered a first-order neighbor, 0.5 if the road entity is considered second-order neighbor, and otherwise, the weight is zero.

(b) Adjacency-route spatial weight feature: This neighboring structure is considered since the road entities along the same corridor are likely to share through-traffic volume and some roadway characteristics, so they are more likely to be spatially correlated (*2*, *3*, *6*). This approach allows adjacency-based second-order neighbors to be promoted to first-order if they are on the same corridor as the road entity of interest. The adjacency-route first-order (AR-FO) model has been considered in this study and in a previous study (*6*). The adjacency-route second-order (AR-SO) model has also been considered, where third-order neighbors that are located on the same corridor as the road entity of interest were promoted to second-order neighbors. The weight in this structure is the inverse of the order. Figure 1 shows a schematic of the adjacency-based and adjacency-route neighboring structures for intersections and roadway segments.

(c) Distance-order spatial weight feature: Three different neighboring structures are explored, and a midpoint of roadway segments is used to calculate the distance between road entities. Aguero-Valverde and Jovanis (*6*) found that the spatial correlation is significantly important if the

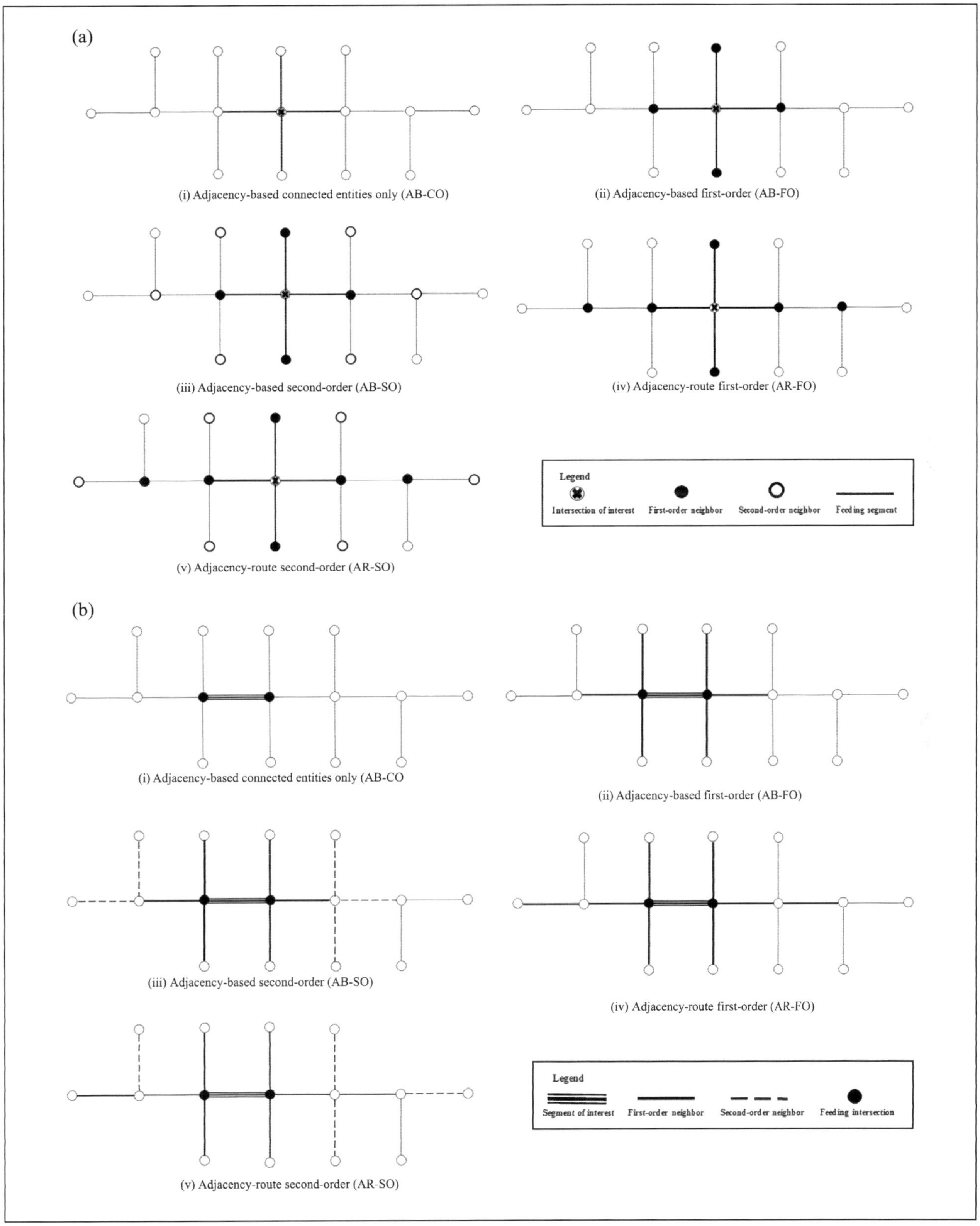

Figure 1. Adjacency-based and adjacency-route neighboring structures for intersections and roadway segments: (*a*) intersections; (*b*) roadway segments.

distance is one mile or less. As a result, the distance-order first-order (DO-FO) neighboring structure considers as the road entity of interest the same road entities that belong to the same corridor and are within half a mile of the road entity of interest in addition to the feeding road entities of the other type to the road entity of interest. The second proposed structure here is the distance-order second-order (DO-SO) neighboring structure, where the same road entities that are located along the same corridor and with half to one mile are additionally added as second-order neighbors. Lastly, a third neighboring structure is suggested in this study. An aerial distance of half a mile of the road entity of interest has been explored (DO-HM), containing road entities that are the same as the road entity of interest and located within half a mile of it in addition to the feeding road entities of the other type to the road entity of interest. The assigned weight in this structure is the inverse of the order.

(d) Distance-based spatial weight feature: The distance between road entities of the same type along the same corridor is tested here. Road entities on different corridors are treated independently. Three distance functions are adopted in this study. First, a commonly used inverse distance (DB-I) is explored. Second, an inverse distance squared (DB-IS) is investigated. The last distance function is the negative exponential of the distance (DB-NE). Many studies have employed one of these three forms of distance-based neighboring structures (*2*, *6*, *15*).

Model Evaluation

Examining the significance of the variables included in the model is the first step in model evaluation, so the estimated coefficient for each independent variable has to be statistically significant. P-value is used to examine the significance of the coefficient. Goodness-of-fit measures were also calculated to evaluate the models. The goodness-of-fit measures include deviance information criterion (DIC), mean absolute deviance (MAD), and root mean square error (RMSE).

DIC is used for model comparison and selection (*26*). The DIC is defined as the posterior mean ($\bar{D}$) of the Bayesian deviance (D) plus the effective number of parameters (P_D), as shown in Equation 7.

$$DIC = \bar{D} + P_D \tag{7}$$

MAD and RMSE are dependent on the difference between the predicted and observed number of crashes, and their equations are as follows:

$$MAD = \frac{\sum_{i=1}^{n} \left| N_{Pred\,ij} - N_{Obsij} \right|}{n} \tag{8}$$

$$RMSE = \sqrt{\frac{\sum_{i=1}^{n} \left(N_{Pred\,ij} - N_{Obsij} \right)^2}{n}} \tag{9}$$

where $N_{Pred\,ij}$ is the predicted number of crashes, and N_{Obsij} is the observed number of crashes.

Analysis and Results

The models in this study were estimated using the Bayesian approach, which draws samples from the targeted posterior distribution of the parameters by using a Markov chain Monte Carlo algorithm (*27*). WinBUGS statistical software was used to estimate the Bayesian models. The posterior summaries were obtained by a chain with 100,000 iterations, 10,000 of which were discarded as a burn-in sample. The convergence of the model was evaluated by inspection of the trace plots and kernel density (*28*).

A total of 13 HPLN joint models were estimated, and the spatial models fall into four main categories: adjacency-based, adjacency-route, distance-order, and distance-based spatial weight features. The first model is HPLN with corridor RE, and this is considered as the base model (Model 1). Four adjacency-based models have been estimated: AB-CO (Model 2), AB-FOWC (Model 3), AB-FO (Model 4), and AB-SO (Model 5). Two adjacency-route models have been estimated: AR-FO (Model 6) and AR-SO (Model 7). Three distance-order models have been estimated: DO-FO (Model 8), DO-SO (Model 9), and DO-HM (Model 10). Lastly, three distance-based models have been estimated: DB-I (Model 11), DB-IS (Model 12), and DB-NE (Model 13). Table 2 shows the model comparison using DIC, MAD, and RMSE, and Table 3 shows the modeling results of the 13 models.

Table 2. Model Comparison Using DIC, MAD, and RMSE

Model	DIC	MAD	RMSE
Corridor RE (Model 1)	3073.60	19.791	38.592
AB-CO (Model 2)	3063.26	20.070	37.568
AB-FOWC (Model 3)	3070.62	25.306	38.069
AB-FO (Model 4)	3059.87	18.593	29.360
AB-SO (Model 5)	3064.14	19.740	34.542
AR-FO (Model 6)	3064.42	19.918	35.643
AR-SO (Model 7)	3060.64	19.512	33.471
DO-FO (Model 8)	3066.05	20.213	42.773
DO-SO (Model 9)	3063.68	19.452	36.130
DO-HM (Model 10)	3062.13	20.343	48.035
DB-I (Model 11)	3069.91	19.853	38.510
DB-IS (Model 12)	3069.77	19.731	40.002
DB-NE (Model 13)	3066.96	19.806	37.702

Table 3. Posterior Summary of Bayesian Hierarchical Poisson-Lognormal Joint Non-Spatial and Spatial Models

(*a*) Non-Spatial Model and Basic Spatial Weight Features

Variable		Corridor RE (Model 1)		AB-CO (Model 2)		AB-FOWC (Model 3)	
		Mean (SD)	95% BCI	Mean (SD)	95% BCI	Mean (SD)	95% BCI
Intersection variables	Intercept	−3.101# (1.474)	(−5.988, 0.010)	−3.273 (1.272)	(−5.225, −0.593)	−6.097 (1.356)	(−8.681, −3.764)
	Log (Major AADT)	0.659 (0.093)	(0.435, 0.827)	0.425 (0.119)	(0.187, 0.648)	0.437 (0.096)	(0.254, 0.629)
	Log (Minor AADT)	0.442 (0.056)	(0.337, 0.553)	0.420 (0.066)	(0.294, 0.545)	0.449 (0.059)	(0.332, 0.572)
	Absolute speed limit difference	−0.028 (0.007)	(−0.042, −0.014)	−0.036 (0.007)	(−0.051, −0.021)	−0.030 (0.007)	(−0.044, −0.016)
	Access points within the influence area of intersection	0.037 (0.017)	(0.002, 0.071)	0.023# (0.017)	(−0.004, 0.057)	0.033 (0.018)	(0.003, 0.070)
Segment variables	Log (AADT)	0.519 (0.209)	(0.164, 0.982)	0.276 (0.137)	(0.022, 0.559)	0.354 (0.177)	(0.029, 0.702)
	Median opening density	0.013 (0.002)	(0.009, 017)	0.013 (0.001)	(0.009, 0.016)	0.013 (0.002)	(0.008, 0.017)
	Driveway density	0.038 (0.013)	(0.012, 0.063)	0.043 (0.012)	(0.018, 0.067)	0.034 (0.014)	(0.005, 0.062)
	Log (Truck AADT)	0.312 (0.148)	(0.005, 0.595)	0.260 (0.130)	(0.012, 0.535)	0.289# (0.008)	(−0.008, 0.609)
Corridor variables	Intercept	−4.909 (1.590)	(−8.224, −1.891)	−2.813 (1.083)	(−6.002, −1.022)	−1.774# (1.247)	(−4.063, 0.346)
	County (Orange = 1, Broward = 0)	0.393 (0.146)	(0.112, 0.698)	–	–	–	–
	Principal arterial versus minor arterial	0.482 (0.168)	(0.130, 0.802)	0.578 (0.219)	(0.149, 1.005)	0.489 (0.174)	(0.140, 0.843)
	Average segment length	–	–	2.499 (0.713)	(1.094, 3.961)	2.000 (0.562)	(0.895, 3.126)
Random effect terms	SD of random effect (ε)	0.600 (0.025)	(0.550, 0.650)	0.516 (0.026)	(0.466, 0.571)	0.487 (0.031)	(0.427, 0.550)
	SD of corridor random effect (u)	0.316 (0.076)	(0.192, 0.492)	–	–	–	–
	SD of spatial effects (φ)	–	–	0.150 (0.025)	(0.103, 0.203)	0.208 (0.037)	(0.137, 0.284)
Proportion of variability (α)		–	–	0.225	(0.161, 0.291)	0.299	(0.207, 0.389)

#Significant at 90% BCI, and all other variables are significant at 95% BCI.

(*b*) Adjacency Spatial Weight Feature

Variable		AB-FO (Model 4)		AB-SO (Model 5)		AR-FO (Model 6)		AR-SO (Model 7)	
		Mean (SD)	95% BCI	Mean (SD)	95% BCI	Mean (SD)	95% BCI	Mean (SD)	95% BCI
Intersection variables	Intercept	−2.984 (1.340)	(−5.889, −0.526)	−2.527# (1.270)	(−4.706, 0.503)	−3.519 (1.121)	(−5.871, −1.499)	−3.232 (1.629)	(−6.045, −0.137)
	Log (Major AADT)	0.499 (0.096)	(0.291, 0.687)	0.479 (0.123)	(0.215, 0.669)	0.427 (0.098)	(0.240, 0.623)	0.480 (0.117)	(0.245, 0.696)
	Log (Minor AADT)	0.409 (0.057)	(0.294, 0.523)	0.423 (0.053)	(0.322, 0.529)	0.420 (0.067)	(0.275, 0.548)	0.423 (0.063)	(0.312, 0.554)
	Absolute speed limit difference	−0.036 (0.007)	(−0.050, −0.022)	−0.036 (0.007)	(−0.049, −0.022)	−0.036 (0.007)	(−0.049, −0.022)	−0.036 (0.007)	(−0.049, −0.022)
	Access points within the inf. area of int.	0.023# (0.017)	(−0.005, 0.057)	0.023# (0.017)	(−0.005, 0.057)	0.023# (0.017)	(−0.005, 0.057)	0.023# (0.017)	(−0.005, 0.057)
Segment variables	Log (AADT)	0.348 (0.174)	(0.023, 0.670)	0.416 (0.171)	(0.076, 0.730)	0.232# (0.134)	(−0.045, 0.467)	0.334# (0.197)	(−0.100, 0.700)
	Median opening density	0.013 (0.002)	(0.009, 0.017)	0.013 (0.002)	(0.009, 0.017)	0.013 (0.002)	(0.009, 0.017)	0.013 (0.002)	(0.009, 0.017)
	Driveway density	0.034 (0.012)	(0.018, 0.067)	0.043 (0.012)	(0.018, 0.067)	0.043 (0.012)	(0.018, 0.067)	0.043 (0.012)	(0.016, 0.066)
	Log (Truck AADT)	0.283# (0.141)	(−0.024, 0.523)	0.233# (0.133)	(−0.020, 0.486)	0.286 (0.134)	(0.030, 0.556)	0.271# (0.142)	(−0.003, 0.568)
Corridor variables	Intercept	−3.624 (1.211)	(−6.005, −1.251)	−3.897 (1.233)	(−5.937, −1.373)	−2.372 (1.036)	(−4.302, −0.204)	−3.226 (1.464)	(−6.037, −0.294)
	Principal arterial versus minor arterial	0.490 (0.198)	(0.106, 0.880)	0.445 (0.190)	(0.063, 0.815)	0.456 (0.191)	(0.065, 0.819)	0.436 (0.200)	(0.043, 0.829)
	Average segment length	2.021 (0.642)	(0.992, 3.523)	2.102 (0.612)	(0.915, 3.304)	2.153 (0.685)	(0.757, 3.466)	2.145 (0.649)	(0.947, 3.462)
Random effect terms	SD of random effect (ε)	0.506 (0.029)	(0.448, 0.565)	0.465 (0.041)	(0.378, 0.540)	0.440 (0.051)	(0.327, 0.528)	0.370 (0.077)	(0.186, 0.495)
	SD of spatial effects (φ)	0.321 (0.058)	(0.217, 0.443)	0.560 (0.102)	(0.376, 0.774)	0.698 (0.128)	(0.465, 0.967)	0.964 (0.162)	(0.658, 1.288)
Proportion of variability (α)		0.386	(0.217, 0.443)	0.542	(0.420, 0.665)	0.609	(0.477, 0.744)	0.718	(0.576, 0.872)

#Significant at 90% BCI, and all other variables are significant at 95% BCI.

Table 3. Continued

(c) Distance-Order Spatial Weight Feature

Variable		DO-FO (Model 8)		DO-SO (Model 9)		DO-HM (Model 10)	
		Mean (SD)	95% BCI	Mean (SD)	95% BCI	Mean (SD)	95% BCI
Intersection variables	Intercept	−2.872 (1.294)	(−5.273, −0.324)	−2.957# (1.555)	(−5.535, 0.321)	−2.784# (1.252)	(−5.041, 0.139)
	Log (Major AADT)	0.440 (0.0955)	(0.258, 0.638)	0.462 (0.087)	(0.295, 0.629)	0.505 (0.123)	(0.234, 0.759)
	Log (Minor AADT)	0.410 (0.065)	(0.275, 0.536)	0.406 (0.062)	(0.281, 0.530)	0.416 (0.066)	(0.289, 0.554)
	Absolute speed limit difference	−0.036 (0.007)	(−0.050, −0.022)	−0.039 (0.007)	(−0.053, −0.025)	−0.037 (0.007)	(−0.051, −0.022)
	No. of access points within the influence area of intersection	0.024# (0.017)	(−0.004, 0.057)	0.020# (0.017)	(−0.007, 0.054)	0.023# (0.017)	(−0.004, 0.056)
Segment variables	Log (AADT)	0.326 (0.174)	(0.030, 0.692)	0.358# (0.215)	(−0.008, 0.762)	0.420 (0.185)	(0.099, 0.801)
	Median opening density	0.014 (0.001)	(0.009, 0.017)	0.013 (0.002)	(0.009, 0.017)	0.014 (0.002)	(0.010, 0.027)
	Driveway density	0.042 (0.012)	(0.018, 0.067)	0.040 (0.012)	(0.016, 0.064)	0.043 (0.012)	(0.019, 0.067)
	Log (Truck AADT)	0.244# (0.134)	(−0.025, 0.513)	0.213# (0.152)	(−0.040, 0.480)	0.220# (0.131)	(−0.026, 0.481)
Corridor variables	Intercept	−3.121 (1.354)	(−6.463, −0.863)	−3.089 (1.563)	(−6.361, −0.516)	−3.804 (1.301)	(−6.542, −1.58)
	Principal arterial versus minor arterial	0.504 (0.198)	(0.121, 0.893)	0.487 (0.181)	(0.124, 0.835)	0.450 (0.192)	(0.078, 0.829)
	Average segment length	2.224 (0.659)	(0.954, 3.596)	1.954 (0.607)	(0.738, 3.168)	1.971 (0.652)	(0.750, 3.265)
Random effect terms	SD of random effect (ε)	0.520 (0.028)	(0.466, 0.576)	0.483 (0.042)	(0.392, 0.559)	0.523 (0.028)	(0.469, 0.580)
	SD of spatial effects (φ)	0.236 (0.046)	(0.154, 0.335)	0.556 (0.125)	(0.334, 0.820)	0.228 (0.046)	(0.147, 0.326)
Proportion of variability (α)		0.310	(0.221, 0.406)	0.529	(0.383, 0.672)	0.302	(0.212, 0.399)

#Significant at 90% BCI, and all other variables are significant at 95% BCI.

(d) Distance-Based Spatial Weight Feature

Variable		DB-I (Model 11)		DB-IS (Model 12)		DB-NE (Model 13)	
		Mean (SD)	95% BCI	Mean (SD)	95% BCI	Mean (SD)	95% BCI
Intersection variables	Intercept	−2.067 (1.725)	(−5.160, 1.279)	−3.245 (1.867)	(−6.568, 0.377)	−1.735 (1.654)	(−4.980, 0.865)
	Log (Major AADT)	0.553 (0.102)	(0.365, 0.733)	0.531 (0.122)	(0.305, 0.749)	0.511 (0.126)	(0.278, 0.779)
	Log (Minor AADT)	0.419 (0.064)	(0.285, 0.540)	0.419 (0.067)	(0.292, 0.546)	0.422 (0.063)	(0.304, 0.548)
	Absolute speed limit difference	−0.033 (0.008)	(−0.048, −0.018)	−0.034 (0.008)	(−0.049, −0.019)	−0.033 (0.007)	(−0.048, −0.019)
	No. of access points within the influence area of intersection	0.032 (0.017)	(0.002, 0.066)	0.032 (0.017)	(0.002, 0.066)	0.032 (0.018)	(0.002, 0.066)
Segment variables	Log (AADT)	0.498 (0.241)	(0.077, 0.991)	0.353# (0.251)	(−0.063, 0.789)	0.482# (0.282)	(−0.020, 0.975)
	Median opening density	0.013 (0.002)	(0.011, 0.062)	0.013 (0.002)	(0.009, 0.017)	0.014 (0.002)	(0.010, 0.018)
	Driveway density	0.037 (0.013)	(0.012, 0.062)	0.038 (0.012)	(0.013, 0.063)	0.0365 (0.013)	(0.011, 0.062)
	Log (Truck AADT)	0.296# (0.164)	(-0.029, 0.606)	0.310 (0.144)	(0.026, 0.590)	0.305 (0.147)	(0.025, 0.592)
Corridor variables	Intercept	−4.434 (1.767)	(−7.748, −1.275)	−3.014 (2.037)	(−6.824, 0.282)	−4.350 (2.375)	(−8.802, −0.177)
	Principal arterial versus minor arterial	0.492# (0.284)	(−0.078, 1.051)	0.493# (0.293)	(−0.094, 1.063)	0.512# (0.230)	(−0.109, 1.079)
Random effect terms	SD of random effect (ε)	0.503 (0.044)	(0.417, 0.589)	0.544 (0.030)	(0.487, 0.604)	0.469 (0.058)	(0.346, 0.578)
	SD of spatial effects (φ)	1.285 (0.300)	(0.632, 1.824)	1.170 (0.277)	(0.627, 1.718)	0.687 (0.167)	(0.328, 0.987)
Proportion of variability (α)		0.708	(0.525, 0.809)	0.673	(0.522, 0.770)	0.585	(0.368, 0.736)

#Significant at 90% BCI, and all other variables are significant at 95% BCI.

Model Comparison

In this study, DIC, MAD, and RMSE have been used for model comparison and selection. Table 2 shows the goodness-of-fit measures values of the 13 models. There are several observations that can be made from this table. First, across the 13 estimated models, it can be seen that the best model is Model 4 (AB-FO) since it has the lowest DIC, MAD, and RMSE values. Model 4 considers the feeding road entities in addition to the directly adjacent road entities of the same type as the road entity of interest. Second, Model 7 (AR-SO) is the second best model in terms of DIC and RMSE. However, in terms of MAD, Model 9, which is DO-SO, (MAD = 19.452) performs slightly better than Model 7 (MAD = 19.512). Model 7 (AR-SO) emphasizes that the road entities on the same corridor are likely to be spatially correlated up to three neighbors. Third, it can be seen that taking into consideration the feeding road entities to the road entity of interest is important since Model 2 (AB-CO), where only the connected road entities of the other type have been considered, outperformed Model 3 (AB-FOWC), where only the directly adjacent road entities of the same type as the road entity of interest have been considered. This proves the importance of including the feeding road entities to the road entity of interest. As a result, the feeding road entities to the road entity of interest have been considered in the whole spatial models in this study. Lastly, Model 9 (DO-SO) performs the best across the distance-order and distance-based models. This also emphasizes that the same corridor road entities are likely to be spatially correlated up to one mile.

With respect to RE terms in Table 3, the standard deviation of RE (ε) is significant at the 95% Bayesian credible interval (BCI) under all the models, which explains the presence of over-dispersion in the data. However, the value in the corridor RE model is 0.600, and it has been reduced by 9.3% to 38.3% in the other models. This indicates that the spatial models reduce the estimates of extra-Poisson variation. Regarding corridor RE, it was significant, which confirms the presence of within-corridor correlations. In terms of the spatially correlated RE, the standard deviation of spatially correlated RE (α) is significant at the 95% BCI under all the models, which confirms the presence of spatial correlations under the proposed spatial weight features. Lastly, the proportion of variability in the REs due to spatial autocorrelation (α) is significant under all spatial models. The value of α varies from 22.5% (Model 2, Table 3*a*) to 71.8% (Model 7, Table 3*b*) under all spatial models. The two models with the lowest explained variation have the smallest number of spatial units. In this study, the spatial clustering was the strongest ($\alpha = 71.8\%$) in Model 7, adjacency-route second-order spatial weight feature.

Interpretation of Explanatory Variables

Every model contains ten significant variables except the distance-based models. These contain nine significant variables, four of which are intersection-related variables: natural log of major AADT, natural log of minor AADT, absolute speed limit difference between major and minor roads, and total number of access points within the influence area of intersections. A total of four roadway-segment significant variables are in the model: natural log of AADT, natural log of truck AADT, driveway density, and median opening density. Lastly, there are two significant corridor-related variables in each model, except the three distance-based models, which have only roadway classification (principal arterial versus minor arterial), which is a significant variable in all models. The significant corridor variables are different in the HPLN with RE from the HPLN with spatial effects. County variable (Orange versus Broward) is significant in the RE model, and it is insignificant in the spatial models. Also, the corridor variable average segment length of corridor is significant in the spatial models, but insignificant in the corridor RE model. In terms of parameter estimates values, the signs across the models are consistent; however, there are some slight differences in the magnitude of the parameter estimates between the non-spatial model and the spatial models. This could be due to the presence of spatial dependence in the crash data among neighboring road entities along corridors, and not accounting for the spatial dependence may result in biased estimates of the model parameters, which may lead to erroneous interpretation and misleading conclusions (*9*, *29*, *30*).

In terms of significant intersection variables, natural log of major and minor AADT were found to be positively associated with crashes as expected. It was also found that as the absolute speed limit difference between major and minor roads increases the total crashes decreases, and this is expected since the corridors have relatively high posted speeds, and high absolute speed limit difference indicates that the minor road is a small road. Therefore, it is expected that intersections with high posted speeds on the major and minor roads experience more crashes than intersections with low or medium posted speeds on the minor roads. The result shows that a unit increase in the absolute speed limit difference would decrease crashes by a factor of 0.965. Lastly, it was found that as the total number of access points within the influence area of intersection increases the crash occurrence increases, and this is consistent with a previous study (*31*). A unit increase in the total number of access points within the influence area of intersection was associated with an increase in crash frequencies by a factor of 1.023.

Regarding roadway segment-related variables, the results show that as the natural log of AADT increases the total number of crashes increases, as expected. It was also found that the natural log of truck AADT is positively associated with crash occurrence. Driveway density is also positively associated with crash occurrence, and this is consistent with previous studies (*32*, *33*). Each unit increase in the driveway density would increase the crashes by a factor of 1.044. Lastly, it was found that as the median opening density increases the total number of crashes increases, as expected (*34*). The coefficient shows that one unit increase in median opening density was associated with an increase in crash frequencies by a factor of 1.013.

With respect to corridor-related variables, it was found that principal arterial roads tend to have more crashes than the minor arterials, and this may be due to the fact that the traffic volume and posted speed are usually higher on principal arterials than on minor arterials. It was also found that Orange County is associated with more crashes than Broward County. Lastly, as the average segment length of corridor increases, the crash occurrence increases, which is consistent with a previous study (*7*). A unit increase in the average segment length of corridors was associated with an increase in crash frequencies by a factor of 8.542.

Summary and Conclusion

This study developed a Bayesian hierarchical joint model with spatial effects for intersections and roadway segments along corridors using a dataset of 255 signalized intersections and 220 roadway segments along 20 corridors from two counties in Florida. This joint model provides a better understanding of risk factors at intersections and roadway segments along corridors. The study also explored the effect of different spatial weight features to identify the best fit of the model to the data. A total of 13 models were estimated in this study.

The results of this study prove the importance of incorporating spatial effects since it provides a better fit of the data compared with the HPLN with corridor RE model. The spatial models can be categorized into four categories: (a) adjacency-based, (b) adjacency-route, (c) distance-order, and (d) distance-based spatial weight features. The results show the benefits of using the joint model since the model (Model 2) that considers the spatial correlation between a roadway segment and its feeding intersections and vice versa outperforms the adjacency-based first-order model (Model 3) without considering the feeding road entities to the road entity of interest. However, one of the best models is the combination of the previous two models. That is, the adjacency-based first-order model (Model 4), which considers the feeding road entities to the road entity of interest and the directly adjacent road entity of the same type as the road entity of interest. Moreover, the results support that the spatial correlation is significantly important between road entities along the same corridor since one of the best models is the adjacency-route second-order model (Model 7).

Detailed roadway characteristics were collected for the intersections and roadway segments in addition to some corridor information. The modeling results show that four intersection-related, four roadway segment-related, and two corridor-related variables were significant in most of the models. The significant intersection and roadway segment variables are the same across the estimated models but with different magnitude in some variables. When considering the risk factors at the intersection level, the number of access points within the influence area of the intersection is positively associated with the number of crashes. Therefore, the minimum distance between an access point and an intersection should be carefully analyzed. With respect to roadway segments, as median opening density and driveway density increase, the total number of crashes increases. This finding emphasizes the importance of choosing the appropriate number and location of driveways and median openings along corridors. The finding of this study will help road designers, planners, and managers as it provides a better understanding of the factors contributing to crashes at intersections and roadway segments along corridors.

Finally, this study increased the understanding of the variables that contribute to crashes along corridors, and there are several possible future developments. First, the effect of the way the distance is measured, aerial versus traveled distance, is worthy of further investigation. Other combinational spatial weight features could also be tried. Since we are modeling roadway segments and intersections along corridors, there is large variation in AADT, and it is suggested that the impact of different ranges of AADT on the model should be investigated. Lastly, this paper could be improved by having more variables related to signal information.

Acknowledgments

The authors thank the Florida Department of Transportation (FDOT) for providing the data.

Author Contributions

The authors confirm contribution to the paper as follows: study conception and design: Mohamed Abdel-Aty, Jaeyoung Lee, Saif Alarifi, and Xuesong Wang; data collection: Jaeyoung Lee and Saif Alarifi; analysis and interpretation of results: Mohamed Abdel-Aty, Jaeyoung Lee, Saif Alarifi, and Xuesong Wang; draft manuscript preparation: Saif Alarifi. All authors reviewed the results and approved the final version of the manuscript.

References

1. Abdel-Aty, M., and X. Wang. Crash Estimation at Signalized Intersections Along Corridors: Analyzing Spatial Effect and Identifying Significant Factors. *Transportation Research Record: Journal of the Transportation Research Board*, 2006. 1953: 98–111.
2. Guo, F., X. Wang, and M. A. Abdel-Aty. Modeling Signalized Intersection Safety with Corridor-Level Spatial Correlations. *Accident Analysis and Prevention*, Vol. 42, No. 1, 2010, pp. 84–92.
3. Xie, K., X. Wang, H. Huang, and X. Chen. Corridor-Level Signalized Intersection Safety Analysis in Shanghai, China using Bayesian Hierarchical Models. *Accident Analysis and Prevention*, Vol. 50, 2013, pp. 25–33.
4. Xie, K., X. Wang, K. Ozbay, and H. Yang. Crash Frequency Modeling for Signalized Intersections in a High-Density Urban Road Network. *Analytic Methods in Accident Research*, Vol. 2, 2014, pp. 39–51.
5. El-Basyouny, K., and T. Sayed. Accident Prediction Models with Random Corridor Parameters. *Accident Analysis and Prevention*, Vol. 41, No. 5, 2009, pp. 1118–1123.
6. Aguero-Valverde, J., and P. Jovanis. Spatial Correlation in Multilevel Crash Frequency Models. *Transportation Research Record: Journal of the Transportation Research Board*, 2010. 2165: 21–32.
7. Wang, X., T. Fan, M. Chen, B. Deng, B. Wu, and P. Tremont. Safety Modeling of Urban Arterials in Shanghai, China. *Accident Analysis and Prevention*, Vol. 83, 2015, pp. 57–66.
8. Cressie, N. A. C. *Statistics for Spatial Data*. Wiley Interscience, New York, 2015.
9. Lesage, J. P., and R. K. Pace. *Introduction to Spatial Econometrics*. CRC Press, Boca Raton, Fla., 2009.
10. Rodegerdts, L. A. *Signalized Intersections: Informational Guide*. Federal Highway Administration, McLean, Va., 2004.
11. Hadayeghi, A., A. S. Shalaby, and B. N. Persaud. Development of Planning Level Transportation Safety Tools Using Geographically Weighted Poisson Regression. *Accident Analysis and Prevention*, Vol. 42, No. 2, 2010, pp. 676–688.
12. Barua, S., K. El-Basyouny, and M. T. Islam. A Full Bayesian Multivariate Count Data Model of Collision Severity with Spatial Correlation. *Analytic Methods in Accident Research*, Vol. 3, 2014, pp. 28–43.
13. Aguero-Valverde, J. Direct Spatial Correlation in Crash Frequency Models: Estimation of the Effective Range. *Journal of Transportation Safety and Security*, Vol. 6, No. 1, 2013, pp. 21–33.
14. Wang, X., Y. Song, R. Yu, and G. G. Schultz. Safety Modeling of Suburban Arterials in Shanghai, China. *Accident Analysis and Prevention*, Vol. 70, 2014, pp. 215–224.
15. Mitra, S. Spatial Autocorrelation and Bayesian Spatial Statistical Method for Analyzing Intersections Prone to Injury Crashes. *Transportation Research Record: Journal of the Transportation Research Board*, 2009. 2136: 92–100.
16. Castro, M., R. Paleti, and C. R. Bhat. A Latent Variable Representation of Count Data Models to Accommodate Spatial and Temporal Dependence: Application to Predicting Crash Frequency at Intersections. *Transportation Research Part B: Methodological*, Vol. 46, No. 1, 2012, pp. 253–272.
17. Dong, N., H. Huang, and L. Zheng. Support Vector Machine in Crash Prediction at the Level of Traffic Analysis Zones: Assessing the Spatial Proximity Effects. *Accident Analysis and Prevention*, Vol. 82, 2015, pp. 192–198.
18. Zeng, Q., and H. Huang. Bayesian Spatial Joint Modeling of Traffic Crashes on an Urban Road Network. *Accident Analysis and Prevention*, Vol. 67, 2014, pp. 105–112.
19. Evans, J. D. *Straightforward Statistics for the Behavioral Sciences*. Brooks/Cole, Pacific Grove, Calif., 1996.
20. El-Basyouny, K., and T. Sayed. Urban Arterial Accident Prediction Models with Spatial Effects. *Transportation Research Record: Journal of the Transportation Research Board*, 2009. 2102: 27–33.
21. Lawson, A. B., W. J. Browne, and C. L. Vidal Rodeiro. *Disease Mapping with WinBUGS and MLwiN*. Wiley, Chichester, 2003.
22. Miaou, S. P., J. J. Song, and B. K. Mallick. Roadway Traffic Crash Mapping: A Space-Time Modeling Approach. *Journal of Transportation and Statistics*, Vol. 6, No. 1, 2003, pp 33–57.
23. Besag, J., J. York, and A. Mollie. Bayesian Image Restoration, with Two Applications in Spatial Statistics. *Annals of the Institute of Statistical Mathematics*, Vol. 43, No. 1, 1991, pp. 1–20.
24. Aguero-Valverde, J., and P. Jovanis. Analysis of Road Crash Frequency with Spatial Models. *Transportation Research Record: Journal of the Transportation Research Board*, 2008. 2061: 55–63.
25. Wang, X., J. Yang, C. Lee, Z. Ji, and S. You. Macro-Level Safety Analysis of Pedestrian Crashes in Shanghai, China. *Accident Analysis and Prevention*, Vol. 96, 2016, pp. 12–21.
26. Spiegelhalter, D., N. Best, B. P. Carlin, and A. Linde. Bayesian Measures of Model Complexity and Fit. *Journal of the Royal Statistical Society B*, Vol. 64, Part 4, 2002, pp. 583–639.
27. Carlin, B. P., and T. A. Louis. *Bayes and Empirical Bayes Methods for Data Analysis*. Chapman and Hall/CRC, Boca Raton, Fla., 2000.
28. Cowles, M. K. *Applied Bayesian Statistics: With R and OpenBugs Examples*. Springer, New York, 2013.
29. Quddus, M. A. Modelling Area-Wide Count Outcomes with Spatial Correlation and Heterogeneity: An Analysis of London Crash Data. *Accident Analysis and Prevention*, Vol. 40, No. 4, 2008, pp. 1486–1497.
30. Anselin, L. *Spatial Econometrics: Methods and Models*. Kluwer Academic, Dordrecht, 2010.
31. Kim, D.-G., and S. Washington. The Significance of Endogeneity Problems in Crash Models: An Examination of Left-Turn Lanes in Intersection Crash Models. *Accident Analysis and Prevention*, Vol. 38, No. 6, 2006, pp. 1094–1100.
32. Eisele, W., and W. Frawley. Estimating the Safety and Operational Impact of Raised Medians and Driveway

Density: Experiences from Texas and Oklahoma Case Studies. *Transportation Research Record: Journal of the Transportation Research Board*, 2005. 1931: 108–116.
33. Mouskos, K., W. Sun, S. Chien, A. Eisdorfer, and T. Qu. Effect of Midblock Access Points on Traffic Accidents on State Highways in New Jersey. *Transportation Research Record: Journal of the Transportation Research Board*, 1999. 1665: 75–83.
34. Cribbins, P., J. M. Aray, and J. K. Donaldson. Effects of Selected Roadway and Operational Characteristics on Accidents on Multi-lane Highways. *Highway Research Record*, Vol. 188, 1967, pp. 8–25.

The Standing Committee on Safety Data, Analysis and Evaluation (ANB20) peer-reviewed this paper (18-01739).

The opinions expressed in this paper are exclusively those of the authors.

Article

TRR
JOURNAL OF THE TRANSPORTATION RESEARCH BOARD

Transportation Research Record
2018, Vol. 2672(38) 223–232

DOI: 10.1177/0361198118782009
journals.sagepub.com/home/trr
SAGE

Time-Based Calibration of the Inertial Operating Speed to Enhance the Assessment of the Geometric Design Consistency

David Llopis-Castelló[1], Francesco Bella[2], Francisco Javier Camacho-Torregrosa[1], and Alfredo García[1]

Abstract

Road crashes are mainly caused by three concurrent factors: infrastructure, vehicle, and human factors. The interaction between the infrastructure and human factors leads to the concept of geometric design consistency. Recently, a global consistency model was developed based on the difference between the inertial operating speed profile and the operating speed profile. The first was defined as the weighted average operating speed of the previous road section based on distance, and represents drivers' expectancies, whereas the second represents road behavior. However, drivers' expectancies are related to short-term memory which declines gradually and depends on time. Thus, a time-based inertial operating speed would allow a more accurate estimation of the phenomenon. This research analyzes different periods of time and weighting distributions to identify how drivers' expectancies should be estimated. A set of 71 homogeneous road segments located in Italy were considered in the study. As a result, 25 seconds and a convex parabolic distribution should be used to calculate the inertial operating speed profile. This new way of estimating drivers' expectancies showed better results than those obtained based on distance. Finally, the proposed consistency model was compared with the previous models and was found to be able to assess more accurately the geometric design consistency. Therefore, the proposed consistency model is a useful tool for engineers to estimate the number of crashes so that they can incorporate road safety considerations into the geometric design of either new two-lane rural roads, or improving the existing ones.

Road safety is a major concern in our society. Around 1.2 million people die and 50 million are injured in road crashes every year (*1*). Most fatalities occur on rural roads. In Italy, between 2011 and 2013, 48% of all road crashes occurred on rural highways (*2*).

Road crashes are mainly caused by three concurrent factors: infrastructure, vehicle, and human factors. In particular, infrastructure is responsible for over 30% of road crashes (*3*). In fact, crashes tend to concentrate at certain road elements. This is why the interaction between the infrastructure and the human factors has been studied extensively in recent years. This interaction can be partially explained by using the concept of geometric design consistency, which can be defined as how road behavior meets drivers' expectancies. In this regard, a consistent road minimizes surprises to road users driving along it, whereas an inconsistent road presents numerous surprises for drivers, leading to anomalous behavior and increasing the likelihood of crashes.

There are several methods for assessing geometric design consistency: operating speed, vehicle stability, alignment indices, and driver workload (*4*). However, most of the consistency models are based on the analysis of the operating speed profile. Operating speed is frequently defined as the 85th percentile of the speed distribution for passenger cars under free-flow conditions with no external restrictions (V_{85}). One important advantage of its use is the possibility to estimate it through operating speed models.

There are two types of consistency model: local and global. Local models focus on localized issues, such as

[1]Highway Engineering Research Group (HERG), Universitat Politècnica de València, Valencia, Spain
[2]Department of Engineering, Roma Tre University, Rome, Italy

Corresponding Author:
Address correspondence to David Llopis-Castelló: dallocas@doctor.upv.es

sudden speed reductions or high differences between design and operating speeds. Such models are ideal for identifying where road crashes are more likely to take place. On the other hand, global consistency models examine the overall speed variation along an entire road segment. Although they do not indicate where crashes are likely to take place, they can be introduced into the Safety Performance Function (SPF) to predict the number of crashes on an entire road segment.

The first global consistency model was developed by Polus and Mattar-Habib (*5*). In their research, two parameters for estimating geometric design consistency were proposed: relative area (R_a) and operating speed dispersion (σ). The first parameter was defined as the area bounded by the operating speed profile and the average operating speed, divided by the length of the road segment. Higher values of R_a and σ thus produced lower consistency values. This model was later enhanced by including the speed dispersion induced by heavy vehicles, as a surrogate measure for the vertical alignment (*6*).

Related to this, Garach et al. (*7*) developed a new consistency model based on the same parameters in Spain. As a result, Polus and Mattar-Habib's model displayed quite conservative behavior, with some road sections classified as poor according to the global model while presenting fair consistency according to R_a and σ.

Finally, Camacho-Torregrosa (*8*) developed another global consistency model considering two operational parameters: the average operating speed and the average deceleration rate. In addition, this study highlights the importance of using homogeneous road segments. The author also presents a classification of road segments attending to their "boundary constraints", i.e., the conditions under which road segments are connected to the rest of the road network. Several SPFs were developed accordingly.

Recently, Garach et al. (*9*) calibrated different SPFs considering different geometric, operating, and consistency variables. Among the consistency models presented above, the model which more accurately estimated the number of crashes on Spanish two-lane rural roads was that proposed by Camacho-Torregrosa. Likewise, the average operating speed also had an important influence on crash occurrence, with more crashes as the average operating speed increases.

In addition, there are several authors who have studied the influence of geometric design consistency on road safety (*10–17*). All of them have concluded that there is a close relationship between consistency and road crashes. In this context, the operating speed reduction and the deviation of the operating speed along a road segment were identified as the most important variables. Furthermore, the selection of the road is critical for the application of global consistency models. The road segments must be homogeneous, since the length along which the a priori expectancies of drivers are acquired plays a major role (*8, 13, 18*).

However, none of the previous consistency models included the definition of the "expectancies acquisition process" phenomenon in their formulation. Here, García et al. (*19*) developed a novel approach for calculating drivers' expectancies and behavior. A new speed concept, the inertial operating speed (V_i), was proposed to estimate drivers' expectancies at a certain location. This speed was defined as the average operating speed of the previous 1,000 m road segment. This distance was determined with a naïve comparison to crash rates. Conversely, road behavior was associated with the operating speed (V_{85}). The Inertial Consistency Index (ICI) was defined as the difference between V_i and V_{85}. Therefore, the greater this index, the greater the difference between drivers' expectancies and road behavior, and thus crash occurrence is higher.

This inertial operating speed was also studied by Montella et al. (*16*) on motorways. Traditionally, this speed has been based on distance, considering different lengths and extracting a simple average. However, this does not match the drivers' expectation acquisition process, which is related to Short-Term Memory (STM).

STM is the memory system that contains our moment-to-moment conscious thoughts and perceptions. The capacity of STM increases with a person's age until it reaches a maximum in young adulthood. As long as we are able to rehearse it, or pay attention to information in our STM, that information can reside there indefinitely. However, without rehearsal, STM declines gradually and the information is lost in about 18 s (*20*).

Drivers do not recall with the same intensity the previous road section. Therefore, the initial and final zones of the preceding section should not be considered in the same way when determining the inertial operating speed. In this regard, Llopis-Castelló et al. (*21*) defined V_i as the weighted average operating speed based on distance to develop a new global consistency model based on the difference between the inertial operating speed profile and the operating speed profile. As expected, this way of calculating V_i showed better results than estimating this speed as the simple average of the operating speed. However, given two homogeneous road segments with a different average operating speed, the periods of time needed to cover the same distance differ. Thus, a time-based inertial operating speed profile determination might lead to a more reliable estimation of the phenomenon.

This paper shows the analysis of different periods of time and weighting distributions in the calculation of the inertial operating speed to identify how drivers' expectancies should be estimated and, consequently, enhance the assessment of the geometric design consistency.

Objectives and Hypotheses

The main objective of this research is to determine how the inertial operating speed (V_i) should be calculated to estimate in a more accurate way drivers' expectancies. The period of time needed to accurately reflect these expectancies in V_i remains unknown, so different trials will be performed using different periods of time, comparing their difference with V_{85} and their impact on the number of crashes.

The underlying hypothesis is that the inertial operating speed profile based on time will allow a more accurate estimation of the number of crashes than the distance-based one. In addition, weighted averages will be used for this determination to achieve a more reliable measure of driver expectancies. The difference between V_i and V_{85} will be considered as a surrogate measure for geometric design consistency. Then, the higher the difference between V_i and V_{85}, the lower the consistency.

Methodology and Data Description

Methodology

This study was developed by examining the relationship between the operating speed behavior and road crashes. Different two-lane rural road sections located in Italy were selected. Next, the geometry for each road section was recreated by means of the methodology proposed by Camacho-Torregrosa et al. (*22*); and the operating speed profiles were estimated considering the speed models calibrated by Marchionna and Perco (*23*). From this, different inertial operating speed profiles were calculated for each homogeneous road segment considering different periods of time and weighting distributions. Crash and traffic data were also obtained. Finally, the relationship between crashes and consistency was studied by calibrating several SPFs. As a result, the inertial operating speed profile that better describes drivers' expectancies was identified and consistency thresholds were proposed.

Road Segments

A total of 48 road sections located in Italy were selected for the study. These resulted in 71 homogeneous road segments, which were identified with the following procedure.

First, road segments were divided into sections with similar traffic volume and cross-sections. Major intersections also influence on drivers' expectancies, so they were considered for segmentation. Finally, each road section was divided according to its geometric behavior using the Curvature Change Rate (CCR), which is defined as the rate between the sum of the absolute deflection angles per length unit.

The homogeneous road segments had Annual Average Daily Traffic (AADT) volumes ranging from 1,319 to 19,577 vehicles per day (vpd). Their length ranged from 1,915 m to 19,325 m, and their longitudinal grade did not exceed 5%. Lane width ranged from 3.00 m to 3.50 m, and the shoulder width varied from 0.50 m to 1.50 m.

Traffic and Crash Data

Traffic volume and crash data were provided by the "Azienda Nazionale Autonoma delle Strade" (ANAS) and the "Automobile Club Italia" (ACI), respectively. Thus, AADT and the number of fatal-and-injury crashes were identified for each homogeneous road segment.

AADT was defined as the average traffic volume from 2012 to 2015. Only fatal-and-injury crashes were considered between 2005 and 2014. The cause of every crash was reviewed so that only those related to geometry were included (e.g., crashes caused by vehicles entering the road from minor roads or driveways were removed from the analysis, since their cause is not the road geometry per se). As a result, a total of 2,080 crashes were reported, involving 202 fatalities and 3,701 injuries.

Speed Profiles

Operating Speed Profiles. The operating speed profile for each road segment was estimated using the model by Marchionna and Perco (*23*) for Italian two-lane rural roads (Figure 1). The model takes into account the general character of the horizontal alignment, since the desired speed is calculated as a function of the CCR.

Inertial Speed Profiles. The inertial operating speed profile was calculated for every road segment from its operating speed profile. This speed attempts to define drivers'

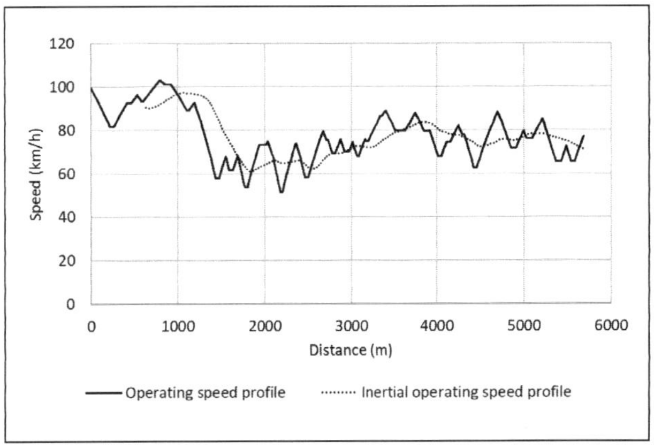

Figure 1. Speed profiles.

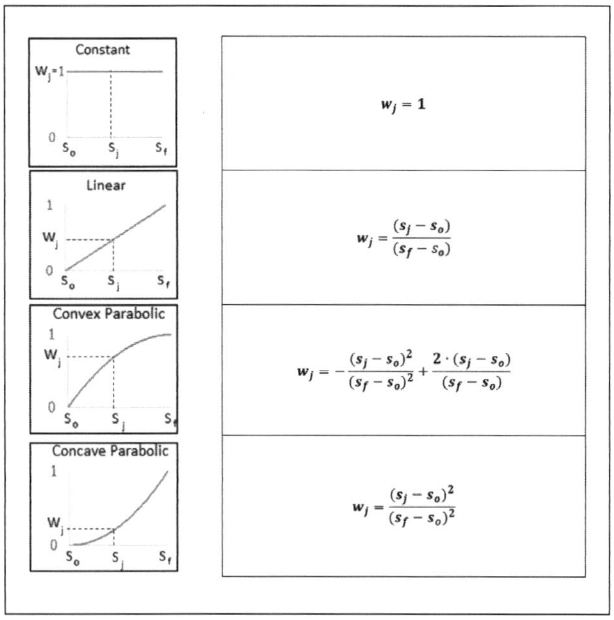

Figure 2. Weighting distributions.

expectancies, which are related to the Short-Term Memory (STM). As noted above, this memory depends on time and is declines gradually (*20*).

Therefore, the inertial operating speed for each point of the alignment was defined as the weighted average operating speed of the preceding road section.

The time for which the inertial operating speed should be calculated was unknown, so periods of time (t) between 10 and 60 s with a step of 5 s and four weighting distributions were analyzed (Figure 2).

A constant distribution provides the average speed, since the operating speed for all different stations under consideration has the same weight. To calculate the other distributions, a weighting factor was proposed, ranging from 0 to 1. Except for the constant distribution, the weighting factor is always 0 for the furthest point, and 1 for the closest point. In addition, the vertexes of the convex and concave parabolic distributions were in 1 and 0, respectively. The equations of these distributions are shown in Figure 2, where s_j is the actual station in meters, w_j is the weighting factor in this station and s_o and s_f are the initial and final stations in meters.

It is worth highlighting that the distance ($s_f - s_o$) to consider for every inertial speed calculation varies along the road, as a function of the speed. For instance, a road segment in which there are two stations presenting an operating speed of 40 and 70 km/h will compute the inertial operating speed using different distances, that for 70 km/h being the longest.

The inertial operating speed is thus determined as the weighted average speed of the last t seconds according to the weighting distribution:

$$V_{i,k} = \frac{\sum w_j \cdot V_{85,j}}{\sum w_j} \quad (1)$$

where $V_{i,k}$: inertial operating speed (km/h) at the point k; $V_{85,j}$: operating speed at the point j; and w_j: weighting factor at the point j.

As a result, 44 (11 periods of time × 4 weighting distributions) inertial operating speed profiles were developed for each road segment. Figure 1 shows V_{85} and V_i for one of the road segments under study in the forward direction, considering 25 s and a convex parabolic distribution.

Consistency Parameters

Different consistency parameters were proposed considering several variables based on the difference between V_i and V_{85} (see *21*). Figure 3 shows the speed differences between the speed profiles depicted in Figure 1. According to this definition, a positive speed difference means that drivers' expectancies are not met, since their speed is lower than the speed they expect to reach. The likelihood of crashes increases with the magnitude of these differences.

The consistency parameters will be composed as a combination of the following, simple parameters (Figure 3).

- A (m·km/h): area bounded by the difference between V_i and V_{85}, and the x axis.
- L (m): length of the road segment.
- σ (km/h): standard deviation of the difference between V_i and V_{85}.
- $A(+)$ (m·km/h): area bounded by the difference between V_i and V_{85} considering only the positive differences.
- $L(+)$ (m): length of the road segment considering only the positive differences.
- $\sigma(+)$ (km/h): standard deviation of the difference between V_i and V_{85} considering only the positive differences.
- $A(> x$ km/h) (m·km/h): area bounded when the difference between V_i and V_{85} is higher than x km/h.

A higher value for either of these variables (dispersion or area bounded) will lead to a lower consistency. This will make the interpretation of the final consistency parameter slightly easier.

Table 1 summarizes the proposed consistency parameters (*21*). All parameters are expressed in terms of

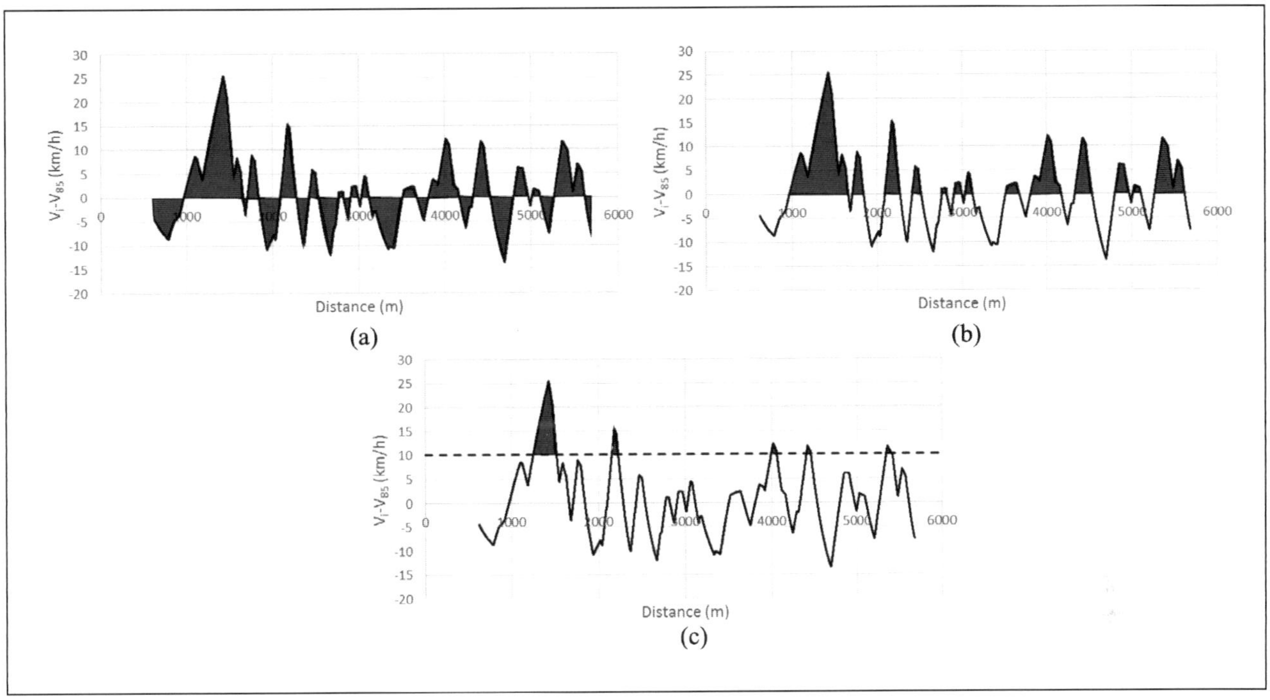

Figure 3. Consistency variables: (*a*) *A*, *L*, and σ; (*b*) *A*(+), *L*(+), and σ(+); and (c) *A* (> *x* km/h).

Table 1. Consistency Parameter

Consistency parameter	Equation
1	$\sqrt{\frac{A(+)\cdot\sigma}{L}}[km/h]$
2	$\sqrt{\frac{A\cdot\sigma}{L}}[km/h]$
3	$\frac{A(+)}{L(+)}[km/h]$
4	$\frac{A(>10km/h)}{L}[km/h]$
5	$\frac{A(>15km/h)}{L}[km/h]$
6	$\frac{A(>20km/h)}{L}[km/h]$
7	$\sqrt{\frac{A(+)\cdot\sigma(+)}{L(+)}}[km/h]$
8	$\sqrt{\frac{A(+)\cdot\sigma}{L(+)}}[km/h]$

speed. This easy interpretation of the consistency parameter is an advantage compared with other consistency models. In all cases, a higher value of the parameter indicates a lower consistency level.

Results

The best consistency parameter was identified by examining its relationship to road crashes. Following common practice, generalized linear modelling techniques were used to fit an SPF that relates risk exposure and consistency to the number of crashes (Equation 2). A negative binomial distribution was assumed, since it is an appropriate solution with overdispersed count data (*24*).

$$Y_{i,10} = e^{\beta_0} \cdot L^{\beta_1} \cdot \mathrm{AADT}^{\beta_2} \cdot e^{\beta_3 \cdot C} \qquad (2)$$

where $Y_{i,10}$: crashes with injuries on the road segment in 10 years; β_i: regression coefficients; *L*: length of the road segment (km); AADT: Average Annual Daily Traffic (vpd); and *C*: consistency parameter (km/h).

The AIC (Akaike Information Criterion) was given for all regressions as a measure of goodness of fit. The smaller the AIC value, the better the model.

The goodness of fit was also studied from the Cumulative Residuals (CURE) Plots (*25*, *26*). This method consists of plotting the cumulative residuals for each independent variable. The aim is to graphically observe how well the function fits the data set. The CURE method has the advantage of not being dependent on the number of observations, as are many other traditional statistical procedures. In general, a good cumulative residuals plot is one that oscillates around 0. Thus, a good fit is given when the residuals do not stray beyond the $\pm 2\sigma^*$ boundaries.

Risk Exposure Influence

It is well known that crashes are significantly affected by the risk exposure. Indeed, several previous researchers have developed SPFs that depend only on the risk

Table 2. Ranking of the Models According to the AIC Value

Model	Parameter	Time (s)	Weighting distribution	AIC
10PX4	Parameter 4	10	Convex	546.87
25PX7	Parameter 7	25	Convex	547.01
10PX7	Parameter 7	10	Convex	547.62
10L4	Parameter 4	10	Linear	547.77
25PX3	Parameter 3	25	Convex	548.02
30PV7	Parameter 7	30	Concave	548.12
25PX8	Parameter 8	25	Convex	548.4
20L7	Parameter 7	20	Linear	548.44
30PV3	Parameter 3	30	Concave	548.46
30PV8	Parameter 8	30	Concave	548.48
15PX7	Parameter 7	15	Convex	548.65
20PX7	Parameter 7	20	Convex	548.81
20L3	Parameter 3	20	Linear	549
10C5	Parameter 5	10	Constant	549.01
30PX7	Parameter 7	30	Convex	549.22
20L8	Parameter 8	20	Linear	549.23
30PV1	Parameter 1	30	Concave	549.3
20C7	Parameter 7	20	Constant	549.56
10L7	Parameter 7	10	Linear	549.59
30L1	Parameter 1	30	Linear	549.6
15C7	Parameter 7	15	Constant	549.6
20PX8	Parameter 8	20	Convex	549.6
40PV7	Parameter 7	40	Concave	549.62
15PV7	Parameter 7	15	Concave	549.64
20PX3	Parameter 3	20	Convex	549.78

exposure (*13*, *27*). An SPF that only considers exposure was calibrated:

$$Y_{i,10} = e^{-7.2212} \cdot L^{0.7070} \cdot \mathbf{AADT}^{1.0307} \qquad \mathbf{AIC} = 557.92 \tag{3}$$

This model is not of major interest, but it is useful for determining how important the inclusion of the consistency term is for crash estimation. As expected, all parameters are statistically significant. The AADT estimate is close to 1, indicating that the number of crashes is linearly affected by the traffic volume under consideration. On the other hand, the length estimate is lower than 1, so longer homogeneous road segments induce lower crash rates.

Consistency Influence

A total of 352 Safety Performance Functions were calibrated by combining 44 inertial operating speed profile types and eight consistency parameters.

All models were sorted as a function of their AIC value. In this regard, Table 2 shows the 25 models with the lowest AIC values. We can see that parameter 7 was the most important. This parameter includes the positive difference between the inertial operating speed and operating speed (Table 1). According to the inception of the inertial operating speed profile, the linear or parabolic weighting distributions produced the best results, validating the hypothesis that the last seconds have a major influence on drivers' expectancies, since the constant weighting distribution was not the best. As expected, the AIC values were lower than for the single-exposure SPF (AIC = 557.92).

The evolution of the AIC for every SPF was also analyzed according to every parameter and weighting distribution. The objective was to examine the sensitivity of the SPF to the type of consistency parameter. Figure 4 shows the trend of the AIC value considering parameter 7 and the convex parabolic distribution. In this regard, the period of time between 20 and 30 seconds was identified for all weighting distributions and consistency parameters as the best. This value is in accordance with previous research.

The lowest AIC value was found for 25 s and the consistency parameter 7 (model 25PX7, where PX indicates a convex parabolic distribution). The corresponding SPF is:

$$Y_{i,10} = e^{-8.57584} \cdot L^{1.03083} \cdot \mathrm{AADT}^{1.02707} \cdot e^{0.17098 \cdot C} \mathrm{AIC} = 547.01 \tag{4}$$

Finally, the model was validated by means of CURE plots (Figure 5). It can be observed that the plots against each explanatory variable do not stray beyond the $\pm 2\sigma^*$

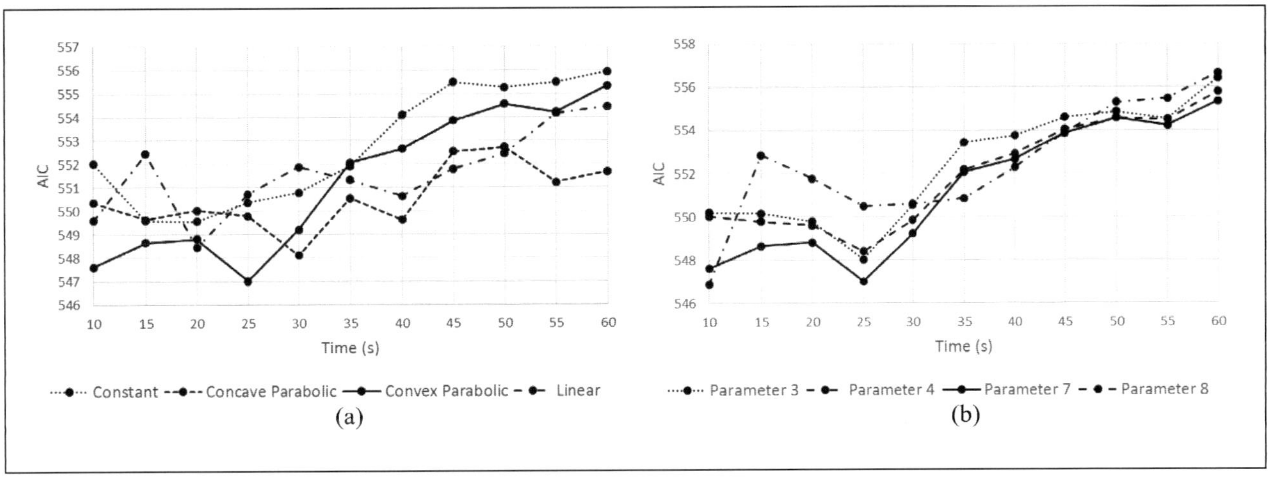

Figure 4. Evolution of the AIC value: (*a*) consistency parameter 7 and (*b*) convex parabolic distribution.

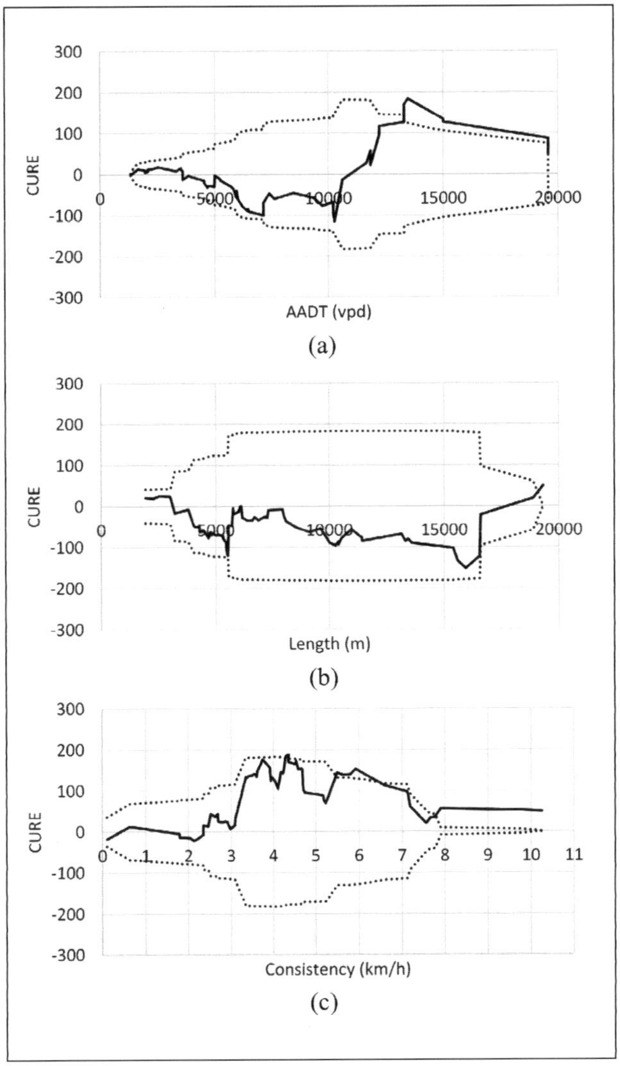

Figure 5. CURE plots: (*a*) AADT, (*b*) length, and (*c*) consistency.

boundaries, apart from a few points when the AADT or C are high. This is mainly due to the limited available data for large traffic volumes and road segments with very poor consistency. In these situations, the proposed model tends to underestimate the number of crashes. Thus, the use of the proposed consistency model for road segments which present a traffic volume lower than 13,500 vpd is recommended. Despite this, the consistency model is a useful tool to estimate the number of fatal-and-injury crashes in Italian two-lane rural roads.

Discussion

Inertial Operating Speed

The inertial operating speed was defined as the weighted average operating speed based on time, which attempts to better reflect the behavior of the STM (*20*), instead of considering the instant operating speed like other consistency models do. As a result, this speed should be calculated at each station as the weighted average operating speed for the last 25 seconds, according to a convex parabolic distribution.

It is important to highlight that V_i was calculated considering a certain period of time – and not a certain distance. This assumption makes sense, since the human mind tends to retain information as a function of time, not of distance covered. In fact, the authors calculated V_i in terms of distance in a previous research (*21*), but time-based models shows a better goodness of fit (AIC of 547.01 Vs. 548.15).

An additional advantage of considering time to calculate the inertial operating speed is its higher stability for different consistency parameters. All consistency parameters showed the minimum AIC values for 25 s, whereas the distance-based model presented their best

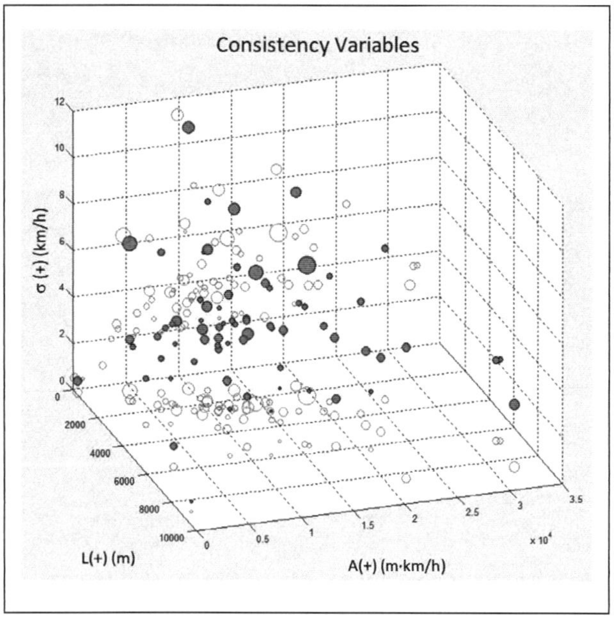

Figure 6. Relationship between consistency variables and crash rate.

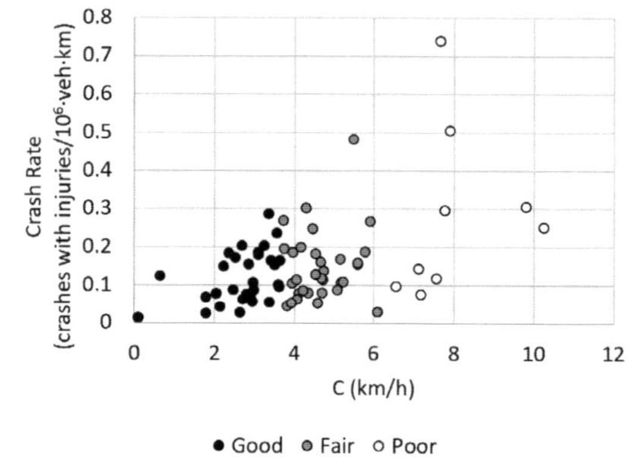

Good	Fair	Poor
C ≤ 3.65 km/h	3.65 < C ≤ 6.5 km/h	C > 6.5 km/h

Figure 7. Relationship between the crash rate and the consistency parameter.

results for a range of distances. This is not surprising, since different lengths can be reached in the same period of time depending on the speed.

Effect of the Consistency Parameter on Road Crashes

The proposed consistency parameter was parameter 7, which is defined considering only the positive differences between V_i and V_{85} (Table 1).

A positive difference between these speed profiles means that drivers' expectancies are not met. Therefore, a higher crash rate is expected. Thus, for a given $A(+)$, a higher length and a lower $\sigma(+)$ leads to a lower crash rate, i.e., a good consistency. Likewise, for a given $L(+)$, a higher $A(+)$ or $\sigma(+)$ leads to a higher crash rate, i.e., a poor consistency. These conclusions can be observed in Figure 6, where the volume of the circles depicts crash rates. In this figure, transparent circles are the projection of the red ones.

As a result, crash rates increase with the consistency parameter. In this way, different consistency thresholds were defined by means of a cluster analysis. Thus, the level of consistency of a homogeneous road segment can be defined as good, fair or poor depending on the value of this consistency parameter (Figure 7).

Therefore, this consistency model can be used to compare and sort different design proposals, maximizing road safety. The proposed SPF (Equation 4) is a useful tool for engineers to estimate the number of crashes, and to determine the potential for improvement offered by a given road solution, or set of solutions.

Comparison with Previous Global Consistency Models

The proposed consistency model was compared with the global consistency models developed by Polus and Mattar-Habib (*5*), Garach et al. (*7*), and Camacho-Torregrosa (*8*).

Different SPFs were calibrated considering the consistency parameters developed by these authors (Table 3). It can be observed that these models offered a worse statistical adjustment than the proposed model, since their AIC values were higher.

In addition, the root-mean-square error (RMSE) and the mean absolute error (MAE) were calculated for each model (Table 3). As a result, the proposed model showed slightly lower values than the previous models.

Thus, the consistency parameter of the enhanced consistency model can better represent the phenomenon than the previous models.

Conclusions

A more accurate way to estimate drivers' expectancies by means of the inertial operating speed profile has been proposed to enhance the assessment of geometric design consistency through the model proposed by Llopis-Castelló et al. (*21*). This model was defined through the difference between the inertial operating speed profile (V_i), which represents drivers' expectancies, and the operating speed profile (V_{85}), which represents road behavior.

Different periods of time and weighting distributions were studied to identify how inertial operating speeds should be calculated. Thus, a total of 352 SPFs were calibrated as a combination of 44 inertial operating speed profile types and eight consistency parameters. Most of

Table 3. Statistical Adjustment – Global Consistency Models

		β_0	β_1	β_2	β_3				
		-	ln *L*	ln AADT	*C*	AIC	α	RMSE	MAE
Polus and Mattar-Habib	Estimate	−6.95815	0.9458	0.9929	−0.3352	553.48	0.3174	22.43	14.03
	Pr(>|z|)	$< 2{\cdot}10^{-16}$	$4.26{\cdot}10^{-8}$	$< 2{\cdot}10^{-16}$	0.00943				
Garach et al.	Estimate	−7.08048	0.9050	1.0146	−0.2917	554.87	0.3247	23.03	14.10
	Pr(>|z|)	$< 2{\cdot}10^{-16}$	$9.58{\cdot}10^{-8}$	$< 2{\cdot}10^{-16}$	0.0169				
Camacho-Torregrosa	Estimate	−6.77612	0.9206	1.0335	−0.2221	555.75	0.3254	21.88	14.10
	Pr(>|z|)	$5.21{\cdot}10^{-16}$	$1.94{\cdot}10^{-7}$	$< 2{\cdot}10^{-16}$	0.0379				
New model	Estimate	−8.57584	1.03083	1.02707	0.17098	547.01	0.287	21.64	13.75
	Pr(>|z|)	$< 2{\cdot}10^{-16}$	$2.28{\cdot}10^{-10}$	$< 2{\cdot}10^{-16}$	0.00015				

them showed lower AIC values than the SPF considering only the risk exposure, so the level of consistency significantly influences crash occurrence.

The best model was 25PX7 model. In this regard, the inertial operating speed profile was estimated considering 25 seconds and a convex parabolic distribution. Likewise, the consistency parameter was obtained from the positive differences between V_i and V_{85}. This model was consistent with STM behavior. Regarding this, an inertial operating speed profile based on time can better represent drivers' expectancies than profiles based on distance.

Addionally, the proposed model was consistent with STM behavior. Regarding this, an inertial operating speed profile based on time can better represent drivers' expectancies than profiles based on distance.

Addionally, the proposed consistency model was compared with the previous ones. As a result, the developed model showed the lowest AIC value and a closer relationship with the observed crashes. Therefore, the proposed global consistency model describes the phenomenon better than the previous models.

Finally, different thresholds were defined to identify the consistency level of a homogeneous road segment. Thus the new SPF is a useful tool for engineers to estimate the number of crashes and brings a more objective assessment of road safety to the geometric road design process.

Acknowledgments

This research was subsidized by the Spanish Ministry of Economy, Industry, and Competitiveness through "Ayudas a la movilidad predoctoral para la realización de estancias breves en centros de I + D 2015." The study presented in this paper is also part of the research project entitled "CASEFU - Estudio experimental de la funcionalidad y seguridad de las carreteras convencionales" (TRA2013-42578-P), subsidized by the Spanish Ministry of Economy, Industry, and Competitiveness and the European Social Fund. In addition, the authors would like to thank the "Azienda Nazionale Autonoma delle Strade" (ANAS) and the "Automobile Club Italia" (ACI), who provided traffic and crash data, respectively.

Author Contributions

The authors confirm contribution to the paper as follows: study conception and design, Llopis-Castelló, D., Camacho-Torregrosa, F.J. and García, A.; data collection, Llopis-Castelló, D. and Bella, F.; analysis and interpretation of results, Llopis-Castelló, D., Bella, F. and Camacho-Torregrosa, F.J.; draft manuscript preparation, Llopis-Castelló, D. All authors reviewed the results and approved the final version of the manuscript.

References

1. WHO. *Global Status Report on Road Safety 2015*. World Health Organization, 2015 http://www.who.int/violence_injury_prevention/road_safety_status/2015/en/.
2. European Transport Safety Council. *Ranking EU Progress on Improving Motorway Safety*. PIN Flash Report, Vol. 28, March 2015.
3. Treat, J. R., N. S. Tumbas, S. T. McDonald, D. Shinar, and R. D. Hume. *Tri-level Study of the Causes of Traffic Accidents: Executive Summary*. Technical Report No. DOT HS-805 099. National Technical Information Services, Bloomington, University of Indiana, 1979.
4. Gibreel, G. M., S. M. Easa, Y. Hassan, and I. A. El-Dimeery. State of the Art of Highway Geometric Design Consistency. *Journal of Transportation Engineering*, Vol. 125, No. 4, 1999, pp. 305–313.
5. Polus, A., and C. Mattar-Habib. New Consistency Model for Rural Highways and Its Relationship to Safety. *Journal of Transportation Engineering*, Vol. 130, No. 3, 2004, pp. 286–293.
6. Mattar-Habib, C., A. Polus, and H. Farah. Further Evaluation of the Relationship between Enhanced Consistency Model and Safety of Two-Lane Rural Roads in Israel and Germany. *European Journal of Transport and Infrastructure Research*, Vol. 4, No. 8, 2008.
7. Garach, L., F. Calvo, M. Pasadas, and J. de Oña. Proposal of a New Global Model of Consistency: Application in Two-Lane Rural Highways in Spain. *Journal of Transportation Engineering*, Vol. 140, No. 8, 2014, pp. 04014030.
8. Camacho Torregrosa, F. J. *Development and Calibration of a Global Geometric Design Consistency Model for Two-Lane Rural Highways, Based on the Use of Continuous*

Operating Speed Profiles. Universitat Politècnica de València, Valencia (Spain), 2015.
9. Garach, L., J. de Oña, G. López, and L. Baena. Development of Safety Performance Functions for Spanish Two-Lane Rural Highways on Flat Terrain. *Accident Analysis & Prevention*, Vol. 95, 2016, pp. 250–265.
10. Anderson, I., K. Bauer, and D. Harwood. Relationship to Safety of Geometric Design Consistency Measures for Rural Two-Lane Highways. *Transportation Research Record: Journal of the Transportation Research Board*, 1999. 1658: 43–51.
11. Ng, J., and T. Sayed. Effect of Geometric Design Consistency on Road Safety. *Canadian Journal of Civil Engineering*, Vol. 31, No. 2, 2004, pp. 218–227.
12. Awatta, M., Y. Hassan, and T. Sayed. Quantitative Evaluation of Highway Safety Performance Based on Design Consistency. *Advances in Transportation Studies*, Vol. 9, 2006.
13. Cafiso, S., A. Di Graziano, G. Di Silvestro, G. La Cava, and B. Persaud. Development of Comprehensive Accident Models for Two-Lane Rural Highways Using Exposure, Geometry, Consistency and Context Variables. *Accident Analysis & Prevention*, Vol. 42, No. 4, 2010, pp. 1072–1079.
14. Quddus, M. Exploring the Relationship between Average Speed, Speed Variation, and Accident Rates Using Spatial Statistical Models and GIS. *Journal of Transportation Safety & Security*, Vol. 5, No. 1, 2013, pp. 27–45.
15. Montella, A., L. Colantuoni, and R. Lamberti. Crash Prediction Models for Rural Motorways. *Transportation Research Record: Journal of the Transportation Research Board*, 2008. 2083: 180–189.
16. Montella, A., and L. L. Imbriani. Safety Performance Functions Incorporating Design Consistency Variables. *Accident Analysis & Prevention*, Vol. 74, 2015, pp. 133–144.
17. Wu, K.-F., E. T. Donnell, S. C. Himes, and L. Sasidharan. Exploring the Association between Traffic Safety and Geometric Design Consistency Based on Vehicle Speed Metrics. *Journal of Transportation Engineering*, Vol. 139, No. 7, 2013, pp. 738–748.
18. Resende, P., and R. Benekohal. *Effect of Roadway Section Length on Accident Modeling Traffic Congestion and Traffic Safety*. The 21st Century Conference, ASCE, Chicago, IL, 1997.
19. García, A., D. Llopis-Castelló, F. J. Camacho-Torregrosa, and A. M. Pérez-Zuriaga. New Consistency Index Based on Inertial Operating Speed. *Transportation Research Record: Journal of the Transportation Research Board*, 2013. 2391: 105–112.
20. Revlin, R. *Cognition: Theory and Practice*. New York, NY: Palgrave Macmillan, 2012.
21. Llopis-Castelló, D., F. Bella, F. J. Camacho-Torregrosa, and A. García. New Consistency Model Based on Inertial Operating Speed Profiles for Road Safety Evaluation. *Journal of Transportation Engineering*, 2017.
22. Camacho-Torregrosa, F. J., A. M. Pérez-Zuriaga, J. M. Campoy-Ungría, A. García, and A. P. Tarko. Use of Heading Direction for Recreating the Horizontal Alignment of an Existing Road. *Computer-Aided Civil and Infrastructure Engineering*, Vol. 30, No. 4, 2015, pp. 282–299.
23. Marchionna, A., and P. Perco. Operating Speed-Profile Prediction Model for Two-Lane Rural Roads in the Italian Context. *Advances in Transportation Studies*, Vol. 2008, No. 14, 2008, pp. 57–68.
24. Lord, D., and F. Mannering. The Statistical Analysis of Crash-Frequency Data: A Review and Assessment of Methodological Alternatives. *Transportation Research Part A: Policy and Practice*, Vol. 44, No. 5, 2010, pp. 291–305.
25. Hauer, E., and J. Bamfo. Two Tools for Finding What Function Links the Dependent Variable to the Explanatory Variables. *Proc., International Co-operation on Theories and Concepts in Traffic Safety 1997 Conference*, Lund, Sweden, 1997.
26. Lord, D., and B. N. Persaud. Accident Prediction Models With and Without Trend: Application of the Generalized Estimating Equations Procedure. *Transportation Research Record: Journal of the Transportation Research Board*, 2000. 1717: 102–108.
27. Harwood, Douglas W., et al. *Prediction of the Expected Safety Performance of Rural Two-Lane Highways*. No. FHWA-RD-99-207, MRI 4584-09, Technical Report. United States. Federal Highway Administration, 2000.

The Standing Committee on Safety Data, Analysis and Evaluation (ANB20) peer-reviewed this paper (18-00813).

Article

TRR
JOURNAL OF THE TRANSPORTATION RESEARCH BOARD

Transportation Research Record
2018, Vol. 2672(38) 233–244

DOI: 10.1177/0361198118797211
journals.sagepub.com/home/trr
SAGE

Real-Time Prediction of Vehicle Trajectories for Proactively Identifying Risky Driving Behaviors at High-Speed Intersections

Chaopeng Tan[1], Nan Zhou[2], Fen Wang[3], Keshuang Tang[1], and Yangbeibei Ji[4]

Abstract

At high-speed intersections in many Chinese cities, a traffic-light warning sequence at the end of the green phase—three seconds of flashing green followed by three seconds of yellow—is commonly implemented. Such a long phase transition time leads to heterogeneous decision-making by approaching drivers as to whether to pass the signal or stop. Therefore, risky driving behaviors such as red-light running, abrupt stop, and aggressive pass are more likely to occur at these intersections. Proactive identification of risky behaviors can facilitate mitigation of the dilemma zone and development of on-board safety altering strategies. In this study, a real-time vehicle trajectory prediction method is proposed to help identify risky behaviors during the signal phase transition. Two cases are considered and treated differently in the proposed method: a single vehicle case and a following vehicle case. The adaptive Kalman filter (KF) model and the K-nearest neighbor model are integrated to predict vehicle trajectories. The adaptive KF model and intelligent driver model are fused to predict the following vehicles' trajectories. The proposed models are calibrated and validated using 1,281 vehicle trajectories collected at three high-speed intersections in Shanghai. Results indicate that the root mean square error between the predicted trajectories and the actual trajectories is 5.02 m for single vehicles and 2.33 m for following vehicles. The proposed method is further applied to predict risky behaviors, including red-light running, abrupt stop, aggressive pass, speeding pass, and aggressive following. The overall prediction accuracy is 95.1% for the single vehicle case and 96.2% for the following vehicle case.

At high-speed intersections in Chinese cities with traffic-light signal controls, at the end of the green phase a long phase transition time is commonly applied, that is, a flashing green indication of 3 s followed by a yellow indication of 3 s (*1*). Such a long phase transition time leads to heterogeneous decision-making by approaching drivers as to whether to pass or stop at the end of the green phase (*2*, *3*). Therefore, risky driving behaviors such as red-light running, abrupt stop, aggressive pass, and inconsistent decisions of the leading vehicle and the following vehicle are more likely to occur at these intersections, potentially leading to right-angle and rear-end traffic accidents (*4*). Proactive identification of these risky behaviors can facilitate mitigation of the dilemma zone and development of on-board safety altering strategies (*5*).

Recent developments of vehicle-to-infrastructure systems, connected vehicles, and video detection technologies provide many possibilities to obtain real-time and high-resolution vehicle trajectory data. A few studies have attempted to use characteristic parameters extracted from vehicle trajectory data, for example, speed, acceleration, and jerk, for the purpose of identifying risky driving behaviors (*6*–*8*). Though those methods performed well in the identification of actual risky behaviors, they were unable to predict future risky behaviors and did not consider the signal phase change and clearance intervals at signalized intersections. Other studies have aimed to predict vehicle trajectories for the development of dilemma zone mitigation, or safety-altering and preventing red-light running, by focusing on signal phase

[1]College of Transportation Engineering, Tongji University, Shanghai, China
[2]Intelligent Transportation Research Institute, Hangzhou Comprehensive Transportation Research Center, Hangzhou, China
[3]School of Urban Rail Transportation, Shanghai University of Engineering Science, Shanghai, China
[4]School of Management, Shanghai University, Shanghai, China

Corresponding Author:
Address correspondence to Yangbeibei Ji: jybb@shu.edu.cn

change and clearance intervals (*5, 9–11*). However, most of the studies were based on segment speed data detected by advance detectors, and the vehicle trajectories thus had to be discretized and sketched. In addition, a general hypothesis was commonly made in those studies that the subject vehicles would keep the same speed as the detected speeds at the advance detectors. Such a hypothesis could greatly constrain the applicability of those methods as approaching drivers might change their decisions and speeds based on their perceived signal states and surrounding traffic environment (*1*).

Hence, the primary objective of this study is to propose a novel method that can predict approaching vehicles' trajectories in a real-time manner during the phase transition time intervals at rural high-speed intersections, based on high-resolution vehicle trajectory data. The secondary objective of this study is to explore the applicability of the proposed vehicle trajectory prediction method in the identification of risky driver behaviors that might occur during the phase transition time intervals. Two cases are considered and treated differently in the proposed method: the single vehicle case and the following vehicle case. The adaptive Kalman filter (KF) model and the K-nearest neighbor (K-NN) model are integrated to predict the single vehicles' trajectories. Meanwhile, the adaptive KF model and the intelligent driver model (IDM) are fused to predict the following vehicles' trajectories. The proposed method is calibrated and validated using 1, 281 real vehicle trajectories with a resolution of 0.12 s, collected at three typical rural high-speed intersections in Shanghai by videorecording. The proposed method is further applied to predict risky behaviors, including red-light running, abrupt stop, aggressive pass, speeding pass, and aggressive following.

The rest of the paper is organized as follows. The second section summarizes the relevant literature on the identification of risky driving behaviors and vehicle trajectory prediction. The third section introduces the proposed vehicle trajectory prediction method for the single vehicle case and the following vehicle case, respectively. The fourth section describes the process of model calibration and validation. The fifth section presents the application of the proposed method in the identification of risky driving behaviors. The final section highlights the conclusions and future works.

Literature Review

Risky driving behaviors usually include adoption of inadequate headway, speeding, excessive reliance on the vehicle's brakes, nearside and reckless overtaking, and passing traffic signals at yellow or even red (*12*). Numerous studies have been done to identify risky driving behaviors based on real-time trajectory data. Dingus et al. (2006) adopted jerk and the peak-to-peak jerk as measures to identify emergency braking behavior from a highly capable instrumentation system including five channels of video and many vehicle state and kinematic sensors (*6*). Khattak et al. (2015) used the absolute jerk and the deviation of jerk to identify extreme driving decision patterns based on the vehicle trajectory data with a resolution of 1.0 s (*7*). Liu et al. (2016) used the 95 percentile acceleration for each speed range to judge the extreme acceleration events through analysis of each vehicle's 10-Hz position and motion data transmitted through vehicle-to-vehicle and vehicle-to-infrastructure applications (*8*). Generally, the above studies have achieved good performance in the identification of risky driving behaviors that have already occurred, but were unable to predict future risky behaviors and did not account for the signal phase transition time period at signalized intersections.

Meanwhile, many studies have focused on the development of dilemma zone (DZ) protection algorithms and systems, in which the probability of risky behaviors occurring is considered as a critical performance indicator. Vehicle trajectory prediction also had a vital role in the prediction of risky behaviors in those studies. In the detection-control system, Zimmerman et al. (2003) used loop detectors to detect the speed and length information of the vehicles in order to predict when the subject vehicles will fall into the DZ (*9*). Zhang et al. (2012) proposed a probabilistic framework to model and predict red-light running hazards based on single upstream loop detector for dynamic all-red extension (*5*). Li et al. (2010) presented a new DZ protection algorithm based on the Markov process, in which the trajectories of the approaching vehicles were predicted using advance detectors with a detection interval of 1.0 s and the times when the vehicles would reach the DZ were ultimately determined (*10*). A common shortcoming of the previous studies is that the segment-based speed data detected by the advance detectors means that the vehicle trajectories are discretized and sketched. A DZ control algorithm proposed by Kong et al. (2014) used real-time trajectory data with a resolution of 1.0 s to predict the number of vehicles in DZ and optimize the green extension time (*11*). Nevertheless, the trajectory prediction in these studies is generally based on a simple hypothesis that vehicles will maintain the same speed as when they are detected by advance detectors. This hypothesis could greatly constrain the applicability of those methods in the real world, as the approaching drivers might change their speeds and even their decisions based on the perceived signal state and surrounding traffic (*12*).

There are other studies on trajectory prediction for improving the safety of intelligent transportation systems in the literature. A study by Talebpour et al. (2014)

found that, with information on interdriver heterogeneity provided by connected vehicles considered, the near-crash identification and prediction could be significantly improved based on trajectory data with a resolution of 0.1 s (*13*).Wu et al. (2014) proposed a collision avoidance system which can predict collision risk by comprehensively considering vehicle motion, driving behavior, and road geometry information from a "Vehicular Cyber Physical System" (*14*). However, the above studies predicted the vehicle motion state solely for the purpose of accident prevention based on connected vehicles or vehicle-to-infrastructure technologies, and did not look into the mechanism of risky behaviors occurring as well as the signal phase transition intervals at high-speed intersections.

In summary, proactive identification of risky behaviors during the phase transition time intervals at high-speed signalized intersections has not yet received enough research attention. In addition, existing methods for vehicle trajectory prediction used either the segment speed data from the upstream loop detectors or high-resolution vehicle motion data from the connected vehicles, and have certain limitations in their objectives or applicability. To fill the research gap, this study proposes a real-time vehicle trajectory prediction method to help proactively identify risky behaviors during the phase transition time intervals at high-speed intersections, based on actual trajectory data.

Methodology

General Descriptions

As mentioned earlier, two distinct cases are considered and treated differently in the proposed method: one is the single vehicle case in which the subject vehicle has sufficient space from its leading vehicle and its movements are not influenced by the leading vehicle; the other is the following vehicle case in which the subject vehicle is in a car-following status and its movements are constrained by its leading vehicle.

The adaptive KF model and the K-NN model are integrated to predict individual vehicle trajectories. Meanwhile, the adaptive KF model and the IDM (*15*) are fused to predict the following vehicles' trajectories. The flow chart of the proposed vehicle trajectory prediction algorithm in this study is shown in Figure 1.

First, at the onset of flashing green, the approaching vehicles are divided into single vehicle and following vehicle, based on their following distances from the leading vehicles as well as the presence of leading vehicles. For further application to signal control, the subject vehicle's speed and position 1.0 s before the onset of flashing green are considered as its initial state for trajectory prediction and act as the inputs of the preliminary trajectory estimation process, given that 1.0 s is the response time of the traffic signal to accept an order and change its strategy. Secondly, the preliminary trajectories for the single vehicles and the following vehicles are acquired by the K-NN model and the IDM, respectively. Finally in the trajectory prediction process, the adaptive KF is used to acquire the predicted trajectories from the initial state to the red onset.

The reasons for choosing the aforementioned models are explained as follows. The adaptive KF, which has many applications in time-varying traffic state estimation (*16–19*), is a mathematical way to control the prediction accuracy when lacking the accurate vehicle motion model and statistical characteristic of system noise given in the

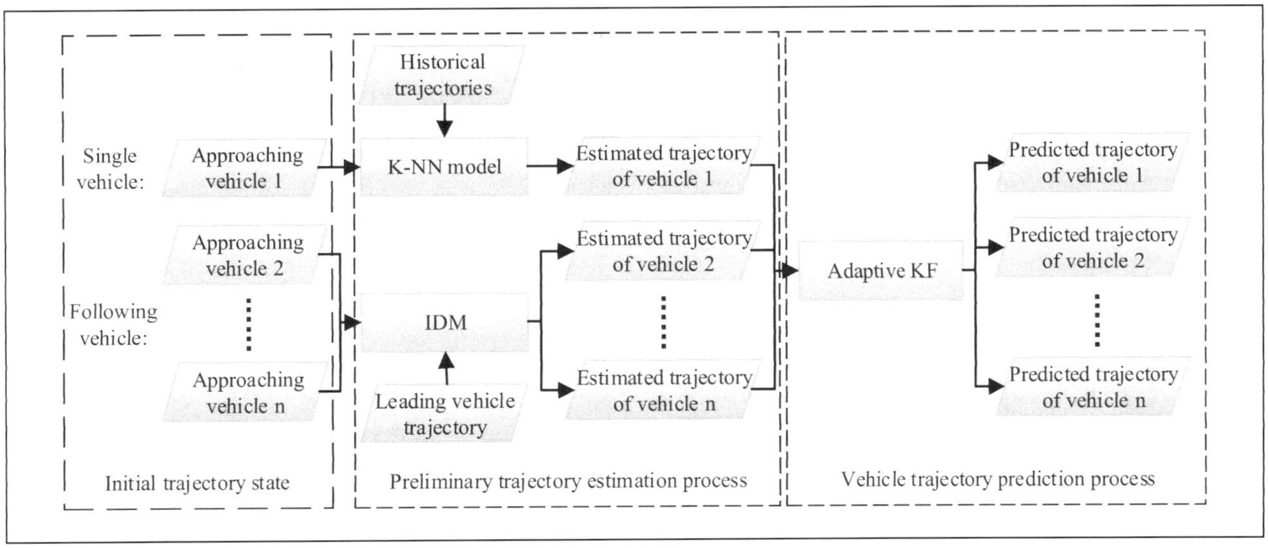

Figure 1. Flow chart of the proposed vehicle trajectory prediction method.

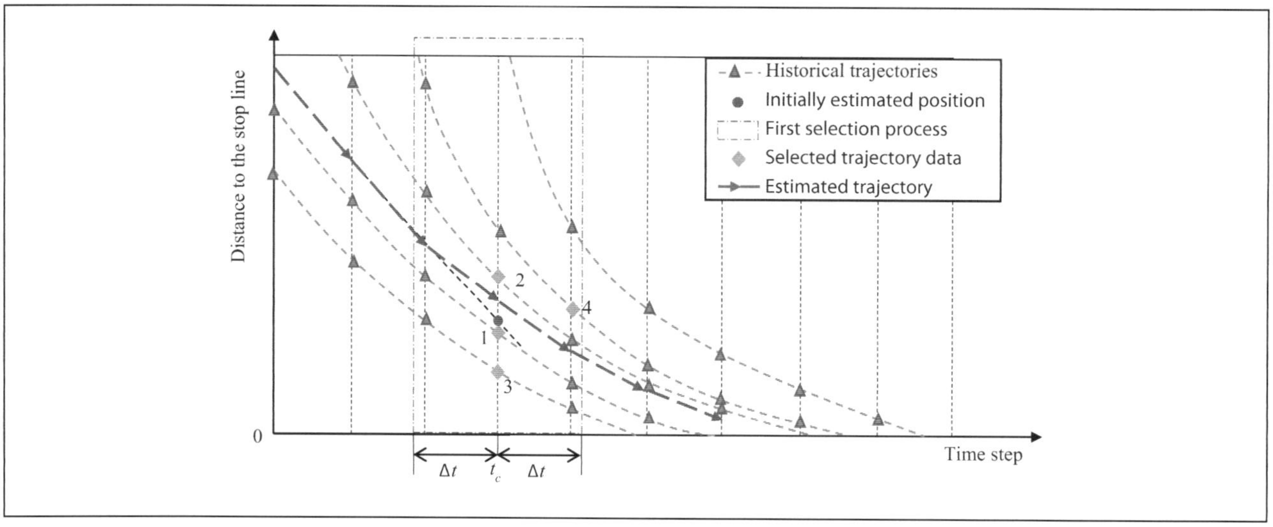

Figure 2. Illustration of trajectory estimation process by the K-NN model.

KF model. In the adaptive KF, the development of the system state is necessary for prediction. While past studies have usually derived it from laws of physics (*14*), the uncertainty of accuracy grows significantly in the step-by-step prediction. Thus, the development of system state is directly acquired by the K-NN model and the IDM in this paper, which makes trajectory prediction of a long time duration possible. For the single vehicle case, the K-NN model, which is a non-parameter estimation method, is used to output the development of the system state in the adaptive KF merely based on historical data, regardless of the potential development law. Compared with the single vehicle case, the development of the following vehicle trajectory state is hard to acquire from historical data directly. Thus, the IDM is used for the following vehicle case.

Preliminary Trajectory Estimation Process

The Single Vehicle Case. Firstly, the parameter K needs to be determined for the application of the K-NN model. The relationship between the mean square error and the number of nearest neighbors K is shown as Equation 1 (*20*).

$$MSE \propto \frac{1}{K} + \alpha^{-\frac{4}{n}\left(\frac{K}{N}\right)^{\frac{4}{n}}} \quad (1)$$

where:

α = change rate of potential density distribution, the system is linear if $\alpha = 0$,

n = dimension of the eigenvector,

K = the number of nearest neighbors,

N = the number of samples in the historical database.

As the system in this paper is nonlinear, according to Equation 1 the K value needs calibration based on historical data, which will be explained later.

Then, K-nearest neighbors are selected for estimating the trajectory state at every time step. Speed is selected as the object estimated by the K-NN model, in which a certain time and position corresponds to a speed value in historical trajectory data. An example of trajectory state estimation process for the calculation of speed at time t_c is shown in Figure 2 while assuming $K = 4$. For less calculation, the univariate searching method was used. The historical trajectory data located in $t_c \pm \Delta t$ will be selected and Δt is a time range for the first selection. The initially estimated position is calculated based on a constant speed. Then K trajectory data with the smallest difference in initially estimated position will finally be selected.

Weight coefficients of the selected historical data are calculated based on the squared weight function, which ensures that the nearest eigenvalue reaches the maximum and then decreases in turn. Weight coefficients in this method are only related to the ordering, but not the distance. The calculation formula of the weight coefficients w_i is given below.

$$w_i = \begin{cases} \frac{K^2 - (i-1)^2}{K(K+1)(4K-1)/6}, & i = 1, 2, ..., K \\ 0, & i = K+1, K+2, ..., N \end{cases} \quad (2)$$

where:

i = the ordering of the selected historical data,

K = the number of selected historical data,

N = the total number of historical data in the time range.

With the weight coefficients calculated, the driving speed $v(t_c)$ at time t_c can be estimated:

$$v(t_c) = \frac{\sum_{i=1}^{K} w_i v_i}{\sum_{i=1}^{K} w_i} \tag{3}$$

where

v_i = the driving speed of *i*-th vehicle selected at time t_c.

Finally, the distance to stop-line z_{s_k} can be calculated by $v(t_c)$. The estimated trajectory is calculated by repeating the above process.

The Following Vehicle Case. With the predicted trajectory of the leading vehicle, the estimated trajectory of the following vehicle can be acquired by IDM. Its basic formula is explained below (*15*):

$$a_n(t) = a_m \left[1 - \left(\frac{v_n(t)}{v_0} \right)^{\delta} - \left(\frac{s^*(v_n(t), \Delta v_n(t))}{\Delta x_n(t)} \right)^2 \right] \tag{4}$$

$$s^*(v_n(t), \Delta v_n(t)) = s_0 + s_1 \sqrt{\frac{v_n(t)}{v_0}} + T_s v_n(t) + \frac{v_n(t) \Delta v_n(t)}{2\sqrt{a_m b_n}} \tag{5}$$

where:

$a_n(t)$ = the acceleration of the following vehicle at time t,

s^* = the desired minimum gap of the following vehicle at time t,

v_0 = desired velocity,

s_0 = safe stopping distance,

s_1 = jam distance, $s_1 = 0$,

T_s = safe time headway,

a_m = maximum acceleration,

b_n = desired deceleration,

δ = acceleration exponent, $\delta = 4$,

$v_n(t)$ = the velocity of the following vehicle at time t,

$\Delta v_n(t)$ = the velocity difference between the following vehicle and the leading vehicle at time t,

$\Delta x_n(t)$ = the distance difference between the following vehicle and the leading vehicle at time t.

In the IDM, five parameters: v_0, s_0, T_s, a_m, and b_n, are constants which need calibration based on historical data. With the difference in position $\Delta x_n(t)$ and velocity $\Delta v_n(t)$ between the following vehicle and the leading vehicle at time t, the acceleration a_n(t) and the desired minimum gap s^* of the following vehicle in the next time step can be calculated by Equations 4 and 5, respectively. The velocity and position of the following vehicle in the next time step can then be determined accordingly. Eventually, the estimated trajectory of the following vehicle can be acquired by repeating the above calculations given the leading vehicle's trajectory.

Trajectory Prediction Process

According to the primary form of the KF, the system state model is presented in Equation 6:

$$\boldsymbol{X}_{k+1} = \boldsymbol{A}_k \boldsymbol{X}_k + \boldsymbol{B}_k \boldsymbol{U}_k + \boldsymbol{W}_k \tag{6}$$

Furthermore, the observation model is set as follows:

$$\boldsymbol{Z}_k = \boldsymbol{H}_k \boldsymbol{X}_k + \boldsymbol{V}_k \tag{7}$$

The system state vector $\boldsymbol{X}_k$ is defined as

$$\boldsymbol{X}_k = \begin{pmatrix} s_k \\ v_k \end{pmatrix} \tag{8}$$

where s_k and v_k are the predicted position and speed at time step k. $\boldsymbol{B}_k$ is a 2×1 control weighting matrix assumed as zero matrix. $\boldsymbol{Z}_k$ is the observation state at time step k, and $\boldsymbol{A}_k$ is a 2×2 state transition matrix at time step k, which take the form of Equations 9 and 10. $\boldsymbol{U}_k$ is the system control input, specified here as acceleration rate. $\boldsymbol{H}_k$ = a 2×2 observation matrix at time step k, as shown in Equation 11. $\boldsymbol{W}_k$ and $\boldsymbol{V}_k$ are the system process noise and the observation noise respectively.

$$\boldsymbol{Z}_k = \begin{pmatrix} z_{s_k} \\ z_{v_k} \end{pmatrix} \tag{9}$$

$$\boldsymbol{A}_k = \begin{pmatrix} \frac{z_{s_{k+1}}}{z_{s_k}} & 0 \\ 0 & \frac{z_{v_{k+1}}}{z_{v_k}} \end{pmatrix} \tag{10}$$

$$\boldsymbol{H}_k = \begin{pmatrix} 1 & 0 \\ 0 & 1 \end{pmatrix} \tag{11}$$

where z_{s_k} and $z_{s_{k+1}}$ are the estimated position at time step k and time step $k+1$. z_{v_k} and $z_{v_{k+1}}$ are the estimated speed at time step k and time step $k+1$. The above position and speed are based on the estimated trajectory by the K-NN model or the IDM.

For prediction of system state at time step k, the equations of the adaptive KF can be expanded as Equation 12 ~ Equation 16, which includes the predicting stage and updating stage. In the predicting stage, the predicted trajectory state at last time step $k-1$(if k is equal to 1, the initial trajectory state is used), $\boldsymbol{X}_{k\text{-}1}$ and $\boldsymbol{P}_{k\text{-}1}$, is used to predict the trajectory state preliminarily at time step k, shown in Equations 12 and 13. In the updating stage, as observation state, the estimated trajectory state is used to optimize the preliminary predicted trajectory, which acquires a more accurate predicted trajectory state, as shown in Equation 14 ~ Equation 16.

The preliminarily predicted trajectory state vector X_k^- is denoted as,

$$X_k^- = \begin{pmatrix} s_k^- \\ v_k^- \end{pmatrix} = A_{k-1}X_{k-1} = \begin{pmatrix} \frac{z_{s_k}}{z_{s_{k-1}}} & 0 \\ 0 & \frac{z_{v_k}}{z_{v_{k-1}}} \end{pmatrix} \begin{pmatrix} s_{k-1} \\ v_{k-1} \end{pmatrix} \quad (12)$$

The preliminarily predicted covariance:

$$P_k^- = A_{k-1}P_{k-1}A_{k-1}^T + \hat{Q}_{k-1} = \begin{pmatrix} \frac{z_{s_k}}{z_{s_{k-1}}} & 0 \\ 0 & \frac{z_{v_k}}{z_{v_{k-1}}} \end{pmatrix} P_{k-1} \begin{pmatrix} \frac{z_{s_k}}{z_{s_{k-1}}} & 0 \\ 0 & \frac{z_{v_k}}{z_{v_{k-1}}} \end{pmatrix}^T + \hat{Q}_{k-1} \quad (13)$$

The Kalman gain matrix is:

$$K_k = P_k^- H_k^T (H_k P_k^- H_k^T + \hat{R}_{k-1})^{-1} = P_k^- \begin{pmatrix} 1 & 0 \\ 0 & 1 \end{pmatrix}^T \left[\begin{pmatrix} 1 & 0 \\ 0 & 1 \end{pmatrix} P_k^- \begin{pmatrix} 1 & 0 \\ 0 & 1 \end{pmatrix}^T + \hat{R}_{k-1} \right]^{-1} \quad (14)$$

The optimized trajectory state is:

$$X_k = \begin{pmatrix} s_k \\ v_k \end{pmatrix} = X_k^- + K_k(Z_k - H_k X_k^-) = \begin{pmatrix} s_k^- \\ v_k^- \end{pmatrix} + K_k \left[\begin{pmatrix} z_{s_k} \\ z_{v_k} \end{pmatrix} - \begin{pmatrix} 1 & 0 \\ 0 & 1 \end{pmatrix} \begin{pmatrix} s_k^- \\ v_k^- \end{pmatrix} \right] \quad (15)$$

The updated covariance is:

$$P_k = [I - K_k H_k] P_k^- \quad (16)$$

where s_k^- and v_k^- are the preliminarily predicted position and speed at time step k. I is a 2×2 identity matrix. K_k is a 2×2 matrix known as the Kalman gain. P_k^- and P_k are the preliminarily predicted and the updated covariance matrix at time step k, respectively. $\hat{Q}_k$ and $\hat{R}_k$ are estimation of the process noise variance and observation noise variance at time step k, respectively, as shown in Equations 17 and 18.

where d_{k-1} is weight coefficient, $d_{k-1} = (1-b)/(1-b^k)$; b is the forgetting factor $(0<b<1)$ generally ranging from 0.95 to 0.99, by which the memory length of the filter can be limited and the weighing factor of new observations can be increased.

The above process is repeated until the target vehicle passes the stop-line or stops. Due to the low volatility of the KF, the predicted trajectory needs to be further processed by using kernel smoothing (*21*).

Model Calibration and Validation

Data Collection and Reduction

Four approaches at three typical high-speed intersections located on Cao'an Road, a rural highway in Shanghai, were selected for data collection in this study. A summary of site conditions is presented in Table 1. Two high-resolution video cameras were used to record the trajectory of the vehicles at the upstream area of the approach and the signal state, respectively. The collected trajectory data, including global time, speed, acceleration/deceleration, and distance of trajectory data, are extracted directly from the image-processing software George 2.0 with a resolution of 0.12 s. For further information about the data collection process and data reduction method, please refer to Tang et al. (*1*).

For the single vehicle case, only the last-to-stop and the first-to-go-through vehicles after the onset of flashing green were selected for the analysis to avoid the influence of the presence of leading vehicles. For the following vehicle case, a vehicle is considered to be in a following status if its time headway to the leading vehicle is less than 6 s or its following distance is less than 100 m, in accordance with previous studies (*22*, *23*). In total, 832 single vehicle trajectories (617 passenger cars and 215 trucks) and 449 following vehicle trajectories (419 passenger cars and 30 trucks) that encountered both the onset of flashing green and yellow indications were eventually used in the model development and validation, as shown in Figure 3.

$$\hat{Q}_k = (1 - d_{k-1})\hat{Q}_{k-1} + d_{k-1}\left\{ K_k \left[\begin{pmatrix} z_{s_k} \\ z_{v_k} \end{pmatrix} - \begin{pmatrix} 1 & 0 \\ 0 & 1 \end{pmatrix} \begin{pmatrix} s_k^- \\ v_k^- \end{pmatrix} \right] \left[\begin{pmatrix} z_{s_k} \\ z_{v_k} \end{pmatrix} - \begin{pmatrix} 1 & 0 \\ 0 & 1 \end{pmatrix} \begin{pmatrix} s_k^- \\ v_k^- \end{pmatrix} \right] K_k^T + P_k - \begin{pmatrix} \frac{z_{s_k}}{z_{s_{k-1}}} & 0 \\ 0 & \frac{z_{v_k}}{z_{v_{k-1}}} \end{pmatrix} P_{k-1} \begin{pmatrix} \frac{z_{s_k}}{z_{s_{k-1}}} & 0 \\ 0 & \frac{z_{v_k}}{z_{v_{k-1}}} \end{pmatrix} \right\}^T \quad (17)$$

$$\hat{R}_k = (1 - d_{k-1})\hat{R}_{k-1} + d_{k-1}\left\{ \left[\begin{pmatrix} z_{s_k} \\ z_{v_k} \end{pmatrix} - \begin{pmatrix} 1 & 0 \\ 0 & 1 \end{pmatrix} \begin{pmatrix} s_k^- \\ v_k^- \end{pmatrix} \right] \left[\begin{pmatrix} z_{s_k} \\ z_{v_k} \end{pmatrix} - \begin{pmatrix} 1 & 0 \\ 0 & 1 \end{pmatrix} \begin{pmatrix} s_k^- \\ v_k^- \end{pmatrix} \right]^T - \begin{pmatrix} 1 & 0 \\ 0 & 1 \end{pmatrix} P_k^- \begin{pmatrix} 1 & 0 \\ 0 & 1 \end{pmatrix}^T \right\} \quad (18)$$

Table 1. Summary of Site Conditions at the Study Intersection Approaches

Intersections	Cao'an Road and Jiasongbei Road	Cao'an Road and Xiangjiang Road	Cao'an Road and Caofeng Road	
Approaches	East-bound	East-bound	West-bound	East-bound
Speed limit	80 km/h	80 km/h	80 km/h	80 km/h
Lane configuration	L-T-T-T-R	L-T-T-T-R	L-T-T-T-R	L-T-T-T-T
Intersection size	54 m	58 m	32 m	32 m
Cycle length	161 s	160 s	104 s	104 s
Number of phases	4	4	3	4
Green time	38 s	45 s	45 s	45 s
Flashing green	3 s	3 s	3 s	3 s
Yellow	3 s	3 s	3 s	3 s
All red	1 s	1 s	1 s	1 s
Percentage of trucks	10.5%	17.3%	24%	8.25%
Traffic volume during the AM peak	2900~3300 veh/h	3000~3500 veh/h	1700~1900 veh/h	1800~2200 veh/h
Observation times	6 AM peak and 3 PM off-peak, 18 h in total	2 AM peak and 2 PM off-peak, 18 h in total	4 AM peak, 8 h in total	4 AM peak, 8 h in total
Observation dates	16 normal weekdays from July to October 2013 under sunny weather conditions			

Note: L = exclusive left-turn lane; T = through-ahead lane; R = exclusive right-turn lane; Intersection size = the distance between the opposite approach stop-lines; AM peak = 07:30~09:30; PM off-peak = 15:00~17:00.

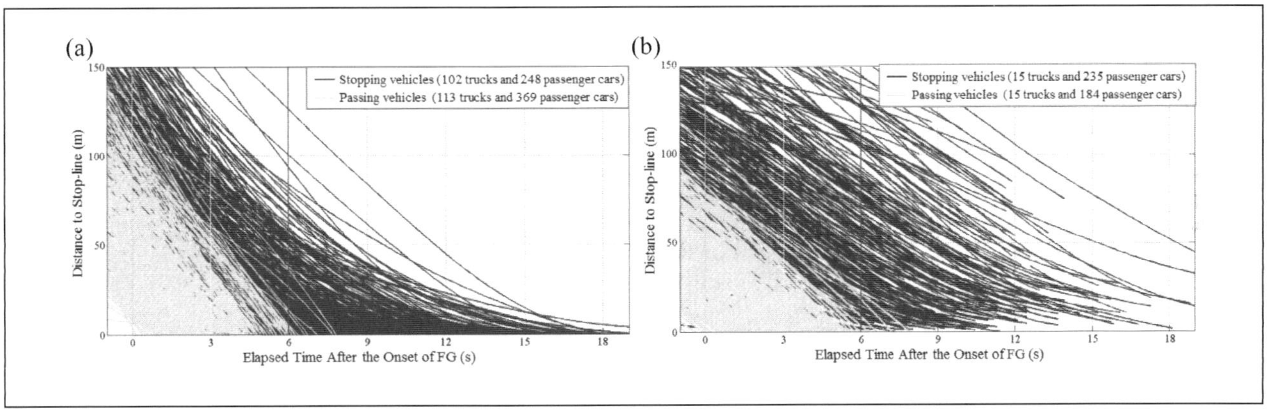

Figure 3. Collected vehicle trajectories: (*a*) single vehicles, (*b*) following vehicles.

Model Calibration

Three quarters of the observed vehicle trajectories were used for model calibration. In the K-NN model, only one parameter, *K*, needs to be calibrated. As mentioned above, different weights are distributed to different states. Under different *K* values, the change of weight (i.e., w_i) in Equation 2 results in the change of root mean square error (RMSE) between the predicted and actual trajectory, as shown in Figure 4. The calculation formula of RMSE is as shown in Equation 19. In general, the prediction error is smallest when *K* is equal to 5, according to the analysis results.

$$RMSE = \sqrt{\frac{\sum_{i=1}^{n} d_i^2}{n}} \tag{19}$$

where d_i is the distance error between the predicted trajectory and the actual trajectory at time step *I* and *n* is the total number of time steps.

As for the IDM, five parameters: safe stopping distance s_0, desired velocity v_0, safe time headway T_s, maximum acceleration a_m, and desired deceleration b_n need calibration. These parameters were determined by using a genetic algorithm proposed by Chen et al. (*24*); the

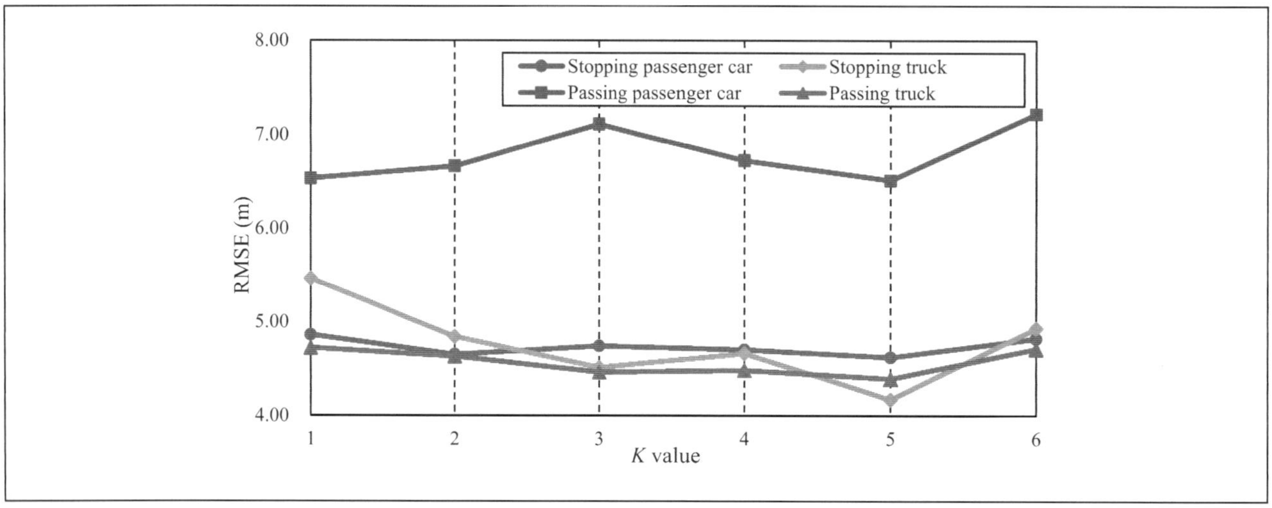

Figure 4. Prediction errors under different *K* values.

Table 2. Calibration Results of the IDM Parameters

Model parameters	Passenger cars	Trucks
Safe stopping distance, s_0 (m)	2.13	1.89
Desired velocity, v_0 (m/s)	25.53	20.36
Safe time headway, T_s (m)	1.98	2.38
Maximum acceleration, a_m (m/s^2)	1.76	1.82
Desired deceleration, b_n (m/s^2)	2.98	3.20

objective of the optimization is to minimize the difference between the measured gap and the simulated gap between the subject vehicle and the leading vehicle. Thus, a performance measure combining the relative error and the absolute error was used in the optimization, as shown in Equation 20.

$$F_{error} = \sqrt{\frac{1}{\langle |s_{data}| \rangle} \left\langle \frac{(s_{sim} - s_{data})^2}{|s_{data}|} \right\rangle} \quad (20)$$

where the operator $\langle s \rangle$ is defined as in Equation 21

$$\langle s \rangle = \frac{1}{N} \sum_{i=1}^{N} s(i) \quad (21)$$

where s is the gap between the subject vehicle and the leading vehicle; s_{sim} denotes the simulated gap, while s_{data} denotes the measured gap; N is the number of chromosomes in the population.

The results based on three quarters of the observed vehicle trajectories for passenger cars and trucks, respectively, are shown in Table 2.

Model Validation

RMSE between the predicted trajectories and the actual trajectories is adopted to evaluate the accuracy of the proposed model. The results are provided in Table 3. As can be seen, the overall average RMSE of the single vehicles is 5.02 m, while it is 2.33 m for the following vehicles.

Figure 5 demonstrates some illustrative examples for trajectory prediction results. It can be seen that, when the estimated trajectory is far from the actual trajectory (Figure 5*b* and *d*), the adaptive KF can correct the predicted result markedly. Note that the following vehicle case for trucks is absent here because of the lack of statistically sufficient trajectory samples.

Application To Identification of Risky Driving Behavior

The proposed method is further applied to identify risky driving behaviors that might occur during the phase transition time intervals at high-speed intersections. Five types of possible risky behaviors are defined first, according to the characteristic parameters extracted from the trajectory data such as speed, acceleration, jerk, etc. Then, the dataset for the model calibration explained previously is used to determine the thresholds of the selected characteristic parameters. Finally, the dataset for model calibration is used to evaluate the performance of the identification, by comparing the actual number of risky behaviors based on the real trajectory and the predicted number of risky behaviors based on the predicted vehicle trajectories.

Table 3. Prediction Accuracy of the Proposed Models

	Average RMSE (m)	Sample size (vehicle)	Overall (m)
Single vehicle case			
Stopping passenger car	4.62	78	5.02
Stopping truck	4.17	22	
Passing passenger car	6.51	42	
Passing truck	4.39	20	
Following vehicle case			
Stopping passenger car 1	2.28	43	2.33
Stopping passenger car 2	2.95	26	
Passing passenger car	1.97	36	

Note: *Stopping car 1, the leading vehicle stopped; stopping car 2, the leading vehicle passed.

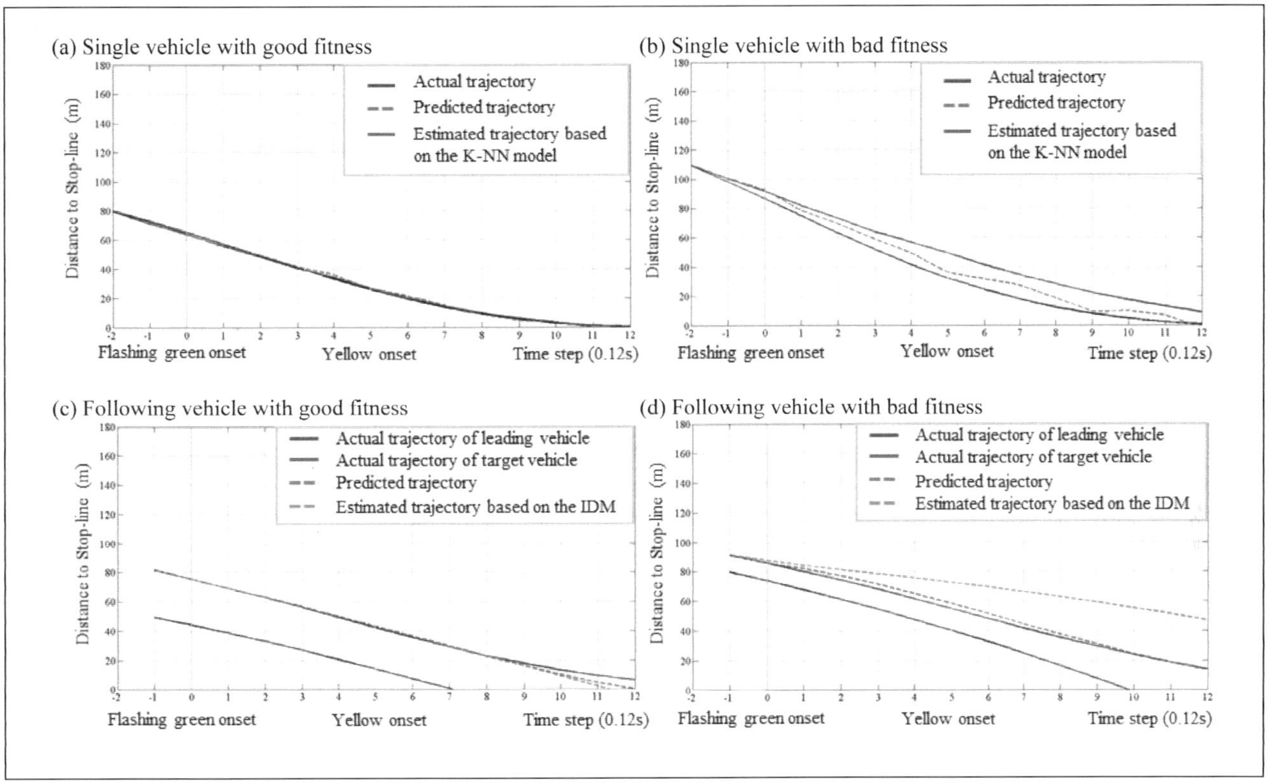

Figure 5. Illustrative examples of predicted trajectories: (*a*) single vehicle with good fitness; (*b*) single vehicle with bad fitness; (*c*) following vehicle with good fitness; (*d*) following vehicle with bad fitness.

Classification of Risky Driving Behaviors

For the identification of risky behaviors by single vehicles, stopping vehicles and passing vehicles need to be separately treated. Typical types of risky behaviors of the single passenger cars are as follows:

- **Type 1: Abrupt Stop**. In this condition, the 85 percentile of deceleration (d_{85}) of the subject vehicle is less than $-3.5\,m/s^2$ (and $-2\,m/s^2$ for trucks) from the onset of flashing green to the moment when its speed is reduced to 10 km/h.
- **Type 2: Red-Light Running**. The subject vehicle enters the intersection after the onset of the red light.
- **Type 3: Speeding Pass**. The subject vehicle crosses the stop-line at a speed greater than 80 km/h, i.e., the posted speed limit at the intersections studied, during time intervals of the flashing green or yellow phases.
- **Type 4: Aggressive Pass**. The subject vehicle passes the stop-line at a speed less than 80 km/h during the flashing green or yellow time intervals, but

Table 4. Identification Results for Risky Behaviors

	Type	Observed (vehicle)	Predicted (vehicle)	Accuracy (%)	Overall (%)
Single vehicle case					
Passenger cars	Type 1	0	0	-	95.1
	Type 2	1	1	100	
	Type 3	3	3	100	
	Type 4	3	2	66.7	
	Not risky	113	111	98.2	
Trucks	Type 1	1	0	0	
	Type 2	1	1	100	
	Type 3	1	0	0	
	Type 4	1	1	100	
	Not risky	38	35	92.1	
Following vehicle case					
Passenger cars	Type 5	41	41	100	96.2
	Not risky	64	60	93.8	

its 85 percentile acceleration (a_{85}) is greater than $2\,m/s^2$ (and $1.5\,m/s^2$ for trucks).

For the identification of risky behaviors by following vehicles, we applied the time to collision (TTC) recommended by an earlier study (*25*).

- **Type 5: Aggressive Following**. In this condition, the 15 percentile TTC between the target vehicle and the leading vehicle is less than 5 s during the phase transition intervals.

Note that all the above threshold values for judging the types of risky behaviors were determined based on the empirical data collected in this study.

Results Analysis

Out of the model validation dataset, 162 single vehicles (120 passenger cars and 42 trucks) and 105 following vehicles (all passenger cars and no trucks) (the total number of sample size is small for trucks as introduced before) were used for the identification of risky behaviors. The identification results for risky behaviors are presented in Table 4. In addition, for illustrative purposes, comparisons of the actual behaviors with the identified behaviors for the single passing vehicles are presented in Figure 6.

As is shown in Table 4, although the total number of risky behaviors observed is small, the overall prediction accuracy is 95.1% for the single vehicle case and 96.2% for the following vehicle case.

In the single vehicle case, most of both the risky behaviors and not-risky behaviors were correctly identified. For passenger cars, the accuracy of Type 4 is relatively low, at 66.7%, because one of the Type 4 risky behaviors was identified as Type 3, as shown in Figure 6*a* indicated by rectangle 1. Because of the small sample size, one incorrect identification will lead to considerable error. In total, six out of the total number of seven passenger cars with risky behavior were identified accurately, and the identification rate of risky behaviors, regardless of risk type, was actually 100%. Only two out of the total number of four risky trucks were identified, however, as shown in Figure 6*b*. That is because the small number of truck samples made the threshold values of trucks for judging the types of risky behaviors less reliable. Moreover, the predicted number of not-risky vehicles is slightly smaller than the actual number, meaning that a small part of the not-risky behaviors were identified as risky behaviors, indicated by rectangles 2 and 3.

In the following vehicle case, as shown in Figure 6*c*, all the risky behaviors were correctly identified by the proposed method. Regarding the not-risky behavior, 60 out of the total number of 64 were correct, while four vehicles (Nos. 27, 41, 47, 105) were identified as demonstrating risky behaviors. The overall accuracy is comparably as high as for the single vehicle case.

Conclusions and Future Works

In summary, a real-time vehicle trajectory prediction method is proposed to help proactively to identify risky driving behaviors during the traffic-light phase transition intervals at high-speed intersections. The proposed models are calibrated and validated using a large dataset of actual vehicle trajectories. Conclusions support that the prediction accuracy of the proposed method is fairly satisfactory in terms of the RMSE between the predicted trajectories and the actual trajectories as well as the correct rate of risky behavior identification. The proposed method can be used in the development of DZ

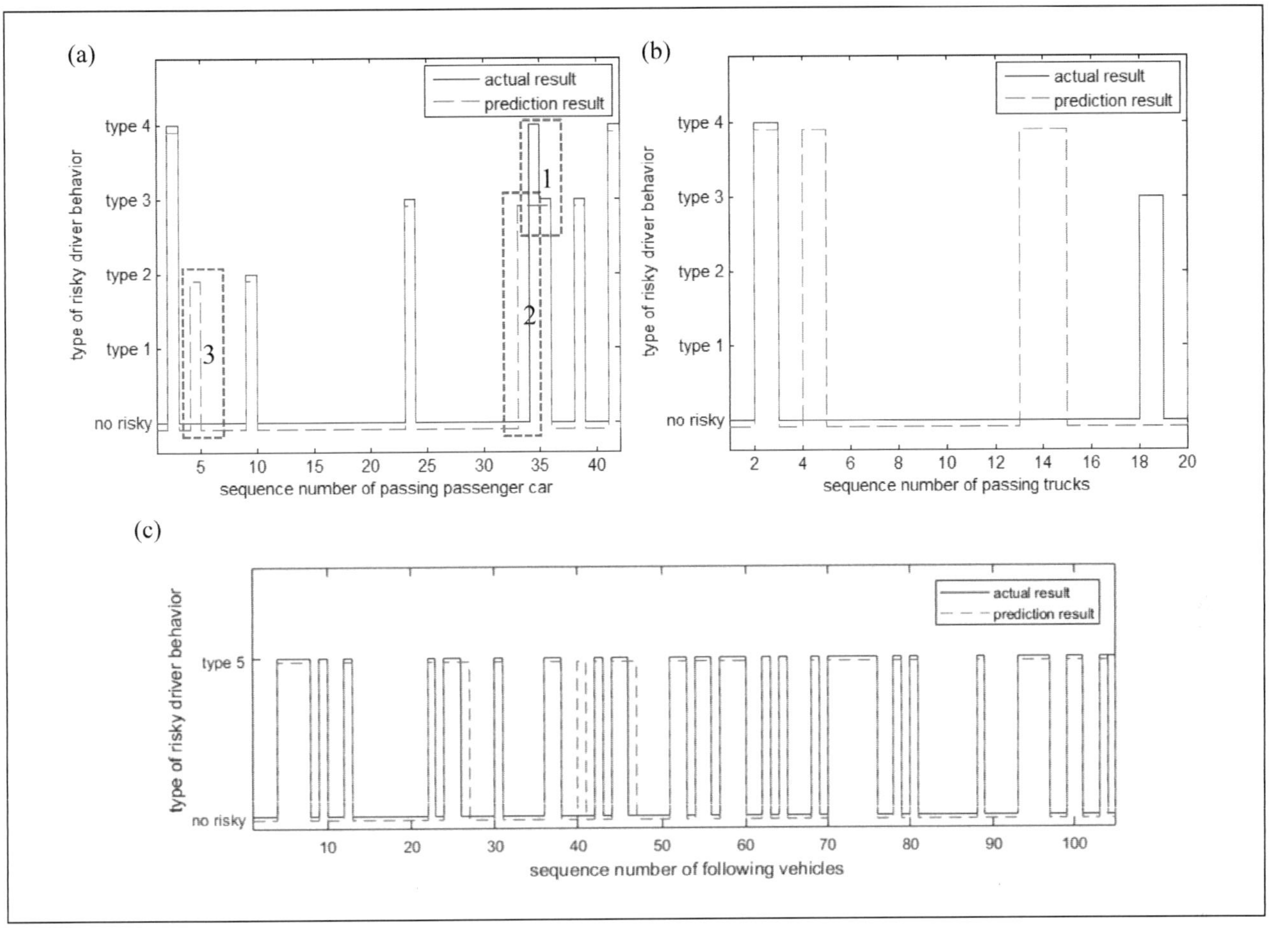

Figure 6. Comparison of actual behaviors and identified behaviors for single passing vehicles and following vehicles: (*a*) comparison of single passing passenger cars; (*b*) comparison of single passing trucks; (*c*) comparison of following vehicles.

mitigation, such as previous works done by Zimmerman et al. (2003), Li et al. (2010) and Kong et al. (2014) (*9–11*), and on-board safety altering and safety-oriented dynamic signal control strategies and systems, such as Zhang et al. (2012), Dingua et al. (2016), Khattak et al. (2015), and Liu and Khattak (2016) (*5–8*), in order to improve safety performance at high-speed intersections with a long phase transition time.

Meanwhile, the presented method has some limitations and demands further studies. First, the sample size of vehicle trajectories for risky behaviors, particularly for trucks, is still limited. Thus, more widespread empirical trajectory data needs to be collected and used for the enhancement and validation of the proposed method. Second, in addition to the adaptive KF model, the K-NN model, and the IDM, other types of models could also be investigated to improve the prediction accuracy. For instance, lane changing behavior of approaching drivers at the upstream of intersection approach was not considered in this study. It needs to be addressed by incorporating a lane changing model into the current models. Third, it is also necessary to develop safety-oriented dynamic signal control algorithms and strategies, based on the proposed method.

Acknowledgments

This study was supported by the Natural Science Foundation of China (No. 61673302). The authors are grateful to Hideki Nakamura at Nagoya University for providing the image-processing software for data reduction in this study, and to Edward Chung at Queensland University of Technology for his insightful advice on this study.

Author Contributions

Study conception and design: Keshuang Tang, Yangbeibei Ji; data collection and reduction: Chaopeng Tan, Nan Zhou; data analysis, modeling and interpretation of results: Chaopeng Tan, Nan Zhou, Fen Wang; draft manuscript preparation: Chaopeng Tan, Keshuang Tang, Yangbeibei Ji. All authors reviewed the results and approved the final version of the manuscript.

References

1. Tang, K., Y. Xu, F. Wang, and T. Oguchi. Exploring Stop-Go Decision Zones at Rural High-Speed Intersections with Flashing Green Signal and Insufficient Yellow Time in China. *Accident Analysis & Prevention*, Vol. 95, Part B, 2016, pp. 470–478.
2. Mahalel, D., D. Zaidel, and T. Klein. Driver's Decision Process on Termination of the Green Light. *Accident Analysis & Prevention*, Vol. 17, No. 5, 1985, pp. 373–380.
3. Newton, C., R. N. Mussa, E. K. Sadalla, E. K. Burns, and J. Matthias. Evaluation of an Alternative Traffic Light Change Anticipation System. *Accident Analysis & Prevention*, Vol. 29, No. 2, 1997, pp. 201–209.
4. Tang, K., Y. Xu, P. Wang, and F. Wang. Impacts of Flashing Green on Dilemma Zone Behavior at High-speed Intersections: Empirical Study in China. *Journal of Transportation Engineering*, Vol. 141, No. 7, 2015, DOI:https://doi.org/10.1061/(ASCE)TE.1943-5436.0000770.
5. Zhang, L., L. Wang, K. Zhou, and W.-B. Zhang. Dynamic All-Red Extension at a Signalized Intersection: A Framework of Probabilistic Modeling and Performance Evaluation. *IEEE Transactions on Intelligent Transportation Systems*, Vol. 13, No. 1, 2012, pp. 166–179.
6. Dingus, T. A., S. G. Klauer, V. L. Neale, A. Petersen, S. E. Lee, J. Sudweeks, M. Perez, J. Hankey, D. Ramsey, and S. Gupta. *The 100-Car Naturalistic Driving Study, Phase II–Results of the 100-Car Field Experiment*. Publication DOT-HS-810-593. NHTSA, U.S. Department of Transportation, Washington, D.C., 2006.
7. Khattak, A., J. Liu, and X. Wang. *Supporting Instantaneous Driving Decisions through Trajectory Data*. Presented at 94th Annual Meeting of the Transportation Research Board, Washington, D.C., 2015.
8. Liu, J., and A. J. Khattak. Delivering Improved Alerts, Warnings, and Control Assistance Using Basic Safety Messages Transmitted between Connected Vehicles. *Transportation Research Part C: Emerging Technologies*, Vol. 68, 2016, pp. 83–100.
9. Zimmerman, K., J. Bonneson, M. Dan, and M. Abbas. Improved Detection and Control System for Isolated High-Speed Signalized Intersections. *Transportation Research Record: Journal of the Transportation Research Board*, 2003. 1856: 212–219.
10. Li, P., M. M. Abbas, R. Pasupathy, and Z. Li. A New Dilemma Zone Protection Algorithm Based on the Vehicle, Trajectories Prediction and Markov Process. Presented at 89th Annual Meeting of the Transportation Research Board, Washington, D.C., 2010.
11. Kong, T., L. Lin, and K. Tang. New Dilemma Zone Protection Algorithm Based on Real-Time Vehicle Trajectory. Presented at 93rd Annual Meeting of the Transportation Research Board, Washington, D.C., 2014.
12. Bonsall, P., R. Liu, and W. Young. Modelling Safety-related Driving Behaviour—Impact of Parameter Values. *Transportation Research Part A: Policy and Practice*, Vol. 39, No. 5, 2005, pp. 425–444.
13. Talebpour, A., H. S. Mahmassani, F. Mete, and S. H. Hamdar. Near-Crash Identification in a Connected Vehicle Environment. *Transportation Research Record: Journal of the Transportation Research Board*, 2014. 2424: 20–28.
14. Wu, C., L. Peng, Z. Huang, M. Zhong, and D. Chu. A Method of Vehicle Motion Prediction and Collision Risk Assessment with a Simulated Vehicular Cyber Physical System. *Transportation Research Part C: Emerging Technologies*, Vol. 47, Part 2, 2014, pp. 179–191.
15. Treiber, M., A. Hennecke, and D. Helbing. Congested Traffic States in Empirical Observations and Microscopic Simulations. *Physical Review E Statistical Physics Plasmas Fluids & Related Interdisciplinary Topics*, Vol. 62, No. 2, 2000, pp. 1805–1824.
16. Batz, T., K. Watson, and J. Beyerer. Recognition of Dangerous Situations within a Cooperative Group of Vehicles. *Presented at 2009 IEEE Intelligent Vehicles Symposium*, Xi'an, Shaansi, China, 2009, pp. 907–912.
17. Barrios, C., and Y. Motai. Improving Estimation of Vehicle's Trajectory Using the Latest Global Positioning System with Kalman Filtering. *IEEE Transactions on Instrumentation & Measurement*, Vol. 60, No. 12, 2011, pp. 3747–3755.
18. Huang, J., and H. S. Tan. Error Analysis and Performance Evaluation of a Future-Trajectory-Based Cooperative Collision Warning System. *IEEE Transactions on Intelligent Transportation Systems*, Vol. 10, No. 1, 2009, pp. 175–180.
19. Yuan, Y., J. W. C. Van Lint, R. E. Wilson, F. Van Wageningen-Kessels, and S. P. Hoogendoorn. Real-Time Lagrangian Traffic State Estimator for Freeways. *IEEE Transactions on Intelligent Transportation Systems*, Vol. 13, No. 1, 2012, pp. 59–70.
20. Fukunaga, K., and L. D. Hostetler. Optimization of K-Nearest Neighbor Density Estimates. *IEEE Transactions on Information Theory*, Vol. 19, No. 3, 1973, pp. 320–326.
21. Speckman, P. Kernel Smoothing in Partial Linear Models. *Journal of the Royal Statistical Society. Series B (Methodological)*, Vol. 50, No. 3, 1988, pp. 413–436.
22. *Highway Capacity Manual 2010*. Transportation Research Board of the National Academies, Washington, D.C., 2010.
23. Weidemann, R., and U. Reiter. *Microscopic Traffic Simulation, The Simulation System-Mission*. University of Karlsruhe, Karlsruhe, Baden-Wurttemberg, Germany, 1992.
24. Chen, C., L. Li, J. Hu, and C. Geng. Calibration of MITSIM and IDM Car-following Model Based on NGSIM Trajectory Datasets. *Presented at 2010 IEEE International Conference on Vehicular Electronics and Safety (ICVES)*, 2010, pp. 48–53.
25. Hayward, J. C. Near-Miss Determination Through Use of a Scale of Danger. *Highway Research Record*, No. 384, 1972, pp. 24–34.

The Standing Committee on Safety Data, Analysis and Evaluation (ANB20) peer-reviewed this paper (18-04827).

Article

TRR
JOURNAL OF THE TRANSPORTATION RESEARCH BOARD

Transportation Research Record
2018, Vol. 2672(38) 245–254

DOI: 10.1177/0361198118772703
journals.sagepub.com/home/trr
SAGE

Crash and Near-Crash Risk Assessment of Distracted Driving and Engagement in Secondary Tasks: A Naturalistic Driving Study

Peter R. Bakhit[1], **BeiBei Guo**[2], **and Sherif Ishak**[3]

Abstract

Distracted driving behavior is a perennial safety concern that affects not only the vehicle's occupants but other road users as well. Distraction is typically caused by engagement in secondary tasks and activities such as manipulating objects and passenger interaction, among many others. This study provides an in-depth analysis of the increased crash/near-crash risk associated with different secondary tasks using the largest real-world naturalistic driving dataset (SHRP2 Naturalistic Driving Study). Several statistical and data-mining techniques were developed to analyze the distracted driving and crash risk. First, a bivariate probit model was constructed to investigate the relationship between engagement in a secondary task and the safety-critical events likelihood. Subsequently, two different techniques were implemented to quantify the increased crash/near-crash risk because of involvement in a particular secondary task. The first technique used the baseline-category logits model to estimate the increased crash risk in terms of conditional odds ratios. The second technique used the a priori association rule mining algorithm to reveal the risk associated with each secondary task in terms of support, confidence, and lift indexes. The results indicate that reaching for objects, manipulating objects, reading, and cell phone texting are the highest crash risk factors among various secondary tasks. Recognizing the effect of different secondary tasks on traffic safety in a real-world environment helps legislators enact laws that reduce crashes resulting from distracted driving, as well as enabling government officials to make informed decisions about the allocation of available resources to reduce roadway crashes and improve traffic safety.

Driving is a daily complex task that requires a driver's full attention. Despite the complexity associated with this task, it is not uncommon to observe drivers perform other secondary tasks while operating a vehicle. These secondary tasks might include reading a newspaper in slow-moving traffic, shaving to be ready for work, and discussing important topics with a passenger, among many others. Although these tasks might seem trivial, they degrade driving performance and increase the likelihood of a crash or near-crash event. Moreover, the technological features embedded in vehicles nowadays, in addition to the advanced wireless communication devices, have brought a new level of distraction to the driving environment (*1*). Thus, it is important to estimate the relative crash/near-crash risk for better understanding of the effect of different types of secondary tasks on traffic safety.

Recognizing the effect of different secondary tasks on traffic safety is imperative. It can help legislators initiate laws that reduce crashes resulting from distracted driving. It can also help government officials make informed decisions about the allocation of available resources to reduce roadway crashes. Most of the previous researches attempted to gather information about secondary tasks through driving simulators and interview studies. Limited research, however, has tapped into comprehensive naturalistic driving datasets. In this study, a large naturalistic driving dataset (SHRP2 Naturalistic Driving Study) is employed to investigate the relationship between engagement in a secondary task

[1]Department of Civil and Environmental Engineering, Louisiana State University, Baton Rouge, LA
[2]Department of Experimental Statistics, Louisiana State University, Baton Rouge, LA
[3]Civil and Environmental Engineering, University of Alabama in Huntsville, Huntsville, AL

Corresponding Author:
Address correspondence to Peter R. Bakhit: pbakhi1@lsu.edu

and the crash/near-crash likelihood using different statistical and data-mining techniques. This dataset includes information collected from more than 3,000 drivers recruited in six different states. By far, this is the largest naturalistic driving study conducted in the United States to date. Thus, this dataset is considered to be one of the most representative datasets of the driving population. Furthermore, in this study, the increased crash risk resulting from distracted driving is quantified so that secondary tasks with a higher crash risk are recognized.

Background

To assess the impact that a particular secondary task has on the crash risk, three different approaches are frequently used, namely, experimental studies, interview studies, and observational studies. Experimental studies are usually conducted in either a driving simulator or in a controlled traffic environment (*2, 3*). During the experiments, participants are asked to perform a specific secondary task at a given time according to the scenarios designed by the experimenter. A comprehensive overview of distracted driving experimental studies can be found in Regan et al. (*3*). Although the experimental studies were successful in recognizing the degradation in driving performance because of the engagement in a secondary task, they were not helpful in making a valid estimate of the actual crash risk for two main reasons. First, the participants did not decide where, when, and how to engage in the secondary task, which is not an accurate representation of real-world secondary task involvement. Second, the transferability of the outcomes from the driving simulators to real life remains questionable. Thus, experimental studies are not considered the best approach to determine the increased crash risk resulting from engagement in different secondary tasks.

Interview studies represent another approach in collecting secondary tasks information (*4–8*). In these studies, information is gathered using telephone surveys, as in McEvoy et al. and Royal (*5, 6*), or online surveys as in Lansdown, and Young and Lenné (*4, 8*). In Sullman and Baas (*7*), a sample of 287 New Zealand drivers were asked about their cell phone use while driving and the perceived risk. The results showed that the percentage of drivers who never used a cell phone or used a cell phone occasionally while driving were 43% and 43%, respectively. The percentage of drivers who used the cell phone frequently was 14%. Although Sullman and Baas's study (*7*) was interested in cell phone use as the sole secondary task, the remaining four studies (*4, 5, 6, 8*) were concerned with all secondary tasks. In McEvoy's study (*5*), the participants were asked to list all secondary tasks that lasted for 5 min or more over the last trip. Young et al. (*8*) conducted another survey asking participants to report all kinds of secondary tasks and how often they engaged in them. Eating/drinking, smoking, clothing/body care, integrated devices, passenger-related, and outside distractors are the most common distractors that contribute to crashes. Despite the valuable information reported in these studies, self-reporting bias (*9*) is a major concern.

Observational studies or naturalistic driving studies (NDS) are the most realistic approaches for gathering secondary tasks data (*10–13*). In these studies, vehicles are equipped with advanced data-collection devices to record normal driving behavior. For example, Klauer (*11*) studied the distracted driving and its relationship to safety-critical events (SCEs). For this 1-year study project, 82 crash events and 761 near-crash events were observed. This data was then analyzed as a part of the 100-Car Naturalistic Driving Study. The results indicated that passenger interaction, embedded devices in a vehicle, and manipulating objects are the most common secondary tasks. Moreover, in three other studies, secondary tasks such as eating/drinking, smoking-related, clothing/body care, integrated devices, passenger-related, outside distractors, and other in-vehicle devices were listed as the most frequent secondary tasks (*10–13*). Another study conducted by Olson et al. investigated the impact of distracted driving on traffic safety (*14*). In Olson et al.'s study, the increased crash risk was reported in terms of odds ratio (OR) estimates. Key findings were that commercial drivers were engaged in secondary tasks in 71% of crashes, 46% of near-crashes, and 60% of all SCEs. The most risky behavior identified was "Cell phone texting" with an OR of 23.2, followed by "Interacting with a device" (OR = 9.9), and "Cell phone dialing" (OR 5.9). Although research approaches derived from NDS data are considered more realistic, a high number of equipped vehicles in addition to long observation periods are required. As a result, in this study, the SHRP2 NDS data (the largest NDS conducted so far) is employed. This real-world driving study project was successful in collecting approximately two petabytes of video and sensor data from more than 3,000 drivers over a 3-year period (2010–2013).

The literature review shows evidence of a relationship between engagement in a secondary task and crash likelihood. However, there are some limitations that need to be addressed. First, most of the statistical models did not take into account the correlation between interrelated variables, such as engagement in a secondary task and the crash likelihood. More specifically, in previous statistical models, the multiple dependent variables were modeled separately, each as a function of a set of independent variables, and therefore, the correlations among the dependent variables were ignored. Second, recent data-mining techniques have captured researchers' attention as they outperform traditional modeling techniques. Thus, this study is the first of its kind to introduce the market basket analysis (association modeling) into the distracted driving area. In this context, the objectives of this study are to (a) construct a statistical model that considers the correlation between engagement in a secondary task and the crash/near-crash occurrence (bivariate probit model); (b) estimate valid crash risk measures for different types of secondary tasks using a large naturalistic driving dataset without sampling bias (baseline-category logits model); and (c) offer a new methodology for investigating the relationship between the different secondary tasks and the crash/near-crash risk using a new data-mining technique.

The rest of the paper is organized as follows. "Data Description" describes the data used in this study. "Model Development" presents model development, detailing the bivariate probit model and results, and the two alternative models used to quantify the increased crash-risk estimate using the baseline-category logits model, and the association rule mining model, respectively. "Discussion" discusses the models' outcomes and their implications. Finally, conclusions are provided in "Conclusion."

Data Description

In this study, the SHRP2 NDS dataset was employed to achieve the research objectives. SHRP2 NDS data is considered the largest study of its kind that has been conducted in the United States to date. The data include observations from more than 3,000 drivers recruited in six different states, namely New York, Washington, Indiana, Florida, Pennsylvania, and North Carolina. The dataset consists of 21,292 driving events (observations) categorized in three groups: crash (1,415 events), near-crash (2,616), and baseline events (18,873) (*15*). For each driving event, different data types are collected, such as event summary data, and driver sociodemographic data. Event summary data include information related to the vehicle and environmental conditions during the event incidence time, such as event type, event severity, surface condition, weather conditions, and whether a secondary task existed or not, among many other factors. Driver sociodemographic data include information related to a driver's socioeconomic and driving characteristics, such as gender, age, education, working status, marital status, and years of driving, among others. Table 1 lists all the different variables used in this research. These variables are selected based on previous distraction-safety-related studies (*11, 14, 16*).

Prior to the models' development, the SHRP2 NDS dataset was reduced to remove any biases that might have affected the crash risk estimates. First, a crash event that did not involve injuries or property damage was excluded from the final dataset. Second, driving events were filtered out to remove any events associated with observable driver impairment. Finally, driving records with missing driver information were excluded; leaving 905 crashes, 2,558 near-crashes, and 18,544 baseline events as a final dataset. It should be noted that the "Secondary Task" variable is the key variable in the rest of the study. This variable shows the type of the secondary task in which the driver was engaged prior to the crash/near-crash time or during the selected normal driving event. The type of the secondary task was manually coded by reviewing video by Virginia Tech Transportation Institute (VTTI) according to the SHRP

Table 1. List of Input Variables

Type	Variable	# of Categories	Categories
Driver characteristics	Age	11	(16–19), (20–24), ... (85–89)
	Gender	2	Male/Female
	Working status	3	Full-time, Part-time, Not working
	Marital status	5	Single, Married, Divorced, Widowed
	Education	5	High School, College degree, Advanced degree
	Driver training	6	Through a private company, Through school, Informal training, No informal training by parents, No training, Other
	Years of driving		Quantitative measure
Event characteristics (summary data)	Event type	3	Crash/Near Crash/Baseline
	Event duration		Quantitative measure
	Secondary task	14	Cell phone, Texting, Passenger interaction, and so forth
	Presence of passenger	2	Yes/No
Roadway characteristics (summary data)	Relation to junction	10	Intersection, Parking entrance, Non-junction, Rail grade crossing, and so forth
	Intersection influence	2	Yes/No
	Alignment	3	Curve left, Curve right, Straight
	Grade	5	Level, Grade up, Grade down, Dip, Hill crest
	Traffic lighting	6	Daylight, Dusk, Darkness lighted, darkness unlighted, Dawn, Other
	Locality	11	Business, School, Interstate, Residential, and so forth

Table 2. Secondary Tasks Classification

Secondary task type	Description
Eating/drinking	Eating with utensils, Eating without utensils, Drinking with lid and straw, Drinking from an open container ... and so forth
Smoking	Smoking cigar/cigarette, Lighting cigar/cigarette, Extinguishing cigar/cigarette
Passenger interaction	Passenger in adjacent seat – interaction, Passenger in rear seat – interaction
Manipulating objects	Object dropped by driver, Object in vehicle, Other, and so forth
Reaching for objects	Reaching for food-related or drink-related item, Reaching for cigar/cigarette, Reaching for personal body-related item, and so forth
Vehicle integral devices	Adjusting/monitoring climate control, Adjusting/monitoring radio, Inserting/retrieving CD (or similar), and so forth
Personal hygiene	Combing/brushing/fixing hair, Applying make-up, Shaving, Brushing/flossing teeth, and so forth
Outside distractors	Looking at previous crash or incident, Distracted by construction, Looking at pedestrian, and so forth
Other secondary tasks	Other non-specific internal eye glance, Other known secondary task, Unknown type (secondary task present)
Dancing	
Reading	
Writing	
Pet in vehicle	
Cell phone, talking/listening handheld	
Cell phone, talking/listening hands-free	
Cell phone, texting	
Cell phone, dialing handheld	
Cell phone, locating/reaching/answering	
Cell phone, other	
No secondary tasks	

2 data dictionary. If no secondary task existed, the variable showed the "No Secondary Task" outcome. In this study, the secondary task activities were classified according to Stutts et al.'s study (*13*) (Table 2).

According to the NHTSA (*17*), distracted driving is responsible for 30% of all crashes. In general, distraction has many sources. Engagement in a secondary task while driving is one of the major sources of distraction. Thus, in this study, distracted driving will be defined as driver engagement in a secondary task. In most of the previous studies, the responsibility of distracted driving as a main cause of accidents was measured by descriptive statistics (for example, mean, standard deviation, chi squared test, etc.). Despite the significant correlation between distracted driving and crash likelihood, these descriptive statistical analyses cannot clearly identify the association among multiple factors in complex relationships. Therefore, this study proposes a new methodology to identify the correlation among responses that are made simultaneously using a discrete choice model.

Model Development

Bivariate Probit Model

One of our objectives was to predict the crash/near-crash likelihood given that the driver was distracted, that is, engaged in a secondary task. Since the occurrence of distraction and crash may both depend on various explanatory variables including the driver's demographic characteristics, vehicle characteristics, and roadway characteristics, the multivariate approach was chosen to link these two variables to the explanatory variables. In particular, a bivariate probit model was constructed to identify the factors affecting these two responses and also capture the correlation between them. Let y_1 be the distraction index with $y_1 = 1$ if the driver is distracted and $y_1 = 0$ otherwise, y_2 be the safety-critical index with $y_2 = 1$ if a crash/near-crash occurs and $y_2 = 0$ otherwise. The bivariate probit model with a latent variable formulation takes the following form:

$$\begin{aligned} z_1 &= \beta_1 X_1 + \epsilon_1, & y_1 = 1 \; \textit{if} \; z_1 \geq 0, \quad y_1 = 0 \; \textit{otherwise} \\ z_2 &= \alpha z_1 + \beta_2 X_2 + \epsilon_2, & y_2 = 1 \; \textit{if} \; z_2 \geq 0, \quad y_2 = 0 \; \textit{otherwise} \end{aligned} \tag{1}$$

where

z_1 is the latent variable that indicates whether the driver is distracted ($y_1 = 1$ if $z_1 \geq 0$) or not ($y_1 = 0$ if $z_1 < 0$)

X_1 is the vector of explanatory variables for the first response

z_2 is the latent variable that indicates whether the driver was involved in a SCE ($y_2 = 1$ if $z_2 \geq 0$) or not ($y_2 = 0$ if $z_2 < 0$)

X_2 is the vector of explanatory variables for the second response

β_1, α, β_2 are the parameters to be estimated
ϵ_1, ϵ_2 are two random errors that follow a normal distribution with mean 0 and variance 1.

If the two responses are interrelated, the coefficient α should be significantly different from 0. By implementing this model, the interrelationship between the distracted driving and the crash likelihood could be investigated but from a different perspective.

Distracted Driving—Crash/Near-Crash Involvement. In this model, SAS® software was employed to investigate the association between distracted driving and the SCE involvement (crash/near-crash). The two response variables are modeled as a function of a set of independent variables, as described in Equation 1. In the first equation to model distraction, the set of independent variables included driver's age, gender, marital status, working status, driver training, education, years of driving, relation to junction, and locality, as shown in Table 1. These independent variables were selected in accordance with previous studies, which examined the driver's willingness to be engaged in a secondary tasks based on different personal and traffic flow factors (*18–21*). Table 3 displays the outcomes of the constructed bivariate probit model (only significant independent variables are shown). In the first equation, it was found that drivers between the ages of 16 and 34 years are more likely to be engaged in a secondary task while driving. The results also showed that the tendency of drivers of either full-time or part-time working status to be distracted while driving is higher than that of non-working drivers. This result is logical, as full-time and part-time drivers are more involved in the driving task. Moreover, Table 3 indicates that drivers are willing to engage in a secondary task when they have passengers on board. The table also depicts that drivers prefer to be engaged in a secondary task while they approach intersections. This might suggest that there is a potential relationship between traffic density and secondary task engagement.

Unlike the first equation, the second equation to model crash/near-crash uses all the event and roadway characteristics shown in Table 1, in addition to driver age and gender as independent variables. The model showed that engagement in a secondary task is significantly correlated to the crash/near-crash likelihood. The positive coefficient implies that secondary task engagement increases the probability of crash/near-crash occurrence. Hence, there is strong statistical evidence of the impact of distracted driving on travel safety. The results also showed that parking and intersection locations are most prone to crash/near-crash occurrences. To summarize, the bivariate probit model found that distracted driving, driver's age, and intersection influence are the most significant predictors of crash/near-crash likelihood. In the next section, further analysis will quantify the increased crash/near-crash risk that results from different types of secondary tasks.

Table 3. Bivariate Probit Model Results

	Coefficient	*t*-statistics	*p*-Value (0.05*)
First Model: Engaged in a secondary task (distracted/not distracted)			
AgeGroup 16–19	0.546	6.11	<0.0001
AgeGroup 20–24	0.653	7.39	<0.0001
AgeGroup 25–29	0.563	4.83	<0.0001
AgeGroup 30–34	0.317	2.22	0.0265
FullTime	0.239	3.54	0.0063
PartTime	0.129	1.96	0.0492
Intersection_Related	0.295	2.19	0.0287
Presence of Passengers	0.419	6.17	<0.0001
Second Model: Involved in crash/near-crash event or not			
Engaged	**1.335**	**14.78**	**<0.0001**
AgeGroup 16–19	1.622	9.258	<0.0001
AgeGroup 20–24	1.17	10.529	<0.0001
AgeGroup 25–29	0.293	2.49	0.009
Parking_Related	0.673	4.23	<0.0001
Intersection	0.26	1.96	0.0492
Intersection Influence	0.629	6.35	<0.0001
Grade Up	–0.318	–2.02	0.037

*Significance level.

Secondary Tasks Risk Assessment

In this section, the increased crash/near-crash risk that results from the different secondary tasks is investigated. For this purpose, two different models were developed: a multinomial logit model, and an association analysis model. The two models attempted to quantify the increased crash/near-crash risk from two different perspectives. The multinomial logit model is a traditional statistical technique that is based on probability theory, whereas association analysis is a new powerful data-mining technique that reveals patterns in big data such as SHRP2 NDS data. The model results will be presented and the merits of each model detailed.

Multinomial Logit Model (Baseline-Category Logits Model). The multinomial logit model is a statistical technique that is employed when the response variable has more than two categories. In this model, the response variable is the event severity (normal, near-crash, or crash event), whereas the explanatory variables are the secondary tasks listed in Figure 1. The baseline-category logits model pairs each response category with a reference response category. As the SHRP2 NDS dataset provides the distribution of secondary tasks in SCEs as well as in

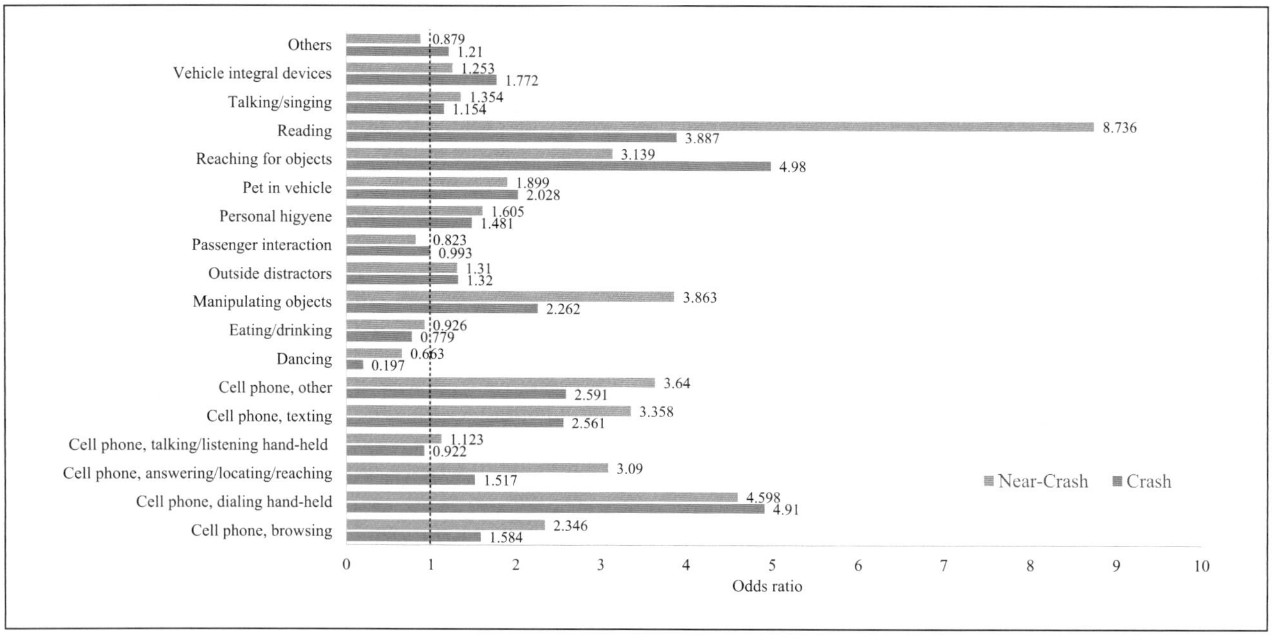

Figure 1. Odds ratios of different secondary tasks.

non-SCEs, the increased crash/near-crash risk could be recognized and quantified. When the "normal" category is the baseline, the baseline-category logits are,

$$\log\left(\frac{\pi_j}{\pi_J}\right), \text{where } j = \text{Near-crash, Crash, and } J = \text{Normal} \tag{2}$$

where π_j is the probability of the jth category. The baseline-category logits model with a set of predictor variables X (secondary tasks in our case) is defined as,

$$\log\left(\frac{\pi_j}{\pi_J}\right) = \beta_j X, \quad j = 1, \ldots, J-1 \tag{3}$$

This model has J-1 equations with separate parameters for each. The effects vary with the category paired with the reference category. If $J = 2$, this model simplifies to an ordinary logistic regression. The main advantage of the baseline-category logits model is the simultaneous fit of all the equations together. This advantage produces parameter estimates with smaller standard errors compared with fitting each equation separately using an ordinary logistic regression. In this regard, a PROC QLIM statement was recalled in SAS platform to achieve the modeling requirements.

According to previous studies (*3, 11, 12*), the OR is frequently used to estimate the relative risk of the secondary tasks while driving. ORs in a baseline-category logits model are defined as in a binary logistic model, except that they describe conditional odds. For instance, Figure 1 shows that the OR for a driver engaged in cell phone texting is 3.358 for near-crash. This means that the odds that drivers who are engaged in cell phone texting will be involved in a near-crash event rather than a normal driving event are about 3.35 times the odds for drivers who are not engaged in cell phone texting, adjusting for all the other secondary tasks. Similarly, the OR for manipulating objects is 2.262 for a crash event. Hence, we may say that the drivers who manipulate objects while driving have the odds of being involved in a crash event vs. normal event that is about 2.262 times the odds for those who are not engaged in manipulating an object, adjusting for all the other secondary tasks. Figure 1 displays the ORs of all secondary tasks in crash and near-crash events.

It should be noted that if the OR for a particular secondary task is less than 1.00 (dashed line), then the secondary task has no harmful effect on traffic safety. Accordingly, passenger interaction, eating/drinking, and dancing show a protective effect rather than a risk effect. However, Figure 1 indicates that the remaining secondary tasks are all within the risk range (OR > 1). For near-crash events, reading while driving showed the highest risk with an OR of 8.739, followed by cell phone dialing handheld, and manipulating objects with ORs of 4.508 and 3.863, respectively. Cell phone texting, cell phone other, and cell phone answering/locating/reaching follow, with ORs of 3.358, 3.64, and 3.00 respectively.

Association Analysis Model (A Priori Algorithm). Data-mining techniques have been receiving increased attention from transportation researchers. These techniques have shown successful implementation in addressing safety problems compared with traditional statistical analyses (*22–27*). Detection of association rules is one of the powerful tools in data-mining techniques. It is considered the most frequent tool employed in web mining within the retail industry (also known as market basket analysis). However, this method has a wide variety of useful applications such as transportation safety. The goal of association analysis is to find rules in the form of conditions (antecedents) and results (consequents). The rules are developed based on the a priori algorithm. More details about the a priori algorithm can be found in Agrawal et al.'s study (*28*). Each developed rule is then evaluated using three performance measures: support, confidence, and lift. For instance, if an extracted rule states that "if ($Var1 = x$), $\rightarrow$ ($Var2 = y$), 30%, 80%", it means that if variable 1 is equal to x, then the probability (Prob) that variable 2 will be equal to y is 80%, and the joint event ($Var1 = x$, $Var2 = y$) occurs in 30% of the observations. Accordingly, Support (S), the first percentage in the rule, is defined as the probability of antecedent and consequent,

$$\text{Support(S)} = \text{Prob(antecedent and consequent)} \quad (4)$$

whereas Confidence (C), the second percentage in the rule, is the conditional probability of the consequent given the antecedent, and is denoted by,

$$\text{Confidence(C)} = \text{Prob(consequent|antecedent)} = \frac{\text{Prob(antecedent and consequent)}}{\text{Prob(antecedent)}} \quad (5)$$

Lift (L) is a performance measure that was presented in Brin et al.'s study (*29*). Lift displays the ratio of confidence for the rule to the marginal probability of having the consequent. To illustrate, suppose that 10% of the entire population buys a product X, then a rule that predicts whether people will buy product X with 20% confidence will have a lift of 20/10 = 2.00. If another rule tells you that people will buy X with 11% confidence, then the rule has a lift close to 1.00, meaning that having antecedent(s) makes little difference to the probability of having a consequent. Therefore, lift is a measure of how helpful the rule is. Rules with a lift index different from 1.00 are more interesting. Equation 6 shows the mathematical expression for the lift measure,

$$\text{Lift(L)} = \frac{\text{Prob(consequent|antecedent)}}{\text{Prob(consequent)}} = \frac{\text{Prob(antecedent and consequent)}}{\text{Prob(antecedent)*Prob(consequent)}} \quad (6)$$

To conclude, support (S) is a measure of frequency, confidence (C) is the measure of belief, and lift (L) is the measure of the improvement brought by the rule. In the marketing industry, sellers are more interested in finding rules with high support levels, high confidence indexes and lifts greater than 1.00. In transportation safety, SCEs (crashes/near-crashes) have much lower frequencies than non-SCEs. As the main objective is to find the association between the SCEs and the associated secondary tasks, support of the rules could be quite low. Therefore, a lift performance measure is used in rules evaluation.

In this research, the authors were more interested in finding rules connecting the secondary task activities (cell phone texting, eating, writing, manipulating objects, etc.) with the event severity. In essence, the study dataset was transformed into a tabular format, in which the columns represented indicator variables for the secondary task activities and the target variable was an SCE or non-SCE. Before interpreting the results, it is important to mention that the minimum support level specified for the proposed model was set at 0.1%. Regardless of the lift value, this means that no rule would have been extracted if it had a support level lower than 0.1%. This low value was selected because of the interest in extracting information related to rare events (crashes/near-crashes). Table 4 shows the rules extracted for secondary tasks and the SCE outcome. The rules are ranked based on the lift index. For better understanding the risk associated with different secondary tasks, rules should be compared with each other. For instance, Table 4 includes the following two rules:

(*Cell phone Texting* = *1*), $\rightarrow$ (*Event* = *SCE*), 2.56%, 27.64%",
(*Vehicle-embedded devices* = *1*), $\rightarrow$ (*Event* = *SCE*), 3.46%, 13.44%"

This means that the risk associated with cell phone texting is higher than that of operating vehicle-embedded devices. In other words, the probability of observing an SCE given that a driver is engaged in cell phone texting, (27.64%), is higher than that of vehicle-embedded devices (13.44%). Following the same criteria, all secondary tasks could be ranked based on how risky they are. The results indicated that reaching for objects, manipulating objects, reading, and other cell phone interaction activities are the riskiest secondary task activities. However, passenger interaction, eating/drinking, and dancing do not indicate a risk factor for SCE occurrence.

Discussion

This study provides quantitative insight into the risk associated with crash/near-crash events when drivers are engaged in secondary task activities. One of the most striking results in Figure 1 is the magnitude of the ORs (risk estimate). The figure shows that some activities can increase the crash or near-crash risk by four- to

Table 4. Association Analysis Results

Consequent	Antecedent	Support (S) %	Confidence (C) %	Lift (L)
EVENTSEVERITY = SCE	Reaching for objects	1.41	40.79	3.35
EVENTSEVERITY = SCE	Manipulating objects	5.78	39.77	3.27
EVENTSEVERITY = SCE	Reading	0.11	39.13	3.21
EVENTSEVERITY = SCE	Cell phone other	0.31	38.81	3.19
EVENTSEVERITY = SCE	Cell phone locating/reaching/answering	0.81	30.06	2.47
EVENTSEVERITY = SCE	Cell phone dialing handheld	0.19	30.00	2.46
EVENTSEVERITY = SCE	Cell phone texting	2.56	27.64	2.27
EVENTSEVERITY = SCE	Pet in vehicle	0.19	26.83	2.20
EVENTSEVERITY = SCE	Cell phone browsing	0.93	23.12	1.90
EVENTSEVERITY = SCE	Personal hygiene	4.01	15.45	1.27
EVENTSEVERITY = SCE	Talking/singing	7.73	13.72	1.13
EVENTSEVERITY = SCE	Vehicle-embedded devices	3.46	13.44	1.10
EVENTSEVERITY = SCE	External distractor	10.94	12.72	1.04
EVENTSEVERITY = SCE	Cell phone talking/listening handheld	3.25	12.02	0.99
EVENTSEVERITY = SCE	Eating/drinking/smoking	4.28	10.34	0.85
EVENTSEVERITY = SCE	Passenger interaction	14.94	9.22	0.76
EVENTSEVERITY = SCE	Other known secondary task	3.72	9.00	0.74
EVENTSEVERITY = SCE	Dancing	1.11	7.11	0.58

eight-fold (such as reaching for objects, reading, and cell phone dialing handheld). These activities are considered high-risk distractors as they not only require multiple steps to be completed but also longer eyes-off-road time (such as, reaching for objects and reading). Surprisingly, other secondary tasks such as passenger interaction showed unexpected impacts. Cooper et al. found that passenger interaction increases the crash risk (*30*). However, this study found that the presence of a passenger has a protective effect rather than a risk effect. This could be explained as the presence of a passenger on board being equivalent to having more eyes on the road, which could reduce the crash or near-crash probability. This result is consistent with the findings in Geyer and Regland's study (*31*).

As shown in Figure 1 and Table 4, some secondary tasks have a similar risk impact in relation to ORs or lift values. Thus, it is preferable to group these secondary tasks together. As a result, the *k*-means clustering algorithm was employed to group secondary tasks with similar risk effects together. *k*-means is a common unsupervised-learning clustering technique, which partitions *n* observations of unlabeled data into *k* clusters in which each observation belongs to the cluster with the nearest mean. Three clustering models were developed using SPSS Modeler with a predetermined number of clusters ($k = 4$). The clustering models used either the OR, obtained from the baseline-category logits model, or the lift index (L), obtained from the association model, as a clustering-based variable. The results of the clustering

Table 5. Secondary Tasks Ranking

Multinomial Logits Model		A Priori Association Model
Crash (*k* = 4)	Near-Crash (*k* = 4)	SCE (Crash/Near-Crash) (*k* = 4)
Reaching for objects	Reading	Reaching for objects
Cell phone, dialing handheld	Cell phone, dialing handheld	Manipulating objects
Reading	Manipulating objects	Reading
Cell phone, texting	Cell phone, texting	Cell phone, other
Cell phone, other	Cell phone, other	Cell phone, answering/reaching
Manipulating objects	Reaching for objects	Cell phone, dialing handheld
Pet in vehicle	Cell phone, answering/reaching	Cell phone, texting
Vehicle-embedded devices	Cell phone, browsing	Pet in vehicle
Cell phone, browsing	Pet in vehicle	Cell phone, browsing
Personal hygiene	Personal hygiene	Personal hygiene
Cell phone, answering/reaching	Talking/singing	Talking/singing
Outside distractors	Outside distractors	Vehicle-embedded devices
Talking/singing	Vehicle-embedded devices	External distractors

Note: Grayscale gradient indicates the *k*-clusters.

analysis are shown in Table 5. Although the highest impact secondary tasks are similar in both models, each model has its own advantages and disadvantages. To have considered more variables in baseline-category logits models would have exposed the developed model to multicollinearity. This problem can produce incorrect parameter estimates and hence, lead to incorrect results and conclusions. However, multicollinearity does not represent a problem for the association analysis model. In the association analysis model, no particular variable is defined as a response variable. Consequently, all rules that describe the association between the SCE/non-SCE event attributes can be extracted. In this study, only the one-product association rules were requested. Rules were then filtered to present only the rules connecting the secondary task activities with the event severity. To the authors' knowledge, this is the first study to employ and adjust the a priori algorithm settings in a distracted driving analysis.

It is essential to understand the impact of distracted driving in the larger context of naturalistic driving to provide useful suggestions for countermeasures. The outcomes of this study can help different sectors (automobile industry, decision makers, safety campaigns, etc.) to address distracted driving behavior. For instance, the automobile industry needs to reduce the in-vehicle features that require visual and physical interaction. This, in turn, will increase driver focus and decrease the eyes-off-road time. One of the possible recommendations is to lock out all the complex in-vehicle features while the vehicle is in motion. Moreover, as cell phone interaction activities are also one of the main internal distraction sources, it would be preferable to develop a new cell phone mode that prohibits all complex features while the vehicle is in motion (similar to airplane mode). Additionally, drivers should be aware of all the relative risks that are associated with the various secondary task activities so that they can adjust their behavior or consider alternatives. Safety campaigns that carry the message "all distractions are bad" are unrealistic and ineffective. Identifying the most serious secondary tasks can help safety campaigns to achieve their goals effectively. Finally, policymakers and legislative institutions should devise their acts (texting bans, handheld cell phone bans, etc.) based on NDS information, not on unrealistic experiments.

Conclusion

This study analyzed the increased crash and near-crash risk associated with multiple secondary tasks using a variety of statistical and data-mining models. First, a bivariate model was constructed using the SHRP2 NDS data to examine the relationship between distracted driving and SCE likelihood from different perspectives. The model indicated that distracted driving is a major contributor to an SCE occurrence. Subsequently, two different models were employed to quantify the increased risk associated with each secondary task: a baseline-category logits model, and a rule mining association model. The baseline-category logits model identified the increased risk in terms of ORs, while the a priori association algorithm detected the associated risks in terms of rules. Each rule was then evaluated based on the lift index (L). The two models succeeded in ranking all the secondary task activities according to the associated increased crash/near-crash risk efficiently. Both models revealed that reading while driving and reaching for objects are the highest crash risk among all secondary tasks. Furthermore, the *k*-means algorithm was implemented to cluster secondary tasks with similar risk impacts. Based on the results, a table was constructed to identify the *k*-means groups and the riskiest secondary tasks within each group. This study's outcomes could help drivers understand the relative risks associated with the various secondary task activities so that they can adjust their behavior or consider alternatives. The study could also help legislators initiate laws that reduce the crashes resulting specifically from distracted driving. Finally, it could help government officials make informed decisions about the allocation of available resources to reduce roadway crashes and improve traffic safety.

Author Contributions

The authors confirm contribution to the paper as follows: study conception and design: P.B., S.I.; analysis and interpretation of results: P.B., B.G; draft manuscript preparation: P.B., S.I., and B.G. All authors reviewed the results and approved the final version of the manuscript.

References

1. Young, K. L., and P. M. Salmon. Examining the Relationship between Driver Distraction and Driving Errors: A Discussion of Theory, Studies and Methods. *Safety Science*, Vol. 50, No. 2, 2012, pp. 165–174.
2. Caird, J. K., C. R. Willness, P. Steel, and C. Scialfa. A Meta-Analysis of the Effects of Cell Phones on Driver Performance. *Accident Analysis & Prevention*, Vol. 40, No. 4, 2008, pp. 1282–1293.
3. Regan, M. A., J. D. Lee, and K. Young. *Driver Distraction: Theory, Effects, and Mitigation*. Boca Raton, Fla., CRC Press, 2008.
4. Lansdown, T. C. Frequency and Severity of In-Vehicle Distractions: A Self-Report Survey. *Proc., 1st International Conference on Driver Distraction and Inattention (DDI 2009)*, 2009.
5. McEvoy, S. P., M. R. Stevenson, and M. Woodward. The Impact of Driver Distraction on Road Safety: Results

from a Representative Survey in Two Australian States. *Injury Prevention*, Vol. 12, No. 4, 2006, pp. 242–247.
6. Royal, D. *National Survey of Speeding and Unsafe Driving Attitudes and Behaviors: 2002*. Volume II-Findings Report No: HS-809 688. National Highway Traffic Safety Administration, Washington, D.C., 2003.
7. Sullman, M. J., and P. H. Baas. Mobile Phone Use amongst New Zealand Drivers. *Transportation Research Part F: Traffic Psychology and Behaviour*, Vol. 7, No. 2, 2004, pp. 95–105.
8. Young, K. L., and M. G. Lenné. Driver Engagement in Distracting Activities and the Strategies Used to Minimise Risk. *Safety Science*, Vol. 48, No. 3, 2010, pp. 326–332.
9. af Wåhlberg, A., L. Dorn, and T. Kline. The Manchester Driver Behaviour Questionnaire as a Predictor of Road Traffic Accidents. *Theoretical Issues in Ergonomics Science*, Vol. 12, No. 1, 2011, pp. 66–86.
10. Johnson, M. B., R. B. Voas, J. H. Lacey, A. S. McKnight, and J. E. Lange. Living Dangerously: Driver Distraction at High Speed. *Traffic Injury Prevention*, Vol. 5, No. 1, 2004, pp. 1–7.
11. Klauer, S. G., T. A. Dingus, V. L. Neale, J. D. Sudweeks, and D. J. Ramsey. *The Impact of Driver Inattention on Near-Crash/Crash Risk: An Analysis Using the 100-Car Naturalistic Driving Study Data*. Report No. DOT HS 810 594. National Highway Traffic Safety Administration, Washington, D.C., 2006.
12. Sayer, J. R., J. M. Devonshire, and C. A. Flannagan. *The Effects of Secondary Tasks on Naturalistic Driving Performance*. Report No. UMTRI-2005-29. The University of Michigan Transportation Research Institute, Ann Arbor, 2005.
13. Stutts, J., J. Feaganes, D. Reinfurt, E. Rodgman, C. Hamlett, K. Gish, and L. Staplin. Driver's Exposure to Distractions in their Natural Driving Environment. *Accident Analysis & Prevention*, Vol. 37, No. 6, 2005, pp. 1093–1101.
14. Olson, R., R. Hanowski, J. Hickman, and J. Bocanegra. Driver Distraction in Commercial Operations Vehicle. *Federal Motor Carrier Safety Administration, DC, FMCSA-RRR-09*, Vol. 42, 2009.
15. Transportation Research Board of the National Academies; Virginia Tech Transportation Institute. The 2nd Strategic Highway Research Program Naturalistic Driving Study InSight Dataset [Data set]. VTTI Dataverse, 2016. https://doi.org/10.15787/VTT1/3YVSY4.
16. Dingus, T. A., F. Guo, S. Lee, J. F. Antin, M. Perez, M. Buchanan-King, and J. Hankey. Driver Crash Risk Factors and Prevalence Evaluation Using Naturalistic Driving Data. *Proceedings of the National Academy of Sciences*, Vol. 113, No. 10, 2016, pp. 2636–2641.
17. National Highway Traffic Safety Adminstration. *Distracted Driving 2014*. Report No. DOT HS 812 260. National Highway Traffic Safety Adminstration, Washington, D.C., 2016.
18. Goldberg, L. R., J. A. Johnson, H. W. Eber, R. Hogan, M. C. Ashton, C. R. Cloninger, and H. G. Gough. The International Personality Item Pool and the Future of Public-Domain Personality Measures. *Journal of Research in Personality*, Vol. 40, No. 1, 2006, pp. 84–96.
19. Lane, W., and C. Manner. The Impact of Personality Traits on Smartphone Ownership and Use. *International Journal of Business and Social Science*, Vol. 2, No. 17, 2011 pp. 22–28.
20. Lansdown, T. C. Individual Differences and Propensity to Engage with In-Vehicle Distractions–A Self-Report Survey. *Transportation Research Part F: Traffic Psychology and Behaviour*, Vol. 15, No. 1, 2012, pp. 1–8.
21. Titchener, K., and I. Y. Wong. Driver Distractions: Characteristics Underlying Drivers' Risk Perceptions. *Journal of Risk Research*, Vol. 13, No. 6, 2010, pp. 771–780.
22. Abdel-Aty, M., and J. Keller. Exploring the Overall and Specific Crash Severity Levels at Signalized Intersections. *Accident Analysis & Prevention*, Vol. 37, No. 3, 2005, pp. 417–425.
23. Abdelwahab, H., and M. Abdel-Aty. Development of Artificial Neural Network Models to Predict Driver Injury Severity in Traffic Accidents at Signalized Intersections. *Transportation Research Record: Journal of the Transportation Research Board*, 2001. 1746: 6–13.
24. Bayam, E., J. Liebowitz, and W. Agresti. Older Drivers and Accidents: A Meta Analysis and Data Mining Application on Traffic Accident Data. *Expert Systems with Applications*, Vol. 29, No. 3, 2005, pp. 598–629.
25. Chang, L.-Y. Analysis of Freeway Accident Frequencies: Negative Binomial Regression Versus Artificial Neural Network. *Safety Science*, Vol. 43, No. 8, 2005, pp. 541–557.
26. Chang, L.-Y., and W.-C. Chen. Data Mining of Tree-Based Models to Analyze Freeway Accident Frequency. *Journal of Safety Research*, Vol. 36, No. 4, 2005, pp. 365–375.
27. Golob, T. F., and W. W. Recker. A Method for Relating Type of Crash to Traffic Flow Characteristics on Urban Freeways. *Transportation Research Part A: Policy and Practice*, Vol. 38, No. 1, 2004, pp. 53–80.
28. Agrawal, R., T. Imieliński, and A. Swami. Mining Association Rules between Sets of Items in Large Databases. *ACM SIGMOD Record*, No. 22, 1993, pp. 207–216.
29. Brin, S., R. Motwani, J. D. Ullman, and S. Tsur. Dynamic Itemset Counting and Implication Rules for Market Basket Data. *ACM SIGMOD Record*, Vol. 26, 1997. pp. 255–264.
30. Cooper, D., F. Atkins, and D. Gillen. Measuring the Impact of Passenger Restrictions on New Teenage Drivers. *Accident Analysis & Prevention*, Vol. 37, No. 1, 2005, pp. 19–23.
31. Geyer, J., and D. Ragland. Vehicle Occupancy and Crash Risk. *Transportation Research Record: Journal of the Transportation Research Board*, 2005. 1908: 187–194.

The Standing Committee on Safety Data, Analysis and Evaluation (ANB20) peer-reviewed this paper (18-01562).

Article

TRR
JOURNAL OF THE TRANSPORTATION RESEARCH BOARD

Transportation Research Record
2018, Vol. 2672(38) 255–265

DOI: 10.1177/0361198118774670
journals.sagepub.com/home/trr
SAGE

Using Naturalistic Driving Study Data to Evaluate the Effects of Intersection Balance on Driver Behavior at Partial Cloverleaf Interchange Terminals

Jin Wang[1] and Huaguo Zhou[2]

Abstract

Past studies showed that poor intersection balances at partial cloverleaf (parclo) interchange terminals significantly impact traffic safety and sight distance of drivers making left turns to entrance ramps. Some state traffic agencies have recommended a "balance" guideline that the length between the left-turn stop line on crossroads to the middle of the intersection should not be greater than 60% of the entire length of the intersection. However, a scarcity of research exists on how the balance of an intersection affects driver behavior, which has been identified as a critical contributing factor to intersection-related crashes. This study utilizes the Naturalistic Driving Study (NDS) data to evaluate the effects of intersection balance on driver behavior at parclo interchange terminals for proof-of-concept. A small but representative data sample was collected from the second Strategic Highway Research Program's (SHRP 2) NDS dataset. It demonstrates statistical characteristics and overall trends of driver speed, acceleration/deceleration rates, and risk perception with the changing of intersection balances. Conclusions provide guidance on optimal intersection balance design that may help drivers make smoother and safer transitions from crossroads to entrance ramps at parclo interchange terminals.

The freeway interchange terminals are essential components in the highway system. The safety performance of the terminals proves critical to the overall system performance. As the access points to freeways, interchange terminals are one main source of accidents and congestion. Several previous studies have found that a majority of freeway crashes occur at interchange terminals, and inappropriate terminal design has caused severe safety problems like wrong-way driving (WWD) crashes. An improper terminal design also increases delay and congestion. Thus, interchange terminals directly affect the safety and operational integrity of freeways. Past studies suggested some interchange types of partial cloverleaf (parclo) interchanges are highly susceptible to WWD movements due to the geometric feature of closely spaced exit ramps and entrance ramps (*1*). One essential design element in the planning, design, and operation of interchange terminals is a proper intersection balance that can provide a safer and smother movement of left-turn vehicles while maintaining vehicular throughput and access to adjoining ramps. It needs to reflect sound traffic engineering principles and driver behavior. Intersection balance is defined as the percentage which the left-turning point is of the way through the intersection. It is the percentage of the distance between the stop line for left turns from a crossroad and the middle of the median separating on and off ramps out of the entire intersection width (Figure 1).

Despite the importance of intersection balance, it is often overlooked in current roadway design and site planning efforts. Driver behavior is one crucial element to consider in the design of signalized intersections, because driver behavior was identified as a critical contributing factor to about 96% of intersection-related crashes (*2*). Therefore, the intersection balance should focus on the element of time and its relationship to the driving task, which includes perception, reaction, navigation, and execution of the necessary maneuvers. However, research on different intersection balances on behavior of left-turn drivers has not been performed because of the lack of good-quality driver behavior data.

[1]Transportation Engineering, Samuel Ginn College of Engineering, Auburn University, University, AL
[2]Department of Civil Engineering, Auburn University, AL

Corresponding Author:
Address correspondence to Jin Wang: jinwang@auburn.edu

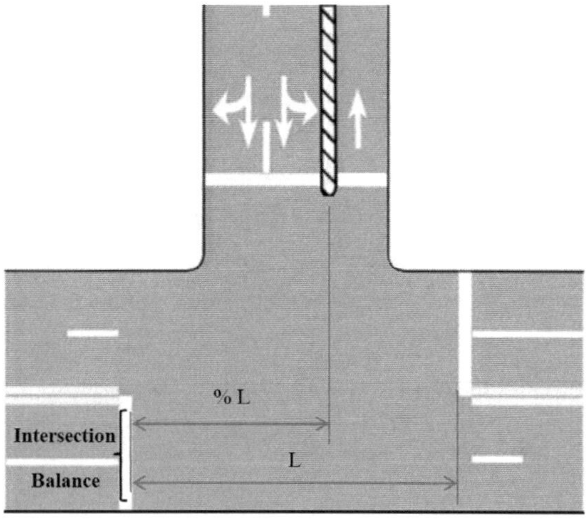

Figure 1. Illustration of current best practice of the intersection balance at parclo interchange terminals.

Thus, this study on the relationship between driver performance and intersection balances proves beneficial for improving freeway traffic safety and operational performance through controlling access location by proper intersection balance design using the second Strategic Highway Research Program's (SHRP 2) Naturalistic Driving Study (NDS) dataset. It also provides rationale for using the naturalistic data to quantify relationships between human and environmental factors for better roadway design, which are unknown from standard crash analysis. Accordingly, recommendation and validation on optimal intersection balance design at parclo interchange terminals can be made.

Literature Review

Intersection Balance

No specific national guideline exists for intersection balances in the Manual on Uniform Traffic Control Devices (MUTCD) (*3*) or AASHTO "Green Book" (*4*). Though regarding stop line positioning, the MUTCD (*3*) indicates that "*if used, stop lines shall consist of solid white lines extending across approach lanes to indicate the point at which the stop is intended or required to be made. If used, stop and yield lines should be placed a minimum of 4 feet in advance of the nearest crosswalk line at controlled intersections, except for yield lines at roundabouts as provided for in Section 3C.04 and at midblock crosswalks. In the absence of a marked crosswalk, the stop line or yield line should be placed at the desired stopping or yielding point, but should not be placed more than 30 feet or less than 4 feet from the nearest edge of the intersecting traveled way.*" However, this guideline on stop line positioning does not consider the intersection balance or driver comfort.

Some state traffic agencies have recommended general guideline on intersection balance. For instance, the Washington State Department of Transportation (WSDOT) (*5*) proposed that the distance between the stop line for left turns from a crossroad and the middle of the median separating on and off ramps be no more than 60% of the entire intersection width. However, there is lack of scientific research to support the guideline by WSDOT.

A recent study by Wang et al. (*6*) using 10-year (2004–2013) Illinois WWD crash data concluded: 1) the stop line positioning of 50% – 60% is used most often at parclo interchange terminals; 2) the possibility of WWD crashes increases significantly when the stop line on crossroads is positioned more than 60% into the intersection at signalized terminals for parclo interchanges.

Another study by Wang (*7*) evaluated the relationships between geometric elements and WWD crashes at parclo interchange terminals based on crash and geometric design data. The study identified vulnerable geometric designs. It also quantified and validated how different geometric elements affect wrong-way crashes with field and crash data, including intersection balances, turning radii, traversable medians, and median widths.

However, despite the effects of different intersection balances on WWD crashes, it is still not clear how the different balances intersection and strategies impact driver behavior. No research was conducted to study driver behavior change by intersection balance due to the lack of sound driver behavior data.

NDS Data

NDS data (including crashes, near-crashes, critical incidents, traffic conflicts, as well as ordinary driving data) enables further finer-grain analysis of the effects of roadway design elements on driver behavior, as outlined in the next section (*8*). The naturalistic driving method offers exposure information, including the frequency of behaviors in normal driving, as well as the larger context of contributing factors. The larger context for exposure enables risk estimates for various driver behaviors and for other contributing factors. Thus, using NDS data allows for empirical support to develop new guidelines and safety countermeasures to reduce crashes. With NDS data, additional analyses can be conducted to evaluate the effects of different roadway design elements or access management technique on driver behavior.

Past safety studies (*9–19*) suggested the use of NDS data can quantify different factors' impacts on driver behavior and crash risk, compared with data from driving simulator, field experiments, or crash reports. These studies addressed road departure, offset left-turn lanes, driver inattention, rear-end crashes on congested freeways, and so forth.

Table 1. Selected Signalized Terminals of Parclo Interchanges

Location	Intersection balance	Interchange	Interchange type	Turning angle	State
Location 1	38.78%	I-75/State Road 26	Parclo	90°	FL
Location 2	51.43%	I-275/E Busch Blvd	Parclo	90°	FL
Location 3	73.33%	I-275/E Busch Blvd	Parclo	90°	FL

Summary

Past research (*6*, *7*) has proved that some minor geometric design improvements and access management techniques can mitigate WWD crashes at parclo interchange terminals. Nonetheless, the crash data analysis cannot determine what caused the differences in the effects of different geometric features, nor does it explain why some interchange types are more vulnerable to WWD maneuvers. It is important to assess how driver behavior is impacted by a particular geometric design. Understanding the detailed effects of the different intersection balances on driver behavior will critically support the development and application of traffic safety countermeasures at intersections.

NDS data may fill the gaps in crash data analysis with real-time continuous naturalistic driving data. In this study, the analysis of NDS data was utilized to investigate the effects of intersection balances at parclo interchange terminals on driver behavior that may result in less wrong-way entries at interchange terminals. Corresponding recommendations can be made to validate the current best practice on the intersection balance guidelines developed by WSDOT.

Data Collection

Study Sites

The study locations were selected from the Roadway Information Database (RID) (*20*), a database containing: 1) roads most frequently driven by NDS participants, and 2) roads of greatest interest to safety researchers. Table 1 includes three signalized terminals of parclo interchanges with different intersection balances (<50%, 50–60%, and >60%). The street view of each location, as shown in Figure 2, is based on high-resolution aerial photography, street views, and the geographic information system (GIS) capabilities of the Google Earth Pro software. In addition, two lengths were measured, that is, L_1: the length from the stop line of the left-turn lane(s) to the centerline of the median on two-way ramps and L: the length of the entire intersection (the distance between the stop lines in two opposite directions on the crossroad). The intersection balance of each interchange terminal was calculated by Equation 1 (*5*). LINK IDs were identified for the approaches of interest at each location for NDS data request and for linkage between NDS data and RID data. Figure 2 illustrates the locations and vehicle movements of interest (indicated by arrows). A manual review was subsequently conducted of each route ID using Google Earth in order to verify whether the segment was actually at the selected interchange terminal.

$$IB(\%) = \frac{L_1}{L} * 100 \tag{1}$$

NDS Data

This research leverages the data from the SHRP 2 NDS database (*8*) to provide a framework for better understanding the relationships between driver behavior and intersection balance. For the purposes of this proof-of-concept study, data were collected at two levels of detail. First, information was obtained as to the general characteristics associated with each trip from the SHRP 2 NDS InSight website. Subsequently, more detailed information is requested from the SHRP 2 NDS database by the Virginia Tech Transportation Institute (VTTI), for example, speed, acceleration/deceleration, and GPS location information, as well as the RID, for example, roadway geometry, traffic control, speed limit, and so forth.

The NDS dataset includes data from participants, vehicles, and trips. The required parameters to collect for each category were determined before the data request, including data of Videos by all cameras during the trips of interest (Forward View and Right-Rear View), Drivers (Driver Physical and Psychological State, Driver Risk Perception Scale, and Driver Demographics and Driving History), Vehicles (Vehicle Type, Vehicle Age and Condition, Vehicle Technologies and Equipment), and Trips (Time Series of Trips which are of a time-series nature, providing details of the geographic location, speed, acceleration/deceleration, turn rate, brake pedal activations, of each trip). The trip data were comprised of a series of comma-separated values (CSV) files (one file for each event and each file includes GPS location information at 1.0-s intervals, as well as speed and acceleration/deceleration data at 0.1-s intervals).

The number of trips and participant data per intersection were specified: for the selected locations with a trip density of less than or equal to 250 trips, all the trips were

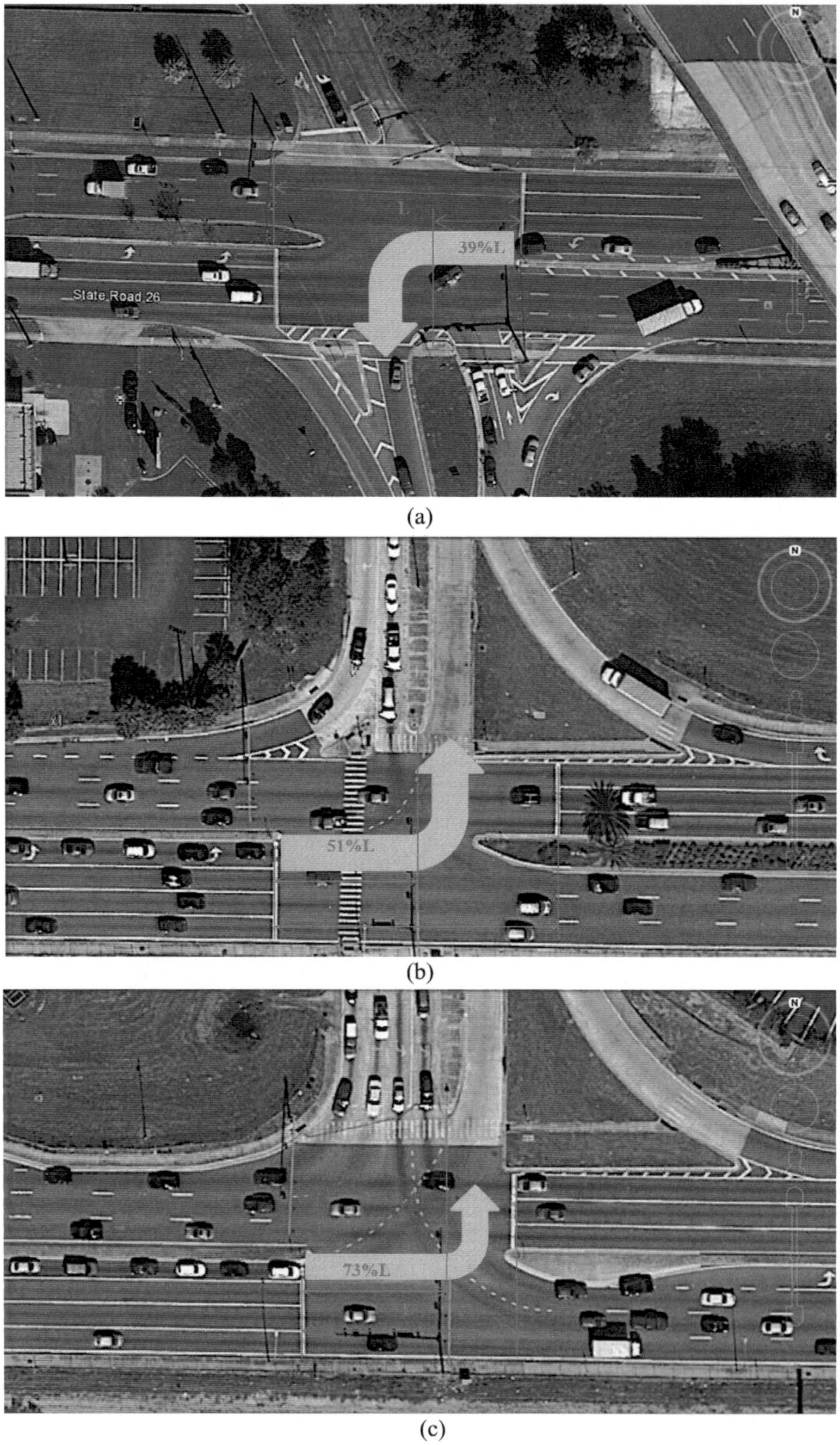

Figure 2. Selected signalized terminals of parclo interchanges: (*a*) Location 1 with marked approaches of interest; (*b*) Location 2 with marked approaches of interest; and (*c*) Location 3 with marked approaches of interest.

requested; for locations with a trip density of larger than 250 trips, 250 trips were requested, of which trips driven by different drivers and trips containing "Events (defined by SHRP 2 NDS)" were preferred.

Thus, information on a total of 545 trips by 137 participant drivers was available and obtained from the NDS dataset. These trips with the traversals of interest were requested from the trip pool passing though the predefined 3 locations: 21 trips by 12 drivers were at Location 1; 496 trips by 111 drivers were at Location 2; 28 trips by 14 drivers were at Location 3. None of the trip data include crash or near-crash events, because data were unavailable for the trips with crash or near-crash events, which were primarily comprised of specifics where the associated GPS data could have potentially allowed for the identification of specific drivers and crash/near-crash locations.

Data Reduction

The collected NDS trip data include front and rear videos, sensor data (speed, acceleration/deceleration, etc.), and supplementary data (driver risk perception based on questionnaires, driver driving history based on questionnaires, vehicle information, etc.). The researchers reviewed all these data and retrieved predefined events. Video review was conducted for each trip to verify they have occurred on the study sites and included the maneuver of interest, as well as to extract the portion of left-turn maneuver between the stop line and the entrance ramps out of each event. Cases with missing data were removed. The retrieved data were then validated by conducting data type checks, image checks, consistency checks, range checks, and format checks. After validation, the validated events were compiled further into the three study sites for analysis. The number of events at each study site is different. To apply a cross-section comparison method, an equal number of events were extracted from the requested data for each study site. Thus, after data reduction, data of 11 events by passenger cars were selected for each study sites for analysis.

Data Analysis

Methodology

Vehicle speed, acceleration/deceleration rates, and videos of front and rear views were used to evaluate drivers' behavior. Statistical analyses were conducted to determine the mean, the variance, and the standard deviation, as well as overall trend of vehicle speed, acceleration/deceleration rates, and risk perception. Statistical hypothesis testing was performed to evaluate potential differences in average speed, deceleration rates, and braking distance between the interchange terminals with different intersection balances. Effectiveness of traffic control devices were quantified by video coding vehicles traveling from the crossroad and then entering the entrance ramps.

The collected NDS speed data are available at 0.1-s intervals from the NDS. Past analyses (*21–27*) showed that one-second intervals were sufficient for speed data. Consequently, the data were aggregated and complete speed, acceleration/deceleration, and latitude/longitude information was determined at one-second intervals, which allow for an investigation of differences in driver speed selection by the intersection balance while controlling for other pertinent factors associated with each driving event. Thus, the collected data were utilized to investigate how driver speed and acceleration/deceleration selections vary among interchange terminals with different intersection balances; how intersection balances affect the standard deviation of speeds and acceleration/deceleration for drivers/vehicles during each event; and how driver risk perception, intersection balances, and speed and acceleration/deceleration related with each other.

To examine these questions, a cross-sectional analysis was used to compare compliant driver interactions from one study location with that of another study location to assess the safety effectiveness of the balanced intersection versus the unbalanced intersection: the higher the proportion of compliant behaviors observed, the better the safety performance of the intersection feature. Three primary metrics of interest were examined: 1) the average speed of vehicles during the time traveling at interchange terminals; 2) the variation in travel speeds for vehicles leading up to each event as quantified by the standard deviation of speed over this period; and 3) the risk perception among study participants included in the events.

Results and Discussion

Speed and Acceleration/Deceleration. First, an explanatory analysis of the data was conducted to ascertain general trends in driver speed selection through the turning maneuver from the stop line on crossroads to entrance ramps. Speed and acceleration/deceleration were profiled for each selected trip at the three study sites. Before profiling, because all acceleration/deceleration data were in the unit of "g" in the NDS database, they were converted to the standard unit "m/s^2" for consistency with speed data. Sample speed and acceleration/deceleration profiles for five trips by five drivers are provided in Figures 3 and 4. Different lengths of the profiles were due to the different lengths of travel time as the speeds were different. The starting speeds of some trips were 0, because the vehicle stopped at red lights. Visual examination of these figures indicates significantly more variability in both the speed

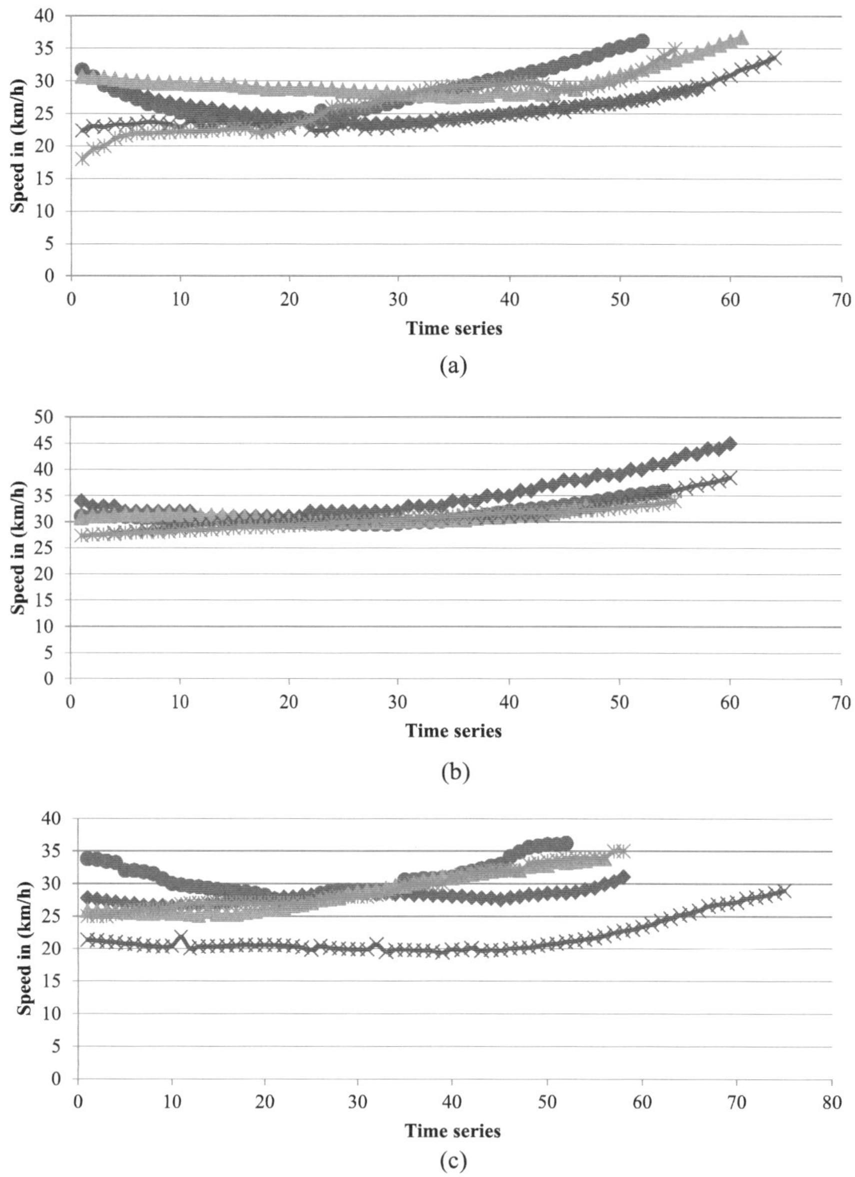

Figure 3. Sample speed profiles: (*a*) sample speed profiles at Location 1; (*b*) sample speed profiles at Location 2; and (*c*) sample speed profiles at Location 3.

and the acceleration/deceleration profiles at unbalanced intersections, that is, Location 1 (a signalized interchange terminal with the intersection balance of less than 50%) and Location 3 (a signalized interchange terminal with the intersection balance of more than 60%), compared with a balanced intersection, Location 2 (a signalized interchange terminal with the intersection balance in 50–60%). Past research (*28–32*) shows that drivers who maintain constant speeds tend to exhibit lower crash/near-crash risk. Combining the speed and acceleration/deceleration selections, which are the main indicators of drivers' comfort and confidence in driving, it shows that a balanced intersection of 50–60% accommodates traffic operation and safety.

Table 2 provides summary statistics data related to the mean speeds among the 33 unique events in the final dataset using a 95% confidence interval as the relative measure at a statistical significance level of 0.05. It provides details of the average values and the standard deviations of the mean speeds across the entire sample of

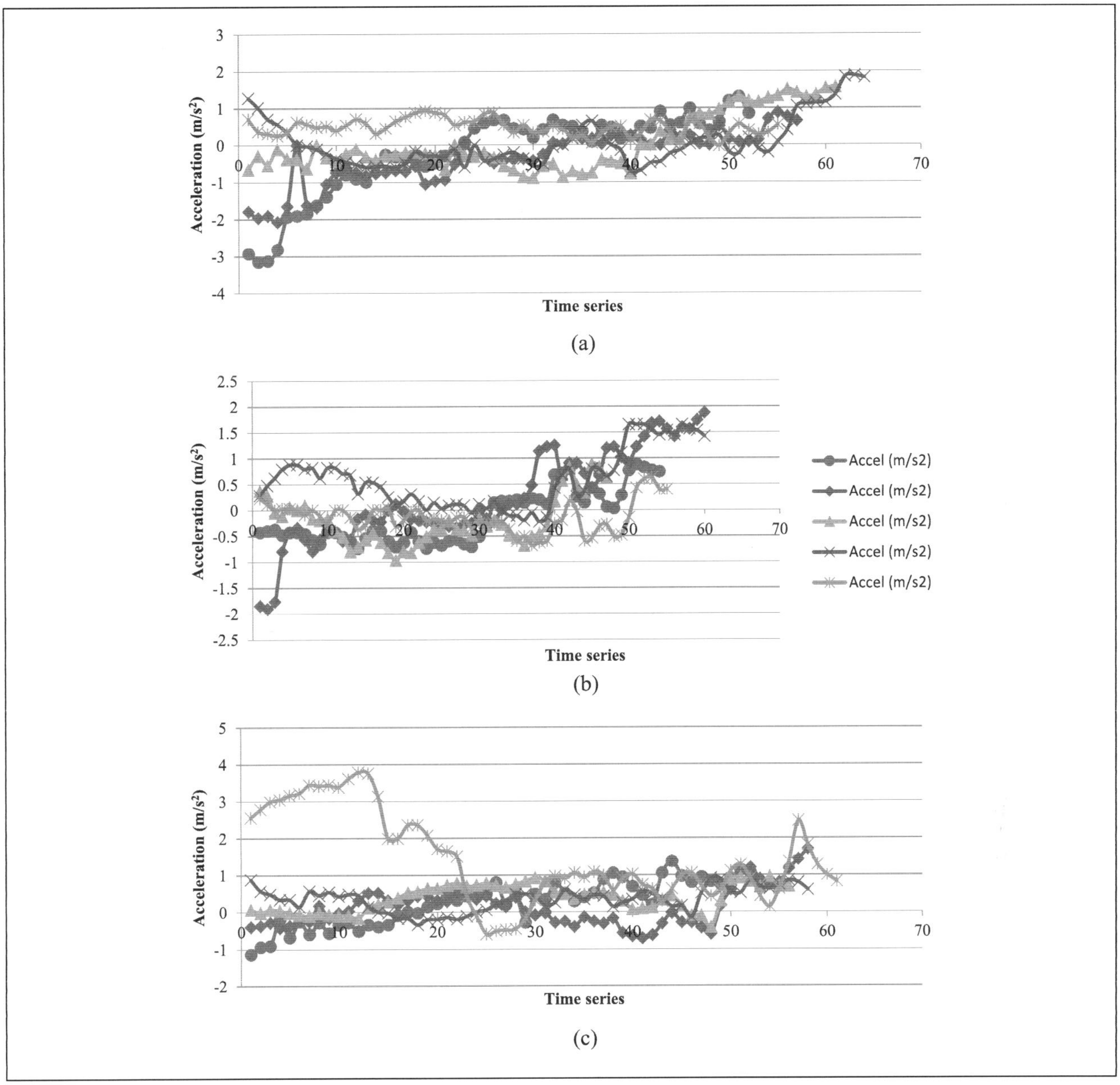

Figure 4. Sample acceleration/deceleration profiles: (*a*) sample acceleration/deceleration profiles at Location 1; (*b*) sample acceleration/deceleration profiles at Location 2; and (*c*) sample acceleration/deceleration profiles at Location 3.

events at each location. To clarify, the mean speed over the process transition from stop line on crossroad to entrance ramps of each trip was calculated.

When examining these aggregate-level data, a few points stand out. First, the mean speeds tended to be relatively consistent when the intersection balance is larger than 50% and most concentrated when it is within 50–60%, for the standard deviation is the smallest. These mean speeds are also not found to be monotonically increasing, for example, the mean speed at intersection balance of 51.43% is greater than at 73.33% of intersection balance. Secondly, it is also noted that the average and standard deviation of mean accelerations/decelerations of intersection balance of 50–60% are significantly smaller than the others, which indicates drivers were able to travel at a relatively smooth transition through the intersection to entrance ramps, hence less crash risk. Thus, again, it proves an intersection balance of 50–60% best accommodates drivers' mobility and safety.

Table 2. Summary Statistics of Mean Speed and Accelerations/Decelerations by Intersection Balance

Intersection balance	Average of mean speeds	Std. dev. of mean speeds	Average of mean accelerations/decelerations	Std. dev. of mean accelerations/decelerations
38.78%	25.621	4.553	0.184	0.413
51.43%	32.280	2.601	0.080	0.242
73.33%	33.280	8.987	0.670	0.565

Note: Std. dev. = standard deviation.

Table 3. Standard Deviations of Mean Speeds over Intersection Balance

Intersection balance	Average of mean speed std. dev. data
38.78%	3.718
51.43%	2.303
73.33%	8.968

Note: Std. dev. = standard deviation.

Table 3 provides similar summary statistics for the standard deviations in travel speeds over the duration of each left-turn trip. In contrast to the overall variability in speeds presented in Table 2, the standard deviations for individual trips are much smaller. This is reflective of the fact that driver speeds tend to be quite consistent. In general, speeds are shown to be more variable at unbalanced intersections when it is lower than 50% or larger than 60%, which is consistent with the speed profiles.

Risk Perception. To gain a better understanding as to how driver risk perceptions affect driver speed and acceleration/deceleration selections, statistical analyses were conducted for speed and acceleration/deceleration as well as the standard deviations, as already described. The driver risk perceptions were obtained from SHRP 2 NDS Risk Perception Questionnaire, a questionnaire designed to gauge the participant's perception of dangerous or unsafe driving behaviors or scenarios. The bigger the risk perception scale, the greater risk the driver is willing to take. For example, as defined in the data dictionary, a score ranging from 4 to 7 means "Moderately Greater Risk" to "Much Greater Risk". Afterwards, the relationship indicators of driver speeds and accelerations/decelerations were calculated using

$$Y = \frac{1}{m}\sum_{i=0}^{m}\frac{1}{n}\sum_{i=0}^{n} v \quad (2)$$

$$Z = \frac{1}{m}\sum_{i=0}^{m}\frac{1}{n}\sum_{i=0}^{n} a \quad (3)$$

where a is the vehicle acceleration/deceleration in the longitudinal direction versus time, v is the vehicle velocity indicated on the speedometer collected from the network, m is the number of drivers with the same risk perception, n is the number of trips with the same risk perception, and Y and Z are the relationship indicators of driver speeds and accelerations/decelerations respectively.

Figure 5 illustrates the relationships of speed vs. risk perceptions (Figure 5*a*), and accelerations/decelerations vs. risk perceptions (Figure 5*b*) based on calculations. It demonstrates that there is no significant variance between drivers' speed and acceleration/deceleration with an increase of driver risk perception. Table 4 provides detailed results. The difference in speed and acceleration among the four groups with four different risk perception scales was tested using a 95% confidence interval at a statistical significance of 0.05. It proves that no significant difference exists among the four groups. Thus, the risk perceptions of different drivers do not significantly influence the drivers' speed and accelerations/decelerations in the collected dataset, that is, there is no significant variance in speed and accelerations/decelerations due to driver risk perceptions.

Conclusion

This paper analyzed the statistical characteristics and overall trends of vehicles speed, acceleration/deceleration rates, and risk perception. It determines how intersection balances affect driver speed and acceleration, as well as the relationship between driver risk perception and speed and acceleration/deceleration. A statistically significant difference was observed in mean speed, acceleration/deceleration rates, and the corresponding standard deviations at the study interchange terminals with different intersection balances. Drivers were found to adapt their speed based upon changes in the roadway environment. Turning to the primary factor of interest, a balance intersection was found to result at smoother speeds and accelerations/decelerations, which exhibit lower crash/near-crash risks. Driver risk perceptions in the dataset have no significant influence on the variability in driver travel speed and acceleration/deceleration.

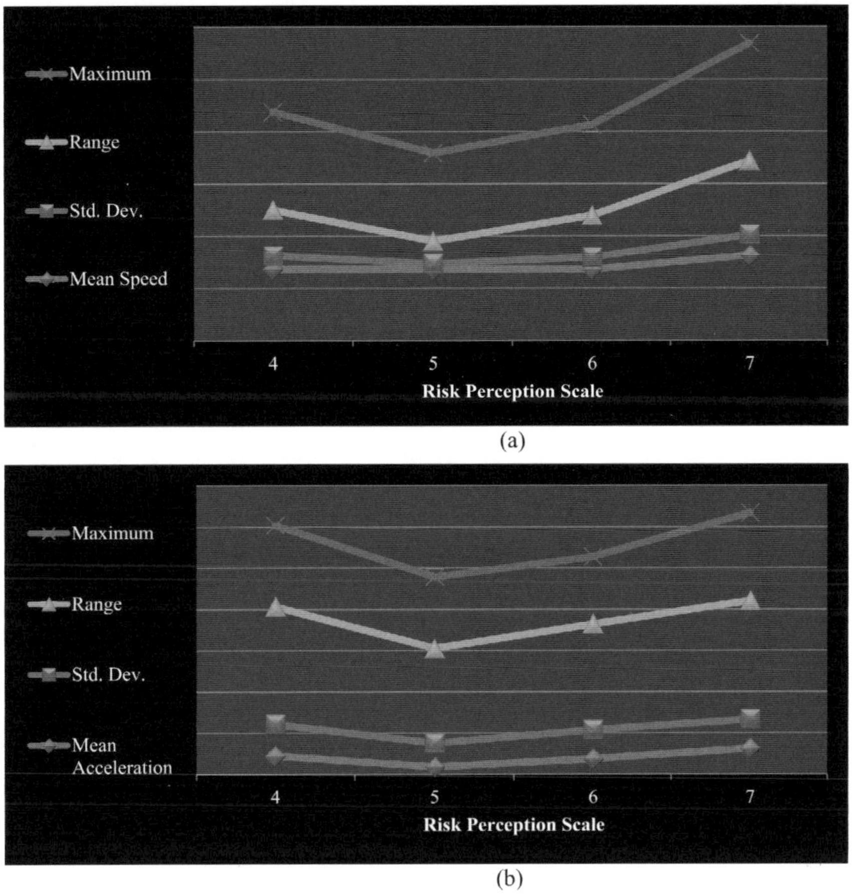

Figure 5. Relationships of (*a*) speed versus risk perception and (*b*) acceleration/deceleration versus risk perception.

Table 4. Relationships of Speed vs. Risk Perception and Acceleration/Deceleration vs. Risk Perception

	Relationship indicator							
	Speed				Acceleration/deceleration			
Risk perception scale	Mean speed	Std. dev.	Range	Maximum	Mean acceleration/ deceleration	Std. dev.	Range	Maximum
4	27.412	5.061	17.548	37.559	0.453	0.762	2.851	1.973
5	27.285	2.160	8.674	33.619	0.181	0.575	2.304	1.725
6	27.178	4. 700	16.028	34.741	0.377	0.697	2.567	1.630
7	32.593	7.876	28.192	45.073	0.626	0.703	2.874	2.133

Note: Std. dev. = standard deviation.

These findings are particularly noteworthy given the difficulty that is normally associated with relating driver perceptions with speed profile data. Most of the extant research literature has relied on traffic detector and crash data, which is often difficult to link directly to the time a maneuver happens due to time lags in these systems. These results provide compelling evidence that is based on real-time speed and acceleration/deceleration profiles.

Conclusively, at signalized interchange terminals of two-way ramps and crossroads, the stop line of left turns on cross roads should be located at no more than 60% of the way through the intersection to provide an appropriate intersection balance for better driver comfort and safety. The findings of this study are consistent with the extant literature. Guidelines on design and application guidelines of intersection balance can be developed accordingly.

Additionally, the paper demonstrated a proof-of-concept for how naturalistic driving data could be leveraged to answer important safety questions of interest as to how drivers adapt their behavior in response to intersection balances. However, it reveals some limitations of the SHRP 2 NDS data: 1) the process for acquiring NDS data can be quite long; 2) turning movements are much less recorded than through movements in the NDS database; 3) the actual number of data applicable to a specific research question may be considerably smaller than the number of trips shown available on the NDS InSight Website.

Author Contributions

The authors confirm contribution to the paper as follows: study conception and design: Jin Wang; data collection: Jin Wang; analysis and interpretation of results: Jin Wang, Huaguo Zhou; draft manuscript preparation: Jin Wang. All authors reviewed the results and approved the final version of the manuscript.

References

1. Morena, D. A., and T. J. Leix. Where These Drivers Went Wrong. *Public Roads*, Vol. 75, No. 6, 2012, pp. 33–41.
2. Tijerina, L., J. D. Chovan, J. Pierowicz, and D. L. Hendricks. *Examination of Signalized Intersection, Straight Crossing Path Crashes and 13 Potential IVHS Countermeasures*. Final Report. No. HS-808 143. National Highway Traffic Safety Administration, Washington, D.C., 1994.
3. Federal Highway Administration. *Manual on Uniform Traffic Control Devices (MUTCD)*. Washington, D.C., 2009.
4. American Association of State Highway and Transportation Officials. *A Policy on Geometric Design of Highways and Streets*, 5th ed. American Association of State Highway and Transportation Officials. ISBN. 49:1-56051, 2004.
5. Washington State Department of Transportation. *Design Manual*. Washington State Department of Transportation. M 22-01.10. 2013.
6. Wang, J., H. Zhou, and Y. Zhang. Improve Sight Distance at Signalized Exit Ramp Terminals of Partial Cloverleaf Interchanges to Reduce Wrong-Way Entries. *ASCE's Journal of Transportation Engineering, Part A: Systems, American Society of Civil Engineers*. 143(6): 04017017. DOI: 10.1061/JTEPBS.0000039. 2017
7. Wang, J. Relationship between Geometric Elements and Wrong-Way Crashes at Partial Cloverleaf Interchange Terminals. *ITE Journal, Institute of Transportation Engineers*. Vol. 88, April - No. 4. 2018.
8. Virginia Tech Transportation Institute. *SHRP 2 InSight Data Access Website – Naturalistic Driving Study (NDS)*. Virginia Tech Transportation Institute. https://insight.shrp2nds.us/. Accessed July, 2017.
9. Hallmark, S. L., N. Oneyear, S. Tyner, B. Wang, and C. Carney. *SHRP 2 Report S2-S08D-RW-1: Analysis of Naturalistic Driving Study Data: Roadway Departures on Rural Two-Lane Curves*. Transportation Research Board of the National Academies, Washington, D.C., 2015.
10. Wu, K., and P. T. Craig. Utilizing Naturalistic Driving Study Data to Study Freeway Rear-end Crash Sequences. Presented at 94th Annual Meeting of the Transportation Research Board, Washington, D.C., 2015.
11. Victor, T., J. Bärgman, C. Boda, M. Dozza, J. Engström, C. Flannagan, J. D. Lee, and G. Markkula. *SHRP 2 Report S08A: Analysis of the SHRP 2 Naturalistic Driving Study Data: Safer Glances, Driver Inattention, and Crash Risk*. Transportation Research Board of the National Academies, Washington, D.C., 2014.
12. Hallmark, S. L., D. McGehee, K. M. Bauer, J. M. Hutton, G. A. Davis, J. Hourdos, I. Chatterjee, T. Victor, J. Bärgman, M. Dozza, H. Rootzén, J. D. Lee, C. Ahlström, O. Bagdadi, J. Engström, D. Zholud, and M. Ljung-Aust. *SHRP 2 Safety Project S08: Initial Analyses from the SHRP 2 Naturalistic Driving Study Addressing Driver Performance and Behavior in Traffic Safety*. Transportation Research Board of the National Academies, Washington, D.C., 2013.
13. Hutton, J. M., K. M. Bauer, C. A. Fees, and A. Smiley. *SHRP 2 Final Report S08B: Analysis of the SHRP 2 Naturalistic Driving Study Data: Evaluation of Offset Left-Turn Lanes*. Transportation Research Board of the National Academies, Washington, D.C., 2014.
14. Lin, P. S., Z. Wang, R. Guo, and A. Kourtellis. A Pilot Study on Interactions between Drivers and Pedestrian Features at Signalized Intersections-Using the SHRP2 Naturalistic Driving Study Data. In *Bridging the East and West*, 2016, pp. 70–77.
15. Carney, C., D. McGehee, K. Harland, M. Weiss, and M. Raby. *Using Naturalistic Driving Data to Assess Vehicle-to-Vehicle Crashes Involving Fleet Drivers*. AAA Foundation for Traffic Safety, Washington, D.C., 2015.
16. Dingus, T. A., F. Guo, S. Lee, J. F. Antin, M. Perez, M. Buchanan-King, and J. Hankey. Driver Crash Risk Factors and Prevalence Evaluation Using Naturalistic Driving Data. *Proceedings of the National Academy of Sciences*, Vol. 113, No. 10, 2016, pp. 2636–2641.
17. Simmons, S. M., A. Hicks, and J. K. Caird. Safety-Critical Event Risk Associated With Cell Phone Tasks As Measured In Naturalistic Driving Studies: A Systematic Review And Meta-Analysis. *Accident Analysis & Prevention*, Vol. 87, 2016, pp. 161–169.
18. Harland, K. K., C. Carney, and D. McGehee. Analysis of Naturalistic Driving Videos of Fleet Services Drivers to Estimate Driver Error and Potentially Distracting Behaviors As Risk Factors For Rear-End Versus Angle Crashes. *Traffic Injury Prevention*, DOI: 10.1080/15389588.2015.1118655. 2016.
19. Kidd, D. G., and A. T. McCartt. The Relevance of Crash Type and Severity When Estimating Crash Risk Using the SHRP2 Naturalistic Driving Data. *Proc., 4th International Conference on Driver Distraction and Inattention*, Sydney, New South Wales, Australia, 2015.

20. The Center for Transportation Research and Education at Iowa State University. *SHRP 2 Roadway Information System (RID)*. http://www.ctre.iastate.edu/shrp2-rid/index.cfm. Accessed July, 2017.
21. Oneyear, N., S. Hallmark, and B. Wang. Analyzing Stopping Behavior at Rural Intersections Using SHRP2 Naturalistic Driving Data. *Proc., 17th International Conference Road Safety on Five Continents (RS5C 2016)*, Rio de Janeiro, Brazil, Statens väg-och transportforskningsinstitut, 2016.
22. Machiani, S. G., A. Medina, R. Gibbons, and B. Williams. Driver Behavior Modeling on Horizontal Curves for Two-Lane Rural Roads Using Naturalistic Driving Data. Presented at 95th Annual Meeting of the Transportation Research Board, Washington, D.C., 2016.
23. Bao, S., H. Xiong, M. L. Buonarosa, and J. R. Sayer. Using Naturalistic Driving Data to Examine Drivers' Seatbelt Use Behavior: Comparison between Teens and Adults. *Journal of Safety Research*, Vol. 54, 2015, pp. 69–73.
24. Johnson, T., R. Chen, R. Sherony, and H. C. Gabler. Investigation of Driver Lane Keeping Behavior in Normal Driving based on Naturalistic Driving Study Data. *SAE International Journal of Transportation Safety*, Vol. 4, No. 2, 2016, pp. 236–244.
25. Guo, F., Y. Fang, and J. F. Antin. 2015. *Evaluation of Older Driver Fitness-to-Drive Metrics and Driving Risk Using Naturalistic Driving Study Data*. NSTSCE; 15-UM-036. National Surface Transportation Safety Center for Excellence, 2015.
26. Oneyear, N. L., S. L. Hallmark, and B. Wang. *SHRP 2 Report: Evaluating the Relationship between the Driver and Roadway to Address Rural Intersection Safety using the SHRP 2 Naturalistic Driving Study Data*. Transportation Research Board of the National Academies, Washington, D.C., 2016.
27. Davis, A. J., E. Hacker, P. T. Savolainen, and T. J. Gates. A Longitudinal Analysis of Rural Interstate Fatalities in Relation to Speed Limit Policies. *Transportation Research Record: Journal of the Transportation Research Board*, 2015. 2514: 21–31.
28. Fitzpatrick, K., P. Carlson, M. A. Brewer, M. D. Wooldridge, and S. P. Miaou. *NCHRP Report 504: Design Speed, Operating Speed, and Posted Speed Practices*. Transportation Research Board of the National Academies, Washington, D.C., 2003.
29. Garber, N. J., and A. A. Ehrhart. Effect of Speed, Flow, and Geometric Characteristics on Crash Frequency for Two-Lane Highways. *Transportation Research Record: Journal of the Transportation Research Board*, 2000. 1717: 76–83.
30. Aljanahi, A. A. M., A. H. Rhodes, and A. V. Metcalfe. Speed, Speed Limits and Road Traffic Accidents under Free Flow. *Accident Analysis & Prevention*, Vol. 31, No. 1–2, 1999, pp. 161–168.
31. Garber, N. J., and R. Gadiraju. *Speed Variance and Its Influence on Accidents*. University of Virginia, Charlottesville, 1988, p. 56.
32. Aarts, L., and I. Van Schagen Driving Speed and the Risk of Road Crashes: A Review. *Accident Analysis & Prevention*, Vol. 38, No. 2, 2006, pp. 215–224.

The Standing Committee on Safety Data, Analysis and Evaluation (ANB20) peer-reviewed this paper (18-01012).

TRR
JOURNAL OF THE TRANSPORTATION RESEARCH BOARD

Article

Transportation Research Record
2018, Vol. 2672(38) 266–276
© National Academy of Sciences:
Transportation Research Board 2018
Article reuse guidelines:
sagepub.com/journals-permissions
DOI: 10.1177/0361198118786829
journals.sagepub.com/home/trr
SAGE

A Novel Technique to Identify Hot Zones for Active Commuters' Crashes

Ahmed Osama[1,3], Tarek Sayed[1], and Emanuele Sacchi[2]

Abstract

This paper presents an approach to identify and rank accident-prone (hot) zones for active transportation modes. The approach aims to extend the well-known empirical Bayes (EB) potential for safety improvement (PSI) method to cases where multiple crash modes are modeled jointly (multivariate modeling). In this study, crash modeling was pursued with a multivariate model, incorporating spatial effects, using the full Bayes (FB) technique. Cyclist and pedestrian crash data for the City of Vancouver (British Columbia, Canada) were analyzed for 134 traffic analysis zones (TAZs) to detect active transportation hot zones. The hot zones identification (HZID) process was based on the estimation of the Mahalanobis distance, which can be considered an extension to the PSI method in the context of multivariate analysis. In addition, a negative binomial model was developed for cyclist and pedestrian crashes, where the EB PSI for each mode crash was quantified. The cyclist and pedestrian PSIs were combined to detect active transportation hot zones. Overall, the Mahalanobis distance method is found to outperform the PSI method in terms of consistency of results; and discrepancy is observed between the hot zones identified using both approaches.

Studies have shown that active modes of transportation such as walking and cycling have considerable social, health, and environmental benefits. They can improve livability, increase activity levels, and reduce congestion as well as the related emissions and crashes. However, active commuters are vulnerable to an elevated risk of severe injury (*1*). With the increasing trend of adopting policies to encourage active transportation in many cities around the world, improving the safety of active commuters is becoming a priority. The identification and ranking of hot zones is an important component in any road safety improvement program, especially when highway agencies can afford to examine and improve only a limited number of sites. Highway agencies strive to narrow down the selected hotspots and maintain only the ones that are likely to generate the greatest return on safety investments. To achieve this objective, considerable research has focused on the development of effective hotspot identification and ranking criteria. Generally, safety literature has focused on methods where identification and ranking are based on a single crash outcome; usually the total number of crashes, for example (*2*, *3*). However, there is a growing interest in accounting for possible correlation among different crash outcomes (i.e., different crash types or severity). For instance, evaluating the safety of active transportation modes requires accounting for both cyclist and pedestrian crashes at the same time. Therefore, the joint modeling of different crash modes as well as the formulation of a method that is able to detect hot zones due to the joint effect of those crashes is a topic that requires further investigation.

This study presents a technique to identify and rank hazardous locations for active transportation modes. The research aims at extending the empirical Bayes (EB) potential for safety improvement (PSI) method to cases where multiple crash severity/types are modeled jointly (multivariate modeling). The procedure proposed in this study is based on the approach adopted in Sacchi et al. (*4*) and involves the following steps: (a) applying multivariate Poisson-lognormal collision prediction models (CPMs) to the data using the full Bayes (FB) method, (b) using the estimates of the Poisson posterior means for each site to compute the Mahalanobis distance from what is the normal Poisson mean for similar sites, and (c) preparing an ordered list of potentially hazardous sites.

The Mahalanobis distance was previously used to account for outliers in the development of safety

[1]Department of Civil Engineering, The University of British Columbia, Vancouver, BC, Canada
[2]Department of Civil, Geological & Environmental Engineering, University of Saskatchewan, Saskatoon, SK, Canada
[3]Department of Civil Engineering, Ain Shams University, Cairo, Egypt

Corresponding Author:
Address correspondence to Ahmed Osama: ahmed.osama@eng.asu.edu.eg; ahmed.osama@alumni.ubc.ca

performance functions (*5*). It was then utilized by Sacchi et al. (*4*) to investigate accident-prone intersections using severity and property-damage-only crash data modeled jointly. Sacchi et al. (*4*) investigated the within-method consistency, and its consistency against other methods such as the EB PSI and the posterior mean of collision frequency methods. Their work demonstrated that the Mahalanobis distance method outperformed other hotspot identification methods and was capable of efficiently extending the univariate concept of PSI in a multivariate setting.

Overall, the objective of this study is to provide traffic safety professionals and transportation planners with a novel technique to identify and prioritize active transportation hot zones for safety improvements. The study made use of a data set based on 134 traffic analysis zones (TAZs) in the City of Vancouver (British Columbia, Canada). Data related to cycling and walking crashes and traffic exposure measures were collected to identify and rank the active transportation hot zones within the city. Multivariate Poisson-lognormal CPMs incorporating spatial effects were developed at the traffic zone level using the FB approach. Moreover, the EB method was used to develop univariate CPMs for the cyclist and pedestrian crashes. Active transportation hot zones were detected using both the FB Mahalanobis distance method and the state-of-the-practice EB PSI method. Lastly, the consistency of both methods in hot zone identification (HZID) was investigated, and this paper concludes with recommendations to make the EB method more consistent with the multivariate FB method.

Literature Review

Hotspot Identification

Several previous studies have shown that hotspot ranking techniques based on Bayesian methodologies are superior to those that simply rely on the observed collision count, for example (*3*). In particular, the EB and FB methods assume that a probability distribution can be found before any data becomes available (prior distribution). Once information becomes available, the prior distribution is converted into a posterior distribution through the Bayes theorem. The EB approach is used to refine the estimate of the expected collision frequency of a location (θ_i) by combining the observed number of collisions (Y_i) with the predicted number of collisions (μ_i) from a safety performance function (SPF) of the same type of road facility (*6*). Alternatively, the FB approach integrates the estimation of the SPF with the posterior distribution estimate of the expected collision frequency (θ_i) through a single step (*7–9*). The EB method typically deals with the entire study period as a single data point (either total or calculated as per year) and does not account for the spatio-temporal and multivariate model structures. The FB approach is more appealing since it treats each time period as an individual data point, allows inference at more than one level for hierarchical models, and allows multivariate and spatial/temporal analysis, which is the focus of this research.

As mentioned in the introduction, a good indicator of the expected safety benefits following a treatment is the PSI or accident reduction potential (*10*). The PSI indicator is able to identify what locations are likely to generate the greatest return on investments. Using this indicator, hotspots are ranked according to their PSI value, which is estimated as the difference between the observed collision frequency θ_i and the normal collision frequency for similar sites (μ_i). The use of PSI to rank sites for required safety improvements is commonly adopted in the literature. This measure is considered among those that produce the most reliable performances for network screening, according to the American Highway Safety Manual, and it is currently known as "excess expected average crash frequency with EB adjustment" (*2*). The EB PSI method continues to be the state-of-the-practice method for identification and ranking, and has been applied in many recent studies (e.g. *11*, *12*).

However, this approach relies on univariate modeling, which may not lead to accurate estimations as collisions of different severity or types might be highly correlated. This correlation would be ignored when separate rankings by collision severity/type are made for each outcome level. Multivariate extensions of the simple Poisson distribution, such as the multivariate Poisson log-normal (MVPLN) model (*8*, *13*), have been proposed in the literature to account for this correlation. Consequently, some hotspot identification techniques that were developed for univariate analyses have been adapted for the multivariate setting. For instance, Aguero-Valverde and Jovanis (*13*) proposed to rank hotspots by excess crash cost using the expected collision frequency as estimated by the MVPLN models. A comparison of this method with the excess crash cost estimated by the independent models indicated significant differences between the multivariate and univariate approaches. El-Basyouny and Sayed (*8*) extended the probability-of-excess method, which is able to identify locations with a very high probability of exceeding the long-term mean of all sites, to the multivariate framework. Moreover, El-Basyouny and Sayed (*9*) proposed a multivariate ranking of hotspots based on statistical depth functions, which are tools for nonparametric multivariate analysis and provide center-out ordering of multivariate data.

Contribution to the Literature

Recently, there has been research that aims at detecting hot zones for active transportation modes on the macro-

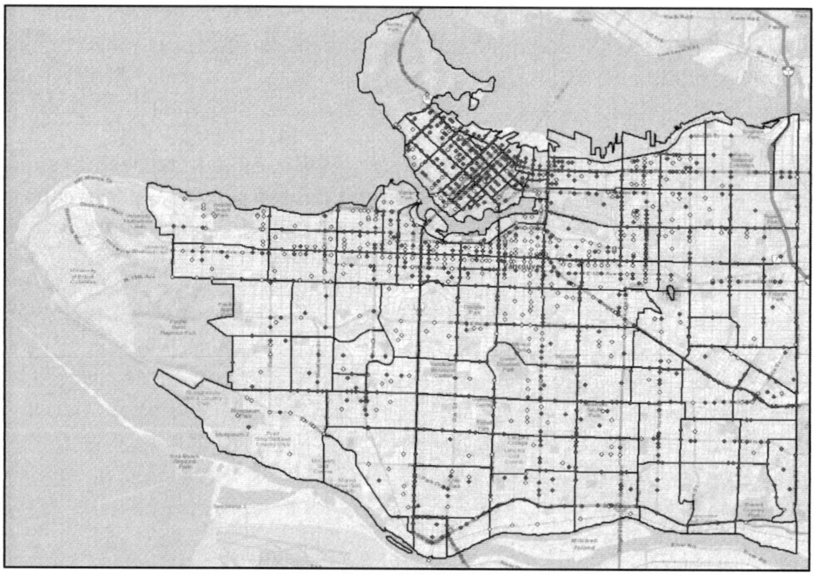

Figure 1. Cyclist collisions (yellow points) and pedestrian collisions (red points).

level. However, these efforts focused on either analyzing the total active transportation crashes without considering possible correlation between crash outcomes (*14*), or analyzing the active modes' crashes taking in consideration the correlation between them, but the hot zones for each active mode are identified independently (*15*, *16*).

This study contributes to the existing literature in several aspects:

1. A new methodology for hotspot identification (HZID), based on the concept of Mahalanobis distance, is proposed for macro-level safety analysis.
2. The proposed identification method incorporates spatial and mode correlations while detecting the active transportation hot zones.
3. A mixed modeling approach is used to allow more flexibility for the multivariate CPM when dealing with the exposure and explanatory variables.

Data Collection

Data Sources

In this study, zone-level CPMs were developed based on the City of Vancouver's 134 TAZs. The data needed for the variables were compiled using ArcGIS software for processing and visual representation, after being extracted from three main sources.

The first source was the Insurance Corporation of British Columbia (ICBC), a public automobile insurance company in Canada, which provided the crash data for a five-year period (2009–2013). Pedestrian-motorist and cyclist-motorist crashes are included in the analysis, as shown in Figure 1 (one point may represent more than one crash of the same type). It should be noted that a main concern about claim crash data would be how precise it is in locating the crash incident. Nevertheless, claim crash data is accurate enough when used in urban roads, as the location can be determined easily and precisely (due to the presence of many witnesses, road names, cameras, etc.). On the other hand, the police reported data would be more accurate in the context of rural highways since claimed crash data does not provide a precise location for such crashes.

The second source was Translink, which provided the output of the Emme2 transportation planning model for travel demand in Metro Vancouver in the year 2011.

Lastly, Acuere Analytics provided the Vancouver Cycling Data Model (VCDM 2011). The VCDM used the bicycle count from the years 2005 to 2011 to estimate the annual average daily bicycle traffic (AADB) over the City of Vancouver bicycle network (*17*). The model was efficient in estimating the AADB on most of the bicycle network links (3180 links, or more than 70% of the network).

Data Variables

Table 1 provides the definition and descriptive statistics of the variables related to crashes and traffic exposure.

Table 1. Definition of Variables and Data Summary (n = 134 TAZs)

Variable	Description	Mean	StdDev	Min	Max
Crashes					
CColl	Cyclist-motorist crashes over 5 years	12.71	13.48	0	78
PColl	Pedestrian-motorist crashes over 5 years	15.45	11.45	0	54
Exposure					
VKT	Vehicle kilometers travelled	4290.43	3315.10	189.46	22,288.79
BKT	Bicycle kilometers travelled	1047.78	2102.07	0	21,462.77
W	Walk trips	3971.64	2677.49	247.11	13,906.56

Crashes were aggregated for each TAZ according to their geo-spatial locations. Boundary crashes were found to be 28% of the total pedestrian crashes and 40% of the total cyclist crashes. The way of those crashes are distributed can have a significant impact on the analysis results due to their high number as they would affect the proportion of cyclist/pedestrian crashes at each zone. Boundary crashes in this study were distributed between the adjacent TAZs according to the relative proportion of bicycle kilometers travelled (BKT) or walk trips (W) at these zones, respectively, depending on whether the crash was a cyclist or pedestrian one. This way of distributing crashes was selected due to the direct association between traffic exposure and crash risk, as discussed in previous research (*18, 19*).

Three exposure measures were used in this study: vehicle kilometers travelled (VKT), BKT, and W. As for the VKT and W variables, they were readily provided by the Emme2 model and were estimated on the TAZ level. A link based method is then used to calculate the zonal BKT. Road segments were represented by links, and the trip count at each link was multiplied by the link length to obtain link BKT. The resulting value was aggregated to obtain the total BKT at each TAZ.

Methodology

This section discusses the methodologies of model specification and HZID using both the univariate EB approach and the multivariate FB approach.

Model Specification

EB Univariate Models. The generalized linear modeling (GLM) approach, which assumes a non-normal distribution error structure, has been widely used for the development of CPMs (*20, 21*). A negative binomial error distribution assumption has become the standard for CPMs developed using the GLM approach (*6, 21*). Based on empirical studies (*20, 21*), a commonly used model form includes an exposure measure (e.g., VKT) raised to some power and multiplied by an exponential function including the other non-exposure explanatory variables. The model can be expressed mathematically as shown in Equation 1.

$$E(Y) = a_0 EXP_1^{a1} EXP_2^{a2} exp(\sum_j b_j x_j) \quad (1)$$

where E(Y) is the predicted collision frequency, EXP_1 and EXP_2 are the measures of traffic volume such as VKT and BKT for cyclist CPMs or VKT and W for pedestrian CPMs, x_j represents any other explanatory variables, and a_0, a_1, a_2, and b_j are the model parameters. The recommended statistical methodology to add explanatory variables into a CPM is a forward stepwise procedure (*21*). Variables are added one by one, and their significance is tested. Variables representing exposure must be included first. Two statistical measures are used to assess the goodness of fit of the GLM models, including Pearson chi square (χ^2) and scaled deviance (SD) statistics (*21*). For a well-fitted model and a relatively large number of observations, the expected value of Pearson χ^2 and SD will be approximately equal to the number of degrees of freedom (df) (*21*).

Full Bayesian Multivariate Model. FB hierarchical modeling is quite flexible and can explicitly use prior information, which results in improved parameter estimates. Incorporating random effects in CPMs was found more suitable through full Bayesian analysis due to its ability to implement complex correlation structures (*22*). Full Bayesian analysis can also provide exact measures of uncertainty on the posterior distributions of the parameters' estimates (*8*).

Recently, there have been increasing efforts to utilize the correlation between adjacent sites as well as the correlation among crash types in order to mitigate some problems (e.g., regression to the mean and low sample mean) resulting from unobserved heterogeneity (*23*). Poisson-lognormal multivariate models incorporating random effects were used in this research. This modeling framework was chosen as it is able to handle overdispersion in count data and account for unobserved heterogeneity, spatial correlation, and crash mode correlation. In order to undergo the multivariate modeling of the pedestrian and cyclist crashes, there are two approaches: the standard approach and the mixed approach (*24*).

The standard approach is to include the same set of covariates for the two modeled collision types. This can be achieved by either including all the possible explanatory variables that would impact the pedestrian or cyclist collisions; or by including only the explanatory variables that are expected to impact both the pedestrian and cyclist collisions. Another approach for multivariate modeling was proposed, which allowed the inclusion of a different set of covariates for each modeled crash type (*25*). This approach, the mixed approach, is already widely used for fitting mixed effects, and is also a valuable tool for multivariate analysis. Capabilities of the mixed approach, which are lacking in the standard multivariate procedures, include, but are not limited to, the ability to use observations that have incomplete responses; and the ability to handle non-standard (e.g., multiple designs) multivariate models (*24*).

Y_i is assumed to be the number of pedestrian/cyclist crashes at zones i, and Y is assumed to follow a Poisson distribution with parameter λ which is itself considered a random variable.

$$Y_i^k \sim Poisson\left(\lambda_i^k\right) \tag{2}$$

$$ln\left(\lambda_i{}^k\right) = a_0{}^k + a_1{}^k ln\left(EXP_{1i}\right) + a_2{}^k ln\left(EXP_{2i}{}^k\right) + \Sigma_n \beta_n{}^k x_{ni}{}^k + u_i^k + s_i^k \tag{3}$$

For k as the mode type, $a_0{}^k$ is the intercept value, $a_1{}^k$, $a_2{}^k$, and $\beta_n{}^k$ are model parameters, EXP_{1i} with EXP_{2i}^k are the traffic exposure variables (VKT with BKT in the case of cyclist collision model and VKT with W in the case of pedestrian collision model), x_{ni}^k represents explanatory variables. The subscript "K" is added to the exposure and explanatory variables to indicate that the independent variables can vary from one mode to another (i.e., one or more variables can be missing in the model of one mode but present in another). Therefore, using this model structure, the advantage is the ability to change the covariates included in the CPM for each crash type. It should be noted that there are no explanatory variables in this study. The objective of this paper is just to demonstrate the approach; we put the explanatory parameters indications for future research. The overdispersion caused by the unobserved or unmeasured heterogeneity is accounted for using u_i^k which is related to zone-specific attributes. It represents unstructured random effects and follows a multivariate normal distribution. s_i^k is the spatial component that represents structured correlated random effects suggesting that zones that are close to each other should have been correlated. The spatial effects can be accounted for by multivariate Gaussian CAR techniques. Spatial effects were included in the model because several univariate CMs were done using this study data, and spatial effects were found to improve the model specification as their proportion of total variance was significantly higher than 0.5, as for example (*18*, *19*).

The procedure of defining the multivariate normal and multivariate CAR error structures is similar to that presented in Islam et al. (*26*) and is shown in Equations 4 and 5. For the multivariate k-dimensional (only two modes in this model) normal error, the diagonal elements of the variance–covariance matrix $\sum$ represent the variances, and the off-diagonal elements represent the covariances. For model estimation, the following prior is used: $\sum^{-1} \sim Wishart\ (I, K)$, where I is the $K \times K$ identity matrix.

$$u_i^k \sim Normal\left(0, \sum\right), \sum = \begin{pmatrix} \sigma 11 & \sigma 12 \\ \sigma 21 & \sigma 22 \end{pmatrix} \tag{4}$$

For the multivariate CAR model, the vector of the spatially correlated i neighborhood is

$$\left(S^1{}_i, S^2{}_i\right) | \left(S^{-1}{}_i, S^{-2}{}_i\right) \sim Normal\left(\bar{s}_{ki}, \frac{\omega}{n_i}\right), \omega = \begin{pmatrix} \sigma s11 & \sigma s12 \\ \sigma s21 & \sigma s22 \end{pmatrix} \tag{5}$$

where, $(S^{-1}{}_i, S^{-2}{}_i)$ denotes the neighborhoods of the $k \times n$ matrix s_i^k, excluding the i-th neighborhood; n_i is the number of neighborhoods adjacent to a neighborhood, ω is the variance–covariance matrix for spatial correlation. The diagonal elements of the covariance matrix ω represent spatial variance. The off-diagonal elements represent the spatial covariance of different severity levels. For model estimation, it is assumed that $\omega^{-1} \sim Wishart\ (I, K)$, where I is the $K \times K$ identity matrix.

The Markov chain Monte Carlo (MCMC) method was implemented using WinBUGS tool to sample the posterior distribution as well as to estimate the parameters. A burn-in sample is used for the purpose of monitoring convergence and then excluded. Parameter estimation, performance evaluation, and inference are obtained by the following iterations. Two chains are used to run the model in WinBUGS, and 20,000 MCMC iterations are discarded as burn-in samples. Afterward, 20,000 iterations were run for each chain. The summary statistics of each chain are then estimated from WinBUGS and the convergences of the developed models are thoroughly checked to ensure that the posterior distribution has been found in order to begin parameter sampling. Convergence can be checked in several ways, as shown by El-Basyouny and Sayed (*8*). In addition to convergence, the significance of the parameter estimates is tested at the 95% confidence level using the confidence intervals.

Hot Zones

There are several ways to detect and rank the hot zones (i.e., collision prone zones). The HZID methodology depends mainly on the technique used for building the macro-level CPM. The two approaches discussed in this study are the EB PSI method and the FB Mahalanobis distance method, as follows:

EB PSI Identification/Ranking Method. The first step in this approach is to calculate the EB safety estimate according to Equation 6.

$$\theta_i = \frac{\mu_i(k + count)}{k + \mu_i} \tag{6}$$

μ_i: predicted collision frequency from the developed CPM.

k: over dispersion parameter of the developed CPM.

θ_i: EB expected safety estimate.

Afterwards, the zones are ranked to ensure that the zones in most need of treatment are given priority. Ranking criteria typically applies the PSI measure, which is based on the difference between the expected and the predicted collision frequency at each site, as shown in Equation 7. The zones with higher PSI are more collision prone.

$$PSI_i = \theta_i - \mu_i \tag{7}$$

The PSIs are calculated for both cyclist and pedestrian crashes. In order to obtain a hot zone ranking for active transportation, the PSIs from both modes need to be combined. The cyclist and pedestrian crashes in Vancouver, British Columbia, are found to have approximately similar costs (*27, 28*). This agrees with a study by Miller at al. (*29*) that comprehensively compared the cyclist and pedestrian crash costs. Therefore, for a univariate method such as EB PSI, it is plausible to add the PSIs due to cyclist and pedestrian crashes to come up with an active transportation hot zone ranking, as shown in Equation 8. The resulting measure "$PSI_{Active\ Transportation}$" is then used to rank the active transportation hot zones.

$$PSI_{ActiveTransportation} = PSI_{Pedestrian} + PSI_{cyclist} \tag{8}$$

FB Mahalanobis Distance Identification/Ranking Method. For univariate data, a road site with an expected collision frequency (θ_i) higher than its long-term collision mean at similar road facilities (μ_i) will also have a higher PSI. Geometrically, the PSI value can be seen as the standard Euclidean distance between these two points on the measuring axis. This concept can be extended to multivariate data by taking into account the likely correlation among collision counts at different (severity/type) levels (*K*).

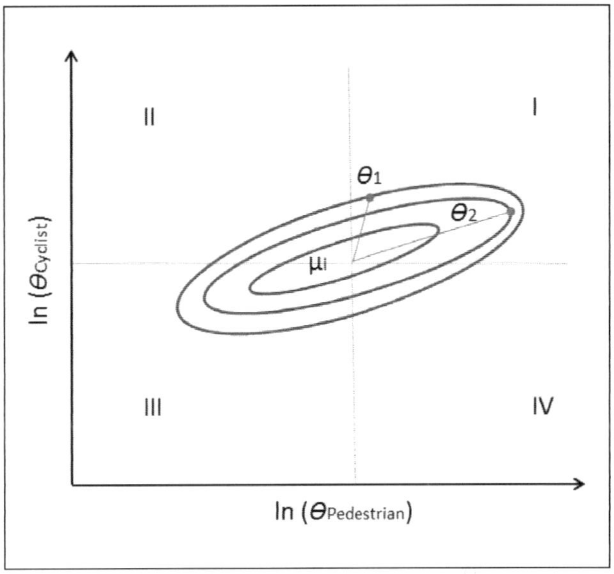

Figure 2. Multivariate distribution of the pedestrian and cyclist collisions.

A multivariate SPF, which takes into account the covariance of different crash modes, for example, cyclist and pedestrian crashes, leads the probability density function to elliptic boundaries (see illustration in Figure 2). The probability density is high for ellipses near the multivariate center (μ_i) and low for ellipses further away. For instance, the Euclidean distance from the multivariate center of the distribution in Figure 2 to the expected collision frequency of site 1 (θ_1) is shorter than the distance to site 2 (θ_2). If μ_i is assumed to be the long-term average collision frequency for sites similar to site 1 and 2, then by applying the PSI concept we would assume that site 2 is more collision prone than site 1. However, for this distribution, the variance for cyclist collisions (y-axis) is less than the variance for pedestrian collisions (x-axis), therefore θ_2 would be "closer" to μ_i than θ_1 in the sense that it is more likely to observe a site with collision frequency equal to θ_2 than one equal to θ_1 (the contour with higher probability that contains θ_2 is nested within the contour that contains θ_1 with lower probability density).

To address the distance between θ_i and the distribution of collision frequency at similar sites, in a multivariate context, another measure is required. The measure, known as the Mahalanobis distance, provides a way to account for the variance of each variable and the covariance between variables (*30*). The Mahalanobis distance, for multivariate normal models, is given by Equation 9.

$$di^2 = (\theta_i{}^* - \mu_i{}^*)\Sigma^{-1}(\theta_i{}^* - \mu_i{}^*) \tag{9}$$

where $\theta_i^* = (ln(\theta_i^1), ln(\theta_i^2), \ldots, ln(\theta_i^K))$ and $\mu_i^* = (ln(\mu_i^1), ln(\mu_i^2), \ldots, ln(\mu_i^K))$

It should be noted that if Σ is the identity matrix, then the Mahalanobis distance reduces to the standard Euclidean distance between θ_i^* and μ_i^*.

The standard method for identifying collision hotspots is to rank them in descending order of the performance metric considered for the analysis (e.g., PSI for univariate data) and the threshold is typically determined according to the amount of funding available for safety improvements. Usually the development threshold is limited to the top 10% of sites (*3, 31*). In this research, locations are ranked in the descending order of their Mahalanobis distances and the top 10% hot zones are investigated. It is worth mentioning that only the zones with both cyclist and pedestrian collisions exceeding the corresponding coordinates of μ_i are considered active transportation hot zones (i.e., zones in quadrant 1). In fact, as shown in Figure 2, the cyclist/pedestrian plane can be divided into four quadrants denoted I, II, III, and IV. The zones in quadrant I represent the zones with both cyclist and pedestrian collisions exceeding both coordinates of μ_i. The zones in quadrant II represent the zones with only cyclist collisions exceeding the cyclist coordinate of μ_i and so on for the remaining quadrants.

Consistency Tests

Two methods of testing HZID consistency are applied in this paper: a within-method consistency test and a between-methods consistency test. The within-method consistency test involves determining the false identifications in each method (*31*). A critical concern of this test is to identify the "truly" hazardous and safe locations, which requires that the truth to be known as a priori. To satisfy this requirement, as has been done in some previous studies (e.g., *31, 32*), the mean of the five years of crash data is used as the true Poisson mean (TPM) of the TAZ. As such, TAZs with higher TPMs (highest 10%) are assumed to be the truly hazardous TAZs in this simulation. The false identification percentage is then calculated according to Equation 10.

$$False\ Identification = \frac{Sum\ of\ False\ Positive\ and\ False\ Negative\ Idintified\ TAZs}{Total\ Number\ of\ TAZs} \quad (10)$$

The between-methods consistency test (Tc), as proposed in the literature (*31*), involves comparing and evaluating two ranked lists of the same dataset from different HZID methods using the evaluation criteria shown in Equation 11.

$$Tc = \{HZ_1, HZ_2, \ldots, HZ_{n\delta}\}_{i,j1} \cap \{HZ_1, HZ_2, \ldots, HZ_{n\delta}\}_{i,j2} \quad (11)$$

where HZ_i is the i-th ranked site identified as a hot zone, n is the total number of sites in the dataset, δ is the threshold of high-risk sites identified (e.g., $\delta = 0.1$ corresponds to the top 10% of zones which are identified as hot zones), and $j1$ and $j2$ represent the ranking method(s) being compared.

The test is designed to assess the homogeneity of the HZID among the different methods applied. The greater the number of hot zones that are similarly identified in both methods, the more consistent the HZID methods are with each other.

Results

Developed Multivariate and Univariate Models

Table 2 shows the developed multivariate and univariate CPMs respectively. These CPMs are used afterwards to identify the active transportation hot zones using the Mahalanobis and PSI methods. The association of the traffic exposure variables (VKT, BKT, and W) with the cyclist and pedestrian crashes is non-linearly positive in both models. The exponents of traffic exposure parameters are less than one, which strengthens the safety in numbers hypothesis (*33*). The parameters in both models are statistically significant at the 5% level or higher. There is no considerable difference between the parameter estimates of the EB and FB CPMs. The correlation (due to the spatial error term) between the pedestrian and cyclist crashes is found to be significant (Corr_Spatial = 0.73), affirming the aforementioned importance of the multivariate CPMs, incorporating spatial effects, for modeling pedestrian and cyclist crashes. The significant correlation between cyclist and pedestrian crashes is plausibly due to the similar behaviors of pedestrians and cyclists, similar vulnerability to the surrounding driving environment, and similar characteristics of the bicycle and pedestrian networks' indicators.

The variance–covariance matrix (SD in parenthesis) for the multivariate model is shown in Table 3.

Hot Zone Ranking Results

Figure 3 illustrates the ranking lists of the top 10% hot zones according to the Mahalanobis distance and PSI methods, respectively. It is obvious that although there are some common zones among the two lists, significant difference in the ranking results is still observed. This agrees with several previous studies (e.g., *13*) which showed that a multivariate approach can yield considerably different HZID and ranking results from a univariate approach.

Table 2. Results of Models

	EB univariate models					
	Model coll=	*K*	*df*	*SD*	X^2	*P*-value
Cyclist collisions	$0.040BKT^{0.59}VKT^{0.22}$	2.43	130	142.60	118.71	a_0, BKT, VKT<0.01
Pedestrian collisions	$0.0008W^{0.61}VKT^{0.54}$	1.42	130	146.35	149.70	a_0, W, VKT<0.01

	FB multivariate model					
	Variable	Mean	StdDev	MC error	Credible interval 2.5 %	Credible interval 97.5 %
Cyclist collisions	Intercept 1	−3.91	0.92	0.040	−3.88	−2.08
	BKT	0.44	0.07	0.003	0.30	0.58
	VKT	0.39	0.11	0.005	0.18	0.60
Pedestrian collisions	Intercept 2	−5.96	0.86	0.037	−7.75	−4.32
	W	0.63	0.10	0.005	0.42	0.83
	VKT	0.41	0.08	0.003	0.26	0.56
	Corr _Spatial	0.73	0.13	0.007	0.41	0.94
	Total DIC	1524.06				

Table 3. Variance–Covariance Matrix for the Multivariate Model

	Pedestrian crashes (SD)	Cyclist crashes (SD)
U error term		
Pedestrian crash	0.014 (0.03)	−0.009 (0.018)
Cyclist crash	−0.009 (0.018)	0.016 (0.02)
S error term		
Pedestrian crash	0.63 (0.21)	0.74 (0.19)
Cyclist crash	0.74 (0.19)	1.63 (0.31)

The findings as presented in Figure 3 can be used by the City of Vancouver to further investigate the factors contributing to active commuters' crashes at these zones. Seven hot zones (six in the case of the EB PSI method), out of the thirteen identified, are within the downtown area (surrounded by a circle in Figure 3). The rest of the zones are spread among the different neighborhoods within the city. The identified hot zones would require the application of countermeasures to improve the safety of active commuters.

Consistency Test Results

This section assesses the consistency of the HZID results that are obtained using the Mahalanobis distance multivariate approach and the PSI univariate approach. The within-methods consistency test shows that the Mahalanobis distance method outperforms the EB PSI method, which agrees with the conclusion of Sacchi et al. (*4*). The percentage of false identification of the Mahalanobis distance method is 7.5% compared with 9% for the PSI method.

The ranked list obtained using the PSI method is then matched to the list from the Mahalanobis distance method in order to estimate the between-methods consistency at different δ values. Figure 4 shows the degree of consistency between the two methods, where substantial inconsistency is noticed in the different percentages of hot zones identified. One approach to increase the consistency between the two methods is to standardize the EB PSIs. This is done using Equation 12.

$$PSI_{Active\ Transportation} = \frac{PSI_{Pedestrian} - Average(PSI_{Pedestrian})}{StdDev(PSI_{Pedestrian})} + \frac{PSI_{Cyclist} - Average(PSI_{Cyclist})}{StdDev(PSI_{Cyclist})} \quad (12)$$

Standardizing the EB PSI values led to more consistent ranking results with the Mahalanobis distance method as revealed by Figure 4. The consistency in the results reached as high as 62%. However, inconsistency still exists between the two approaches especially for low percentages of hot zones identified. In fact, the typical threshold of 10% of sites used by highway agencies seems to provide different ranked lists depending on whether the univariate or the multivariate approach is used, as well as whether the EB or FB technique is used. This agrees with previous studies (*4*, *12*, *13*).

Conclusion

Traffic safety analysis methods need to consider correlations among different crash types. Fortunately, there is a

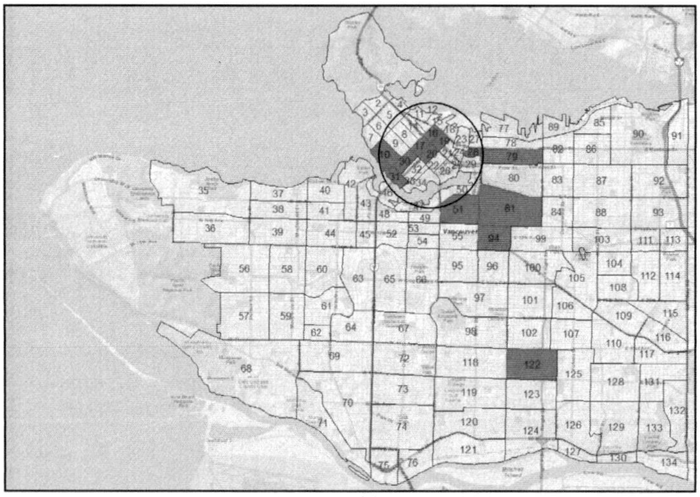

Rank	Zone
1	79
2	51
3	28
4	122
5	81
6	19
7	20
8	94
9	30
10	17
11	31
12	10
13	16

(a)

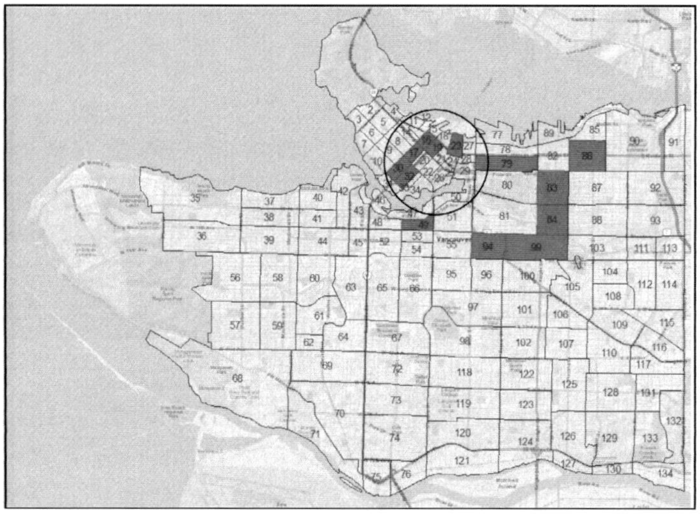

Rank	Zone
1	99
2	17
3	30
4	94
5	79
6	84
7	16
8	86
9	83
10	49
11	19
12	23
13	32

(b)

Figure 3. Vancouver's active transportation hot zones: (*a*) Mahalanobis distance method and (*b*) EB PSI method.

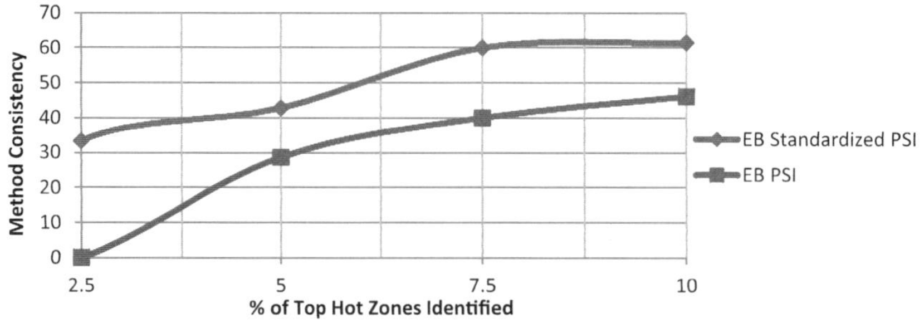

Figure 4. Consistency test results.

wealth of statistical tools that permit more precise safety analyses, taking into account the correlations that exist among different crash classifications. These tools can be used to assist transportation engineers to better identify and rank hotspots.

In this study, a new multivariate method, based on the Mahalanobis distance concept, was presented for identifying and ranking active transportation hot zones. A mixed multivariate CPM incorporating spatial effects was developed, in a FB context, as a base for the novel HZID method. The method was applied to pedestrian and cyclist crash and traffic exposure data extracted from 134 TAZs within the City of Vancouver.

Univariate EB CPMs were also developed to compare the HZID results of the state-of-the-practice EB PSI method with those of the new FB Mahalanobis distance method. When consistency tests were applied, the Mahalanobis distance method outperformed the PSI method, and a substantial inconsistency in ranking of hot zones was observed between the two methods. This affirms the results from the few previous studies that suggested shifting to multivariate methods for HZID and ranking rather than relying on the univariate methods. Using standardized PSIs has improved the consistency between the results of the EB PSI method and the Mahalanobis distance method regarding active transportation HZID. However, inconsistency is still observed, especially at low percentage of hot zones identified.

In conclusion, in a multivariate setting, the Mahalanobis distance can efficiently extend the univariate concept of PSI for HZID of active transportation modes. The Mahalanobis distance method would be more reliable when dealing with active transportation safety analysis, since it can directly provide an accurate identification/ranking of active transportation hot zones without relying on speculative assumptions that control the relation between the incorporated active modes.

There are several areas of further research. First, research can be directed into obtaining more conclusive results by simulating crash data to establish priori sites that are hazardous, and then assess whether the proposed multivariate method can correctly identify them. Second, the results are based on a single dataset, therefore; additional research is required in different active transportation environments to confirm the findings. Third, explanatory variables can be added to improve the CM's predictability and the HZID performance. Fourth, the Mahalanobis distance spatial CM could be compared with multivariate CM, multivariate spatial CM, and multivariate Mahalanobis distance CM in order to yield more insights about HZID methods. Lastly, different levels of severity could be incorporated in the active transportation crash data to provide more comprehensive analysis.

Author Contributions

The authors confirm contribution to the paper as follows: study conception and design: Ahmed Osama and Tarek Sayed; data collection: Ahmed Osama and Tarek Sayed; analysis and interpretation of results: Ahmed Osama, Tarek Sayed, and Emanuele Sacchi; draft manuscript preparation: Ahmed Osama, Tarek Sayed, and Emanuele Sacchi. All authors reviewed the results and approved the final version of the manuscript.

References

1. De Hartog, J. J., H. Boogaard, H. Nijland, and G. Hoek. Do the Health Benefits of Cycling Outweigh the Risks? *Environmental Health Perspectives*, Vol. 118, No. 8, 2010, p. 1109.
2. National Research Council (US), Transportation Research Board, Task Force on Development of the Highway Safety Manual and Transportation Officials, Joint Task Force on the Highway Safety Manual. *Highway Safety Manual*, Vol. 2, AASHTO, 2010.
3. Lan, B., and B. Persaud. Fully Bayesian Approach to Investigate and Evaluate Ranking Criteria for Black Spot Identification. *Transportation Research Record: Journal of the Transportation Research Board*, 2011. 2237: 117–125.
4. Sacchi, E., T. Sayed, and K. El-Basyouny. Multivariate Full Bayesian Hot Spot Identification and Ranking: New Technique. *Transportation Research Record: Journal of the Transportation Research Board*, 2015. 2515: 1–9.
5. El-Basyouny, K., and T. Sayed. A Method to Account for Outliers in the Development of Safety Performance Functions. *Accident Analysis & Prevention*, Vol. 42, No. 4, 2010, pp. 1266–1272.
6. Hauer, E. *Observational Before-After Studies in Road Safety: Estimating the Effects of Highway and Traffic Engineering Measures on Road Safety*. Pergamon, Tarrytown, NY, USA, 1997, p. 10.
7. Li, W., A. Carriquiry, M. Pawlovich, and T. Welch. The Choice of Statistical Models in Road Safety Countermeasure Effectiveness Studies in Iowa. *Accident Analysis & Prevention*, Vol. 40, No. 4, 2008, pp. 1531–1542.
8. El-Basyouny, K., and T. Sayed. Collision Prediction Models using Multivariate Poisson-Lognormal Regression. *Accident Analysis & Prevention*, Vol. 41, No. 4, 2009, pp. 820–828.
9. El-Basyouny, K., and T. Sayed. Depth-Based Hotspot Identification and Multivariate Ranking using the Full Bayes Approach. *Accident Analysis & Prevention*, Vol. 50, 2013, pp. 1082–1089.
10. Sayed, T., and F. Rodriguez. Accident Prediction Models for Urban Unsignalized Intersections in British Columbia. *Transportation Research Record: Journal of the Transportation Research Board*, 1999. 1665: 93–99.
11. Yang, B. Z., and B. P. Loo. Land Use and Traffic Collisions: A Link-Attribute Analysis using Empirical Bayes Method. *Accident Analysis & Prevention*, Vol. 95, 2016, pp. 236–249.
12. Cheng, W., W. H. Lin, X. Jia, X. Wu, and J. Zhou. Ranking Cities for Safety Investigation by Potential for Safety

Improvement. *Journal of Transportation Safety & Security*, 2017, Vol. 10, No. 4, pp. 1–22.
13. Aguero-Valverde, J., and P. Jovanis. Bayesian multivariate Poisson lognormal models for crash severity modeling and site ranking. *Transportation Research Record: Journal of the Transportation Research Board*, 2009. 2136: 82–91.
14. Tasic, I., and R. J. Porter. Modeling Spatial Relationships between Multimodal Transportation Infrastructure and Traffic Safety Outcomes in Urban Environments. *Safety Science*, Vol. 82, 2016, pp. 325–337.
15. Lee, J., M. Abdel-Aty, and X. Jiang. Multivariate Crash Modeling for Motor Vehicle and Non-Motorized Modes at the Macroscopic Level. *Accident Analysis & Prevention*, Vol. 78, 2015, pp. 146–154.
16. Nashad, T., S. Yasmin, N. Eluru, J. Lee, and M. A. Abdel-Aty. Joint Modeling of Pedestrian and Bicycle Crashes: Copula-Based Approach. *Transportation Research Record: Journal of the Transportation Research Board*, 2016. 2601: 119–127.
17. El Esawey, M., C. Lim, and T. Sayed. Development of a Cycling Data Model: City of Vancouver Case Study. *Canadian Journal of Civil Engineering*, Vol. 42, No. 12, 2015, pp. 1000–1010.
18. Osama, A., and T. Sayed. Evaluating the Impact of Socioeconomics, Land Use, Built Environment, and Road Facility on Cyclist Safety. *Transportation Research Record: Journal of the Transportation Research Board*, 2017. 2659: 33–42.
19. Osama, A., and T. Sayed. Evaluating the Impact of Connectivity, Continuity, and Topography of Sidewalk Network on Pedestrian Safety. *Accident Analysis and Prevention*, Vol. 107, 2017, pp. 117–125. In Press.
20. Miaou, S. P., and H. Lum. Modeling Vehicle Accidents and Highway Geometric Design Relationships. *Accident Analysis & Prevention*, Vol. 25, No. 6, 1993, pp. 689–709.
21. Sawalha, Z., and T. Sayed. Evaluating Safety of Urban Arterial Roadways. *Journal of Transportation Engineering*, Vol. 127, No. 2, 2001, pp. 151–158.
22. El-Basyouny, K., and T. Sayed. Application of Generalized Link Functions in Developing Accident Prediction Models. *Safety Science*, Vol. 48, No. 3, 2010, pp. 410–416.
23. Jonathan, A. V., K. F. K. Wu, and E. T. Donnell. A Multivariate Spatial Crash Frequency Model for Identifying Sites with Promise Based on Crash Types. *Accident Analysis & Prevention*, Vol. 87, 2016, pp. 8–16.
24. Wright, S. P. Multivariate Analysis using the MIXED Procedure. In *Proc. 38th Annual SAS Users Group International Conference*, Nashville, Tenn., SAS Institute, Cary, NC, 1998, pp. 1238–1242.
25. Osama, A., and T. Sayed. Investigating the Effect of Spatial and Mode Correlations on Active Transportation Safety Modeling. *Analytic Methods in Accident Research*, Vol. 16, 2017, pp. 60–74.
26. Islam, M. T., K. El-Basyouny, S. E. Ibrahim, and T. Sayed. Before–After Safety Evaluation Using Full Bayesian Macroscopic Multivariate and Spatial Models. *Transportation Research Record: Journal of the Transportation Research Board*, 2016. 2601: 128–137.
27. Button, S. *The Cost of Cyclist Collision in Vancouver*. Unpublished Report, 2013.
28. City of Vancouver. *Pedestrian Safety Study*. Technical Report. Urban Systems, 2012.
29. Miller, T. R., E. Zaloshnja, B. A. Lawrence, J. Crandall, J. Ivarsson, and A. E. Finkelstein. Pedestrian and Pedalcyclist Injury Costs in the United States by Age and Injury Severity. In *Annual Proceedings/Association for the Advancement of Automotive Medicine*, Vol. 48, Association for the Advancement of Automotive Medicine, 2004, p. 265.
30. De Maesschalck, R., D. Jouan-Rimbaud, and D. L. Massart. The Mahalanobis Distance. *Chemometrics and Intelligent Laboratory Systems*, Vol. 50, No. 1, 2000, pp. 1–18.
31. Cheng, W., and S. Washington. New Criteria for Evaluating Methods of Identifying Hot Spots. *Transportation Research Record: Journal of the Transportation Research Board*, 2008. 2083: 76–85.
32. Washington, S., and W. Cheng. *High Risk Crash Analysis*. No. FHWA-AZ-05-558. Arizona Department of Transportation, Pheonix, Az., 2005.
33. Jacobsen, P. L. Safety in Numbers: More Walkers and Bicyclists, Safer Walking and Bicycling. *Injury Prevention*, Vol. 9, No. 3, 2003, pp. 205–209.

The Standing Committee on Safety Data, Analysis and Evaluation (ANB20) peer-reviewed this paper (18-01280).

Article

TRR
JOURNAL OF THE TRANSPORTATION RESEARCH BOARD

Transportation Research Record
2018, Vol. 2672(38) 277–289

DOI: 10.1177/0361198118793497
journals.sagepub.com/home/trr
SAGE

Exploring the Effects of Important Predictors of Ramp Speed Choice

Bahar Dadashova[1], Karen Dixon[1], and Raul Avelar[1]

Abstract

Traditional measures of speed obtained through traffic observations are not based on detailed information about the related drivers and vehicles. Data from naturalistic studies, such as SHRP2 – NDS, can mitigate this issue by combining the key data on driver, roadway, and speed choice behavior. The objective of this study is to assess drivers' speed choice on freeway ramps as a function of ramp design, trip summary, and driver characteristics. The data analysis provides insights into various spatial and temporal factors. To conduct the data analysis time series reduction, matching, and clustering methods were implemented to define a new speed choice behavior response variable denoted as driving state. Using the resulting response variable and the three groups of predictors, neural network analysis was conducted to identify the most influential predictors and their effects on the speed choice behavior of drivers during on-ramp and off-ramp travels. Results of this analysis of speed choice behavior on freeway ramps indicate that the speed choice at these locations is indeed a complex process and is mainly influenced by the temporal and traffic conditions. Personal characteristics of drivers were also found to influence speed choice in these locations.

Freeway crashes frequently occur at on-ramp and off-ramp locations. These crashes account for 18% of all Interstate crashes, 17% of injury crashes, and 11% of fatal crashes at interchange locations (*1*). A 2004 study by McCartt et al. found that about 50% of ramp-related crashes occurred while the vehicles were exiting the freeway, 36% occurred while the vehicles were entering the freeway, and 16% occurred at the midpoints of access roads (*2*). They also observed that 48% of the crashes were run-off-road crashes, and 36% of them were rear-end collisions. These findings suggest that speed adjustments that occur at interchange locations, such as freeway-to-ramp transitions, may be associated with an increase in crashes. Kim et al. found that 85% of all freeway rear-end crashes occurred within 2,000 feet of the on-ramp gore (*3*). This study found that there was a strong association between rear-end crash rate and deceleration rate. Overall, speed indicators as well as acceleration rates have been found to be important predictors of highway safety (*4–8*).

Several researchers have analyzed speed choice behavior at ramps as an important roadway design factor, but few of these studies have analyzed the merging and diverging process of drivers as the result of their personal characteristics (*3, 9*). Speed choice is a complex process that can be influenced by several factors: roadway characteristics, traffic conditions, driver characteristics, vehicle dynamics, and weather. Analysis of speed choice behavior on freeway ramps might become even more challenging due to the sophisticated design of these sites. The roadway characteristics, speed limits, and other drivers have been found to be among the most influential predictors attributed to speed choice behavior (*10, 11*). Personality characteristics that have been found to be predictive of excessive speeding behavior include conscientiousness, reward sensitivity, sensation seeking tendencies, road rage, inattention, and risk perception (*12–16*).

Research Objectives and Methodological Approach

The goal of this paper is to evaluate the speed choice behavior of drivers during freeway merging and diverging activities. The secondary objective of this paper is to explore how well the speed choice behavior reflects the driver's personal characteristics and surrounding factors. For this purpose, the authors used trip time series data

[1]Texas A&M Transportation Institute, College Station, TX

Corresponding Author:
Address correspondence to Bahar Dadashova: b-dadashova@tti.tamu.edu

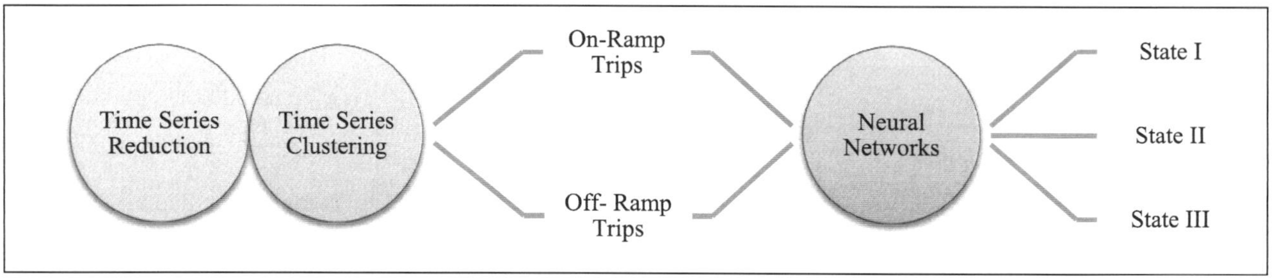

Figure 1. Ramp speed choice behavior analysis framework.

from the Naturalistic Driving Study (NDS) of the Second Strategic Highway Research Program (SHRP 2).

In the present study, it is hypothesized that there are three types of speed choice, labelled as states (State I, State II and State III). Drivers are assumed to adopt one of the three types of driving state during each trip including both on-ramp (street to freeway, SF) and off-ramp (freeway to street, FS) trips. Driver's speed choice state depends on multiple factors such as the ramp design, direction of travel, driver characteristics, time and day of trip, and other trip details including speed variance, speed mean, and number of trips per driver at each study location. A three-step procedure was used to conduct time series classification and analyze the impact of external and personal factors on the speed choice behavior (Figure 1).

Several methods exist for conducting the time series feature extraction and classification analysis. These include wavelet transformation, Fourier transformation K-nearest neighbor, piecewise linear approximation, support vector machines, and artificial neural networks (*17–21*). In this study wavelet transformation, dynamic time warping and neural networks were applied to conduct the data analysis, as explained below.

As the first step, wavelet transformation was applied to reduce the time series data dimension to obtain a relatively balanced times series data. Using dynamic time warping tools, the on-ramp and off-ramp speed traversals were then clustered into three states of driving behavior. As the final step, joint analysis of driving states was conducted using neural networks architecture.

The rest of this paper is organized as follows. In the next section the SHRP 2 dataset is described briefly. This section is followed by the description of the statistical methods and tools implemented in this study. In the third section the data analysis results are discussed. The paper ends with conclusions of the study.

SHRP 2 Data Description

The SHRP 2 – NDS dataset includes data that represent more than 3,500 volunteer drivers aged from 16 to 98. The data collection duration extended over a four-year study period (2010–2013). The SHRP 2 – NDS study includes data from various locations in six states in the United States: Indiana (IN), New York (NY), North Carolina (NC), Washington (WA), Pennsylvania (PA), and Florida (FL). NDS data is complemented by the Roadway Information Database (RID) which contains detailed information for roadways in the study regions. NDS trip data can be linked to the RID roadway data using the Link ID—a unique road segment identifier, variable. The integrated SHRP 2 data can provide researchers with an abundance of information such as driver and roadway characteristics, vehicle speed information over distance and time, and other congestion-related factors including the time and day of participant trips.

NDS Traversals

In this study, NDS trip information (or *traversals*) was collected from ramps at four interchanges on two freeway sections in Altoona, PA. For the analysis of speed choice behavior, a state where the enforcement of the speed traffic law was not very strict was chosen in order to diminish the impact of this factor on drivers' speed choice behavior (*22*). Data were obtained from four interchanges on two freeways (Figure 2):

- Urban freeway: William Penn, Blair County
- Rural freeway: Bud Shuster Freeway, Blair County

Each ramp located on these freeway interchanges is approximately two miles long, which include locations on the intersecting freeway and street, and the ramp connecting the two. To fulfill the objectives of this study, the authors identified trips where each driver had traversed the same ramp in both directions of travel. These trips are identified as:

- Merging: street to freeway (SF)
- Diverging: freeway to street (FS)

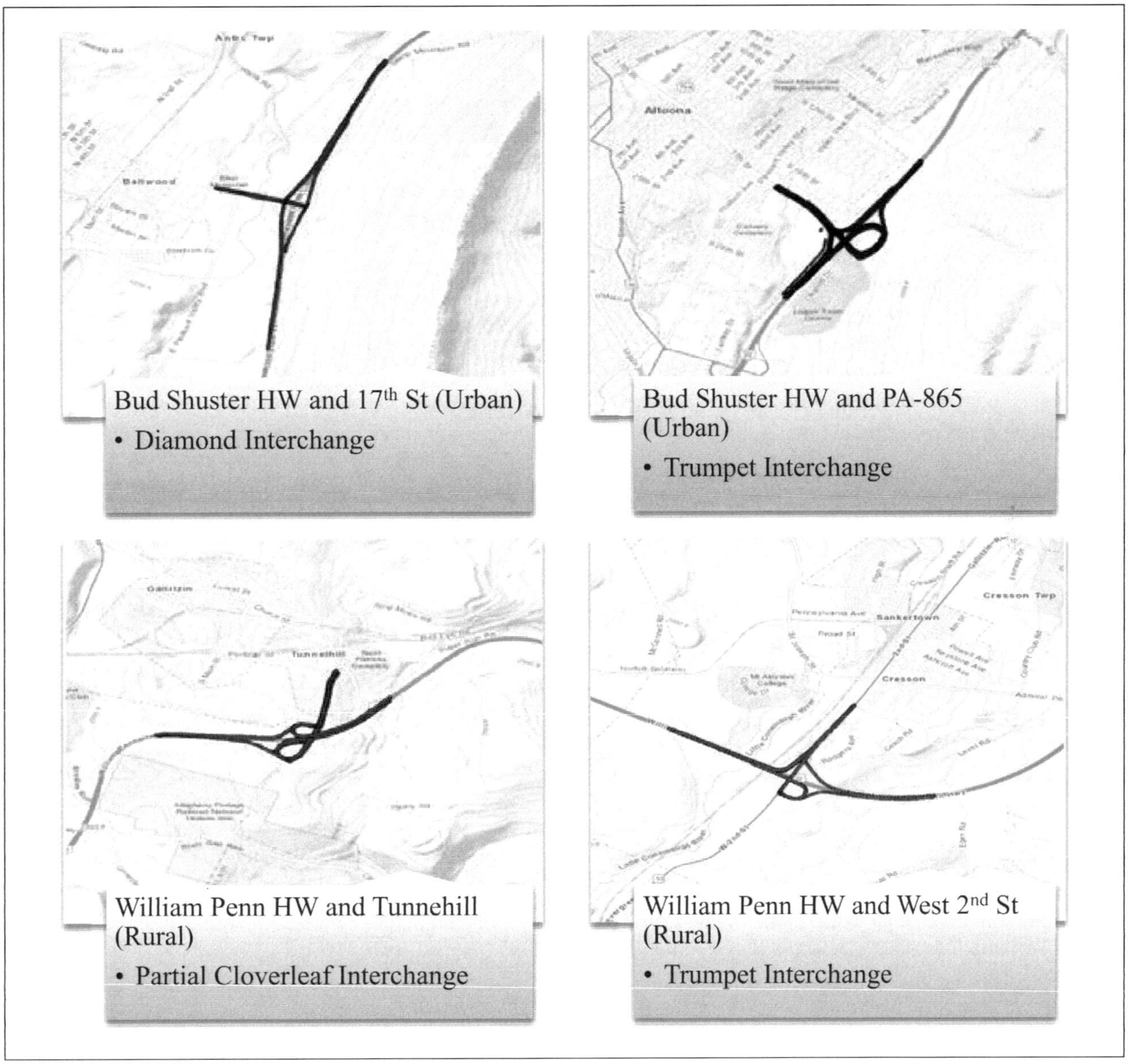

Figure 2. Ramp trajectories.

For the analysis of speed choice behavior, speeds were recorded by a global position system (GPS) transponder located in each study vehicle. The NDS trip time series provides both GPS speed and the speed recorded by the vehicle's own network. However, since the speeds recorded by the GPS also have the matching latitude and longitude information for every second, this data source was used as the primary indicator of speed choice behavior.

To obtain a consistent study sample, only trip data that spanned the entire duration of the target freeway ramps with travel times of 60 to 70 seconds or longer were selected. This produced 859 trips taken by 32 participants. In many cases, drivers travelled the study route multiple times over the study period. The number of trips per driver for most of the drivers ranged from two to 70 trips, with the average of 30 trips per driver. Two of the study participants regularly used one of these routes; they had accumulated 341 trips (117 and 224 trips) which accounted for almost 40% of all trips. Such an overrepresentation in the trip numbers can present a bias in the final results where the speed choice will be analyzed as the function of driver characteristics among other explanatory factors. Therefore, due to the potential for undue and biased influence in the results from these two drivers if all their trips were used, it was decided to reduce the number of trips by random sampling. For this purpose, the authors randomly assigned 0 and 1 to each trip by these two drivers and selected the trips that were assigned value 1. As a result 53 and 77 trips were left. The final study data set used in this study included 256 SF and 392 FS trips.

Next, the authors assembled a trip summary for each trip. The associated data included the year, month, day of week, and time bin as well as each trip's maximum speed, mean speed, and speed variance. There is no exact hour for the trips, instead each trip was assigned a time bin of three hours.

NDS Driver Characteristics

To explore how the individual characteristics of drivers may influence their speed choice behavior, the following driver information, obtained through interviews and psychological testing of the SHRP2-NDS participants, was used:

- Driver demographics (age and gender)
- Barkley's Attention Deficit Hyperactivity Disorder (ADHD) screening
- Risk perception score (RPS)
- Sleeping habits
- Depth perception

Barkley's ADHD Screening Test. Individuals with attention deficit disorder (ADD) and ADHD are prone to frequent inattention and distraction while performing tasks. In addition to inattention being a factor that is potentially associated with speed choice (*14*), a study by Quinn et al. found that individuals with ADD and ADHD have a higher likelihood to be ticketed for speeding (*23*). This observation leads to the potential hypothesis that those who score high on Barkley's ADHD screening test may have higher incidence of speeding than those who score low on the test. The six items included in the Barkley's ADHD screening are:

1. Easily distracted
2. Difficulty organizing
3. Loses objects
4. Difficulty waiting for turn
5. Feels restless
6. Difficulty enjoying leisure activities

Each of these six items is scored by the participant by using one of the following three answers: Never or Rarely (1), Sometimes (2), and Often (3). The Barkley's ADHD score was then calculated using the answers provided to these items (*24*).

Risk Perception Score (RPS). Risk perception is well documented in the literature to have a strong impact on speed choice behavior (*25*). Those with low risk perception tend to have a high perception of their individual driver control. As part of this characteristic, these drivers tend to dismiss risks, exude a high level of self-confidence (especially with regard to their driving ability), and demonstrate unrealistic optimism (*26*). It is hypothesized that those with low risk perception (which equates to a low RPS score) will have higher number of speeding incidents than those who have a high RPS score. The list of the questions used for the RPS is shown in Table 1. Respondents answered the questionnaire by choosing No Greater Risk, Moderately Greater Risk, or Greater Risk for each question. RPS was then quantified by allocating one point to No Greater Risk, four points to Moderately Greater Risk; and seven points to Greater Risk.

Sleeping Habits. Previous research has found personality differences between long and short sleepers. Specifically, short sleepers were found to be efficient persons who handle stress by keeping busy and by denial. Long sleepers, on the other hand, had higher instances of depression and anxiety and scored higher on most pathology tests (*27*). This portion of the evaluation is an exploratory analysis on both sleep schedule (i.e., whether the participant keeps a fairly regular sleep schedule, Yes or No) and sleeper type (i.e., light, normal, or heavy) to determine if these factors encompass a variety of traits related to speeding propensity.

Depth Perception. Depth perception is the ability to visually perceive the world in three-dimensional space and is necessary to accurately determine the distance to an object. Depth perception is a "personal" characteristic that directly affects an individual's visual perception, and therefore may be connected to speeding behavior (or lack thereof).

Participants were shown a picture of four rings (top, bottom, left and right) and were asked if the bottom ring seems to be floating towards them. If the participant answered "Yes", they moved to a second picture and were asked which ring seemed to be floating toward them. Drivers who could not see the ring floating towards them received no score. All others were scored in seconds of arc, where the smaller the seconds of arc, the better the participant's depth perception.

The list of all the predictors used in the study is shown in Table 2.

Table 1. Elements of Risk Perception Questionnaire

Questionnaire elements
• Running red light
• Risks for fun
• Sudden lane changes
• Running stop sign
• Speeding for thrill
• Tailgating
• Illegal turn
• In a hurry
• Risk of passing on right
• Yellow light acceleration
• Driving after taking drug or alcohol
• Road rage
• Driving to reduce tension
• Passenger interaction
• Racing
• Speeding <20 mph over limit
• Speeding >20 mph over limit
• Not wearing safety belt
• Risk perception score
No greater risk (= 1)
Moderately greater risk (= 4)
Greater risk (= 7)

Statistical Methodology

As indicated above, the spatial distance completed by all traversals is similar between study sites (approximately two miles) since all the ramps included in the study had approximately the same length. However, because of the speed differentials, the trip durations vary, hence resulting in unbalanced time series data. Therefore, as an initial step for the time series dimension reduction evaluation, the authors used the discrete wavelet transformation (DWT) method so as to obtain a relatively balanced time series.

The second step of the statistical analysis involves the matching and clustering of the reduced traversal sequences. Time series clustering can be conducted in several ways. In this study, the dynamic time warping (DTW) algorithm was used to cluster the reduced time series data into three driving state clusters: State I, State II, and State III. Given that the on-ramp and off-ramp speeds have two different shapes (e.g. when travelling from freeway to street) the time series clustering might become challenging. Therefore, the two sets of speed time series data were clustered based on the direction of the trips: freeway to street trips (FS) and street to freeway trips (SF).

The clusters or driving states are assumed to be the three levels of the speed choice state response variable. After identifying the response variables, the authors then used neural networks (NN) to evaluate the impact of the following group of factors on the speed choice behavior:

- Ramp characteristics
- Driver characteristics

Table 2. Predictors Used to Analyze Driving Behaviors

Predictor group	Variable name	Variable type	Descriptive statistics: Min	Max	Mean	St. D.
Ramp characteristics	Area type	Categorical	1 = rural, 2 = urban			
	Travel direction	Categorical	1 = freeway to street, 2 = street to freeway			
	Interchange ID	Categorical	1 = Bud Shuster HW and 17th St, 2 = Bud Shuster HW and PA-865, 3 = Will Penn HW and Tunnehill St, 4 = Will Penn HW and West 2nd St			
	Ramp design	Categorical	1 = diamond; 2 = trumpet; 3 = partial cloverleaf			
Trip summary	Trip year	Numerical	1 = 2011; 2 = 2012; 3 = 2013			
	Trip month	Categorical	1 = January, 2 = February, ..., 12 = December			
	Trip week day	Categorical	1 = Monday, ..., 7 = Sunday			
	Trip time	Categorical	Three-hour bin (8 bins)			
	Maximum speed	Numerical	41	80	61.9	5.8
	Mean speed	Numerical	5.3	60.4	41.3	6.7
	Speed variance	Numerical	3.5	28.1	12.8	5.1
	Vehicle classification	Categorical	1 = car, 2 = SUV, 3 = cross-over, 4 = minivan			
Driver characteristics	Age	Categorical	1 = 16–19; 2 = 20–24, 3 = 25–29, 4 = 30–34, 5 = 35–39, 6 = 45–49, 7 = 50–54, 8 = 55–59, 9 = 60–64, 10 = 65–69, 11 = 70–74			
	Gender	Categorical	1 = female, 2 = male			
	Barkley score	Numerical	1	6	2.02	1.06
	Risk perception score	Numerical	1	126	74.05	38.6
	Sleeper type	Categorical	1 = light; 2= normal; 3 = heavy			
	Has a sleep schedule	Categorical	0 = no; 1 = yes			
	Depth perception	Categorical	1 = B/400, 2 = L/200, 3 = B/100, 4 = T/70, 5 = T/50, 6 = L/40, 7 = R/30, 8 = L/25, 9, = R/20			

- Trip summary

Although the clustering was conducted for the on-ramp and off-ramp time series separately, joint NN analysis was conducted using the results from both sets of clusters with travel direction used as one of the predictors.

Discrete Wavelet Transformation

As previously indicated, due to the speed differentials, the length of the series (i.e. number of seconds) included in this study ranged between 70 and 850 seconds. Such a significant inequality among the series lengths could be a cause for concern. This problem can be dealt with by reducing the time series dimension (length-wise) using algorithms such as discrete wavelet and discrete Fourier transformation.

In this study, DWT was used to reduce the dimensions of relatively longer series. Namely, the Haar wavelet is used to conduct the dimension reduction. Haar wavelet allows the time series of length T to be represented in terms of its orthonormal basis by calculating a set of averages and coefficients (usually $\sqrt{2}$ to ensure energy conservation) (*28*, *29*). After the first iteration, the time series length reduces to $T/2$. The resulting time series is referred to as the first level of wavelet transformation. The DWT method can be applied recursively until a single coefficient and average is obtained.

Dynamic Time Warping

In order to match and cluster the reduced trip time series of varying lengths, the DTW algorithm was used. DTW is essentially a point-to-point matching method under some boundary and temporal consistency constraints. DTW was originally developed for speech recognition (*30*). The algorithm aims to match two time series vectors by warping the time axis iteratively until the optimal alignment is achieved. Optimal alignment in this case is measured by the time warping distance (*31*).

DTW clustering can be conducted based on the distance or shape of the series. In this study the authors have implemented the partitioning around the medoids (PAM) algorithm. The PAM algorithm is based on the principle of finding the sequence that is located in the cluster center. The members of the given cluster are then selected based on the cutoff distance from the medoids. This method essentially optimizes the matching by using both shape and distance measures.

Neural Networks

To describe the relationships between the predictors (inputs) and the speed choice state (output), multi-layer perception (MLP) NN architecture trained by a backpropagation algorithm was used. NN methods are known for their ability to deal with a relatively large number of predictors. The NN framework or architecture has three elements: input, hidden layer, and output (Figure 3). Input and output refer to the predictors and response variable respectively. Hidden layers are the collection of neurons organized and connected to each other using the arrows that are referred to as the weights. Weights can also be understood as the parameter estimates although they should not be interpreted as such (*32*).

In the NN architecture, the relative contribution of the inputs to the output depends on the magnitude and the direction of the connection weights (*33*). Connection weights are computed using the weights of individual inputs in each hidden layer. Greater connection weight indicates the higher intensity of the association. Negative connection weights represent an inhibitory (reducing) effect while the positive connection weight represents an excitatory (increasing) effect of the neurons on the output. The NN architecture is depicted in Figure 4.

Result and Discussion

Time Series Reduction and Matching

To reduce the dimension of the speed series and create a relatively balanced set of traversals, the authors recursively applied the DWT method. Each series was saved together with all levels of wavelet decompositions. After reviewing the resulting reduced sequences of the trip time series, the authors selected the reduced series of lengths ranging from $20 < \text{seconds} \leq 41$. An example of the SF and FS trips completed by two drivers (D-171263 and D-639928) is depicted in Figure 4.

All four trips (2 SF and 2 FS) were completed on Ramp #3. After applying the recursive wavelet transformation in two iterations, the FS speed series of the two drivers was reduced from 126 and 167 seconds to 31 and 41 seconds respectively. The SF trips, after two iterations, were reduced from 109 and 113 seconds to 27 and 28 seconds respectively. The transformed values represent the trip speeds after they have been transformed twice. For example, if driver D-171263 has the following speeds at the initial four seconds of the SF trip: $S^0{}_{t,t=1,2,3,4} = (30.5, 30.02, 31.31, 33.7)$ where the superscript 0 indicates the initial level (Figure 4, upper-left graph), this time series will be reduced to a sequence of three values after the first-level wavelet transformation : $S^1{}_{t,t=1,2,3,} = \left(30.26\sqrt{2}, 30.66\sqrt{2}, 32.52\sqrt{2}\right) = (42.79, 43.36, 45.99)$ where $30.26 = \frac{30.5+30.02}{2}$ and so on. The rationale of applying the coefficient $\sqrt{2}$ is to "conserve the energy". After the next iteration, the second level wavelet transformation for this sequence will be equal

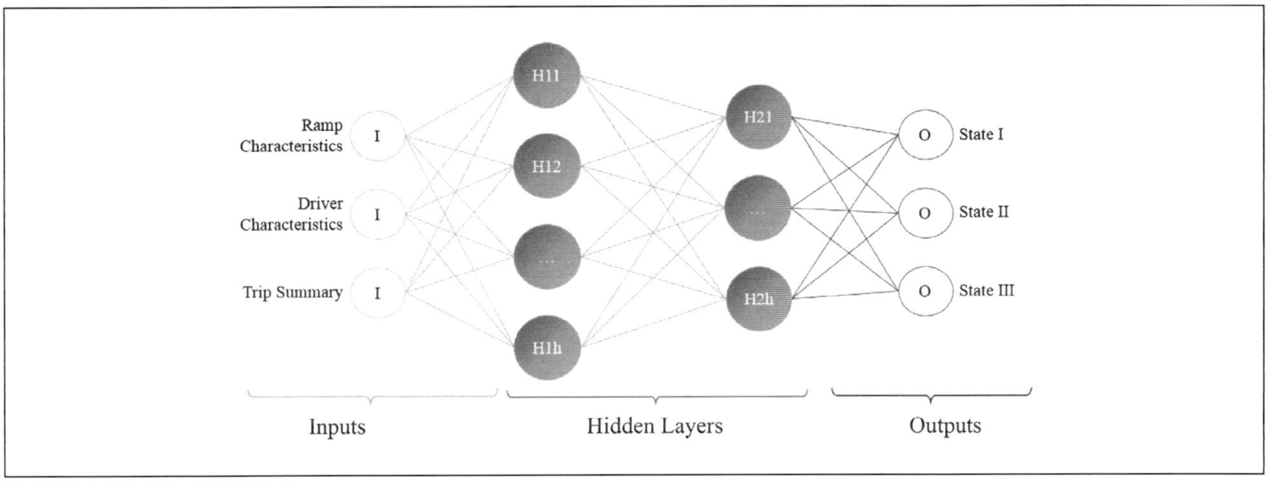

Figure 3. Neural networks framework.

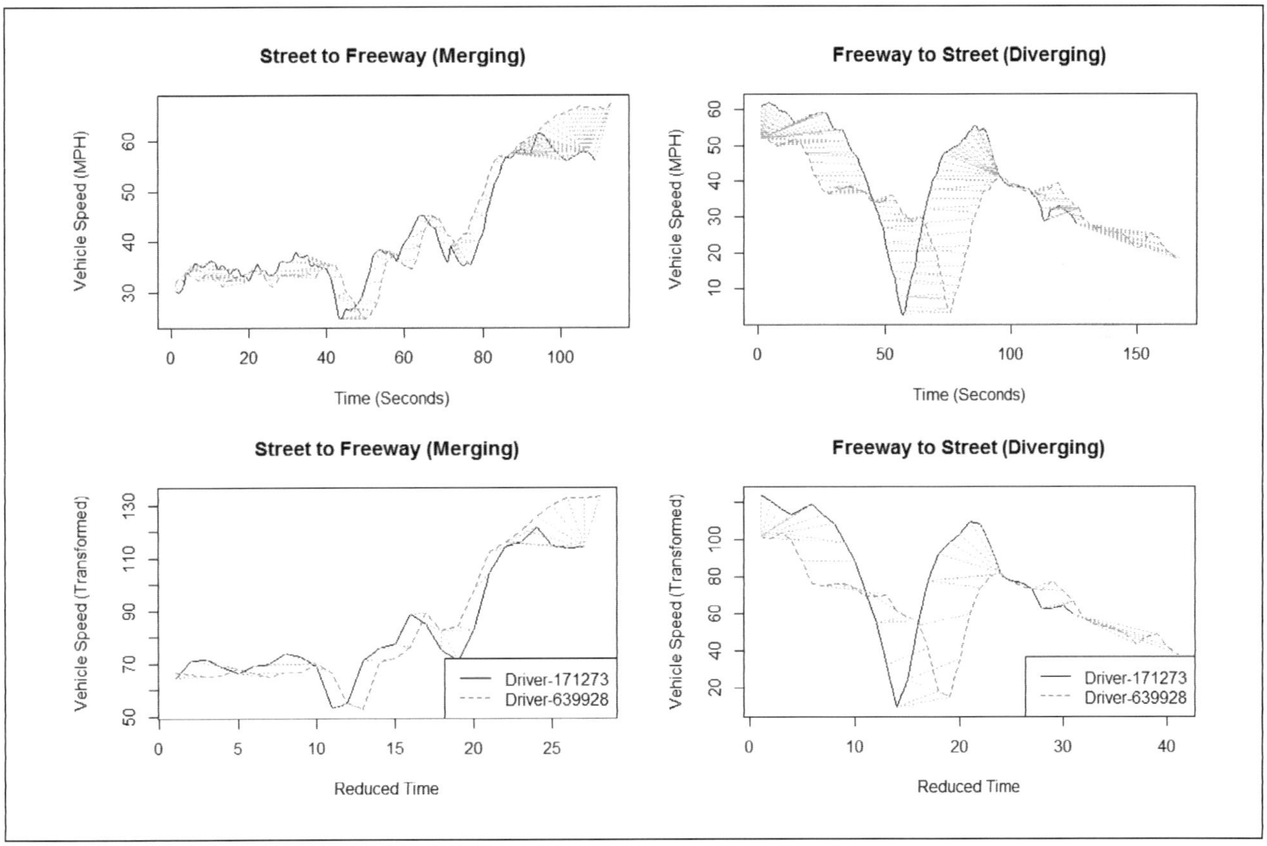

Figure 4. Matching two traversals: original versus DWT series.

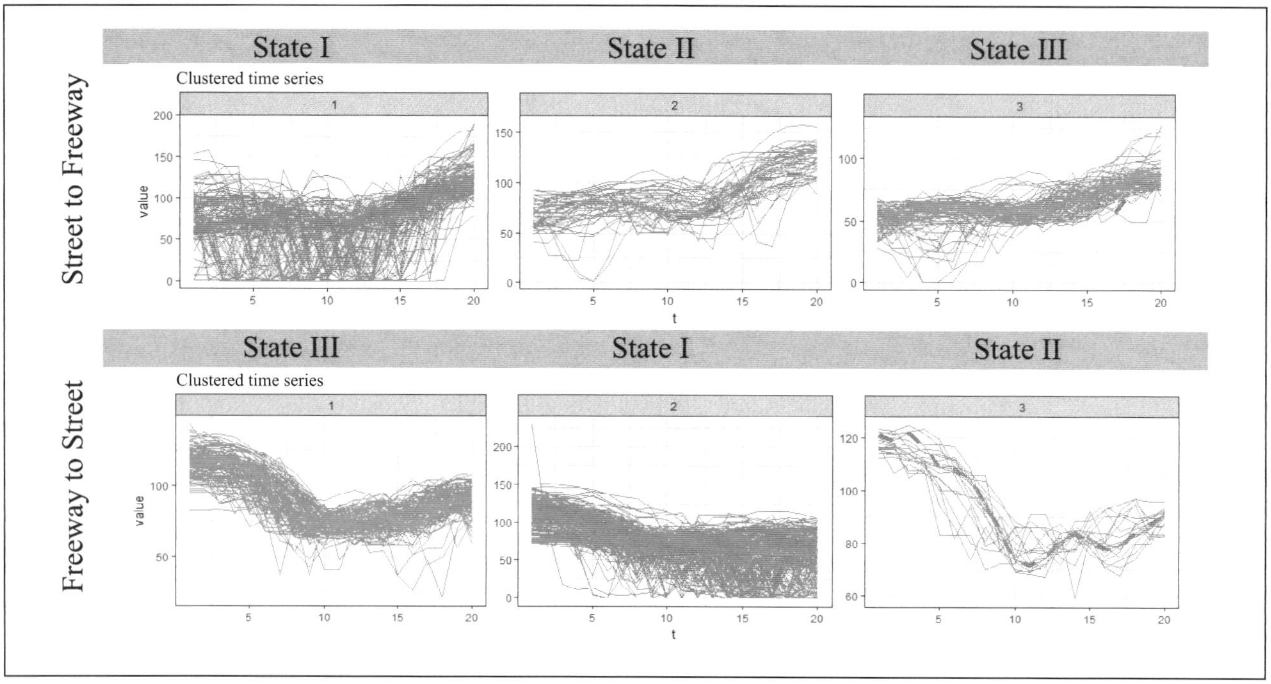

Figure 5. Speed profile clusters.

to: $S^2_{t,\,t\,=\,1,2} = \left(43.08\sqrt{2}, 44.68\sqrt{2}\right) = (60.92, 63.19)$ where $43.08 = \frac{42.79 + 43.36}{2}$, and so on (Figure 4, lower left graph). As can be observed, no information is lost, and the shape of the series is not altered after the wavelet transformations.

Speed Profile Clusters

The reduced speed profiles were clustered using the PAM algorithm into three driver states: State I, State II and State III. The clustering was conducted using R-CRAN open software (*34*). As shown in Table 3, 242 FS trips and 129 SF trips were assigned to State I, 19 FS trips and 45 SF trips were assigned to State II and 131 FS trips and 82 SF trips were assigned to State III.

The speed profile clusters can be visualized as in Figure 5.

Since the time series data were reduced using wavelets, in Figure 5 the values shown on the *y*-axes do not represent actual speeds but rather transformed speeds. Also, it should be noted that the clusters show individual trip series, meaning that the same driver could belong to all three clusters. However, it is assumed that over a longer period of time, the true behavior of the driver will surface as additional trips occur. This is also one of the reasons why all of the trips need to be included in the clustering analysis. On the contrary, the analysis might not be able to capture the true nature of the driver's behavior, as there is no way to assure convergence with "long term behavior."

Table 3. Summary of Speed Profile Clusters

	Freeway to street		Street to freeway	
Clusters	# of series	% of total	# of series	% of total
State I	242	62%	129	50%
State II	19	5%	45	18%
State III	131	33%	82	32%
Total	392	100%	256	100%

At first inspection, it can be observed that the speed series belonging to State I (purple colored) are highly volatile, with some of the values reaching zero (coming to a complete stop). Speed profiles assigned to State II (green and yellow colored) are still relatively volatile; however, the speeds rarely reach a value of zero. The speed series assigned to State III (green and turquoise colored), on the other hand, exhibit smoother speed transitions when travelling from one location to another.

Observing the overall volatility of the series belonging to each cluster, one explanation might be that at State I the drivers are less cautious because there are abrupt changes and sudden stops in the speed profiles. However, their driving behavior could also be attributed to congestion, that is, the vehicles might be stopping due to congestion. It appears that at States II and III the drivers could be more cautious (or under less congested conditions), although at State II the drivers still continue making abrupt speed changes. At State III the drivers seem very cautious because the transition from freeway to street and from street to freeway appears to be quite smooth. Yet, this smooth transition could also be indicating minimal or nonexistent congestion or opposing traffic. For the following analysis, the authors review and interpret the results under the hypothesis that an increasing state level indicates more cautious or safe driving behavior by the drivers.

Results of Network Analysis

The authors used 47 inputs to assess the three states of speed choice behavior on the ramps using the NN architecture (Table 4). These inputs belong to the following categories: ramp characteristics, driver characteristics, and trip characteristics. To conduct the NN analysis, the data was sampled into training and testing datasets. The NN is trained using 70% of observations connected through 15 hidden neurons. The testing data (remaining 30% of observations) was then predicted using the trained algorithm. The mean squared errors showing the difference between the predicted and observed data was estimated to be 2.5%, 2.0%, and 3.4%, for the States I, II and III, respectively. This observation indicates a highly accurate prediction of each output.

The relative importance of the variables was found by using the Garson's algorithm (*35*). Garson's algorithm uses the absolute values of connection weights to determine the relative importance of inputs in the network. This relative importance was calculated for the entire network and not for each driver state individually. It shows the individual contributions of each input to the network. The sum of all contributions will be equal to 100%. The importance of each variable is shown in Table 4. As shown, speed variance is the most important predictor of the speed choice state response variable with almost 7% contribution. Day of week, mean speed, maximum speed, trip time, and travel direction each contribute 3% or more to the entire network. In general, the importance diagram implies that the trip time and date and the locations of the trips are the main contributors to the speed choice behavior. Therefore, it can be concluded that speed choice is mainly influenced by traffic conditions rather than by driver characteristics. To explore the direction and the magnitude of the input effects on the outputs, it is helpful to examine the connection weights. The connection weights are calculated for each input by using the weights from all hidden neurons (15 in total). The results were obtained for all three outputs which are the speed choice states.

Figure 6 graphically represents the relationships between the driving states and some of the most important factors.

The magnitude of speed variance effect on speed choice behavior is 37.04, 21.83, and 1.95 for States I, II and III respectively. Connection weights for speed variance indicates that this variable has an increasing impact (positive sign) on speed choice, although as drivers shift from State I to State III, the magnitude of this effect starts to diminish. This finding is consistent with the initial assumption that as the drivers move from State I to State III, they start driving more cautiously thus the magnitude of the effect of this variable is decreasing as the states progress in increasing order. The two other speed related inputs, mean and maximum speeds, are also important. It can be observed that in State I the mean speed is decreasing, this is expected because some drivers come to a complete stop. As the drivers move from State I to State III, the mean speed is increasing. This finding, together with the effect of speed variance on the speed choice states, yields very interesting results. As shown, the effect of speed variance is smaller in State III compared with State II, however the effect of mean speed in State III is observed to be much higher compared with State II. This would imply that the speed variance is perhaps a reliable indicator of cautious driving behavior.

The second most important variable group is the time and day of the trips. For State I, the effect of day of week is increasing while for the States II and III the effect is decreasing. Note that the effect of day week is in increasing order from Monday to Sunday. This implies that as the week progresses more drivers start driving in State I and fewer drivers drive in State III. Using the previous assumptions about the driving states, this result can be interpreted as follows: during weekdays, from Monday through Friday, drivers will be driving more cautiously and on the weekends they will drive less cautiously.

Another interesting relationship is presented by the ramp design. The ramp design is ordered from least sophisticated (or complex) to more sophisticated. The importance of the influence of ramp design is moderate to high (ranked 11 out of 47 factors in Table 4). The effect of the ramp design on the State I is positive and its magnitude is the highest, suggesting that as the ramps become more challenging, drivers would tend to adjust their driving to the characteristics of the State I pattern. In State II, the effect is negative, implying that fewer drivers will be driving in this state as the interchanges become more sophisticated. The relationship between State III driving and ramp design is positive although the magnitude of the effect is very small. This could imply that the speed choice behavior of State III drivers is not affected significantly by the ramp design.

Table 4. NN Connection Weights and Importance of Predictors of Speed Choice

Variable name	Importance (% contribution)	Overall connection weights			Variable name	Importance (% contribution)	Overall connection weights		
		State I	State II	State III			State I	State II	State III
Speed variance	6.3%	37.04	21.83	1.95	Risks for fun	2.0%	7.96	4.70	2.93
Trip week day	4.0%	7.15	−7.68	−14.00	Difficulty enjoying leisure activities	2.0%	18.04	−2.76	−3.79
Mean speed	3.9%	−43.98	17.03	21.73	Sudden lane change	2.0%	−8.70	0.76	−4.78
Maximum speed	3.4%	15.22	26.80	−1.95	Loses things	2.0%	14.02	−8.08	4.92
Trip time	3.0%	−16.26	−15.34	6.23	Feels restless	1.9%	5.91	6.06	9.15
Travel direction	3.0%	−11.41	−26.51	22.14	Risk perception score	1.9%	0.49	−5.24	26.23
Not wearing seatbelt	2.9%	−10.77	12.78	8.22	Area type	1.9%	−3.94	−5.65	−2.37
Trip year	2.8%	14.48	−14.92	−10.40	Vehicle classification	1.9%	−12.25	6.91	−7.86
Sleeper type	2.8%	11.95	15.77	−5.43	Road rage	1.9%	−0.25	3.95	4.45
Interchange ID	2.7%	−12.12	−4.39	−3.48	Passenger interaction	1.9%	−1.88	3.17	4.97
Ramp design	2.5%	3.38	−7.61	0.03	Speed choice < 20 mph	1.8%	−9.18	−7.66	15.32
Trip month	2.4%	−13.20	8.60	−28.88	Passing on right	1.8%	−0.10	5.43	−0.80
Accelerates at yellow light	2.4%	5.56	9.42	−15.27	Driving to reduce tension	1.7%	−0.69	14.70	17.59
Age	2.4%	−1.39	−4.44	8.65	Difficulty awaiting turn	1.7%	−0.04	11.17	−2.30
Speed choice >20 mph	2.3%	0.82	12.75	2.88	Rolling stop sign	1.7%	−8.33	2.50	1.90
Driving under influence	2.3%	4.56	5.42	−4.53	Number of trips per driver	1.7%	21.35	−10.84	−5.69
Road racing	2.2%	16.54	10.01	4.81	Speed choice for thrill	1.5%	−1.12	0.20	1.74
Depth perception	2.2%	−9.09	4.76	−1.36	In a hurry	1.5%	3.11	4.48	−10.62
Sleep schedule	2.2%	14.38	19.27	11.42	Running red light	1.5%	−14.24	8.60	4.89
Tailgating	2.1%	0.52	11.36	−8.95	Illegal turns	1.3%	−8.82	0.80	−1.44
Difficulty organizing	2.1%	−9.44	14.48	−4.12	Gender	1.3%	2.30	3.06	20.53
Easily distracted	2.1%	28.11	−15.44	−10.07	Barkley score	1.1%	12.60	0.54	5.69

Among the driver characteristics, driver's age, RPS, and sleeping habits are observed to contribute the most to the driving states. The results show that driver age is negatively related to State I and II and positively related to State III, implying that as drivers grow older their driving patterns become more like State III, and less like States I and II. However, one can observe that more senior drivers also drive less in State II and more in State I. This could indicate that the relationship between age and speed choice is in fact U-shaped and not linear. Further analysis will be necessary to explore the driving patterns in terms of young, middle aged, and older drivers.

RPS is positively related to State I and State III driving, although the magnitude for State I driving is very small, implying an almost non-existent impact. The magnitude of the effect for State III is very high, which would imply that State III drivers have very high risk perception. For State II drivers this effect is negative, implying that the drivers in this group have lower risk perception. Analyzing the speed-related elements of the risk perception survey (i.e., accelerates at yellow light, speeding for thrill, or driving over the speed limit), one can observe that the State III drivers seem to perceive the risks related to these behaviors poorly when compared with State I drivers. For example, State III drivers do not perceive

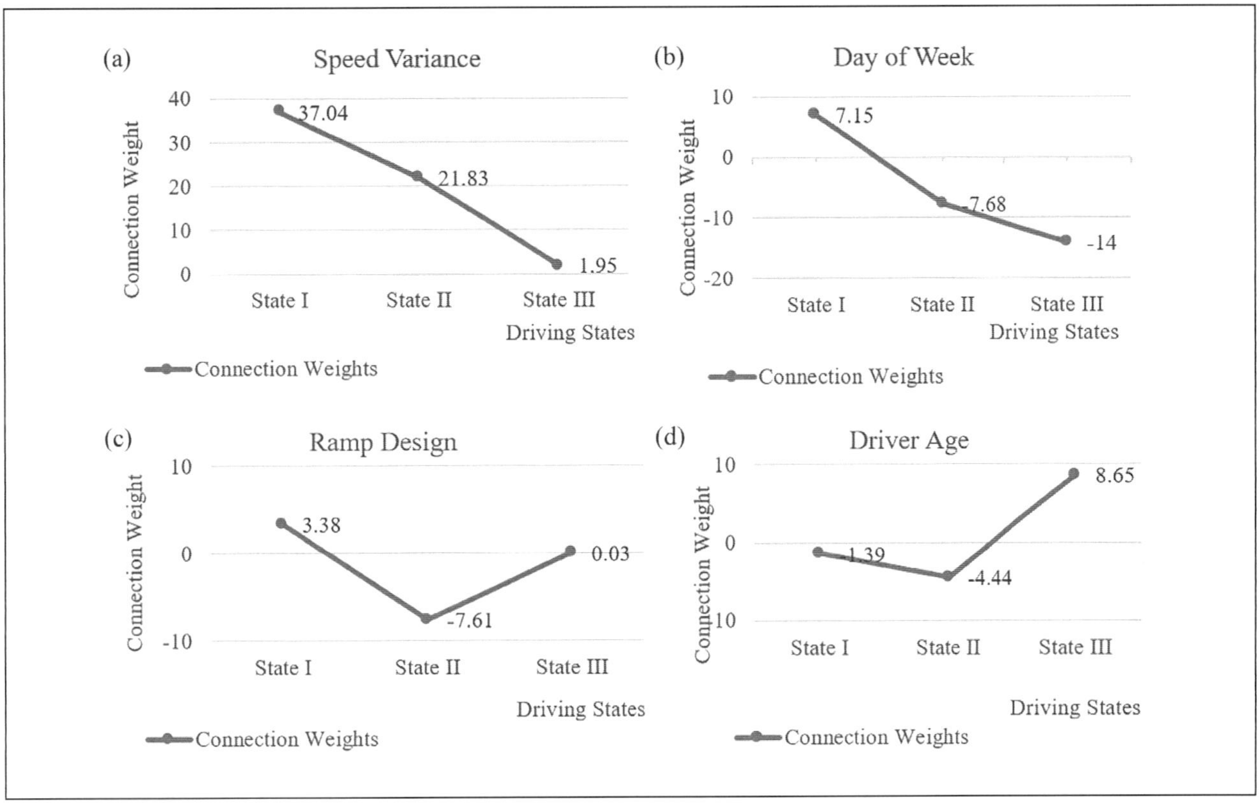

Figure 6. Description of relationships between the driving states and important variables: (*a*) speed variance; (*b*) day of week; (*c*) ramp design; (*d*) age of driver.

the risk of accelerating at a yellow light as highly as the State I and State II drivers. It can also be observed that State II drivers show the highest concern to this behavior. Assuming that State II driving can be interpreted as "moderate" driving, results show that most drivers might assume that accelerating at a yellow light as a risky behavior. These results indicate that there might be a U-shaped relationship between risk perception elements and speed choice. More risky behaviors (driving under influence, tailgating, sudden lane changes, and passing on right), on the other hand, are perceived as highly risky by State III drivers. Further research will be needed in order to explore the relationships between the risk perception and speed choice.

The Barkley's ADHD score was not found to be important although individual elements of the survey were found to be quite important (easily distracted and difficulty organizing). It can be observed that drivers belonging to State II and State III tend to be less prone to being distracted, perhaps because they tend to be more aware of their surroundings, while drivers at State I are more prone to being distracted. Moreover, the State III drivers are found to have lower Barkley scores than the State I drivers. This is consistent with the initial hypothesis that drivers scoring less in the Barkley's ADHD test were expected to have less incidence of speeding.

Conclusions

In this study, the authors conducted a speed profile analysis of a sample of NDS participants at freeway ramps in Pennsylvania. The authors focused the data analysis using three years of trip data from four ramp locations. To analyze the speed choice behavior, 47 predictors indicating the ramp characteristics, driver characteristics, and trip summary were used.

To perform this evaluation, the data analysis was conducted in three stages. The authors implemented the data reduction and clustering of the speed time series to distinguish the three types of driving behaviors, named as states. Using the states as the three levels of speed choice, the authors then developed NN architecture to explore the relationships between the 47 inputs and three outputs (one output per state). Overall, the results of the NN analysis uncovered interesting relationships between the ramp speed choice behavior and the other explanatory factors. The results suggest that traffic conditions and geographic location are the main predictors of a driver's speed choice behavior. Based on the influence of these factors, drivers adjust their speed in a more or less cautious manner.

The effect of driver characteristics was not found to be as important, although the results of the NN analysis did reveal some interesting facts. For example, the analysis of the drivers' RPS revealed that the drivers who drove in a more cautious manner and adjusted their speeds more smoothly (State III) did not perceive driving at higher speeds or accelerating as risky behaviors. They did, however, perceive some other behaviors, such as the driving under influence or sudden lane changes, as risky. When examining the answers of these drivers to the risk perception questionnaire, they were observed to have higher risk perceptions as far as the speeding-related questions were concerned. They were also observed to be easily distracted based on their Barkley ADHD scores. These results might explain why the drivers observed in this driving state stopped more often compared with the other clusters; they would be more likely to be distracted and might have had to stop or reduce the speed significantly more often when encountering highly risky situations. As future improvement on this work, the authors would like to evaluate more ramp trips to define and explain speed choice behavior of drivers at these locations with more accuracy. This future work will also consider analysing the effects of risk perception elements in batches (e.g. speed related questions) rather than evaluating the impact of each element separately.

Acknowledgments

This work has been carried out in the framework of the Strategic Research Plan and Advancing Transportation Leadership and Safety (ATLAS) Program at the Texas A&M Transportation Institute (TTI). The authors thank Susan Chrysler, Stephanie Quinn, and three anonymous referees for providing their inputs and comments that helped to improve this paper.

Author Contributions

The authors confirm contribution to the paper as follows: study conception and design: Bahar Dadashova; data collection: Bahar Dadashova; analysis and interpretation of results: Bahar Dadashova, Karen Dixon, and Raul Avelar; draft manuscript preparation: Bahar Dadashova, Karen Dixon, and Raul Avelar. All authors reviewed the results and approved the final version of the manuscript.

References

1. Firestine, M., H. McGee, and P. Toeg. Improving Truck Safety at Interchanges. Report No. FHWA-IP-89-024. Federal Highway Administration, Washington, D.C., 1989.
2. McCartt, A.T., V.S. Northrup, and R.A. Retting. Types and Characteristics of Ramp-Related Motor Vehicle Crashes on Urban Interstate Roadways in Northern Virginia. *Journal of Safety Research*, Vol. 35, No. 1, 2004, pp. 107–114.
3. Kim, S. K., T. Song, N. M. Rouphail, S. Aghdashi, A. Amaro, and G. Goncalves. Exploring the Association of Rear-End Crash Propensity and Micro-Scale Driver Behaviour. *Safety Science*, Vol. 89, 2016, pp. 45–54.
4. Hauer, E. Speed and Safety. *Transportation Research Record: Journal of the Transportation Research Board*, 2009. 2103: 10–17.
5. Elvik, R., P. Christensen, and A. Amundsen. *Speed and Road Accidents. An Evaluation of the Power Model.* Institute of Transport Economics Report 740. Norway, 2004.
6. Solomon, D. *Accidents on Main Rural Highways Related to Speed, Driver, and Vehicle.* ARRB Group Limited, Victoria, Australia, 1964.
7. Lave, C. A. Speed Choice, Coordination, and the 55 mph Limit. *The American Economic Review*, Vol. 75, No. 5, 1985, pp. 1159–1164.
8. Garber, N. J., and R. Gadiraju. Factors Affecting Speed Variance and Its Influence on Accidents. *Transportation Research Record: Journal of the Transportation Research Board*, 1989. 1213: 64–71.
9. Kondyli, A., and L. Elefteriadou. Modeling Driver Behavior at Freeway-Ramp Merges. *Transportation Research Record: Journal of the Transportation Research Board*, 2011. 2249: 29–37.
10. Haglund, M., and I. Åberg. Speed Choice in Relation to Speed Limit and Influences from Other Drivers. *Transportation Research Part F: Traffic Psychology and Behaviour*, Vol. 3, No. 1, 2000, pp. 39–51.
11. Aarts, L., and I. Van Schagen. Driving Speed and the Risk of Road Crashes: A Review. *Accident Analysis & Prevention*, Vol. 38, No. 2, 2006, pp. 215–224.
12. Jovanović, D., K. Lipovac, P. Stanojević, and D. Stanojević. The Effects of Personality Traits on Driving-Related Anger and Aggressive Behavior in Traffic Among Serbian Drivers. *Transportation Research Part F: Traffic Psychology and Behavior*, Vol. 14, No. 1, 2011, pp. 43–53.
13. Scott-Parker, B., B. Watson, M. J. King, and M. K. Hyde. A Further Exploration of Sensation Seeking Propensity, Reward Sensitivity, Depression, Anxiety, and the Risky Behavior of Young Novice Drivers in a Structural Equation Model. *Accident Analysis & Prevention*, Vol. 50, 2013, pp. 465–471.
14. National Highway Traffic Safety Administration. *Risky Driving.* https://www.nhtsa.gov/risky-driving. Accessed May 26, 2017.
15. Beanland, V., M. Selbom, and A. K. Johnson. Personality Domains and Traits that Predict Self-Reported Aberrant Driving Behaviors in a Southeastern US University Sample. *Accident Analysis & Prevention*, Vol. 72, 2014, pp. 184–192.
16. Bailey, S., A. Lennon, and B. Watson. Getting Mad May Not Mean Getting Even: The Influence of Drivers' Ethical Ideologies on Driving Anger and Related Behavior. *Transportation Research Part F: Traffic Psychology and Behavior*, Vol. 36, 2016, pp. 104–116.

17. Byun, H., and S. W. Lee. Applications of Support Vector Machines for Pattern Recognition: A Survey. *Proc., First International Workshop Pattern Recognition with Support Vector Machines,* Niagara Falls, ON, Canada, 2002, pp. 213–236.
18. Bracewell, R. N. The Fourier Transform and Its Applications (2nd ed.). McGraw-Hill, New York, NY, 1978.
19. Li, Z., X. Jin, and X. Zhao. Drunk Driving Detection Based on Classification of Multivariate Time Series. *Journal of Safety Research*, Vol. 54, 2015, pp.61–e29.
20. Liang, Y., L. M. Reyes, and J. D. Lee. Real-time Detection of Driver Cognitive Distraction Using Support Vector Machines. *IEEE Transactions on Intelligent Transportation Systems*, Vol. 8, No. 2, 2007, pp. 340–350.
21. Delen, D., R. Sharda, and M. Bessonov. Identifying Significant Predictors of Injury Severity in Traffic Accidents Using a Series of Artificial Neural Networks. *Accident Analysis & Prevention*, Vol. 38, No. 3, 2006, pp. 434–444.
22. *Strictest and Most Lenient States on Speed Choice and Reckless Driving*, https://wallethub.com/edu/strictest-and-most-lenient-states-on-speed choice/14211/.
23. Quinn, P. *Road Safety: Overcoming Driving Distractions. Attitude-Strategies and Support for ADHD and LD.* http://www.additudemag.com/adhd/article/904.html.
24. Antin, J. F. *SHRP 2 Report S2-S05-RR-1: Design of the In-vehicle Driving Behavior and Crash Risk Study: In Support of the SHRP 2 Naturalistic Driving Study*. Transportation Research Board of the National Academies, Washington D.C., 2011.
25. Jonah, B. A. Sensation Seeking and Risky Driving: A Review and Synthesis of the Literature. *Accident Analysis & Prevention*, Vol. 29, No. 5, 1997, pp. 651–665.
26. Thielen, I. P., R. C. Hartmann, and D. P. Soares. Risk Perception and Speed Choice. *Cadernos de Saude Publica*, Vol. 24, No. 1, 2008, pp. 131–139.
27. Hartmann, E., F. Baekland, and G. R. Zwilling. Psychological Differences between Long and Short Sleepers. *Archives of General Psychiatry*, Vol. 25, No. 5, 1972, pp. 463–468.
28. Chaovalit, P., A. Gangopadhyay, G. Karabatis, and Z. Chen. Discrete Wavelet Transform-based Time Series Analysis and Mining. *ACM Computing Surveys (CSUR)*, Vol. 43, No. 2, 2011, 6p.
29. Percival, D. B., and A. T. Walden. *Wavelet Methods for Time Series Analysis*, Cambridge University Press, Cambridge, MA, 2000.
30. Sakoe, H., and S. Chiba. Dynamic Programming Algorithm Optimization for Spoken Word Recognition. *IEEE Transactions on Acoustics, Speech, and Signal Processing*, Vol. 26, 1978, pp. 43–49.
31. Li, S. Z., and A. Jain. Dynamic Time Warping (DTW). In S. Z. Li, and A. Jain (eds.), *Encyclopedia of Biometrics*. Springer, Boston, MA, 2009, pp. 213–231.
32. Karlaftis, M. G., and E. I. Vlahogianni. Statistical Methods Versus Neural Networks in Transportation Research: Differences, Similarities and Some Insights. *Transportation Research Part C: Emerging Technologies*, Vol. 19, No. 3, 2011, pp. 387–399.
33. Olden, J. D., and D. A. Jackson. Illuminating the "Black Box": A Randomization Approach for Understanding Variable Contributions in Artificial Neural Networks. *Ecological Modelling*, Vol. 154, No. 1, 2002, pp. 135–150.
34. Giorgino, T. Computing and Visualizing Dynamic Time Warping Alignments in R: The dtw Package. *Journal of Statistical Software*, Vol. 31, No. 7, 2009, pp. 1–24.
35. Garson, G. D. Interpreting Neural Network Connection Weights. *Artificial Intelligence Expert*, Vol. 6, 1991, pp. 47–51.

The Standing Committee on Safety Data, Analysis and Evaluation (ANB20) peer-reviewed this paper (18-01431).

Article

TRR
JOURNAL OF THE TRANSPORTATION RESEARCH BOARD

Transportation Research Record
2018, Vol. 2672(38) 290–301

DOI: 10.1177/0361198118773869
journals.sagepub.com/home/trr
SAGE

Extracting Useful Information from Basic Safety Message Data: An Empirical Study of Driving Volatility Measures and Crash Frequency at Intersections

Mohsen Kamrani[1], Ramin Arvin[1], and Asad J. Khattak[1]

Abstract

With the emergence of high-frequency connected and automated vehicle data, analysts can extract useful information from them. To this end, the concept of "driving volatility" is defined and explored as deviation from the norm. Several measures of dispersion and variation can be computed in different ways using vehicles' instantaneous speed, acceleration, and jerk observed at intersections. This study explores different measures of volatility, representing newly available surrogate measures of safety, by combining data from the Michigan Safety Pilot Deployment of connected vehicles with crash and inventory data at several intersections. For each intersection, 37 different measures of volatility were calculated. These volatilities were then used to explain crash frequencies at intersection by estimating fixed and random parameter Poisson regression models. Given that volatility reflects the degree to which vehicles move, erratic movements are expected to increase crash risk. Results show that an increase in three measures of driving volatility are positively associated with higher intersection crash frequency, controlling for exposure variables and geometric features. More intersection crashes were associated with higher percentages of vehicle data points (speed & acceleration) lying beyond threshold-bands. These bands were created using mean plus two standard deviations. Furthermore, a higher magnitude of time-varying stochastic volatility of vehicle speeds when they pass through the intersection is associated with higher crash frequencies. These measures can be used to locate intersections with high driving volatilities. A deeper analysis of these intersections can be undertaken, and proactive safety countermeasures considered to enhance safety.

High-frequency connected vehicle (CV) data offers an opportunity to detect dispersions in vehicular speeds, accelerations, and jerks. Measures of dispersion attempt to quantify the spread of data. Commonly used dispersion measures include variance, range, minimum, and maximum values. In this paper, the concept of "driving volatility" is expanded, defined as deviation from the norm.

Volatility in driving reflects the degree to which a vehicle moves in three dimensions. If the vehicle's movements are erratic, then the risk of a crash is higher. Higher driving volatility is associated with higher safety risks, more fuel consumption, and increased emissions (*1*). The focus of this paper is to explore different measures of driving volatility, which have not yet been explored systematically in a spatial context.

CVs transmit high-frequency data between vehicles and road infrastructure. Widespread deployment of communication technologies has provided an unprecedented amount of data. Such "big data" combined with new tools can help researchers study, monitor, and evaluate transportation network performance in real-time (*2*, *3*). This study takes advantage of the big data provided by the Safety Pilot Model Deployment (SPMD). SPMD is a field test in Ann Arbor, Michigan, that offers detailed and relevant data demonstrating real-world vehicle-to-vehicle (V2V) and vehicle-to-infrastructure (V2I) communication. In this program, around 3,000 vehicles, equipped with Dedicated Short Range Communications (DSRC) devices, communicate with roadside equipment (4). The SPMD test provides rich information packages transmitted as Basic Safety Messages (BSM) through V2V and V2I communication. BSM contain the vehicles' position and motion information, their component

[1]Department of Civil & Environmental Engineering, The University of Tennessee, Knoxville, TN

Corresponding Author:
Address correspondence to Mohsen Kamrani: mkamrani@vols.utk.edu

status, and other information (4, 5). To explore the relationship between volatility and crash frequency, this study has created a new and unique database that integrates BSM, crashes, and inventory data to extract critical information from large-scale BSM data.

This study defines measures to quantify the driving volatility in a spatial context. Then it explores correlations between the measures of driving volatilities and crash frequencies at 116 intersections in Ann Arbor, MI, where the movements of sufficient instrumented vehicles were recorded. The objectives of the study are to

1. define and calculate several measures of volatility using vehicles' kinematic characteristics; and
2. identify measures of driving volatility (if any) that are strongly associated with crashes at intersections.

Given that driver behavior is the main contributing factor in crashes (*6*, *7*), findings from this study are beneficial in two ways. First, they can help proactively identify locations with high levels of driving volatility but that might not have many crashes as candidates for safety improvements. Second, reduction of driving volatility at high crash locations can reduce future crashes.

Literature Review

There are various definitions for aggressive driving in the literature, but there is little agreement among them. In the current literature, researchers often use the term "aggressive" for describing behaviors that threaten the safety of drivers and occupants in the host vehicle and other vehicles. In the U.S., aggressive driving such as speeding, failure to yield the right of way, and reckless driving account for more than 50% of fatal crashes (*8*). Different definitions of aggressive driving have been presented in the literature. Lajunen et al. (*9*) defined driver aggression as "any form of driving behavior that is intended to injure or harm other road users physically or psychologically." These behaviors vary from less aggressive forms such as flashing lights, verbal threats, tailgating, and cutting other vehicles off, to more extreme behaviors such as physical attacks (*10*). When it comes to instantaneous driver behavior, aggressive driving can be described using different aspects of vehicle kinematics such as speed, acceleration, and vehicular jerk.

Many previous studies used common vehicle kinematics to quantify aggressive behavior or deviation from normal behavior (*11–13*). One of the more favorable variables for describing aggressiveness is maximum acceleration or deceleration of the vehicle. In the urban driving environment, Kim et al. (*14*) suggested the threshold of 1.47 m/s^2 and 2.28 m/s^2 for aggressive and extreme aggressive acceleration. De Vlieger (*15*) defined different thresholds for different driving styles in urban areas, such as a range of 0.85 to 1.10 m/s^2, as aggressive driving. Han et al. (*16*) quantified variations in driving behaviors under different driving conditions by providing different acceleration thresholds that vary with speed of the vehicle. Vehicular jerk, change in acceleration rate with respect to time, has been used to classify aggressive driving style (*17*) by using the ratio of standard deviation to the mean of jerk within a time span for identifying accident-prone drivers (*18*). Feng et al. (*19*) showed that there are unique characteristics of vehicular jerk in gas pedal operations. Also, aggressive drivers are found to be associated with significantly higher values of vehicular jerk (*19*).

More recently, a new term "driver volatility" was introduced to describe the performance of driving behavior. The difference between "volatility" and "aggressiveness" is analogous to "crash" and "accident" (*20*). Different measures for driving volatilities have been used in the previous studies (*21*, *22*). Kamrani et al. (*22*) defined volatility score as the coefficient of variation (ratio of standard deviation to the mean) of acceleration and deceleration. To the best of the authors' knowledge, different measures of driving volatility have not been explored systematically, especially in the transportation context. Therefore, this study comprehensively explores several measures of driving volatility (applied to BSM data) and investigates their associations with intersection crash frequency.

Methodology

Various instantaneous driving measures can be used to quantify driving volatilities such as acceleration, brake position, and steering angle. Volatility in instantaneous driving behavior should be measured by considering both longitudinal and lateral acceleration. Considering speed, acceleration, or jerk solely as the measure of volatility might ignore the importance of information embedded in the data. However, given a significant questionable error in the lateral acceleration data (22), only longitudinal acceleration, speed, and jerk are used in this paper. It should be noted that excluding lateral acceleration does not affect the results drastically for two reasons. First, the lateral acceleration is more critical if there is a noticeable amount of curvature in the trip, whereas the boundary of the intersection in this study is limited to 150 ft from the center toward each approach. Second, in the area of an intersection, the traveled distance is short (called "passing" in this paper), and the geometry of the intersection does not allow drivers to have considerable lane changing space.

A total of 116 intersections were selected in the city of Ann Arbor, MI to extract BSM data consisting of speed, longitudinal acceleration (hereafter acceleration), time and geocodes. For each intersection, appropriate polygons are drawn based on 150 ft from the center of intersection toward all approaches. These polygons are used to filter the BSM data based on the longitude and latitude values available in the data. After the filtration, out of nearly 2,500,000,000 BSM, 215,000,000 were found to be at the selected intersections. Data at this level are used for "level 1" calculations of driving volatilities (discussed later). The time and device ID variables of the BSM are used to identify passings taken by each vehicle. Around 3,300,000 passings have been taken by more than 900 vehicles. Data from this step are used to do "level 2" calculations of driving volatilities (discussed later) at intersections. Crash and inventory data were also collected for individual intersections. The driving volatility and intersection-related data are integrated to form the final dataset. The study uses rigorous modeling techniques that are suitable for the analysis of newly available volatility data.

Measures of Driving Volatility

Although some of the measures used for volatility are common, as shown in Table 1, other measures presented are relatively new in the transportation field. Variations in longitudinal control of a vehicle are reflected in speed, acceleration, and vehicular jerk. The values of vehicular speed and acceleration are available directly from BSM data whereas the jerk values are calculated from the acceleration values, since it is the rate of change of acceleration.

Standard Deviation. A key measure for quantifying volatility is the standard deviation (S_{dev}) which is a simple and desirable statistic used for expressing variation in data.

$$S_{dev} = \sqrt{\frac{1}{n-1}\sum_{i=1}^{n}(x_i - \bar{x})^2} \tag{1}$$

where x_i is the value of observation i, $\bar{x}$ is the mean, and n is the number of observations.

Coefficient of Variation. A basic measure of dispersion is the coefficient of variation which is obtained from the division of S_{dev} by the mean (*22, 23*), providing a relative measure of dispersion

$$C_v = \frac{S_{dev}}{|\bar{x}|} * 100 \tag{2}$$

Mean Absolute Deviation around Central Point. This measure is defined as the average distance between each observation and the central tendency of the dataset (here, mean) which is defined as (*24*)

Table 1. Summary of Measures for Driving Volatility Quantification

		Applied to vehicular						
			Acceleration			Jerk		
Measure of driving volatility	Formula	Speed	+	−	Both	+	−	Both
Standard deviation	$S_{dev} = \sqrt{\frac{1}{n-1}\sum_{i=1}^{n}(x_i - \bar{x})^2}$	√			√			√
Coefficient of variation	$C_v = \frac{S_{dev}}{\lvert\bar{x}\rvert} * 100$	√	√	√		√	√	
Mean absolute deviation	$D_{mean} = \frac{1}{n}\sum_{i=1}^{n}\lvert x_i - \bar{x}\rvert$	√			√			√
Quartile coefficient of variation	$Q_{c_v} = \frac{Q_3 - Q_1}{Q_3 + Q_1} * 100$	√	√	√		√	√	
Percent of extreme values	$\%T = \frac{c > Threshold}{n} * 100$; $Threshold = \bar{x} \pm z * S_{dev}$	√			√			√
Time-varying stochastic volatility	$r_i = \ln\left(\frac{x_i}{x_{i-1}}\right) * 100$; $V_f = \sqrt{\frac{1}{n-1}\sum_{i=1}^{n}(r_i - \bar{r})^2}$	√						

$$D_{\text{mean}} = \frac{1}{n}\sum_{i=1}^{n}|x_i - \bar{x}| \quad (3)$$

Quartile Coefficient of Variation. Another measure for describing dispersion of a dataset is the quartile coefficient of variation, especially when the sample has non-normal distribution. The quartile coefficient of variation is defined as (*25*)

$$Q_{C_V} = \frac{Q_3 - Q_1}{Q_3 + Q_1} * 100 \quad (4)$$

where Q_1 and Q_3 are the sample 25th and 75th percentiles, respectively.

Count of Extreme Values. This measure captures driving volatility by counting the number of observations beyond a defined threshold-band. The function is (*26*)

$$\%T = \frac{c > \text{Threshold}}{n} * 100 \quad (5)$$

where c is the count of observations beyond the threshold. The threshold-band can be defined as (*26*)

$$\text{Threshold} = \bar{x} \pm z * S_{\text{dev}} \quad (6)$$

where z represents the distance between a mean and a point in units of S_{dev}, that is, $z = 1, 2, 3$, and so forth. Application of this measure takes into account the magnitude of vehicular speed when calculating volatility of acceleration (*22*). Figure 1 shows how the speed bin concept is applied to the real-world acceleration data obtained from the BSM. Notably, the ability of a vehicle to accelerate declines with higher speeds. Therefore, instead of having a fixed pair of upper and lower bounds to count the number of acceleration and deceleration extreme points, speed bins of 5 mph are used in this study. The upper and lower bound for each bin are calculated using its mean and S_{dev}. Similarly, vehicular jerk is classified based on corresponding speed bins.

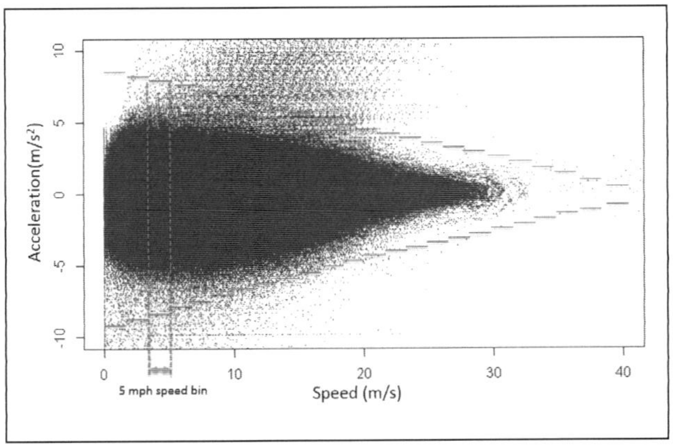

Figure 1. Speed bins for calculating acceleration thresholds at various speeds using BSM data.

Time-Varying Stochastic Volatility. The time-varying stochastic volatility which is commonly used in finance is computed by (*27, 28*)

$$V_f = \sqrt{\frac{1}{n-1}\sum_{i=1}^{n}(\mathrm{r_i} - \bar{r})^2} \text{ from } i = 1 \text{ to } n \quad (7)$$

where

$$\mathrm{r_i} = \ln\left(\frac{x_i}{x_{i-1}}\right) * 100 \quad (8)$$

and

x_{i-1} = previous observation (in this study instantaneous vehicular speed), and

ln is the natural logarithm.

This measure requires positive time-series observations; therefore, it is not applicable to the acceleration and jerk values due to their negative values. Using only the positive values of acceleration and jerk will be inconsistent with the time-series nature of data required by this measure. That said, this measure is applied to speed at the vehicle passing level (level 2), which is discussed next.

Two Levels for Calculating Volatility

Volatility measures can be applied in two ways to obtain driving volatility at intersections as shown in Figure 2.

Level 1 Calculation of Volatility. The level 1 calculation of driving volatility disregards the individual passings and treats all data for each intersection as bulk (at the aggregate level, $N \sim 215{,}000{,}000$). Compared with Level 2, this calculation is simpler, easier, and faster to perform.

Level 2 Calculation of Volatility. In this method, volatility of each passing at the intersections is calculated separately. For this, the time and device ID available in BSM are used to identify the passings. The averages of calculated volatilities for all passings are reported as measures of volatility for each intersection. Nearly 3,300,000 passings were identified for 116 intersections during the two-month period taken by around 900 unique device IDs.

Notation of Variables

Applying each of the measures to the speed, acceleration, and jerk at two levels results in 37 driving volatility values for each intersection. To distinguish them, a notation

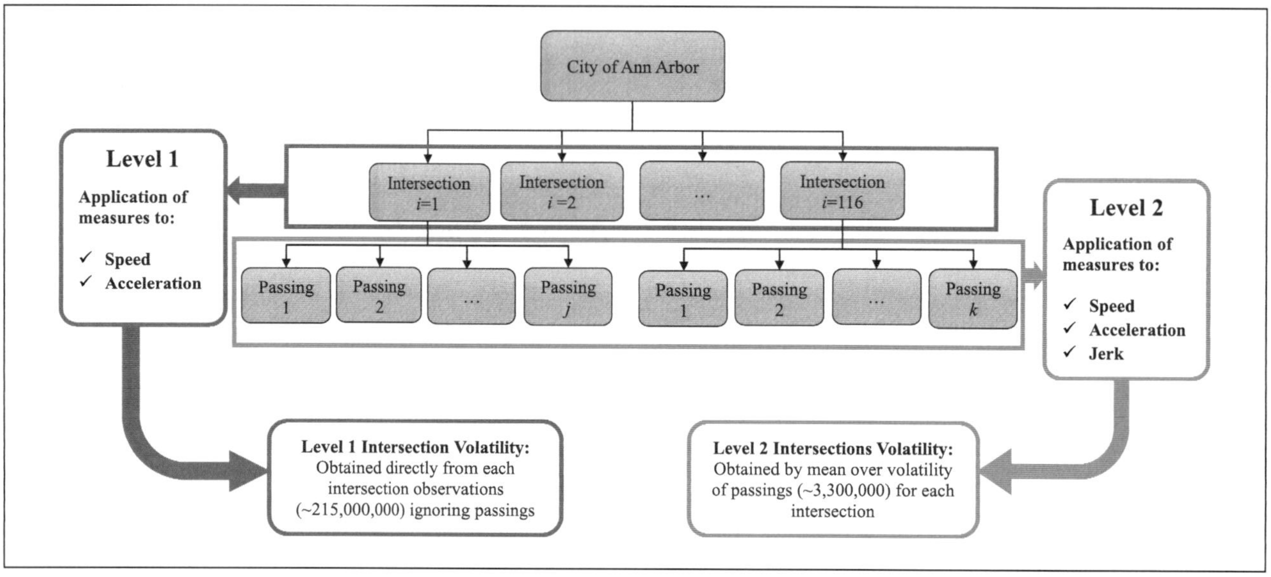

Figure 2. Measures of driving volatility at intersections.

system is used in which the volatilities have three terms in their names separated by a dash.

- The first term is either L_1 for Level 1 or L_2 for Level 2 indicating the method of calculation.
- The second term indicates the element to which the volatility measure is applied. Since some of the measures of volatilities necessitate the separation of positive and negative values, the second term can have the following notation:
  - Speed: vehicular speed
  - AccDec: both positive and negative values of acceleration
  - Accel: positive values of acceleration
  - Decel: negative values of acceleration
  - Jerk: vehicular jerk calculated from acceleration
  - PosJerk: positive values of jerk
  - NegJerk: negative values of jerk
- The last term shows what measure was applied to obtain the volatility. For example, if S_{dev} is applied to the acceleration (both positive and negative values) for individual passings (level 2), the variable will be named: L_2-AccDec-S_{dev}.

Modeling Approach

Count-data models are usually applied to accident frequency because the number of crashes is a non-negative integer (*29*). Poisson, negative binomial and zero-inflated are common count models used for this purpose (*30, 31*). For the Poisson regression model, the probability of having n crashes at intersection i is (*32*)

$$P(n_i) = \frac{\lambda_i^{n_i} \exp(-\lambda_i)}{n_i!} \tag{9}$$

where $P(n_i)$ is the probability of having n crashes at intersection i, and λ_i is the Poisson parameter for the intersection i. These are the expected number of crashes for the intersections in each year. In order to fit the model, λ_i can be expressed in the logarithm form as the function of a set of independent variables (*32*)

$$\ln(\lambda_i) = \beta X_i \tag{10}$$

where X_i is a vector of explanatory variables, and β is a vector of estimated coefficients. The Poisson function defined in Equations 9 and 10 can be maximized by standard maximum likelihood procedures.

Applying Poisson regressions to the data while the mean and variance are not equal might lead to inappropriate results. To address over-dispersion ($E(n_i) < \text{VAR}(n_i)$), or under-dispersion ($E(n_i) > \text{VAR}(n_i)$) in the data, the negative binomial model can be derived as

$$\lambda_i = \exp(\beta X_i + \varepsilon_i) \tag{11}$$

where error term, $\exp(\varepsilon_i)$, is a gamma-distributed with mean 1 and variance α. The additional term, allows variance to be different from the mean,

$$\text{Var}(n_i) = E(n_i) + \alpha E(n_i)^2 \tag{12}$$

where $\text{Var}(n_i)$ and $E(n_i)$ are the variance and the expected number of crashes respectively.

Choosing between Poisson and negative binomial regression depends on the estimated α parameter. If α

significantly does not differ from zero, Poisson regression model should be used. Otherwise, the negative binomial model is appropriate (*33*). Although the presence of over-dispersion can be evaluated by the mean and variance of crash data (33), a Lagrange multiplier (LM) can be used to statistically test the existence of over-dispersion in Poisson model (32).

On the other hand, it is possible that associations between independent variables and the dependent variable are not consistent across all observations. Several observed and unobserved factors associated with crash frequency might lead to unobserved heterogeneity (34–38). To address the heterogeneity with random parameters, using simulated maximum likelihood estimation, Greene (32) developed an approach to model random parameters in the Poisson model. Equation 13 indicates the formulation of estimated coefficients as

$$\beta_i = \beta + \varphi_i \tag{13}$$

where φ_i is a randomly distributed term with any specified distribution (e.g., normal distribution with mean zero and standard deviation of σ). The negative binomial parameter in Equation 10 can be written as

$$\lambda_i|\varphi_i = e^{(\beta X_i + \varepsilon_i)} \tag{14}$$

The log-likelihood function for the random-parameter model can be written as (*29*)

$$\mathrm{LL} = \sum_i \ln \int_{\varphi_i}^{i} g(\varphi_i)P(n_i|\varphi_i)d\varphi_i \tag{15}$$

where $g(.)$ is the pre-specified probability density function for φ_i. To maximize the log-likelihood function, a simulation-based approach using Halton draws can be used. Different studies (*39*, *40*) have shown that Halton draws is more efficient in comparison with random draws. Further details on random parameter models can be found in (*29*).

Data

The data used in this study are the result of integrating BSM from the Michigan Safety Pilot with intersection crash and inventory data. The steps for data preparation are shown in Figure 3. The BSM data were collected, under real-world conditions, at the Ann Arbor test site by equipping around 3,000 vehicles with DSRC devices enabling them to log different variables including their instantaneous speed, acceleration heading, coordinates, and so forth at usually 10 Hz. The data is accessible via ITS Public Data Hub (https://www.its.dot.gov/data/), maintained by the Federal Highway Administration under U.S. DOT. Speed, acceleration, longitude, and latitude values of the complete two-month data (October and April 2012) were utilized in this study. The data examination and error-checking process shows high accuracy in the variables used. For instance,

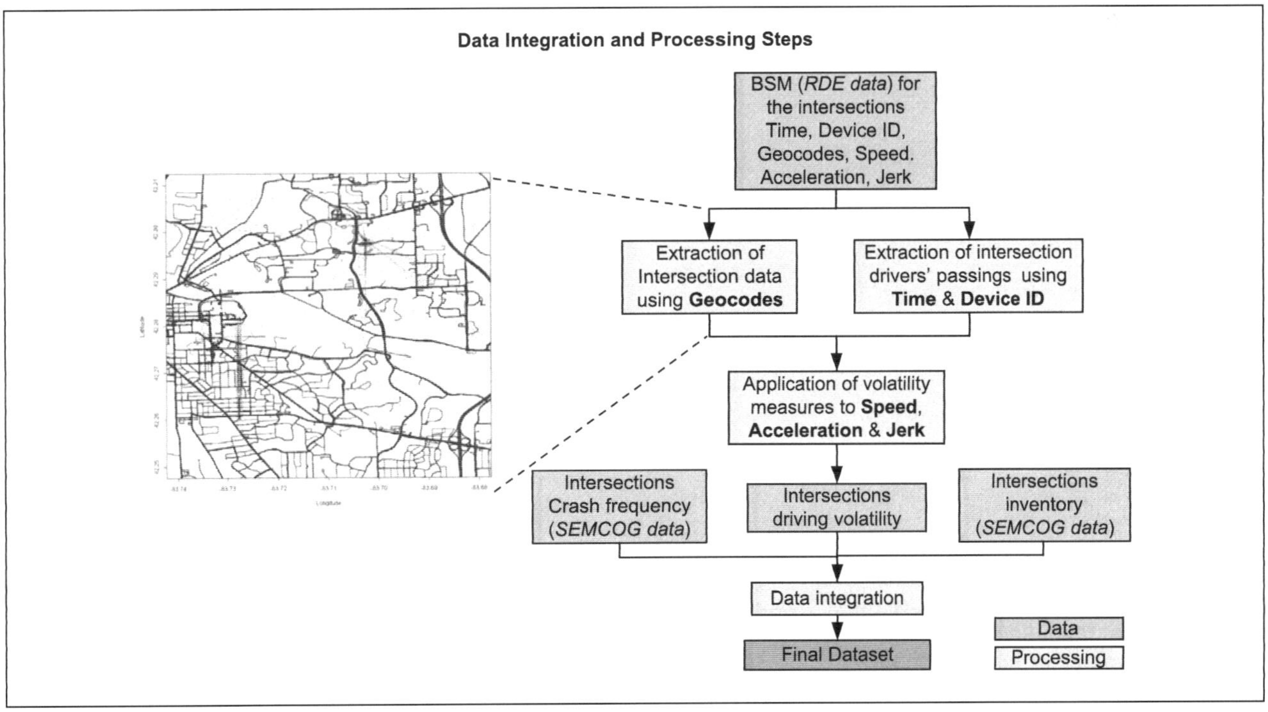

Figure 3. Ann Arbor map created from BSM data, and data preparation steps.

Table 2. Descriptive Statistics of Measures of Volatilities ($n = 116$)

Variables	Mean	S_{dev}	Min	Max
Intersection-related variable				
Average crashes (5 years)	7.56	7.64	0	44
AADT major road	20805	8326	3100	45400
AADT minor road	9396	4138	1100	27400
Speed limit major road	35.34	7.24	25	45
Speed limit minor road	30.47	3.95	25	45
Signalized intersection (yes = 1)	0.46	0.5	0	1
4-legged intersection (yes = 1)	0.4	0.49	0	1
Total through lanes	4.45	1.28	2	8
Total left turn lanes	1.53	1.32	0	6
Total right turn lanes	0.93	0.78	0	4
Volatility of Level 1 variables (ignoring individual vehicle passings)*				
L_1-Speed-S_{dev} (m/s)	11.35	2.4	4.92	16.41
L_1-Speed-C_v (%)	45	16	13	71
L_1-Speed-Q_{c_v} (%)	32	16	6	61
L_1-Speed-D_{mean} (m/s)	7.85	1.96	3.21	12.32
L_1-Speed-%$T(1S_{dev})$ (%)	28	13	11	59
L_1-Speed-%$T(2S_{dev})$ (%)	4	3	0	11
L_1-AccDec-S_{dev} (m/s^2)	0.75	0.17	0.34	1.43
L_1-Accel-C_v (%)	59	6	44	73
L_1-Decel-C_v (%)	65	9	51	103
L_1-Accel-Q_{c_v} (%)	39	6	23	51
L_1-Decel-Q_{c_v} (%)	44	7	23	59
L_1-AccDec-D_{mean} (m/s^2)	0.4	0.09	0.15	0.52
L_1-AccDec-%$T(1S_{dev})$ (%)	23	4	14	36
L_1-AccDec-%$T(2S_{dev})$ (%)	7	1	3	9
Volatility of Level 2 variables (averaged over passings)*				
L_2-Speed-S_{dev} (m/s)	2.02	0.95	0.41	5.28
L_2-Speed-V_f (%)	2	2	0	6
L_2-Speed-C_v (%)	15	10	1	40
L_2-Speed-Q_{c_v} (%)	10	7	1	26
L_2-Speed-D_{mean} (m/s)	1.49	0.7	0.3	3.47
L_2-Speed-%$T(1S_{dev})$ (%)	34	2	29	39
L_2-Speed-%$T(2S_{dev})$ (%)	2	1	1	4
L_2-AccDec-S_{dev} (m/s^2)	0.4	0.13	0.17	1.18
L_2-Accel-C_v (%)	27	6	15	43
L_2-Decel-C_v (%)	29	5	16	44
L_2-Accel-Q_{c_v} (%)	18	4	10	28
L_2-Decel-Q_{c_v} (%)	20	4	12	29
L_2-AccDec-D_{mean} (m/s^2)	0.16	0.06	0.05	0.35
L_2-AccDec-%$T(1S_{dev})$ (%)	36	4	27	49
L_2-AccDec-%$T(2S_{dev})$ (%)	4	1	2	8
L_2-Jerk-S_{dev} (m/s^3)	1.37	0.15	1.04	1.78
L_2-JerkPos-C_v (%)	59	3	52	65
L_2-JerkNeg-C_v (%)	59	3	52	64
L_2-JerkPos-Q_{c_v} (%)	44	3	32	48
L_2-JerkNeg-Q_{c_v} (%)	44	3	32	47
L_2-Jerk-D_{mean} (m/s^3)	0.81	0.11	0.56	1.09
L_2-Jerk-%$T(1S_{dev})$ (%)	26	1	23	28
L_2-Jerk-%$T(2S_{dev})$ (%)	7	1	4	10

Note: *L_1 = level 1 calculation; L_2 = level 2 calculation; S_{dev} = standard deviation; %$T(1S_{dev})$ = % of extreme points beyond mean ± one S_{dev}; %$T(2S_{dev})$ = % of extreme points beyond mean ± two S_{dev}; C_v = coefficient of variation; Q_{c_v} = quartile coefficient of variation; D_{mean} = mean absolute deviation; V_f = time-varying stochastic volatility; Accel = acceleration; Decel = deceleration; AccDec = both acceleration & deceleration; JerkPos = positive jerk; JerkNeg = negative jerk.

the accuracy of the map created from BSM shown in Figure 3 is a good indication of data precision.

Intersection specific data such as the average number of crashes (2010–2014), annual average daily traffic (AADT), and speed limits for all approaches were collected. The dataset was error checked (by a third person randomly double checking 10% of the data) and verified. The data can be obtained via the Metropolitan Planning Organization

Table 3. Fixed and Random Parameter Poisson Model Results

Variables	Fixed parameter			Random parameter		
	Estimate[a]	z value	Marginal effect	Estimate[a]	z value	Marginal effect
Constant	−1.497***	−4.73	–	−1.852***	−5.42	–
Intersection-related						
AADT major approach (1000)	0.033***	7.39	0.25	0.033***	7.84	0.17
Scale parameter	–	–	–	0.007***	5.36	–
AADT minor approach (1000)	0.023***	3.55	0.17	0.024***	3.70	0.12
Signalized intersection (yes = 1)	0.789***	6.01	5.21	0.704***	5.77	3.58
Four-legged intersection (yes = 1)	0.260**	3.11	1.95	0.248***	2.93	1.26
Measures of volatility[b]						
L_1-Speed-%$T(2S_{dev})$	0.050***	3.57	0.38	0.041***	2.97	0.21
Scale parameter	–	–	–	0.065***	8.53	–
L_1-AccDec-%$T(2S_{dev})$	0.225***	4.38	1.70	0.260***	4.63	1.32
L_2-Speed-V_f	0.061	1.92	0.47	0.109***	3.47	0.55
Summary statistics						
AIC	609.65			585.6		
Log-likelihood at zero $L(0)$	−578.32			−578.32		
Log-likelihood at convergence $L(\beta)$	−296.83			−282.81		
McFadden (ρ^2)	0.487			0.517		
Sample size (N)	116			116		

[a]Significance codes: ***0.01%, **1%, *5%, · 10%
[b]L_1: level 1 calculation; L_2: level 2 calculation; %$T(2S_{dev})$: % of extreme points beyond mean ± two S_{dev}; V_f: time-varying stochastic volatility; AccDec: both acceleration & deceleration.

website: http://semcog.org/Crash-and-Road-Data. Among intersections in the Ann Arbor area, 116 intersections were identified, keeping in mind that enough BSM data should be available for the calculation of different measures of driving volatility. Finally, appropriate geocodes were used to filter out BSM data for each intersection. These BSM were used to calculate 37 different measures of driving volatilities. The final dataset was created by integrating intersection inventory data, crash data and computed driving volatilities.

Results

Descriptive Statistics

Table 2 presents the descriptive statistics. For all intersections, the five-year mean of crashes is 7.56 with standard deviation of 7.64. About 46% of the intersections are signalized, 40% of the intersections are 4-legged, and the rest are T-intersections. Table 2 also presents the descriptive statistics of variables calculated from BSM data, that is, measures of volatilities. Please note that the unit of analysis is the intersection.

Correlations

Given the number of computed volatilities, correlation analysis may shed some light on relationships between crash frequency and driving volatilities (Figure 4). Bars in the figure are sorted based on the value of positive correlation. Blue bars show volatilities with a positive correlation between average crashes whereas the red ones indicate negatively correlated volatilities. This figure was used as a guide to insert variables in the model specification and to examine their associations and improvements in model fit. As expected, there is a high level of correlation among some of the computed volatilities. For instance, two highly correlated volatilities at the bottom of the figure (L_2-AccDec-$1S_{dev}$ and L_2-AccDec-$2S_{dev}$) are calculated in a similar way, with the only difference being in the number of standard deviations from the mean. If such highly correlated variables are used simultaneously in estimation, then the model may suffer from multicollinearity. Using engineering judgment and variance inflation factor (VIF > 5), multicollinearity was addressed in the model specification.

Modeling Results and Discussion

Table 3 provides the results for fixed and random parameter Poisson regression. Fixed parameter model is estimated for crash frequency as a function of intersection-related variables and measures of driving volatility. Starting out with intersection-related variables and keeping the significant ones in the model, measures of volatility variables were inserted in the model based on correlations from Figure 4. The model fits were compared using Akaike information criterion (AIC) and log-likelihood.

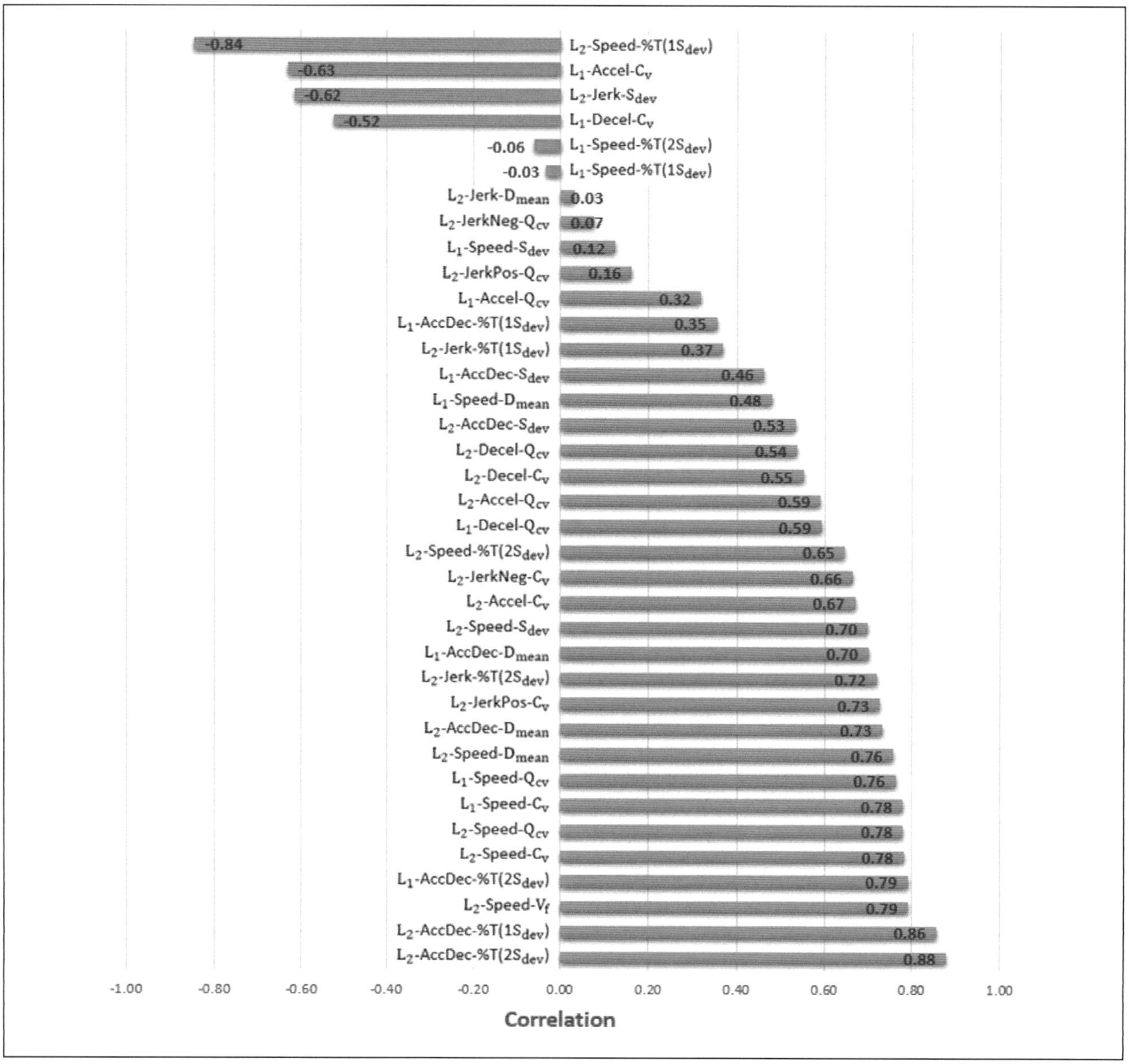

Figure 4. Correlations between crash frequency and measures of volatilities.

The random-parameter Poisson model is estimated (using simulated maximum likelihood) assuming a normal distribution for random parameters (29). Compared to the fixed parameter model, the random parameter model shows a better fit based on log-likelihood, AIC, and McFadden ρ^2 (33). As Figure 5 shows, the random-parameter model outperforms the fixed parameter in terms of crash frequency prediction.

The marginal effects are shown in Table 3. These effects are the average increases in crash frequencies of intersections given one unit increase in the respective independent variable. For instance, a one-percent increase in the time-varying stochastic volatility of speed (L_2-Speed-V_f) is associated with a 0.55 increase in average crash frequency. That means a higher magnitude of time-varying stochastic volatility of vehicle speeds when they pass through the intersection is associated with higher crash frequencies, as expected. In addition, more intersection crashes were associated with higher percentages of vehicle data points (speed & acceleration) lying beyond threshold-bands created using mean plus two S_{dev} at intersections (L_1-Speed-$\%T(2S_{dev})$ and L_1-AccDec-$\%T(2S_{dev})$ variables).

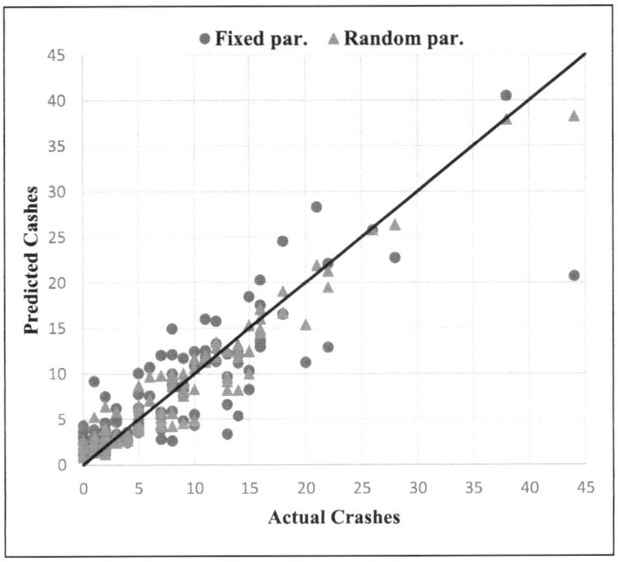

Figure 5. Expected–actual number of crashes for fixed and random parameter models.

Other variables which are used as controls in the model show the expected signs and magnitudes. According to Table 3, 1,000 more vehicles per day on the major approach is associated with a 0.17 increase in crash frequency. As expected, the association of the minor approach AADT is less than the major approach ADDT. An increase of 1,000 vehicles on the minor road is correlated with a 0.12 increase in crash frequency. According to the model, signalized intersections on average have 3.58 more crashes than un-signalized ones. Likewise, 4-legged intersections on average have more crashes than 3-legged intersections.

Limitations

The sample data used in this study does not come from representative drivers. This study did not consider volatility in the lateral direction, which could result in a sideswipe crash. Given that lane change frequency is generally relatively small at intersections, the results might not be considerably different. Furthermore, the data used in this study is the product of averaging 5-year crashes and using two-month BSM data. In other words, a short period of instantaneous driving behavior was used to explore correlations with 5-year average crash frequencies. The authors have used all available data to make the results as accurate as possible, even though handling and processing such large-scale data was difficult. Although the data was error-checked, it is possible that some errors, made during collection of data, remain. This paper considers only crash frequency although it is worthwhile to investigate the associations of driving volatility with crash severity. Finally, it should be noted that only the means of calculated volatilities for passings (level 2 volatility) were used to model volatility at each intersection, whereas the between-passings variation could also be used as measures of volatility.

Conclusions

This study discusses a way to extract useful information in the form of driving volatility from newly available BSM data. Such data are increasingly becoming available, providing a valuable resource for studying vehicle kinematics and microscopic behaviors of drivers, for example, instantaneous vehicle speed, acceleration, and jerk. This study creates a new and unique database (BSM data integrated with crash and inventory data) and mines critical information from large-scale BSM data. More than 2,500,000,000 BSM were processed along with crash and inventory data from 116 intersections in the city of Ann Arbor, Michigan. Volatilities of vehicles passing within 150 ft from the center of each intersection were calculated. Using nearly 215,000,000 observations for nearly 3,300,000 passings, 37 measures of driving volatility were calculated. To explore relationships between measures of driving volatility and crash frequency at intersections, rigorous statistical models were estimated. The models account for unobserved heterogeneity associated with crashes at intersections.

Three measures of driving volatilities show positive and statistically significant association with crash frequencies at the intersections. More intersection crashes are found to be associated with a higher percentage of BSM data points of speed and acceleration lying beyond the threshold-bands created using mean plus two standard deviations at intersections. Furthermore, a higher magnitude of time-varying stochastic volatility of vehicle speeds as they pass through the intersection is associated with higher crash frequency. The findings are significant in the sense that they can be used to identify intersections with high levels of driving volatility. In particular, intersections where crash frequency may be low, but the volatility is high, may be good candidates for further study and future safety treatments. These are likely to be intersections where crashes are waiting to happen due to higher driving volatility. Such intersections can be proactively examined to find the causes of driving volatility to prevent crashes. Higher levels of driving volatility might be due to outdated signal timing, higher speed limits, limited line of sight, inappropriate signal timing, and so forth. In practice, depending on the detected reasons, proactive countermeasures can be taken to reduce drivers' volatility. In addition, appropriate alerts can be given to vehicle drivers when they are approaching locations (*41*) with a high level of driving volatility.

Acknowledgments

This paper is based on work supported by National Science Foundation (Award number: 1538139). Additional support was provided by the U.S. Department of Transportation through the Collaborative Sciences Center for Road Safety, a consortium led by The University of North Carolina at Chapel Hill in partnership with The University of Tennessee.

Author Contributions

The authors confirm contribution to the paper as follows: study conception and design: Asad J Khattak, Mohsen Kamrani, Ramin Arvin; data collection: Mohsen Kamrani, Ramin Arvin, Asad J Khattak; analysis and interpretation of results: Ramin Arvin, Mohsen Kamrani, Asad J Khattak; draft manuscript preparation: Mohsen Kamrani, Ramin Arvin, Asad J Khattak. All authors reviewed the results and approved the final version of the manuscript.

References

1. "Brian" Park, B., I. Yun, and K. Ahn. Stochastic Optimization for Sustainable Traffic Signal Control. *International Journal of Sustainable Transportation*, Vol. 3, No. 4, 2009, pp. 263–284.
2. Arvin, R., M. Kamrani, A. J. Khattak, and J. Rios-Torres. Safety Impacts of Automated Vehicles in Mixed Traffic. Presented at 97th Annual Meeting of the Transportation Research Board, Washington D.C., 2018.
3. Ghiasi, A., O. Hussain, Z. S. Qian, and X. Li. A Mixed Traffic Capacity Analysis and Lane Management Model for Connected Automated Vehicles: A Markov Chain Method. *Transportation Research Part B: Methodological*, Vol. 106, 2017, pp. 266–292.
4. Henclewood, D. *Safety Pilot Model Deployment–One Day Sample Data Environment Data Handbook*. Research and Technology Innovation Administration. Research and Technology Innovation Administration, US Department of Transportation, McLean, VA, 2014.
5. Zhang, M., and A. J. Khattak. Identifying and Analyzing Extreme Lane Change Events Using Basic Safety Messages in a Connected Vehicle Environment. Presented at 97th Annual Meeting of the Transportation Research Board, Washington, D.C., 2018.
6. Kludt, K., J. Brown, J. Richman, and J. Campbell. *Human Factors Literature Reviews on Intersections, Speed Management, Pedestrians and Bicyclists, and Visibility*. Report No. FHWA-HRT-06-034, Federal Highway Administration, Washington, D.C., 2006.
7. Arvin, R., M. Khademi, and H. Razi-Ardakani. Study on Mobile Phone Use While Driving in a Sample of Iranian Drivers. *International Journal of Injury Control and Safety Promotion*, Vol. 24, No. 2, 2017, pp. 256–262.
8. Wali, B., A. J. Khattak, and J. Xu. Contributory Fault and Level of Personal Injury to Drivers Involved in Head-on Collisions: Application of Copula-based Bivariate Ordinal Models. *Accident Analysis & Prevention*, Vol. 110, 2018, pp. 101–114.
9. Lajunen, T., D. Parker, and S. G. Stradling. Dimensions of Driver Anger, Aggressive and Highway Code Violations and Their Mediation by Safety Orientation in UK Drivers. *Transportation Research Part F: Traffic Psychology and Behaviour*, Vol. 1, No. 2, 1998, pp. 107–121.
10. Özkan, T., T. Lajunen, D. Parker, N. Sümer, and H. Summala. Symmetric Relationship Between Self and Others in Aggressive Driving Across Gender and Countries. *Traffic Injury Prevention*, Vol. 11, No. 3, 2010, pp. 228–239.
11. Papazikou, E., M. A. Quddus, and P. Thomas. Detecting Deviation from Normal Driving Using SHRP2 NDS Data. Presented at 96th Annual Meeting of the Transportation Research Board, Washington, D.C., 2017.
12. Ghasemzadeh, A., and M. M. Ahmed. Drivers' Lane-Keeping Ability in Heavy Rain: Preliminary Investigation Using SHRP 2 Naturalistic Driving Study Data. *Transportation Research Record: Journal of the Transportation Research Board*, 2017. 2663: 99–108.
13. Ahmed, M. M., and A. Ghasemzadeh. Exploring the Impacts of Adverse Weather Conditions on Speed and Headway Behaviors Using the SHRP2 Naturalistic Driving Study Data. Presented at 96th Annual Meeting of the Transportation Research Board, Washington, D.C., 2017.
14. Kim, E., and E. Choi. Estimates of Critical Values of Aggressive Acceleration from a Viewpoint of Fuel Consumption and Emissions. Presented at 92nd Annual Meeting of the Transportation Research Board, Washington, D.C., 2013.
15. De Vlieger, I., D. De Keukeleere, and J. Kretzschmar. Environmental Effects of Driving Behaviour and Congestion Related to Passenger Cars. *Atmospheric Environment*, Vol. 34, No. 27, 2000, pp. 4649–4655.
16. Han, I., and K. Yang. Characteristic Analysis for Cognition of Dangerous Driving Using Automobile Black Boxes. *International Journal of Automotive Technology*, Vol. 10, No. 5, 2009, pp. 597–605.
17. Murphey, Y. L., R. Milton, and L. Kiliaris. Driver's Style Classification Using Jerk Analysis. *IEEE Workshop on Computational Intelligence in Vehicles and Vehicular Systems, 2009. CIVVS'09.*, Nashville, TN, IEEE, New York, 2009, pp. 23–28.
18. Bagdadi, O., and A. Várhelyi. Development of a Method for Detecting Jerks in Safety Critical Events. *Accident Analysis & Prevention*, Vol. 50, 2013, pp. 83–91.
19. Feng, F., S. Bao, J. R. Sayer, C. Flannagan, M. Manser, and R. Wunderlich. Can Vehicle Longitudinal Jerk be Used to Identify Aggressive Drivers? An Examination Using Naturalistic Driving Data. *Accident Analysis & Prevention*, Vol. 104, 2017, pp. 125–136.
20. Stewart, A. E., and J. H. Lord. Motor Vehicle Crash Versus Accident: A Change in Terminology is Necessary. *Journal of Traumatic Stress*, Vol. 15, No. 4, 2002, pp. 333–335.
21. Khattak, A. J., and B. Wali. Analysis of Volatility in Driving Regimes Extracted from Basic Safety Messages Transmitted Between Connected Vehicles. *Transportation Research Part C: Emerging Technologies*, Vol. 84, 2017, pp. 48–73.
22. Kamrani, M., B. Wali, and A. J. Khattak. Can Data Generated by Connected Vehicles Enhance Safety? Proactive Approach to Intersection Safety Management. *Transportation Research Record: Journal of the Transportation Research Board*, 2017. 2659: 80–90.

23. Wali, B., A. J. Khattak, H. Bozdogan, and M. Kamrani. How is Driving Volatility Related to Intersection Safety? A Bayesian Heterogeneity-Based Analysis of Instrumented Vehicles Data. *Transportation Research Part C: Emerging Technologies*, 2018. (Revised version submitted).
24. Huber, P. J. *Robust Statistics*. John Wiley & Sons, New York, 2005.
25. Bonett, D. G. Confidence Interval for a Coefficient of Quartile Variation. *Computational Statistics & Data Analysis*, Vol. 50, No. 11, 2006, pp. 2953–2957.
26. Kamrani, M., A. J. Khattak, and T. Li. A Framework to Process and Analyze Driver, Vehicle and Road Infrastructure Volatilities in Real-time. Presented at 97th Annual Meeting of the Transportation Research Board, Washington D.C., 2018.
27. Figlewski, S. *Forecasting Volatility Using Historical Data*. Working Paper, New York University, Stern School of Business, New York, 1994.
28. Kamrani, M., R. Arvin, and A. J. Khattak. Analyzing Highly Volatile Driving Trips Taken by Alternative Fuel Vehicles. Presented at 97th Annual Meeting of the Transporation Research Board, Washington D.C., 2018.
29. Anastasopoulos, P. C., and F. L. Mannering. A Note on Modeling Vehicle Accident Frequencies with Random-parameters Count Models. *Accident Analysis & Prevention*, Vol. 41, No. 1, 2009, pp. 153–159.
30. Zhang, M., A. J. Khattak, and E. Shay. Analysis of Crashes Involving Pedestrians Across the United States: Implications for Connected and Automated Vehicles. Presented at 97th Annual Meeting of the Transportation Research Board, Washington, D.C., 2018.
31. Azizi, L., and A. Sheikholeslami. Safety Effect of U-Turn Conversions in Tehran: Empirical Bayes Observational Before-and-After Study and Crash Prediction Models. *Journal of Transportation Engineering*, Vol. 139, No. 1, 2012, pp. 101–108.
32. Greene, W. H. *Econometric Analysis*. Pearson Education, India, 2003.
33. Washington, S. P., M. G. Karlaftis, and F. Mannering. *Statistical and Econometric Methods for Transportation Data Analysis*. CRC Press, Boca Raton, FL, 2010.
34. Wali, B., A. Ahmed, and N. Ahmad. An Ordered-probit Analysis of Enforcement of Road Speed Limits. *Proceedings of the Institution of Civil Engineers-Transport*, Thomas Telford Ltd, Vol. Ahead of Print 2017, pp. 1–10.
35. Li, X., A. J. Khattak, and B. Wali. Role of Multiagency Response and On-Scene Times in Large-Scale Traffic Incidents. *Transportation Research Record: Journal of the Transportation Research Board*, 2017. 2616: 39–48.
36. Jamali, A., and Y. Wang. Estimating Pedestrian Exposure for Small Urban and Rural Areas. *Transportation Research Record: Journal of the Transportation Research Board*, 2017. 2661: 84–94.
37. Wang, Y., K. Kockelman, and A. Jamali. A Synthesis of Spatial Models for Multivariate Count Responses. *Regional Research Frontiers*, Vol. 2, 2017. pp. 221–237.
38. Wali, B., A. J. Khattak, J. Waters, D. Chimba, and X. Li. Development of Safety Performance Functions: Incorporating Unobserved Heterogeneity and Functional Form Analysis. *Transportation Research Record: Journal of the Transportation Research Board*, 2018. In press.
39. Bhat, C. R. Simulation Estimation of Mixed Discrete Choice Models Using Randomized and Scrambled Halton Sequences. *Transportation Research Part B: Methodological*, Vol. 37, No. 9, 2003, pp. 837–855.
40. Train, K. *Halton Sequences for Mixed Logit*. Department of Economics, UC Berkeley, 2000.
41. Poirier, J., F. Ludwig, and S. Motamedi. Traffic Light Assistant Simulation: Foggy Weather. *Advances in Human* Aspects of Transportation, Vol. 484, 2017, pp. 1129–1138.

The Standing Committee on Safety Data, Analysis and Evaluation (ANB20) peer-reviewed this paper (18-00089).

Any opinions, findings, and conclusions or recommendations expressed in this paper are those of the authors and do not necessarily reflect views of the sponsors.

Article

TRR
JOURNAL OF THE TRANSPORTATION RESEARCH BOARD

Transportation Research Record
2018, Vol. 2672(38) 302–312

DOI: 10.1177/0361198118790843
journals.sagepub.com/home/trr
SAGE

Analysis of Rear-End Conflicts in Urban Networks using Bayesian Networks

Katerina Stylianou[1] and Loukas Dimitriou[1]

Abstract

Crash analysis and modeling studies have provided insightful information on crash-contributing factors and the methodologies utilized provide evidence that they could also be beneficial for conflict analysis, as traffic conflict data share similar traits with crash data. In this study, a Bayesian network (BN) is estimated to comprehensively analyze rear-end conflict likelihood in an urban network, using disaggregate vehicle-by-vehicle data and the time-to-collision indicator to identify conflicts. The variables imported in the BN include (i) individual driver characteristics (e.g., speed); (ii) traffic operational characteristics (e.g., volume); and (iii) general characteristics (e.g., weather conditions). The inference analyses of the BN conducted to quantify the contributions of the variables affecting rear-end conflict likelihood in the urban network showed that rear-end conflict likelihood could be increased when the involved vehicles are of a different type, when the speed of the following vehicle is higher than the speed of the leading vehicle, when the individual speed is high when the individual headway is small, with a higher coefficient of variation of speed values, when the type of intersection nearest to the measuring point was a priority intersection, when the carriageway was of dual design, and when it was rainy. It was also shown that rear-end conflict likelihood increases during congestion and free-flow traffic. The findings of this study could be further developed to provide a good understanding of contributing factors to possible crashes in the urban network.

The number of crashes in the network and the magnitude of their effect, mainly on humans, justify the importance of analyzing the contributing factors to their occurrence. It has been reported that rear-end crashes are one of the most common types of crashes, especially in urban networks. Therefore, identifying various contributing factors to rear-end crashes, such as geometric, environmental, or traffic characteristics and driver behavior has been the objective of various safety studies.

A large number of previous studies has been conducted to investigate rear-end crashes with the main goal to reduce the number of crashes and mitigate their effect. Traditionally, crash frequency analysis has been conducted, which estimates crash frequency by employing aggregated measures of traffic (e.g., annual average daily traffic) and other crash-contributing factors such as road geometry features (e.g., horizontal or vertical curves). In recent years, research efforts have focused on the introduction of the dynamics of traffic, concentrating on real-time safety analysis. Real-time safety analysis focuses on the investigation of crash likelihood over a very short time period, assuming that the occurrence of a crash is mainly due to short-term turbulence right before a crash (*1*). The implementation of real-time analysis is, in proactive systems, capable of timely spotting and evolving hazardous conditions that can be mitigated with various traffic smoothing measures (*2*). In general, both crash frequency and real-time analyses aim to find crash mechanisms and identify the relationship between safety and its contributing factors (*1*). From this perspective, the identification of both design characteristics and traffic operations responsible for traffic crashes in the network could help identify hazardous traffic conditions and hence support the development of effective countermeasures to improve safety levels.

However, the reliance on crashes for traffic safety analysis usually yields problems. A common issue with crash data is the poor quality of the data or the lack of their availability (*3*). Several research attempts have been undertaken to overcome the shortcomings associated with crash data. Multiple authors have advocated the use of alternative indicators, also known as surrogate measures, which reflect road safety. The most used surrogate measure is traffic conflict, defined as "an observable

[1]LaB for Transport Engineering, Department of Civil and Environmental Engineering, University of Cyprus, Nicosia, Cyprus

Corresponding Author:
Address correspondence to Loukas Dimitriou: lucdimit@ucy.ac.cy

situation in which two or more road users approach each other in space and time for such an extent that there is a risk of collision if their movements remain unchanged" (*4*). Several indicators have been developed for the identification of conflicts, and the well-known time-to-collision (TTC) is considered an effective indicator to evaluate traffic safety in the field of traffic conflict (*5*).

This study focuses on furthering understanding of the dynamic traffic characteristics that precede rear-end conflicts and other factors that contribute to these types of conflicts. The objective of this study is to develop a conflict-based methodology to interpret rear-end conflicts in urban networks using real-time data provided by the ubiquitous traffic surveillance system of inductive loop detectors (ILDs).

Background Review

Rear-end crashes have been a recurrent research subject in safety analysis, and many studies have investigated this type of crash under different focuses and modeling approaches. For example Abdel-Aty and Abdelwahab (*6*) investigated the effect of geometric incompatibility of light truck vehicles on other drivers' visibility by using nested logit models. The results of this study show that drivers' visibility and inattention on the following vehicle have the largest effect on being involved in a rear-end collision when an LTV (light transport vehicle) is leading a passenger car. Kim et al. (*7*) developed a microscopic model of freeway rear-end crash risk based on a modified negative binomial regression model to estimate rear-end risk using Washington State data, and captured the dual impacts of explanatory variables in the occurrence of the crashes. Kim et al. (*8*) also explored the association of rear-end crash propensity and micro-scale driver behavioral data gathered by an in-vehicle sensing system and showed that high crash rates were detected with high deceleration rates. Another approach for rear-end crash investigation includes the use of surrogate safety measures instead of crash data. For example, Oh et al. developed a method for identifying rear-end collision risks based on stopping distance using ILD data, and proposed a rear-end collision risk index to quantify the potential of rear-end collisions (*5*). Oh and Kim proposed a methodology to estimate rear-end crash probabilities by using vehicle trajectories obtained from video surveillance mechanisms (*9*). In this study a statistical methodology was developed to estimate rear-end crash potential using TTC measures. Similarly, Meng and Weng used TTC as a surrogate safety measure to estimate rear-end crash potential in the work zone merging areas (*10*). Meng and Qu, applied an inverse Gaussian regression model to TTC data, in order to estimate rear-end vehicle crash frequencies in road tunnels (*11*). Li et al. developed a rear-end collision risk index for evaluating the risk related to kinematic waves near freeway recurrent bottlenecks, utilizing aggregated traffic data from ILDs (*12*). This study reported that the likelihood of a rear-end collision is highest when traffic approaching from upstream is at capacity state while traffic downstream is highly congested.

These studies have provided an insightful understanding on crash modeling and analysis, providing evidence that they could also be beneficial for conflict analysis, as traffic conflict data share similar traits with crash data, as highlighted by Zhang et al. (*13*). Recent research has attempted to transfer crash frequency methodologies to conflict-based analysis. For example, El Basyouny and Sayed proposed a two-phase model to investigate the relationship between conflicts and crashes (*14*). In this study, a lognormal model was employed to predict conflicts based on traffic volume, area type, and geometric variables, and later a negative binomial safety performance function was employed to predict crashes. The results showed that a proportional relationship exists between conflicts and crashes. Similarly, Zhang et al. modeled the frequency of opposing left-turn conflicts at signalized intersections using generalized linear regression models and showed that the frequency of traffic conflicts follows a negative binomial distribution and concluded that the occurrences of conflicts at signalized intersections can be modeled using the generalized linear regression models (*13*). Another approach was the development of conflict-based safety performance functions (SPF) within the framework of Bayesian statistics to predict the number of rear-end conflicts at different intersection approaches (*3*). Similarly, Sacchi and Sayed developed conflict-based safety performance functions using a two-phase nested modeling process in which firstly a Poisson-gamma SPF, which uses traffic volume as an exposure, was used to predict conflicts, which were then used in another Poisson-gamma SPF to predict collisions (*15*). The results of this study demonstrated the proportional relationship between conflicts and collisions.

The aforementioned studies have focused on statistical methods for crash likelihood and frequency or conflict-based analysis, however recently researchers started using Bayesian networks (BN) in modeling crashes. BN, which were introduced by Pearl belong to the artificial intelligence family and are graphical structures used to represent knowledge about an uncertain domain (*16*, *17*). An application of BNs has been in injury severity analysis. For instance, de Ona et al. used a BN to classify traffic accidents according to their injury severity on rural highways and identified the variables that best reveal the factors that are associated with each injury severity level (*18*). Similarly, Mujalli and de Ona developed several

BNs to identify the most significant variables for accident injury severity, presenting that the reduction of model variables does not reduce the performance of the model (*19*). BNs have also been used with simulated data for validating safety requirements. For example Georgiades et al. investigated the integration of an agent-based micro-simulation model with a BN to predict accident risk on roadways (*20*). Additionally, Gerogiades and Mouskos proposed an accident risk quantification through BNs for black spots identification (*21*). The relationship between crashes and traffic characteristics has also been investigated with the use of BNs. Hossain and Muromachi proposed a framework based on BNs for real-time crash prediction of urban expressways (*2*). In this study the random multinomial logit model was employed to identify influencing factors to crashes and then a BN was estimated to predict crash risk in real time. Dias et al. used BNs to understand the impact of congestion on accidents and showed that the probability of an accident increases with the congestion level (*22*). Sun and Sun proposed a dynamic Bayesian network model of time sequence time data to investigate the relationship between crash occurrence and dynamic speed condition data, and reported a high accuracy prediction model (*23*). It is evident that BNs have been proven to be an effective methodology for describing crashes involving many interdependent variables and their most important aspect is that, contrary to the statistical methods, BNs are an effective technique to investigate the interrelationships among variables in any kind of system and provide a graphical representation of these.

Even though a few attempts at transferring crash frequency methodologies to conflict-based analysis have been made, to the best of our knowledge no conflict-based studies have been performed using BNs. This study aims to address this gap by developing a BN model to investigate the effect of design and traffic operations variables on rear-end conflict likelihood, and discover the interrelations between these independent variables.

Methodology

BNs combine principles from graph theory, probability theory, computer science, and statistics (*17*). Models of such class are powerful tools to deal with uncertainty and have been extensively used to represent knowledge, inference, and reasoning. A BN model can capture statistical associations between independent and dependent variables, as well as extract correlations among independent variables to present all interactions among all variables. The discovery of underlying patterns of data, and the investigation of relationships among variables therefore, is relatively easy using BNs and, as such, BNs make it easy to describe crashes involving many interdependent variables (*18*). A BN is developed in this study to analyze rear-end conflict likelihood in an urban network based on both traffic operational characteristics and other contributing factors such as weather and roadway design.

A BN B is an annotated acyclic graph that represents a joint probability distribution (JPD) over a set of random variables V. According to Ben-Gal et al. (*17*) a network is defined by a pair $B = (G,\Theta)$, where G is the directed acyclic graph (DAG) whose nodes $X_1, X_2, \ldots, X_n$ represent random variables, and whose edges represent the direct dependencies between these variables (*17*). In particular, an edge from node X_i to node X_j represents a statistical dependence between the corresponding variables, and node X_i is referred to as "parent" whereas node X_j is referred to as "child." The graph G encodes independence assumptions, by which each variable X_i is independent of its non-descendants given its parents in G. The second component Θ denotes the set of parameters of the network. This set contains the parameter $\theta_{x_i|\pi_i} = P_B(x_i|\pi_i)$ for each realization x_i of X_i conditioned on π_i, the set of parents of X_i in G. Accordingly, B defines a unique JPD over V, namely:

$$P_B(X_1, X_2, \ldots, X_n) = \prod_{i=1}^{n} P_B(X_i|\pi_i) = \prod_{i=1}^{n} \theta_{X_i|\pi_i} \quad (1)$$

Each node of the graph contains the state of the random variable and it represents a conditional probability table that contains the probabilities of the node being in a specific state given the states of its parents.

In most cases the BN is unknown and thus needs to be "learned" from the data. Therefore, both the graph topology (network structure) and the parameters of the JPD in the BN need to be estimated; this is known as the BN learning problem. Learning the structure is considered to be a harder problem than learning the parameters, as the number of possible DAGs on N variables is super exponential in N. There are three ways of estimating the structure of a BN, namely: (i) expert knowledge, (ii) automatically through a learning algorithm, and (iii) a combination of the previous two. In this study the K2 searching algorithm is applied to BN learning combined with expert knowledge to modify the initial DAG (*24*). The K2 algorithm is a score-based algorithm that heuristically searches for the most probable structure given a database of cases. This algorithm assumes first that there is an ordering available for the variables and second, a priori, all structures are equally likely. Technically, the algorithm begins by assuming that a node has no parents and then in every step, the K2 algorithm adds the parent to whose addition most increases the probability of the resulting structure until the addition of a single parent cannot increase the probability (*25*). Secondly, a simple estimator is used here to estimate the conditional probability tables (CPT) of the nodes after the BN structure is determined.

Application and Results

Data

For the estimation of the BN this study utilized disaggregate traffic and weather data. The traffic data used in the current study was collected from the traffic surveillance system of the city of Nicosia in Cyprus. The urban network of Nicosia is equipped with 16 ILDs, providing an adequate coverage of the urban area of 20 km^2. Vehicle-by-vehicle (VBV) data is extracted in real time and stored in the server of the Traffic Center of the Public Works Department of Cyprus. For the purposes of this study VBV data for the day of April 12th, 2016 was used, totaling in 24 h of individual vehicle data. Complete data from 14 ILDs were used, totaling 354,760 vehicles. The surveillance system covers most of the urban network of Nicosia, thus the dataset contains both unique physical characteristics of the vehicle as well as other general roadway design characteristics. Unique vehicle characteristics used in this study include arrival time, vehicle type (passenger car, motorcycle, and truck), vehicle speed, vehicle length, and headway (HW) between consecutive vehicles. General characteristics include the vehicle's lane of travel, the total number of lanes, and the direction of travel. The location of the ILD in regard to the network's geometrical characteristics was also known (e.g., near signalized intersection or in the middle of the link). In addition, aggregated traffic parameters (all in 5-min intervals) were developed from the disaggregated data, including traffic volume, average speed, speed standard deviation, and coefficient of variation of speed (CVS). Information about temperature and precipitation was also collected for the corresponding day by a weather station located at the University of Cyprus. The weather data are recorded every minute, but for the purposes of this study a 5-min binary variable for precipitation was developed: rain, no rain.

As mentioned earlier, the objective of this study is to develop a conflict-based methodology to interpret rear-end conflicts in urban networks, therefore the TTC was used in this analysis to define a traffic conflict. TTC is defined as "the time that remains until a collision between two vehicles would have occurred if the collision course and speed difference are maintained" (*26*). In this study only traffic conflicts with a TTC of 1.5 s or less were used (*14*, *27*, *28*). The TTC indicator was calculated through the disaggregate VBV data following Equation 2:

$$TTC_i = \frac{V_i\left(HW_i - \frac{l_{i-1}}{V_i}\right)}{V_i - V_{i-1}} \quad (2)$$

where

HW is the temporal headway between two vehicles

t_{i-1} the time at which the leading vehicle *(i-1)* passes the ILD

t_i the time at which the following vehicle *(i)* passes the same ILD

V_i the speed of the following vehicle

V_{i-1} the speed of the leading vehicle

L_{i-1} the length of the leading vehicle

Before starting the process of the BN estimation the data has to be prepared in such a way as the learning algorithm can deal with it (*29*). BNs can deal with both discrete and continuous data, however categorical variables are more advantageous in the sense of lower computational complexity. Therefore, all the variables in this study are discretized as presented in Table 1. The independent variables are divided into three main categories: (i) individual driver characteristics; (ii) traffic operational characteristics; and (iii) general contributing factors such as design and weather conditions.

Rear-End Conflict Detection

Rear-end conflicts in the urban network were detected using VBV data from ILDs and following Equation 2 it was shown that 317 conflicts were detected (0.09% of all vehicle observations) for the given observation period. The association of the detected conflicts with the independent variables is presented in Table 1. Indicatively, more rear-end conflicts were detected when the following vehicle had a higher speed than the leading vehicle; when the HW between consecutive vehicles was smaller than 4 s; near signalized intersections; and in dual carriageways. In addition, more rear-end conflicts were detected during peak hours when compared with off-peak hours, and when speed standard deviation was higher. Higher volumes also presented more rear-end conflicts, however after a peak value, fewer rear-end conflicts were detected in higher volumes.

Structure Learning Results

Generally, there are three main approaches in structure learning: score based, in which a score function is used and the algorithm attempts to find the structure that maximizes that score function; constrained based, in which conditional independence tests are carried out and independence relations established provide constraints; and hybrid algorithms, which use both constrained based and score techniques (*30*). This study follows the score-based approach to estimate the BN. According to findings from de Ona et al., who used three major BN scoring metrics, including minimum description length (MDL), Akaike information criterion (AIC), and likelihood-equivalence Bayesian Dirichlet (BDe) in BN learning, the

Table 1. Variable Definitions and Descriptions

Variable	Description	Codes/values	Detected conflicts	
C	Rear-end conflict index	(Yes) if a rear-end conflict is evident	317	0.09%
		(No) if there is no rear-end conflict	354,443	99.91%
Individual characteristics				
Type	Type of vehicle	(pass car) passenger car	269	84.85%
		(motor) motorcycle	33	10.41%
		(other) truck or otherwise	15	4.73%
Inter	Interaction between types of vehicles	(Yes) if vehicles are of the same type (e.g., both passenger cars)	241	76.03%
		(No) if vehicles are of different type (e.g., passenger car–truck)	76	23.97%
DV	Speed difference	(Yes) If DV > 0	124	39.12%
		(No) If DV < 0	193	60.88%
Speed	Individual vehicle speed	(S1) 0–20 km/h	228	71.92%
		(S2) 20–40 km/h	42	13.25%
		(S3) 40–60 km/h	28	8.83%
		(S4) 60–80 km/h	16	5.05%
		(S5) >80 km/h	3	0.95%
HW	Temporal headway between two consecutive vehicles	(HW1) 0–2 s	94	29.65%
		(HW2) 2–4 s	120	37.85%
		(HW3) 4–6 s	53	16.72%
		(HW4) >6 s	50	15.77%
Traffic characteristics				
AVSP	Average speed (aggregated in 5-min intervals)	(AVSP1) 0–20 km/h	2	0.63%
		(AVSP 2) 20–40 km/h	165	52.05%
		(AVSP 3) 40–60 km/h	144	45.43%
		(AVSP 4)60–80 km/h	6	1.89%
		(AVSP 5) >80 km/h	0	0.00%
Vol	Volume (aggregated in 5-min intervals)	(Vol1) 0–50	3	0.95%
		(Vol 2) 50–100	56	17.67%
		(Vol 3) 100–150	80	25.24%
		(Vol 4) 150–200	96	30.28%
		(Vol 5) 200–250	79	24.92%
		(Vol 6) >250	3	0.95%
SSTD	Speed standard deviation (aggregated in 5-min intervals)	(SSTD1) 0–5	0	0.00%
		(SSTD 2) 5–10	44	13.88%
		(SSTD 3) 10–15	149	47.00%
		(SSTD 4) 15–20	124	39.12%
CVS	Coefficient of variation of speed (aggregated in 5-min intervals)	(CVS 1) 0–0.5	230	72.56%
		(CVS 2) 0.5–1	87	27.44%
		(CVS 3) 1–1.5	0	0.00%
		(CVS 4) 1.5–2	0	0.00%
PTI	Peak time indicator	(Yes) during peak hours	166	52.37%
		(No) during off-peak hours	151	47.63%
General characteristics				
Location	Location of measuring point in the network	(signal) near signalized intersection	153	48.27%
		(middle) middle of corridor	70	22.08%
		(priority) near priority intersection	94	29.65%
Design	Carriageway design	(single) single carriageway	65	20.50%
		(dual) dual carriageway	222	70.03%
		(two + one) 2 + 1	30	9.46%
Median	Directions separated by median	(Yes) if there is a median	204	64.35%
		(No) if there is not a median	113	35.65%
Rain	Weather conditions (aggregated in 5-min intervals)	(Yes) rainy	36	11.36%
		(No) not rainy	281	88.64%

MDL scoring criterion produced a better prediction accuracy compared with the other two metrics, and therefore the MDL criterion is also adopted in this study (*18*).

This study was employed using Weka software, for structure and parameter learning of the BN (*31*). The dataset was used for the structure learning, by using the cross validation method. Cross validation involves the partitioning of the data into subsets, performing the analysis on one subset (called the training set) and validating the analysis on the other subset (called the testing set). By using the K2 algorithm, the initial structure of the BN was obtained as presented in Figure 1*a*. The initial structure of the learned BN included 15 nodes and 49 arcs. As noted before, the edges between the nodes represent the direct dependence between the variables, specifically, the direction of the arc indicates which variable is the parent and which is the child. For example, as can be seen from Figure 1*a*, AVSP is the parent node of variables Speed, DV, Vol, and PTI (peak time indicator), whereas it is the child node of Median, SSTD, Design and Location. Following this initial structure, expert knowledge was applied to optimize the model's structure. Operations for the optimization of the learned BN, based on experts' knowledge, included the insertion or deletion of an arc in the network. Finally, the optimized network included 15 nodes and 44 arcs, as presented in Figure 1*b*. Specifically, inaccuracies conflicting with common practice included: (a) the existence of a Median in the roadway was found to be the child of Design, Location, HW and CVS. The design and the location are indeed related with the median, however the other two variables are solely based on human behavior and clearly do not affect network design; (b) Rain was shown to be the child of several traffic operational characteristics, which is inaccurate in practice, and in the current case weather conditions do not have any of the examined variables as parents (root node); (c) Rain was not found to have a relationship with Speed, which is again inaccurate as several studies have shown that speed is highly affected by the weather conditions; (d) roadway Design was shown to be affected by the standard deviation of Speed and the CVS, which in practice cannot be; and finally (e) individual HW, Speed, and Rain were added as parents to the conflict node.

Parameter Learning Results

Following the BN structure created, Bayesian model parameters can then be estimated. The structure of the BN formulates the interdependency among the variables and thus is capable of providing probability inference based on the CPT of each node, as described by Equation 1. Figure 2 presents a perspective of the current situation with respect to rear-end conflict likelihood and the other variables of this study. As an example, the CPT of the node "Headway" that was obtained from the parameter estimation step is shown in Table 2.

In accordance to Figure 2, it can be ascertained that based on the prior probabilities and variables, the average probability of rain, location, speed standard deviation (in 5-min aggregation), individual speed and individual HW being associated with a rear-end conflict is 0.2%. In addition, other evidence from Figure 2 includes that around half of the drivers drive with a speed between 40–60 km/h and 58% of the average speed in 5-min aggregation is also within the same speed range. Aggressive behavioral driving characteristics are also shown: 20% of individual HWs were within 0 and 2 s and a high proportion of following drivers (44%) had a higher speed than the leading driver.

Bayesian Inference Results

The developed BN structure formulates the interdependency between the variables and thus is capable of providing the CPTs for each node. Inference in BNs consists of the computation of the conditional probability of one variable, given that other variables are set to evidence, meaning that for each examined variable its value is set as 1.0, and the probabilities of the affected variables are updated through the CPTs. Inference may be done for a specific variable given evidence on the state of other variables, thus here through setting evidences for the related variables, their contribution on the likelihood of rear-end conflict can be quantified. For example, node "Location" has three alternatives: middle, priority, and signal. If this node was to be set as evidence in order to see its effect on conflict likelihood, it would mean that the probability of one of the alternatives would be set as 1.0 (e.g., P[middle] = 1.0), and the remaining alternatives would have a probability equal to zero (e.g., P[priority] = 0.0, P[signal] = 0.0). Table 3 illustrates the inference results for the variables that increase the likelihood of rear-end conflict, in which the first row (original distribution) corresponds to the outcome of rear-end conflict as extracted from the data set, and the rest of the rows present the results for each node that was set as evidence.

The effect of all the variables was examined on rear-end conflict likelihood, but here only states of variables that increased the likelihood are presented. First, a probability of 1.0 was assigned for the variable Inter with the value of no (i.e., Inter = no), which represents whether the two vehicles examined are of the same type, that is, passenger car–passenger car. The probability of rear-end conflict increased from 0.0021 to 0.0028, compared to the original distributions, when the involved vehicles are not of the same type. This shows that different types of vehicles will increase the propensity of rear-end conflicts in the urban network. This result is consistent with findings in previous studies when modeling rear-end collisions in regard to light truck vehicles (*6*). Speed difference also has an effect on rear-end conflict propensity, as indicated by the inference results of the variable, DV. When the speed difference is set as evidence (i.e., DV = yes) the

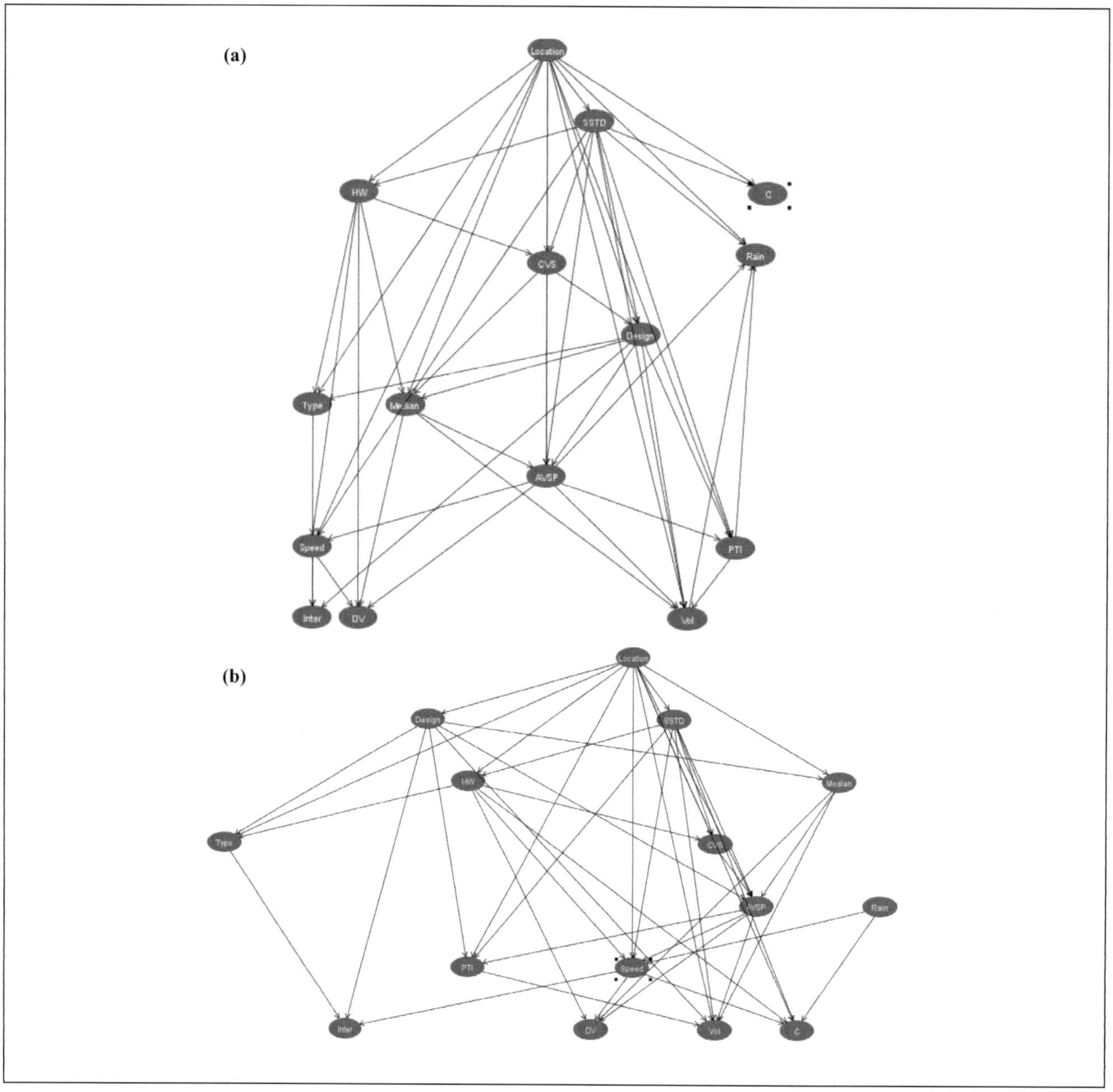

Figure 1. BN structures of rear-end conflicts in an urban network: (*a*) learned structure; (*b*) optimized structure.

probability of rear-end conflict increases from 0.21% to 0.22%. Individual speed also affects rear-end conflict likelihood. The inference results of the variable Speed, when a probability of 1.0 is assigned for S5 (Speed > 80 km/h) indicate that the probability of rear-end conflict is up to seven times higher than the original distribution. The temporal distance between vehicles also affects the probability of rear-end conflict as suggested by the inference results of the variable, HW. When the probability of HW1 (0 < HW < 2 s) is set to 1.0, rear-end conflict probability is increased from 0.21% to 0.30%. This implies that drivers driving with lower HWs have an increased probability of being involved in a rear-end conflict. Similarly, when setting evidence for state HW3 to be true (4 < HW < 6 s), rear-end conflict likelihood probability increased, implying that higher values of HWs are associated with higher probabilities of rear-end conflicts. This may be due to the multiple interrelations between variables. For example, when HW1 is set as evidence, states S1 and S5 of variable Speed reduce from

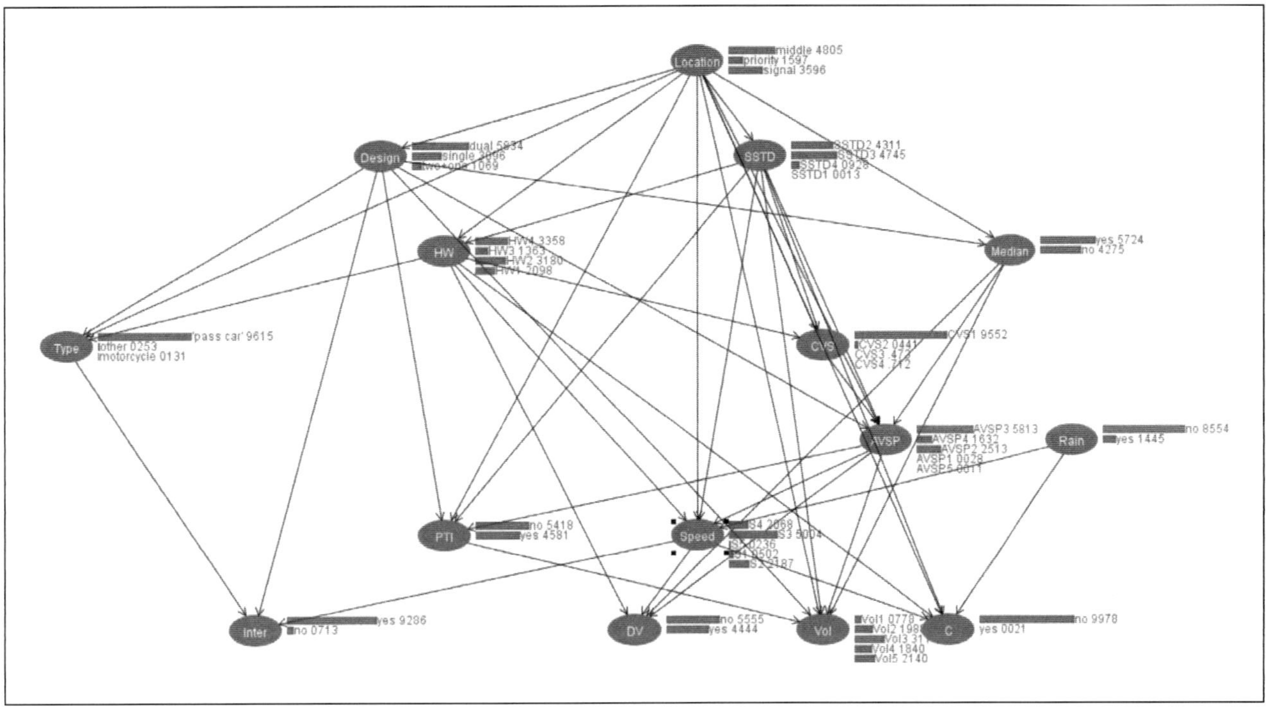

Figure 2. BN structure for rear-end conflicts in urban network with corresponding conditional probability tables.

Table 2. CPT of the Node Headway for Rear-End Conflict in an Urban Network

Parent nodes		Headway			
SSTD	Location	HW1	HW2	HW3	HW4
SSTD1	Signal	0.056	0.049	0.011	0.884
SSTD1	Priority	0.017	0.017	0.017	0.948
SSTD1	Middle	0.052	0.213	0.133	0.601
SSTD2	Signal	0.204	0.284	0.114	0.397
SSTD2	Priority	0.191	0.324	0.134	0.351
SSTD2	Middle	0.238	0.322	0.135	0.305
SSTD3	Signal	0.206	0.348	0.125	0.321
SSTD3	Priority	0.229	0.294	0.135	0.341
SSTD3	Middle	0.191	0.303	0.153	0.354
SSTD4	Signal	0.206	0.356	0.127	0.310
SSTD4	Priority	0.198	0.289	0.185	0.328
SSTD4	Middle	0.182	0.351	0.167	0.300

2.36% to 1.42% and from 5.02 to 0.96% respectively. On the other hand, when HW3 is set as evidence, states S1 and S5 of variable Speed increase from 2.36% to 2.53% and from 5.02% to 8.15% respectively. This implies that when individual HWs are short (0–2 s) there are fewer probabilities of having very low (0–20 km/h) or very high (>80 km) speeds in the urban network, and subsequently when HWs are medium (4–6 s) there are higher probabilities of having very low (0–20 km/h) or very high (>80 km) speeds. These individual speed conditions clearly represent the congestion and free-flow states of traffic, and the aforementioned results imply that rear-end conflict probabilities are increased during these states of traffic.

A similar conclusion can be drawn from the inference results of the first traffic operational variable, AVSP. When the states AVSP1, AVSP2 and AVSP5 are set as evidence, the probability of rear-end conflict increases from 0.21% to 2.82%, and 0.32% and 2.74%. These results underscore the significant impact of traffic states

Table 3. BN Probability Inference

Variable		Rear-end conflict	
Name	State	Yes	No
Original distribution		0.0021	0.9978
Individual characteristics			
Inter	no	0.0028	0.9971
DV	yes	0.0022	0.9977
Speed	S5	0.0147	0.9852
HW	HW1	0.0030	0.9969
	HW3	0.0033	0.9966
Traffic characteristics			
AVSP	AVSP1	0.0282	0.9717
	AVSP2	0.0032	0.9967
	AVSP5	0.0274	0.9725
Vol	Vol1	0.0035	0.9964
	Vol4	0.0030	0.9969
	Vol6	0.0026	0.9973
SSTD	SSTD1	0.0987	0.9012
	SSTD4	0.0116	0.9883
CVS	CVS2	0.0122	0.9877
	CVS3	0.0139	0.9860
	CVS4	0.0176	0.9823
General characteristics			
Location	priority	0.0057	0.9942
Design	dual	0.0022	0.9977
Rain	yes	0.0066	0.9933

on rear-end conflicts in the urban network. Volume (in 5-min aggregation) is another factor affecting rear-end crash likelihood in the urban environment. As shown in Table 3, rear-end conflict probabilities increase with both lower values of volume (Vol1) and higher values (Vol4, Vol6). As explained previously, again this should be checked in tandem with the other nodes of the network. When a probability of 1.0 is assigned to the state Vol1 of variable Vol, the probability of state SSTD1 of variable SSTD indicating the lowest speed standard deviation level here (0–5 km/h), is increased by 10 times (from 0.13% to 1.07%), suggesting that volumes between 0 and 50 vehicles per 5 min are more likely to present low speed variations (between 0 and 5 km/h). On the other hand, when setting evidence for Vol6 (>80 veh/5 min), the state SSTD3 (10–15 km/h) of variable SSTD is increased from 47.45% to 84.18%. These values of speed standard deviation again present the different traffic states in the urban network and their effect on rear-end conflict likelihood. Speed standard deviation in a 5-min aggregation, was also shown to significantly affect rear-end conflict likelihood. As can be seen from Table 3, when the probabilities of state SSTD1 and SSTD4 of variable SSTD are set to 1.0, the probability of rear-end conflict increases from 0.22% to 9.87% and 1.16%, respectively. As already discussed, the values of speed variation reflect the traffic states in the network, and these results show that a rear-end conflict is more likely during congestion rather than free-flow conditions, which is consistent with the conclusions of previous studies developed for crashes (*32*). The final traffic characteristic that was found to affect rear-end conflict likelihood was the CVS. It was shown that the higher the CVS, the higher the probability of a rear-end conflict, which has been reported in a previous study (*33*).

The final variables include the general characteristics affecting rear-end conflicts in the urban network. The variable Location denotes the type of intersection nearest to the measuring spot: signal (near signalized intersection), middle (in the middle of the corridor), and priority (near priority intersection). The reference results indicate that the likelihood of rear-end conflicts increases at locations near to priority intersections, rather than signalized intersections or in the middle of the corridor. This finding may imply that signal control is a mitigating factor in rear-end conflicts. Similar analysis for the design of the carriageway showed that rear-end conflict probability is increased from 0.21% to 0.22% in dual carriageways. Finally, the likelihood of rear-end conflict was increased by three times when rain was evident. As presented in Table 3, the probability of rear-end conflict in the urban network was increased from 0.21% to 0.66% when the variable Rain was assigned a probability of 1.0 (i.e., Rain = yes). This result highlights the findings of previous research, which has shown that weather conditions, and especially precipitation, has been found to influence crash frequency (*34*).

Conclusion

The present study examines the interrelationships between several variables and their association with rear-end conflicts in an urban network, by estimating a BN. The proposed methodology incorporated surrogate safety measures and disaggregated VBV data to estimate conflicts in the urban network of Nicosia, Cyprus and developed the BN model to quantitatively and qualitatively analyze rear-end conflict patterns.

In this study, conflicts were identified using the widely used surrogate measure indicator, TTC, and it was found that 0.09% of all vehicle interactions were identified as rear-end conflicts. A BN was employed to capture the statistical associations between the examined variables and discover the underlying patterns to describe rear-end conflict likelihood in the urban network. Three types of explanatory attributes were used in the model including (i) individual driver characteristics (speed, HW, vehicle type, vehicle interaction, and speed difference); (ii) traffic operational characteristics (average speed, speed standard deviation, volume, CVS, and peak time indicator, all aggregated in 5-min intervals); and (iii) general contributing factors (location, carriageway design, and weather conditions). The structure of the BN was learned using the K2 algorithm and was later optimized using expert knowledge to better represent the association among variables.

Further research is needed to supplement the findings of this study. First the BN model performance should be examined and enhanced by evaluating the different score metrics and the available search algorithms. Additionally, more data over multiple time periods and the investigation of different BN structures is desirable to better understand the statistical associations between the rear-end conflict likelihood and other variables.

Author Contributions

The authors confirm contribution to the paper as follows: study conception and design: LD, KS; data collection: KS; analysis and interpretation of results: LD, KS; draft manuscript preparation: KS. All authors reviewed the results and approved the final version of the manuscript.

References

1. Wang, L., M. Abdel-Aty, and J. Lee. Safety Analytics for Integrating Crash Frequency and Real-Time Risk Modeling for Expressways. *Accident Analysis and Prevention*, Vol. 104, 2017, pp. 58–64. https://doi.org/10.1016/j.aap.2017.04.009.
2. Hossain, M., and Y. Muromachi. A Bayesian Network Based Framework for Real-Time Crash Prediction on the Basic Freeway Segments of Urban Expressways. *Accident Analysis and Prevention*, Vol. 45, 2012, pp. 373–381. https://doi.org/10.1016/j.aap.2011.08.004.
3. Sacchi, E., and T. Sayed. Bayesian Estimation of Conflict-Based Safety Performance Functions. *Journal of Transportation Safety and Security*, Vol. 8, No. 3, 2016, pp. 266–279. https://doi.org/10.1080/19439962.2015.1030807.
4. Sayed, T., and S. Zein. Traffic Conflict Standards for Intersections. *Transportation Planning and Technology*, Vol. 22, No. 4, 1999, pp. 309–323. https://doi.org/10.1080/03081069908717634.
5. Oh, C., S. Park, and S. G. Ritchie. A Method for Identifying Rear-End Collision Risks Using Inductive Loop Detectors. *Accident Analysis and Prevention*, Vol. 38, No. 2, 2006, pp. 295–301. https://doi.org/10.1016/j.aap.2005.09.009.
6. Abdel-Aty, M., and H. Abdelwahab. Modeling Rear-End Collisions Including the Role of Driver's Visibility and Light Truck Vehicles Using a Nested Logit Structure. *Accident Analysis and Prevention*, Vol. 36, No. 3, 2004, pp. 447–456. https://doi.org/10.1016/S0001-4575(03)00040-X.
7. Kim, J. K., Y. Wang, and G. Ulfarsson F. Modeling the Probability of Freeway Rear-End Crash Occurrence. *Journal of Transportation Engineering*, Vol. 133, No. 1, 2007, pp. 11–19. https://doi.org/10.1061/(ASCE)0733-947X(2007)133:1(11).
8. Kim, S. K., T. J. Song, N. M. Rouphail, S. Aghdashi, A. Amaro, and G. Gonçalves. Exploring the Association of Rear-End Crash Propensity and Micro-Scale Driver Behavior. *Safety Science*, Vol. 89, 2016, pp. 45–54. https://doi.org/10.1016/j.ssci.2016.05.016.
9. Oh, C., and T. Kim. Estimation of Rear-End Crash Potential Using Vehicle Trajectory Data. *Accident Analysis and Prevention*, Vol. 42, No. 6, 2010, pp. 1888–1893. https://doi.org/10.1016/j.aap.2010.05.009.
10. Meng, Q., and J. Weng. Evaluation of Rear-End Crash Risk at Work Zone Using Work Zone Traffic Data. *Accident Analysis and Prevention*, Vol. 43, No. 4, 2011, pp. 1291–1300. https://doi.org/10.1016/j.aap.2011.01.011.
11. Meng, Q., and X. Qu. Estimation of Rear-End Vehicle Crash Frequencies in Urban Road Tunnels. *Accident Analysis and Prevention*, Vol. 48, 2012, pp. 254–263. https://doi.org/10.1016/j.aap.2012.01.025.
12. Li, Z., S. Ahn, K. Chung, D. R. Ragland, W. Wang, and J. W. Yu. Surrogate Safety Measure for Evaluating Rear-End Collision Risk Related to Kinematic Waves Near Freeway Recurrent Bottlenecks. *Accident Analysis and Prevention*, Vol. 64, 2014, pp. 52–61. https://doi.org/10.1016/j.aap.2013.11.003.
13. Zhang, X., P. Liu, Y. Chen, L. Bai, and W. Wang. Modeling the Frequency of Opposing Left-Turn Conflicts at Signalized Intersections Using Generalized Linear Regression Models. *Traffic Injury Prevention*, Vol. 15, No. 6, 2014, pp. 645–651. https://doi.org/10.1080/15389588.2013.860526.
14. El-Basyouny, K., and T. Sayed. Safety Performance Functions Using Traffic Conflicts. *Safety Science*, Vol. 51, No. 1, 2013, pp. 160–164. https://doi.org/10.1016/j.ssci.2012.04.015.
15. Sacchi, E., and T. Sayed. Conflict-Based Safety Performance Functions for Predicting Traffic Collisions by Type. *Transportation Research Record: Journal of the Transportation Research Board*, 2016. 2583: 50–55.
16. Pearl, J. Reverend Bayes on Inference Engines: A Distributed Hierarchical Approach. *Proceedings of the AAAI National Conference on AI*, Pittsburgh, Pa., 1982, pp. 133–136. https://doi.org/10.1.1.294.5479.
17. Ben-Gal, I., F. Ruggeri, R. S. Kenett, and F. Faltin. Bayesian Networks. *Encyclopedia of Statistics in Quality & Reliability*, Vol. 1, No. 1, John Wiley & Sons, Hoboken, NJ, 2007, p. 4. https://doi.org/10.1002/9780470061572.eqr089.
18. De Oña, J., R. O. Mujalli, and F. J. Calvo. Analysis of Traffic Accident Injury Severity on Spanish Rural Highways Using Bayesian Networks. *Accident Analysis and Prevention*, Vol. 43, No. 1, 2011, pp. 402–411. https://doi.org/10.1016/j.aap.2010.09.010.
19. Mujalli, R. O., and J. De Oña. A Method for Simplifying the Analysis of Traffic Accidents Injury Severity on Two-Lane Highways Using Bayesian Networks. *Journal of Safety Research*, Vol. 42, No. 5, 2011, pp. 317–326. https://doi.org/10.1016/j.jsr.2011.06.010.
20. Gregoriades, A., A. Sutcliffe, G. Papageorgiou, and P. Louvieris. Human-Centered Safety Analysis of Prospective Road Designs. *IEEE Transactions on Systems, Man, and Cybernetics Part A:Systems and Humans*, Vol. 40, No. 2, 2010, pp. 236–250. https://doi.org/10.1109/TSMCA.2009.2037011.

21. Gregoriades, A., and K. C. Mouskos. Black Spots Identification through a Bayesian Networks Quantification of Accident Risk Index. *Transportation Research Part C: Emerging Technologies*, Vol. 28, 2013, pp. 28–43. https://doi.org/10.1016/j.trc.2012.12.008.
22. Dias, C., and M. Miska. Relationship between Congestion and Traffic Accidents on Expressways: An Investigation with Bayesian Belief Networks. *Proc., 40th Annual Meeting of Infrastructure Planning (JSCE)*, Japan, 2009.
23. Sun, J., and J. Sun. A Dynamic Bayesian Network Model for Real-Time Crash Prediction Using Traffic Speed Conditions Data. *Transportation Research Part C: Emerging Technologies*, Vol. 54, 2015, pp. 176–186. https://doi.org/10.1016/j.trc.2015.03.006.
24. Cooper, G. F., and E. Herskovits. A Bayesian Method for the Induction of Probabilistic Networks from Data. *Machine Learning*, Vol. 9, No. 4, 1992, pp. 309–347. https://doi.org/10.1007/BF00994110.
25. Zong, F., H. Xu, and H. Zhang. Prediction for Traffic Accident Severity: Comparing the Bayesian Network and Regression models. *Mathematical Problems in Engineering*, Vol. 2013, 2013, pp. 1–9. https://doi.org/10.1155/2013/475194.
26. Hayward, J. C. Near-Miss Determination through Use of a Scale of Danger. *Highway Research Record*, Vol. 384, 1972, pp. 24–34.
27. Van Der Horst, R., and J. Hogema. Time-to-Collision and Collision Avoidance Systems. *Proc., 6th Workshop of the International*, Co-operation on Theories and Concepts in Traffic Safety, Salzburg, Austria, 1993, pp. 1–12.
28. Sayed, T., and S. Zein. Traffic Conflict Standards for Intersections. *Transportation Planning and Technology*, Vol. 22, No. 4, 1999, pp. 309–323.
29. Zhao, J., and W. Deng. The Use of Bayesian Network in Analysis of Urban Intersection Crashes in China. *Transport*, Vol. 30, No. 4, 2015, p. 411. https://doi.org/10.3846/16484142.2013.816365.
30. Koski, T. J. T., and J. Noble. A Review of Bayesian Networks and Structure Learning. *Mathematica Applicanda*, Vol. 40, No. 1, 2012, pp. 53–103. https://doi.org/10.14708/ma.v40i1.278.
31. Witten, I. H., E. Frank, M. A. Hall, and C. J. Pal. The WEKA Workbench. Data Mining: Practical Machine Learning Tools and Techniques, 4th ed. *Morgan Kaufmann*, Cambridge, MA, 2016, pp. 553–571
32. Shi, Q., M. Abdel-Aty, and J. Lee. A Bayesian Ridge Regression Analysis of Congestion's Impact on Urban Expressway Safety. *Accident Analysis and Prevention*, Vol. 88, 2016, pp. 124–137. https://doi.org/10.1016/j.aap.2015.12.001.
33. Zheng, Z., S. Ahn, and C. M. Monsere. Impact of Traffic Oscillations on Freeway Crash Occurrences. *Accident Analysis and Prevention*, Vol. 42, No. 2, 2010, pp. 626–636. https://doi.org/10.1016/j.aap.2009.10.009.
34. Yu, R., M. Abdel-Aty, and M. Ahmed. Bayesian Random Effect Models Incorporating Real-Time Weather and Traffic Data to Investigate Mountainous Freeway Hazardous Factors. *Accident Analysis and Prevention*, Vol. 50, 2013, pp. 371–376. https://doi.org/10.1016/j.aap.2012.05.011.

The Standing Committee on Safety Data, Analysis and Evaluation (ANB20) peer-reviewed this paper (18-04602).

Sage Publications USA
sagepub.com

ISBN 978-1-5108-8246-1

CURRAN ASSOCIATES INC.
proceedings.com